# PARTIAL DIFFERENTIAL EQUATIONS

## AN INTRODUCTION

BICENTENNIAL
1807
WILEY
2007
BICENTENNIAL

## THE WILEY BICENTENNIAL—KNOWLEDGE FOR GENERATIONS

*E*ach generation has its unique needs and aspirations. When Charles Wiley first opened his small printing shop in lower Manhattan in 1807, it was a generation of boundless potential searching for an identity. And we were there, helping to define a new American literary tradition. Over half a century later, in the midst of the Second Industrial Revolution, it was a generation focused on building the future. Once again, we were there, supplying the critical scientific, technical, and engineering knowledge that helped frame the world. Throughout the 20th Century, and into the new millennium, nations began to reach out beyond their own borders and a new international community was born. Wiley was there, expanding its operations around the world to enable a global exchange of ideas, opinions, and know-how.

For 200 years, Wiley has been an integral part of each generation's journey, enabling the flow of information and understanding necessary to meet their needs and fulfill their aspirations. Today, bold new technologies are changing the way we live and learn. Wiley will be there, providing you the must-have knowledge you need to imagine new worlds, new possibilities, and new opportunities.

Generations come and go, but you can always count on Wiley to provide you the knowledge you need, when and where you need it!

WILLIAM J. PESCE
PRESIDENT AND CHIEF EXECUTIVE OFFICER

PETER BOOTH WILEY
CHAIRMAN OF THE BOARD

# PARTIAL DIFFERENTIAL EQUATIONS

## AN INTRODUCTION

### WALTER A. STRAUSS
Brown University

BICENTENNIAL
BICENTENNIAL
1807
WILEY
2007
BICENTENNIAL
BICENTENNIAL

John Wiley & Sons, Ltd

| | |
|---|---|
| Publisher | Laurie Rosatone |
| Associate Editor | Shannon Corliss |
| Senior Editorial Assistant | Jeffrey Benson |
| Marketing Manager | Jaclyn Elkins |
| Production Manager | Dorothy Sinclair |
| Senior Production Editor | Sandra Dumas |
| Creative Director | Harry Nolan |
| Senior Designer | Madelyn Lesure |
| Production Management Services | Aptara, Inc. |
| Bicentennial Logo Design | Richard J. Pacifico |

This book was set in 11/12 Times Ten by Aptara, Inc. and printed and bound by Courier(Westford). The cover was printed by Courier(Stoughton).

This book is printed on acid free paper. ∞

To order books or for customer service, please call 1-800-CALL WILEY (225-5945).

ISBN-13     978-0470-05456-7

Printed in the United States of America

10 9 8 7 6 5 4 3

# PREFACE

Our understanding of the fundamental processes of the natural world is based to a large extent on partial differential equations. Examples are the vibrations of solids, the flow of fluids, the diffusion of chemicals, the spread of heat, the structure of molecules, the interactions of photons and electrons, and the radiation of electromagnetic waves. Partial differential equations also play a central role in modern mathematics, especially in geometry and analysis. The availability of powerful computers is gradually shifting the emphasis in partial differential equations away from the analytical computation of solutions and toward both their numerical analysis and the qualitative theory.

This book provides an introduction to the basic properties of partial differential equations (PDEs) and to the techniques that have proved useful in analyzing them. My purpose is to provide for the student a broad perspective on the subject, to illustrate the rich variety of phenomena encompassed by it, and to impart a working knowledge of the most important techniques of analysis of the solutions of the equations.

One of the most important techniques is the method of separation of variables. Many textbooks heavily emphasize this technique to the point of excluding other points of view. The problem with that approach is that only certain kinds of partial differential equations can be solved by it, whereas others cannot. In this book it plays a very important but not an overriding role. Other texts, which bring in relatively advanced theoretical ideas, require too much mathematical knowledge for the typical undergraduate student. I have tried to minimize the advanced concepts and the mathematical jargon in this book. However, because partial differential equations is a subject at the forefront of research in modern science, I have not hesitated to mention advanced ideas as further topics for the ambitious student to pursue.

This is an undergraduate textbook. It is designed for juniors and seniors who are science, engineering, or mathematics majors. Graduate students, especially in the sciences, could surely learn from it, but it is in no way conceived of as a graduate text.

The main prerequisite is a solid knowledge of calculus, especially multivariate. The other prerequisites are small amounts of ordinary differential equations and of linear algebra, each much less than a semester's worth. However, since the subject of partial differential equations is by its very nature not an easy one, I have recommended to my own students that they should already have taken full courses in these two subjects.

The presentation is based on the following principles. Motivate with physics but then do mathematics. Focus on the three classical equations: All the important ideas can be understood in terms of them. Do one spatial dimension before going on to two and three dimensions with their more complicated geometries. Do problems without boundaries before bringing in boundary conditions. (By the end of Chapter 2, the student will already have an intuitive and analytical understanding of simple wave and diffusion phenomena.) Do not hesitate to present some facts without proofs, but provide the most critical proofs. Provide introductions to a variety of important advanced topics.

There is plenty of material in this book for a year-long course. A quarter course, or a fairly relaxed semester course, would cover the starred sections of Chapters 1 to 6. A more ambitious semester course could supplement the basic starred sections in various ways. The unstarred sections in Chapters 1 to 6 could be covered as desired. A computational emphasis following the starred sections would be provided by the numerical analysis of Chapter 8. To resume separation of variables after Chapter 6, one would take up Chapter 10. For physics majors one could do some combination of Chapters 9, 12, 13, and 14. A traditional course on boundary value problems would cover Chapters 1, 4, 5, 6, and 10.

Each chapter is divided into sections, denoted A.B. An equation numbered (A.B.C) refers to equation (C) in section A.B. A reference to equation (C) refers to the equation in the same section. A similar system is used for numbering theorems and exercises. The references are indicated by brackets, like [AS].

The help of my colleagues is gratefully acknowledged. I especially thank Yue Liu and Brian Loe for their extensive help with the exercises, as well as Costas Dafermos, Bob Glassey, Jerry Goldstein, Manos Grillakis, Yan Guo, Chris Jones, Keith Lewis, Gustavo Perla Menzala, and Bob Seeley for their suggestions and corrections.

*Walter A. Strauss*

# PREFACE TO SECOND EDITION

In the years since the first edition came out, partial differential equations has become yet more prominent, both as a model for scientific theories and within mathematics itself. In this second edition I have added 30 new exercises. Furthermore, this edition is accompanied by a solutions manual that has answers to about half of the exercises worked out in detail. I have added a new section on water waves as well as new material and explanatory comments in many places. Corrections have been made wherever necessary.

I would like to take this opportunity to thank all the people who have pointed out errors in the first edition or made useful suggestions, including Andrew Bernoff, Rustum Choksi, Adrian Constantin, Leonid Dickey, Julio Dix, Craig Evans, A. M. Fink, Robert Glassey, Jerome Goldstein, Leon Greenberg, Chris Hunter, Eva Kallin, Jim Kelliher, Jeng-Eng Lin, Howard Liu, Jeff Nunemacher, Vassilis Papanicolaou, Mary Pugh, Stan Richardson, Stuart Rogers, Paul Sacks, Naoki Saito, Stephen Simons, Catherine Sulem, David Wagner, David Weinberg, and Nick Zakrasek. My warmest thanks go to Julie and Steve Levandosky who, besides being my co-authors on the solutions manual, provided many suggestions and much insight regarding the text itself.

# CONTENTS

(The starred sections form the basic part of the book.)

# 1

# WHERE PDEs COME FROM

After thinking about the meaning of a partial differential equation, we will flex our mathematical muscles by solving a few of them. Then we will see how naturally they arise in the physical sciences. The physics will motivate the formulation of boundary conditions and initial conditions.

## 1.1 WHAT IS A PARTIAL DIFFERENTIAL EQUATION?

The key defining property of a partial differential equation (PDE) is that there is more than one independent variable $x, y, \ldots$. There is a dependent variable that is an unknown function of these variables $u(x, y, \ldots)$. We will often denote its derivatives by subscripts; thus $\partial u / \partial x = u_x$, and so on. A PDE is an identity that relates the independent variables, the dependent variable $u$, and the partial derivatives of $u$. It can be written as

$$F(x, y, u(x, y), u_x(x, y), u_y(x, y)) = F(x, y, u, u_x, u_y) = 0. \quad (1)$$

This is the most general PDE in two independent variables of *first* order. The *order* of an equation is the highest derivative that appears. The most general *second*-order PDE in two independent variables is

$$F(x, y, u, u_x, u_y, u_{xx}, u_{xy}, u_{yy}) = 0. \quad (2)$$

A *solution* of a PDE is a function $u(x, y, \ldots)$ that satisfies the equation identically, at least in some region of the $x, y, \ldots$ variables.

When solving an ordinary differential equation (ODE), one sometimes reverses the roles of the independent and the dependent variables—for instance, for the separable ODE $\dfrac{du}{dx} = u^3$. For PDEs, the distinction between the independent variables and the dependent variable (the unknown) is always maintained.

1

Some examples of PDEs (all of which occur in physical theory) are:

1. $u_x + u_y = 0$     (transport)
2. $u_x + yu_y = 0$     (transport)
3. $u_x + uu_y = 0$     (shock wave)
4. $u_{xx} + u_{yy} = 0$     (Laplace's equation)
5. $u_{tt} - u_{xx} + u^3 = 0$     (wave with interaction)
6. $u_t + uu_x + u_{xxx} = 0$     (dispersive wave)
7. $u_{tt} + u_{xxxx} = 0$     (vibrating bar)
8. $u_t - iu_{xx} = 0$     $(i = \sqrt{-1})$     (quantum mechanics)

Each of these has two independent variables, written either as $x$ and $y$ or as $x$ and $t$. Examples 1 to 3 have order one; 4, 5, and 8 have order two; 6 has order three; and 7 has order four. Examples 3, 5, and 6 are distinguished from the others in that they are not "linear." We shall now explain this concept.

*Linearity* means the following. Write the equation in the form $\mathscr{L}u = 0$, where $\mathscr{L}$ is an *operator*. That is, if $v$ is any function, $\mathscr{L}v$ is a new function. For instance, $\mathscr{L} = \partial/\partial x$ is the operator that takes $v$ into its partial derivative $v_x$. In Example 2, the operator $\mathscr{L}$ is $\mathscr{L} = \partial/\partial x + y\partial/\partial y$. ($\mathscr{L}u = u_x + yu_y$.) The definition we want for linearity is

$$\mathscr{L}(u + v) = \mathscr{L}u + \mathscr{L}v \qquad \mathscr{L}(cu) = c\mathscr{L}u \tag{3}$$

for any functions $u$, $v$ and any constant $c$. Whenever (3) holds (for all choices of $u$, $v$, and $c$), $\mathscr{L}$ is called *linear operator*. The equation

$$\mathscr{L}u = 0 \tag{4}$$

is called *linear* if $\mathscr{L}$ is a linear operator. Equation (4) is called a *homogeneous linear equation*. The equation

$$\mathscr{L}u = g, \tag{5}$$

where $g \neq 0$ is a given function of the independent variables, is called an *inhomogeneous linear equation*. For instance, the equation

$$(\cos xy^2)u_x - y^2u_y = \tan(x^2 + y^2) \tag{6}$$

is an inhomogeneous linear equation.

As you can easily verify, five of the eight equations above are linear as well as homogeneous. Example 5, on the other hand, is not linear because although $(u + v)_{xx} = u_{xx} + v_{xx}$ and $(u + v)_{tt} = u_{tt} + v_{tt}$ satisfy property (3), the cubic term does not:

$$(u + v)^3 = u^3 + 3u^2v + 3uv^2 + v^3 \neq u^3 + v^3.$$

The advantage of linearity for the equation $\mathcal{L}u = 0$ is that if $u$ and $v$ are both solutions, so is $(u + v)$. If $u_1, \ldots, u_n$ are all solutions, so is any linear combination

$$c_1 u_1(x) + \cdots + c_n u_n(x) = \sum_{j=1}^{n} c_j u_j(x) \qquad (c_j = \text{constants}).$$

(This is sometimes called the superposition principle.) Another consequence of linearity is that if you add a homogeneous solution [a solution of (4)] to an inhomogeneous solution [a solution of (5)], you get an inhomogeneous solution. (Why?) The mathematical structure that deals with linear combinations and linear operators is the vector space. Exercises 5–10 are review problems on vector spaces.

We'll study, almost exclusively, linear systems with constant coefficients. Recall that for ODEs you get linear combinations. The coefficients are the arbitrary constants. *For an ODE of order m, you get m arbitrary constants.*

Let's look at some PDEs.

## Example 1.

Find all $u(x, y)$ satisfying the equation $u_{xx} = 0$. Well, we can integrate once to get $u_x = \text{constant}$. But that's not really right since there's another variable $y$. What we really get is $u_x(x, y) = f(y)$, where $f(y)$ is arbitrary. Do it again to get $u(x, y) = f(y)x + g(y)$. This is the solution formula. Note that *there are two arbitrary* functions *in the solution*. We see this as well in the next two examples. $\square$

## Example 2.

Solve the PDE $u_{xx} + u = 0$. Again, it's really an ODE with an extra variable $y$. We know how to solve the ODE, so the solution is

$$u = f(y) \cos x + g(y) \sin x,$$

where again $f(y)$ and $g(y)$ are two arbitrary functions of $y$. You can easily check this formula by differentiating twice to verify that $u_{xx} = -u$. $\square$

## Example 3.

Solve the PDE $u_{xy} = 0$. This isn't too hard either. First let's integrate in $x$, regarding $y$ as fixed. So we get

$$u_y(x, y) = f(y).$$

Next let's integrate in $y$ regarding $x$ as fixed. We get the solution

$$u(x, y) = F(y) + G(x),$$

where $F' = f$. $\square$

**Moral**   *A PDE has arbitrary functions in its solution.* In these examples the arbitrary functions are functions of one variable that combine to produce a function $u(x, y)$ of two variables which is only partly arbitrary.

A function of two variables contains *immensely* more information than a function of only one variable. Geometrically, it is obvious that a surface $\{u = f(x, y)\}$, the graph of a function of two variables, is a much more complicated object than a curve $\{u = f(x)\}$, the graph of a function of one variable.

To illustrate this, we can ask how a computer would record a function $u = f(x)$. Suppose that we choose 100 points to describe it using equally spaced values of $x$: $x_1, x_2, x_3, \ldots, x_{100}$. We could write them down in a column, and next to each $x_j$ we could write the corresponding value $u_j = f(x_j)$. Now how about a function $u = f(x, y)$? Suppose that we choose 100 equally spaced values of $x$ and also of $y$: $x_1, x_2, x_3, \ldots, x_{100}$ and $y_1, y_2, y_3, \ldots, y_{100}$. Each pair $x_i, y_j$ provides a value $u_{ij} = f(x_i, y_j)$, so there will be $100^2 = 10{,}000$ lines of the form

$$x_i \qquad y_j \qquad u_{ij}$$

required to describe the function! (If we had a prearranged system, we would need to record only the values $u_{ij}$.) A function of three variables described discretely by 100 values in each variable would require a million numbers!

To understand this book what do you have to know from calculus? Certainly all the basic facts about partial derivatives and multiple integrals. For a brief discussion of such topics, see the Appendix. Here are a few things to keep in mind, some of which may be new to you.

1. Derivatives are *local*. For instance, to calculate the derivative $(\partial u / \partial x)(x_0, t_0)$ at a particular point, you need to know just the values of $u(x, t_0)$ for $x$ *near* $x_0$, since the derivative is the limit as $x \to x_0$.

2. Mixed derivatives are equal: $u_{xy} = u_{yx}$. (We assume throughout this book, unless stated otherwise, that all derivatives exist and are continuous.)

3. The chain rule is used frequently in PDEs; for instance,

$$\frac{\partial}{\partial x}[f(g(x, t))] = f'(g(x, t)) \cdot \frac{\partial g}{\partial x}(x, t).$$

4. For the integrals of derivatives, the reader should learn or review Green's theorem and the divergence theorem. (See the end of Section A.3 in the Appendix.)

5. Derivatives of integrals like $I(t) = \int_{a(t)}^{b(t)} f(x, t)\, dx$ (see Section A.3).

6. Jacobians (change of variable in a double integral) (see Section A.1).

7. Infinite series of functions and their differentiation (see Section A.2).

8. Directional derivatives (see Section A.1).

9. We'll often reduce PDEs to ODEs, so we must know how to solve simple ODEs. But we won't need to know anything about tricky ODEs.

## EXERCISES

1.  Verify the linearity and nonlinearity of the eight examples of PDEs given in the text, by checking whether or not equations (3) are valid.

2.  Which of the following operators are linear?
    (a)  $\mathcal{L}u = u_x + xu_y$
    (b)  $\mathcal{L}u = u_x + uu_y$
    (c)  $\mathcal{L}u = u_x + u_y^2$
    (d)  $\mathcal{L}u = u_x + u_y + 1$
    (e)  $\mathcal{L}u = \sqrt{1 + x^2}\,(\cos y)u_x + u_{yxy} - [\arctan(x/y)]u$

3.  For each of the following equations, state the order and whether it is nonlinear, linear inhomogeneous, or linear homogeneous; provide reasons.
    (a)  $u_t - u_{xx} + 1 = 0$
    (b)  $u_t - u_{xx} + xu = 0$
    (c)  $u_t - u_{xxt} + uu_x = 0$
    (d)  $u_{tt} - u_{xx} + x^2 = 0$
    (e)  $iu_t - u_{xx} + u/x = 0$
    (f)  $u_x(1 + u_x^2)^{-1/2} + u_y(1 + u_y^2)^{-1/2} = 0$
    (g)  $u_x + e^y u_y = 0$
    (h)  $u_t + u_{xxxx} + \sqrt{1 + u} = 0$

4.  Show that the difference of two solutions of an inhomogeneous linear equation $\mathcal{L}u = g$ with the same $g$ is a solution of the homogeneous equation $\mathcal{L}u = 0$.

5.  Which of the following collections of 3-vectors $[a, b, c]$ are vector spaces? Provide reasons.
    (a)  The vectors with $b = 0$.
    (b)  The vectors with $b = 1$.
    (c)  The vectors with $ab = 0$.
    (d)  All the linear combinations of the two vectors $[1, 1, 0]$ and $[2, 0, 1]$.
    (e)  All the vectors such that $c - a = 2b$.

6.  Are the three vectors $[1, 2, 3]$, $[-2, 0, 1]$, and $[1, 10, 17]$ linearly dependent or independent? Do they span all vectors or not?

7.  Are the functions $1 + x$, $1 - x$, and $1 + x + x^2$ linearly dependent or independent? Why?

8.  Find a vector that, together with the vectors $[1, 1, 1]$ and $[1, 2, 1]$, forms a basis of $\mathbb{R}^3$.

9.  Show that the functions $(c_1 + c_2 \sin^2 x + c_3 \cos^2 x)$ form a vector space. Find a basis of it. What is its dimension?

10. Show that the solutions of the differential equation $u''' - 3u'' + 4u = 0$ form a vector space. Find a basis of it.

11. Verify that $u(x, y) = f(x)g(y)$ is a solution of the PDE $uu_{xy} = u_x u_y$ for all pairs of (differentiable) functions $f$ and $g$ of one variable.

12.   Verify by direct substitution that

$$u_n(x, y) = \sin nx \sinh ny$$

is a solution of $u_{xx} + u_{yy} = 0$ for every $n > 0$.

## 1.2   FIRST-ORDER LINEAR EQUATIONS

We begin our study of PDEs by solving some simple ones. The solution is quite geometric in spirit.

The simplest possible PDE is $\partial u / \partial x = 0$ [where $u = u(x, y)$]. Its general solution is $u = f(y)$, where $f$ is any function of *one* variable. For instance, $u = y^2 - y$ and $u = e^y \cos y$ are two solutions. Because the solutions don't depend on $x$, they are constant on the lines $y = $ constant in the $xy$ plane.

### THE CONSTANT COEFFICIENT EQUATION

Let us solve

$$au_x + bu_y = 0,\qquad(1)$$

where $a$ and $b$ are constants not both zero.

**Geometric Method**   The quantity $au_x + bu_y$ is the directional derivative of $u$ in the direction of the vector $\mathbf{V} = (a, b) = a\mathbf{i} + b\mathbf{j}$. It must always be zero. This means that $u(x, y)$ must be constant in the direction of $\mathbf{V}$. The vector $(b, -a)$ is orthogonal to $\mathbf{V}$. The lines parallel to $\mathbf{V}$ (see Figure 1) have the equations $bx - ay = $ constant. (They are called the *characteristic lines*.) The solution is constant on each such line. Therefore, $u(x, y)$ depends on $bx - ay$ only. Thus the solution is

$$u(x, y) = f(bx - ay),\qquad(2)$$

where $f$ is any function of one variable. Let's explain this conclusion more explicitly. On the line $bx - ay = c$, the solution $u$ has a constant value. Call

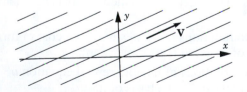

Figure 1

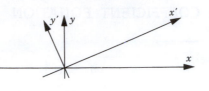

**Figure 2**

this value $f(c)$. Then $u(x, y) = f(c) = f(bx - ay)$. Since $c$ is arbitrary, we have formula (2) for all values of $x$ and $y$. In $xyu$ space the solution defines a surface that is made up of parallel horizontal straight lines like a sheet of corrugated iron.

**Coordinate Method** Change variables (or "make a change of coordinates"; Figure 2) to

$$x' = ax + by \qquad y' = bx - ay. \tag{3}$$

Replace all $x$ and $y$ derivatives by $x'$ and $y'$ derivatives. By the chain rule,

$$u_x = \frac{\partial u}{\partial x} = \frac{\partial u}{\partial x'}\frac{\partial x'}{\partial x} + \frac{\partial u}{\partial y'}\frac{\partial y'}{\partial x} = au_{x'} + bu_{y'}$$

and

$$u_y = \frac{\partial u}{\partial y} = \frac{\partial u}{\partial y'}\frac{\partial y'}{\partial y} + \frac{\partial u}{\partial x'}\frac{\partial x'}{\partial y} = bu_{x'} - au_{y'}.$$

Hence $au_x + bu_y = a(au_{x'} + bu_{y'}) + b(bu_{x'} - au_{y'}) = (a^2 + b^2)u_{x'}$. So, since $a^2 + b^2 \neq 0$, the equation takes the form $u_{x'} = 0$ in the new (primed) variables. Thus the solution is $u = f(y') = f(bx - ay)$, with $f$ an arbitrary function of *one* variable. This is exactly the same answer as before!

### Example 1.

Solve the PDE $4u_x - 3u_y = 0$, together with the auxiliary condition that $u(0, y) = y^3$. By (2) we have $u(x, y) = f(-3x - 4y)$. This is the general solution of the PDE. Setting $x = 0$ yields the equation $y^3 = f(-4y)$. Letting $w = -4y$ yields $f(w) = -w^3/64$. Therefore, $u(x, y) = (3x + 4y)^3/64$.

Solutions can usually be checked much easier than they can be derived. We check this solution by simple differentiation: $u_x = 9(3x + 4y)^2/64$ and $u_y = 12(3x + 4y)^2/64$ so that $4u_x - 3u_y = 0$. Furthermore, $u(0, y) = (3 \cdot 0 + 4y)^3/64 = y^3$. $\square$

## THE VARIABLE COEFFICIENT EQUATION

The equation

$$u_x + yu_y = 0 \tag{4}$$

is linear and homogeneous but has a variable coefficient ($y$). We shall illustrate for equation (4) how to use the geometric method somewhat like Example 1.

The PDE (4) itself asserts that *the directional derivative in the direction of the vector* $(1, y)$ *is zero*. The curves in the $xy$ plane with $(1, y)$ as tangent vectors have slopes $y$ (see Figure 3). Their equations are

$$\frac{dy}{dx} = \frac{y}{1} \tag{5}$$

This ODE has the solutions

$$y = Ce^x. \tag{6}$$

These curves are called the *characteristic curves* of the PDE (4). As $C$ is changed, the curves fill out the $xy$ plane perfectly without intersecting. On each of the curves $u(x, y)$ is a constant because

$$\frac{d}{dx}u(x, Ce^x) = \frac{\partial u}{\partial x} + Ce^x\frac{\partial u}{\partial y} = u_x + yu_y = 0.$$

Thus $u(x, Ce^x) = u(0, Ce^0) = u(0, C)$ is independent of $x$. Putting $y = Ce^x$ and $C = e^{-x}y$, we have

$$u(x, y) = u(0, e^{-x}y).$$

It follows that

$$u(x, y) = f(e^{-x}y) \tag{7}$$

is the *general solution* of this PDE, where again $f$ is an arbitrary function of only a single variable. This is easily checked by differentiation using the chain rule (see Exercise 4). Geometrically, the "picture" of the solution $u(x, y)$ is that it is *constant on each characteristic curve* in Figure 3.

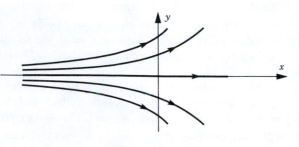

Figure 3

## Example 2.

Find the solution of (4) that satisfies the auxiliary condition $u(0, y) = y^3$. Indeed, putting $x = 0$ in (7), we get $y^3 = f(e^{-0}y)$, so that $f(y) = y^3$. Therefore, $u(x, y) = (e^{-x}y)^3 = e^{-3x}y^3$. $\qquad\qquad\square$

## Example 3.

Solve the PDE

$$\boxed{u_x + 2xy^2u_y = 0.} \qquad\qquad (8)$$

The characteristic curves satisfy the ODE $dy/dx = 2xy^2/1 = 2xy^2$. To solve the ODE, we separate variables: $dy/y^2 = 2x\,dx$; hence $-1/y = x^2 - C$, so that

$$y = (C - x^2)^{-1}. \qquad\qquad (9)$$

These curves are the characteristics. Again, $u(x, y)$ is a constant on each such curve. (Check it by writing it out.) So $u(x, y) = f(C)$, where $f$ is an arbitrary function. Therefore, the general solution of (8) is obtained by solving (9) for $C$. That is,

$$\boxed{u(x, y) = f\left(x^2 + \frac{1}{y}\right).} \qquad\qquad (10)$$

Again this is easily checked by differentiation, using the chain rule: $u_x = 2x \cdot f'(x^2 + 1/y)$ and $u_y = -(1/y^2) \cdot f'(x^2 + 1/y)$, whence $u_x + 2xy^2u_y = 0$. $\qquad\qquad\square$

In summary, the geometric method works nicely for any PDE of the form $a(x, y)u_x + b(x, y)u_y = 0$. It reduces the solution of the PDE to the solution of the ODE $dy/dx = b(x, y)/a(x, y)$. If the ODE can be solved, so can the PDE. Every solution of the PDE is constant on the solution curves of the ODE.

**Moral**  Solutions of PDEs generally depend on arbitrary functions (instead of arbitrary constants). You need an auxiliary condition if you want to determine a unique solution. Such conditions are usually called *initial* or *boundary* conditions. We shall encounter these conditions throughout the book.

## EXERCISES

1.  Solve the first-order equation $2u_t + 3u_x = 0$ with the auxiliary condition $u = \sin x$ when $t = 0$.

2.  Solve the equation $3u_y + u_{xy} = 0$. (*Hint*: Let $v = u_y$.)

3. Solve the equation $(1 + x^2)u_x + u_y = 0$. Sketch some of the characteristic curves.

4. Check that (7) indeed solves (4).

5. Solve the equation $xu_x + yu_y = 0$.

6. Solve the equation $\sqrt{1 - x^2}u_x + u_y = 0$ with the condition $u(0, y) = y$.

7. (a) Solve the equation $yu_x + xu_y = 0$ with $u(0, y) = e^{-y^2}$.
   (b) In which region of the $xy$ plane is the solution uniquely determined?

8. Solve $au_x + bu_y + cu = 0$.

9. Solve the equation $u_x + u_y = 1$.

10. Solve $u_x + u_y + u = e^{x+2y}$ with $u(x, 0) = 0$.

11. Solve $au_x + bu_y = f(x, y)$, where $f(x, y)$ is a given function. If $a \neq 0$, write the solution in the form

$$u(x, y) = (a^2 + b^2)^{-1/2} \int_L f \, ds + g(bx - ay),$$

where $g$ is an arbitrary function of one variable, $L$ is the characteristic line segment from the $y$ axis to the point $(x, y)$, and the integral is a line integral. (*Hint:* Use the coordinate method.)

12. Show that the new coordinate axes defined by (3) are orthogonal.

13. Use the coordinate method to solve the equation

$$u_x + 2u_y + (2x - y)u = 2x^2 + 3xy - 2y^2.$$

## 1.3 FLOWS, VIBRATIONS, AND DIFFUSIONS

The subject of PDEs was practically a branch of physics until the twentieth century. In this section we present a series of examples of PDEs as they occur in physics. They provide the basic motivation for all the PDE problems we study in the rest of the book. We shall see that most often in physical problems the independent variables are those of space $x$, $y$, $z$, and time $t$.

### Example 1. Simple Transport

Consider a fluid, water, say, flowing at a constant rate $c$ along a horizontal pipe of fixed cross section in the positive $x$ direction. A substance, say a pollutant, is suspended in the water. Let $u(x, t)$ be its concentration in grams/centimeter at time $t$. Then

$$\boxed{u_t + cu_x = 0.} \tag{1}$$

(That is, the rate of change $u_t$ of concentration is proportional to the gradient $u_x$. Diffusion is assumed to be negligible.) Solving this equation as in Section 1.2, we find that the concentration is a function of $(x - ct)$

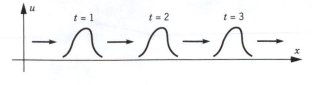

**Figure 1**

only. This means that the substance is transported to the right at a fixed speed $c$. Each individual particle moves to the right at speed $c$; that is, in the $xt$ plane, it moves precisely along a characteristic line (see Figure 1).                                                □

**Derivation of Equation (1).**   The amount of pollutant in the interval $[0, b]$ at the time $t$ is $M = \int_0^b u(x, t)\, dx$, in grams, say. At the later time $t + h$, the same molecules of pollutant have moved to the right by $c \cdot h$ centimeters. Hence

$$M = \int_0^b u(x, t)dx = \int_{ch}^{b+ch} u(x, t + h)\, dx.$$

Differentiating with respect to $b$, we get

$$u(b, t) = u(b + ch, t + h).$$

Differentiating with respect to $h$ and putting $h = 0$, we get

$$0 = cu_x(b, t) + u_t(b, t),$$

which is equation (1).                                                □

## Example 2. Vibrating String

Consider a flexible, elastic homogenous string or thread of length $l$, which undergoes relatively small transverse vibrations. For instance, it could be a guitar string or a plucked violin string. At a given instant $t$, the string might look as shown in Figure 2. Assume that it remains in a plane. Let $u(x, t)$ be its displacement from equilibrium position at time $t$ and position $x$. Because the string is perfectly flexible, the tension (force) is directed tangentially along the string (Figure 3). Let $T(x, t)$ be the magnitude of this tension vector. Let $\rho$ be the density (mass per unit length) of the string. It is a constant because the string is homogeneous. We shall write down Newton's law for the part of the string between any two points at $x = x_0$ and $x = x_1$. The slope of the string at $x_1$ is

**Figure 2**

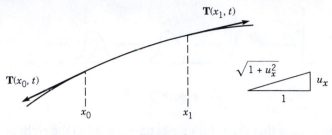

**Figure 3**

$u_x(x_1, t)$. Newton's law $\mathbf{F} = m\mathbf{a}$ in its longitudinal $(x)$ and transverse $(u)$ components is

$$\left.\frac{T}{\sqrt{1 + u_x^2}}\right|_{x_0}^{x_1} = 0 \qquad \text{(longitudinal)}$$

$$\left.\frac{Tu_x}{\sqrt{1 + u_x^2}}\right|_{x_0}^{x_1} = \int_{x_0}^{x_1} \rho u_{tt}\, dx \quad \text{(transverse)}$$

The right sides are the components of the mass times the acceleration integrated over the piece of string. Since we have assumed that the motion is purely transverse, there is no longitudinal motion.

Now we also assume that the motion is small—more specifically, that $|u_x|$ is quite small. Then $\sqrt{1 + u_x^2}$ may be replaced by 1. This is justified by the Taylor expansion, actually the binomial expansion,

$$\sqrt{1 + u_x^2} = 1 + \tfrac{1}{2}u_x^2 + \cdots$$

where the dots represent higher powers of $u_x$. If $u_x$ is small, it makes sense to drop the even smaller quantity $u_x^2$ and its higher powers. With the square roots replaced by 1, the first equation then says that $T$ is constant along the string. Let us assume that $T$ is independent of $t$ as well as $x$. The second equation, differentiated, says that

$$(Tu_x)_x = \rho u_{tt}.$$

That is,

$$\boxed{u_{tt} = c^2 u_{xx} \quad \text{where } c = \sqrt{\frac{T}{\rho}}.} \qquad (2)$$

This is the *wave equation*. At this point it is not clear why $c$ is defined in this manner, but shortly we'll see that $c$ is the *wave speed*. $\qquad\square$

There are many *variations* of this equation:

(i)  If significant air resistance $r$ is present, we have an extra term proportional to the speed $u_t$, thus:

$$u_{tt} - c^2 u_{xx} + r u_t = 0 \quad \text{where } r > 0. \tag{3}$$

(ii)  If there is a transverse elastic force, we have an extra term proportional to the displacement $u$, as in a coiled spring, thus:

$$u_{tt} - c^2 u_{xx} + k u = 0 \quad \text{where } k > 0. \tag{4}$$

(iii)  If there is an externally applied force, it appears as an extra term, thus:

$$u_{tt} - c^2 u_{xx} = f(x, t), \tag{5}$$

which makes the equation inhomogeneous.

Our derivation of the wave equation has been quick but not too precise. A much more careful derivation can be made, which makes precise the physical and mathematical assumptions [We, Chap. 1].

The same wave equation or a variation of it describes many other wavelike phenomena, such as the vibrations of an elastic bar, the sound waves in a pipe, and the long water waves in a straight canal. Another example is the equation for the electrical current in a transmission line,

$$u_{xx} = CL u_{tt} + (CR + GL) u_t + GR u,$$

where $C$ is the capacitance per unit length, $G$ the leakage resistance per unit length, $R$ the resistance per unit length, and $L$ the self-inductance per unit length.

### Example 3. Vibrating Drumhead

The two-dimensional version of a string is an elastic, flexible, homogeneous drumhead, that is, a membrane stretched over a frame. Say the frame lies in the $xy$ plane (see Figure 4), $u(x, y, t)$ is the vertical

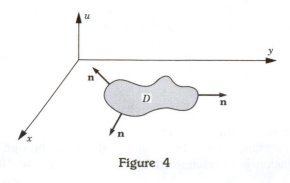

**Figure 4**

displacement, and there is no horizontal motion. The horizontal components of Newton's law again give constant tension $T$. Let $D$ be any domain in the $xy$ plane, say a circle or a rectangle. Let bdy $D$ be its boundary curve. We use reasoning similar to the one-dimensional case. The vertical component gives (approximately)

$$F = \int_{\text{bdy } D} T \frac{\partial u}{\partial n} \, ds = \iint_D \rho u_{tt} \, dx \, dy = ma,$$

where the left side is the total force acting on the piece $D$ of the membrane, and where $\partial u / \partial n = \mathbf{n} \cdot \nabla u$ is the directional derivative in the outward normal direction, $\mathbf{n}$ being the unit outward normal vector on bdy $D$. By Green's theorem (see Section A.3 in the Appendix), this can be rewritten as

$$\iint_D \nabla \cdot (T \nabla u) \, dx \, dy = \iint_D \rho u_{tt} \, dx \, dy.$$

Since $D$ is arbitrary, we deduce from the second vanishing theorem in Section A.1 that $\rho u_{tt} = \nabla \cdot (T \nabla u)$. Since $T$ is constant, we get

$$u_{tt} = c^2 \nabla \cdot (\nabla u) \equiv c^2 (u_{xx} + u_{yy}), \tag{6}$$

where $c = \sqrt{T/\rho}$ as before and $\nabla \cdot (\nabla u) = \operatorname{div} \operatorname{grad} u = u_{xx} + u_{yy}$ is known as the *two-dimensional laplacian*. Equation (6) is the two-dimensional wave equation.   □

The pattern is now clear. Simple three-dimensional vibrations obey the equation

$$u_{tt} = c^2 (u_{xx} + u_{yy} + u_{zz}). \tag{7}$$

The operator $\mathcal{L} = \partial^2/\partial x^2 + \partial^2/\partial y^2 + \partial/\partial z^2$ is called the *three-dimensional laplacian* operator, usually denoted by $\Delta$ or $\nabla^2$. Physical examples described by the three-dimensional wave equation or a variation of it include the vibrations of an elastic solid, sound waves in air, electromagnetic waves (light, radar, etc.), linearized supersonic airflow, free mesons in nuclear physics, and seismic waves propagating through the earth.

### Example 4. Diffusion

Let us imagine a motionless liquid filling a straight tube or pipe and a chemical substance, say a dye, which is diffusing through the liquid. Simple diffusion is characterized by the following law. [It is not to

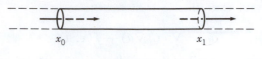

**Figure 5**

be confused with convection (transport), which refers to currents in the liquid.] The dye moves from regions of higher concentration to regions of lower concentration. The rate of motion is proportional to the concentration gradient. (This is known as Fick's law of diffusion.) Let $u(x, t)$ be the concentration (mass per unit length) of the dye at position $x$ of the pipe at time $t$.

In the section of pipe from $x_0$ to $x_1$ (see Figure 5), the mass of dye is

$$M(t) = \int_{x_0}^{x_1} u(x, t)\, dx, \quad \text{so} \quad \frac{dM}{dt} = \int_{x_0}^{x_1} u_t(x, t)\, dx.$$

The mass in this section of pipe cannot change except by flowing in or out of its ends. By Fick's law,

$$\frac{dM}{dt} = \text{flow in} - \text{flow out} = ku_x(x_1, t) - ku_x(x_0, t),$$

where $k$ is a proportionality constant. Therefore, those two expressions are equal:

$$\int_{x_0}^{x_1} u_t(x, t)\, dx = ku_x(x_1, t) - ku_x(x_0, t).$$

Differentiating with respect to $x_1$, we get

$$\boxed{u_t = ku_{xx}.} \tag{8}$$

This is the *diffusion equation*.

In three dimensions we have

$$\iiint_D u_t\, dx\, dy\, dz = \iint_{\text{bdy } D} k(\mathbf{n} \cdot \nabla u)\, dS,$$

where $D$ is any solid domain and bdy $D$ is its bounding surface. By the divergence theorem (using the arbitrariness of $D$ as in Example 3), we get the *three-dimensional diffusion equation*

$$\boxed{u_t = k(u_{xx} + u_{yy} + u_{zz}) = k\, \Delta u.} \tag{9}$$

If there is an external source (or a "sink") of the dye, and if the rate $k$ of diffusion is a variable, we get the more general inhomogeneous equation

$$u_t = \nabla \cdot (k\,\nabla u) + f(x, t).$$

The same equation describes the conduction of heat, brownian motion, diffusion models of population dynamics, and many other phenomena.

□

## Example 5. Heat Flow

We let $u(x, y, z, t)$ be the temperature and let $H(t)$ be the amount of heat (in calories, say) contained in a region $D$. Then

$$H(t) = \iiint_D c\rho u \, dx \, dy \, dz,$$

where $c$ is the "specific heat" of the material and $\rho$ is its density (mass per unit volume). The change in heat is

$$\frac{dH}{dt} = \iiint_D c\rho u_t \, dx \, dy \, dz.$$

Fourier's law says that heat flows from hot to cold regions proportionately to the temperature gradient. But heat cannot be lost from $D$ except by leaving it through the boundary. This is the law of conservation of energy. Therefore, the change of heat energy in $D$ also equals the heat flux across the boundary,

$$\frac{dH}{dt} = \iint_{\text{bdy } D} \kappa(\mathbf{n} \cdot \nabla u)\,dS,$$

where $\kappa$ is a proportionality factor (the "heat conductivity"). By the divergence theorem,

$$\iiint_D c\rho \frac{\partial u}{\partial t} \, dx \, dy \, dz = \iiint_D \nabla \cdot (\kappa\,\nabla u)\,dx\,dy\,dz$$

and we get the *heat equation*

$$\boxed{c\rho \frac{\partial u}{\partial t} = \nabla \cdot (\kappa\,\nabla u).} \tag{10}$$

If $c$, $\rho$, and $\kappa$ are constants, it is exactly the same as the diffusion equation!

□

## Example 6. Stationary Waves and Diffusions

Consider any of the four preceding examples in a situation where the physical state does not change with time. Then $u_t = u_{tt} = 0$. So *both* the wave *and* the diffusion equations reduce to

$$\Delta u = u_{xx} + u_{yy} + u_{zz} = 0. \qquad (11)$$

This is called the *Laplace equation*. Its solutions are called *harmonic functions*. For example, consider a hot object that is constantly heated in an oven. The heat is not expected to be evenly distributed throughout the oven. The temperature of the object eventually reaches a steady (or equilibrium) state. This is a harmonic function $u(x, y, z)$. (Of course, if the heat were being supplied evenly in all directions, the steady state would be $u \equiv$ constant.) In the *one*-dimensional case (e.g., a laterally insulated thin rod that exchanges heat with its environment only through its ends), we would have $u$ a function of $x$ only. So the Laplace equation would reduce simply to $u_{xx} = 0$. Hence $u = c_1 x + c_2$. The two- and three-dimensional cases are *much* more interesting (see Chapter 6 for the solutions). $\qquad \square$

## Example 7. The Hydrogen Atom

This is an electron moving around a proton. Let $m$ be the mass of the electron, $e$ its charge, and $h$ Planck's constant divided by $2\pi$. Let the origin of coordinates $(x, y, z)$ be at the proton and let $r = (x^2 + y^2 + z^2)^{1/2}$ be the spherical coordinate. Then the motion of the electron is given by a "wave function" $u(x, y, z, t)$ which satisfies Schrödinger's equation

$$-ihu_t = \frac{h^2}{2m}\Delta u + \frac{e^2}{r}u \qquad (12)$$

in all of space $-\infty < x,y,z < +\infty$. Furthermore, we are supposed to have $\iiint |u|^2 dx \, dy \, dz = 1$ (integral over all space). Note that $i = \sqrt{-1}$ and $u$ is complex-valued. The coefficient function $e^2/r$ is called the potential. For any other atom with a single electron, such as a helium ion, $e^2$ is replaced by $Ze^2$, where $Z$ is the atomic number. $\qquad \square$

What does this mean physically? In quantum mechanics quantities cannot be measured exactly but only with a certain probability. The *wave function* $u(x, y, z, t)$ represents a possible *state* of the electron. If $D$ is *any* region in $xyz$ space, then

$$\iiint_D |u|^2 \, dx \, dy \, dz$$

is the probability of finding the electron in the region $D$ at the time $t$. The *expected $z$ coordinate of the position* of the electron at the time $t$ is the value of the integral

$$\iiint z|u(x, y, z, t)|^2 \, dx \, dy \, dz;$$

similarly for the $x$ and $y$ coordinates. The *expected $z$ coordinate of the momentum* is

$$\iiint -ih\frac{\partial u}{\partial z}(x, y, z, t) \cdot \bar{u}(x, y, z, t) \, dx \, dy \, dz,$$

where $\bar{u}$ is the complex conjugate of $u$. All other observable quantities are given by operators $A$, which act on functions. The expected value of the observable $A$ equals

$$\iiint Au(x, y, z, t) \cdot \bar{u}(x, y, z, t) \, dx \, dy \, dz.$$

Thus the position is given by the operator $Au = \mathbf{x}u$, where $\mathbf{x} = x\mathbf{i} + y\mathbf{j} + z\mathbf{k}$, and the momentum is given by the operator $Au = -ih\nabla u$.

Schrödinger's equation is most easily regarded simply as an axiom that leads to the correct physical conclusions, rather than as an equation that can be derived from simpler principles. It explains why atoms are stable and don't collapse. It explains the energy levels of the electron in the hydrogen atom observed by Bohr. In principle, elaborations of it explain the structure of *all* atoms and molecules and so all of chemistry! With many particles, the wave function $u$ depends on time $t$ and all the coordinates of all the particles and so is a function of a large number of variables. The Schrödinger equation then becomes

$$-ihu_t = \sum_{i=1}^{n} \frac{h^2}{2m_i}(u_{x_i x_i} + u_{y_i y_i} + u_{z_i z_i}) + V(x_1, \ldots, z_n)u$$

for $n$ particles, where the potential function $V$ depends on all the $3n$ coordinates. Except for the hydrogen and helium atoms (the latter having two electrons), the mathematical analysis is impossible to carry out completely and cannot be calculated even with the help of the modern computer. Nevertheless, with the use of various approximations, many of the facts about more complicated atoms and the chemical binding of molecules can be understood. □

This has been a brief introduction to the sources of PDEs in physical problems. Many realistic situations lead to much more complicated PDEs. See Chapter 13 for some additional examples.

## EXERCISES

1.  Carefully derive the equation of a string in a medium in which the resistance is proportional to the velocity.

2.  A flexible chain of length $l$ is hanging from one end $x = 0$ but oscillates horizontally. Let the $x$ axis point downward and the $u$ axis point to the right. Assume that the force of gravity at each point of the chain equals the weight of the part of the chain below the point and is directed tangentially along the chain. Assume that the oscillations are small. Find the PDE satisfied by the chain.

3.  On the sides of a thin rod, heat exchange takes place (obeying Newton's law of cooling—flux proportional to temperature difference) with a medium of constant temperature $T_0$. What is the equation satisfied by the temperature $u(x, t)$, neglecting its variation across the rod?

4.  Suppose that some particles which are suspended in a liquid medium would be pulled down at the constant velocity $V > 0$ by gravity in the absence of diffusion. Taking account of the diffusion, find the equation for the concentration of particles. Assume homogeneity in the horizontal directions $x$ and $y$. Let the $z$ axis point upwards.

5.  Derive the equation of one-dimensional diffusion in a medium that is moving along the $x$ axis to the right at constant speed $V$.

6.  Consider heat flow in a long circular cylinder where the temperature depends only on $t$ and on the distance $r$ to the axis of the cylinder. Here $r = \sqrt{x^2 + y^2}$ is the cylindrical coordinate. From the three-dimensional heat equation derive the equation $u_t = k(u_{rr} + u_r/r)$.

7.  Solve Exercise 6 in a ball except that the temperature depends only on the spherical coordinate $\sqrt{x^2 + y^2 + z^2}$. Derive the equation $u_t = k(u_{rr} + 2u_r/r)$.

8.  For the hydrogen atom, if $\int |u|^2 \, d\mathbf{x} = 1$ at $t = 0$, show that the same is true at all later times. (*Hint:* Differentiate the integral with respect to $t$, taking care about the solution being complex valued. Assume that $u$ and $\nabla u \to 0$ fast enough as $|\mathbf{x}| \to \infty$.)

9.  This is an exercise on the divergence theorem

$$\iiint_D \nabla \cdot \mathbf{F} \, d\mathbf{x} = \iint_{\text{bdy } D} \mathbf{F} \cdot \mathbf{n} \, dS,$$

valid for any bounded domain $D$ in space with boundary surface bdy $D$ and unit outward normal vector $\mathbf{n}$. If you never learned it, see Section A.3. It is crucial that $D$ be bounded As an exercise, verify it in the following case by calculating both sides separately: $\mathbf{F} = r^2\mathbf{x}$, $\mathbf{x} = x\mathbf{i} + y\mathbf{j} + z\mathbf{k}$, $r^2 = x^2 + y^2 + z^2$, and $D =$ the ball of radius $a$ and center at the origin.

10.  If $\mathbf{f}(\mathbf{x})$ is continuous and $|\mathbf{f}(\mathbf{x})| \le 1/(|\mathbf{x}|^3 + 1)$ for all $\mathbf{x}$, show that

$$\iiint_{\text{all space}} \nabla \cdot \mathbf{f}\, dx = 0.$$

(*Hint:* Take $D$ to be a large ball, apply the divergence theorem, and let its radius tend to infinity.)

11.  If curl $\mathbf{v} = \mathbf{0}$ in all of three-dimensional space, show that there exists a scalar function $\phi(x, y, z)$ such that $\mathbf{v} = \text{grad } \phi$.

## 1.4  INITIAL AND BOUNDARY CONDITIONS

Because PDEs typically have so many solutions, as we saw in Section 1.2, we single out one solution by imposing auxiliary conditions. We attempt to formulate the conditions so as to specify a unique solution. These conditions are motivated by the physics and they come in two varieties, initial conditions and boundary conditions.

An *initial condition* specifies the physical state at a particular time $t_0$. For the diffusion equation the initial condition is

$$u(\mathbf{x}, t_0) = \phi(\mathbf{x}), \tag{1}$$

where $\phi(\mathbf{x}) = \phi(x, y, z)$ is a given function. For a diffusing substance, $\phi(\mathbf{x})$ is the initial concentration. For heat flow, $\phi(\mathbf{x})$ is the initial temperature. For the Schrödinger equation, too, (1) is the usual initial condition.

For the wave equation there is a *pair* of initial conditions

$$u(\mathbf{x}, t_0) = \phi(\mathbf{x}) \quad \text{and} \quad \frac{\partial u}{\partial t}(\mathbf{x}, t_0) = \psi(\mathbf{x}), \tag{2}$$

where $\phi(\mathbf{x})$ is the initial position and $\psi(\mathbf{x})$ is the initial velocity. It is clear on physical grounds that both of them must be specified in order to determine the position $u(\mathbf{x}, t)$ at later times. (We shall also prove this mathematically.)
□

In each physical problem we have seen that there is a *domain D* in which the PDE is valid. For the vibrating string, $D$ is the interval $0 < x < l$, so the boundary of $D$ consists only of the two points $x = 0$ and $x = l$. For the drumhead, the domain is a plane region and its boundary is a closed curve. For the diffusing chemical substance, $D$ is the container holding the liquid, so its boundary is a surface $S = \text{bdy } D$. For the hydrogen atom, the domain is all of space, so it has no boundary.

It is clear, again from our physical intuition, that it is necessary to specify some *boundary condition* if the solution is to be determined. The three most important kinds of boundary conditions are:

(D)  $u$ is specified ("*Dirichlet* condition")

(N)  the normal derivative $\partial u/\partial n$ is specified ("*Neumann* condition")

(R)  $\partial u/\partial n + au$ is specified ("*Robin* condition")

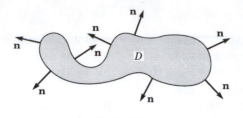

**Figure 1**

where $a$ is a given function of $x$, $y$, $z$, and $t$. Each is to hold for all $t$ and for $\mathbf{x} = (x, y, z)$ belonging to bdy $D$. Usually, we write (D), (N), and (R) as equations. For instance, (N) is written as the equation

$$\frac{\partial u}{\partial n} = g(\mathbf{x}, t) \tag{3}$$

where $g$ is a given function that could be called the boundary datum. Any of these boundary conditions is called *homogeneous* if the specified function $g(\mathbf{x}, t)$ vanishes (equals zero). Otherwise, it is called inhomogenous. As usual, $\mathbf{n} = (n_1, n_2, n_3)$ denotes the unit normal vector on bdy $D$, which points outward from $D$ (see Figure 1). Also, $\partial u/\partial n \equiv \mathbf{n} \cdot \nabla u$ denotes the directional derivative of $u$ in the outward normal direction.

In *one-dimensional* problems where $D$ is just an interval $0 < x < l$, the boundary consists of just the two endpoints, and these boundary conditions take the simple form

(D)   $u(0, t) = g(t)$   and   $u(l, t) = h(t)$

(N)   $\dfrac{\partial u}{\partial x}(0, t) = g(t)$   and   $\dfrac{\partial u}{\partial x}(l, t) = h(t)$

and similarly for the Robin condition.   □

Following are some illustrations of physical problems corresponding to these boundary conditions.

## THE VIBRATING STRING

If the string is held *fixed* at both ends, as for a violin string, we have the homogeneous Dirichlet conditions $u(0, t) = u(l, t) = 0$.

Imagine, on the other hand, that one end of the string is *free* to move transversally without any resistance (say, along a frictionless track); then there is no vertical component of tension $T$ at that end, so $u_x = 0$. This is a Neumann condition.

Third, the Robin condition would be the correct one if one were to imagine that an end of the string were free to move along a track but were attached to a coiled spring or rubber band (obeying Hooke's law) which tended to pull it back to equilibrium position. In that case the string would exchange some of its energy with the coiled spring.

Finally, if an end of the string were simply moved in a specified way, we would have an inhomogeneous Dirichlet condition at that end.

## DIFFUSION

If the diffusing substance is enclosed in a container $D$ so that none can escape or enter, then the concentration gradient in the normal direction must vanish, by Fick's law (see Exercise 2). Thus $\partial u/\partial n = 0$ on $S = $ bdy $D$, which is the Neumann condition.

If, on the other hand, the container is permeable and is so constructed that any substance that escapes to the boundary of the container is immediately washed away, then we have $u = 0$ on $S$.

## HEAT

Heat conduction is described by the diffusion equation with $u(\mathbf{x}, t) = $ temperature. If the object $D$ through which the heat is flowing is perfectly *insulated*, then no heat flows across the boundary and we have the Neumann condition $\partial u/\partial n = 0$ (see Exercise 2).

On the other hand, if the object were immersed in a large *reservoir* of specified temperature $g(t)$ and there were perfect thermal conduction, then we'd have the Dirichlet condition $u = g(t)$ on bdy $D$.

Suppose that we had a uniform rod insulated along its length $0 \leq x \leq l$, whose end at $x = l$ were immersed in the reservoir of temperature $g(t)$. If heat were exchanged between the end and the reservoir so as to obey Newton's law of cooling, then

$$\frac{\partial u}{\partial x}(l, t) = -a[u(l, t) - g(t)],$$

where $a > 0$. Heat from the hot rod radiates into the cool reservoir. This is an inhomogeneous Robin condition.

## LIGHT

Light is an electromagnetic field and as such is described by Maxwell's equations (see Chapter 13). Each component of the electric and magnetic field satisfies the wave equation. It is through the boundary conditions that the various components are related to each other. (They are "coupled.") Imagine, for example, light reflecting off a ball with a mirrored surface. This is a scattering problem. The domain $D$ where the light is propagating is the exterior of the ball. Certain boundary conditions then are satisfied by the electromagnetic field components. When polarization effects are not being studied, some scientists use the wave equation with homogeneous Dirichlet or Neumann conditions as a considerably simplified model of such a situation.

## SOUND

Our ears detect small disturbances in the air. The disturbances are described by the equations of gas dynamics, which form a system of nonlinear equations with velocity $\mathbf{v}$ and density $\rho$ as the unknowns. But *small* disturbances are described quite well by the so-called linearized equations, which are a lot simpler; namely,

$$\frac{\partial \mathbf{v}}{\partial t} + \frac{c_0^2}{\rho_0} \operatorname{grad} \rho = 0 \tag{4}$$

$$\frac{\partial \rho}{\partial t} + \rho_0 \operatorname{div} \mathbf{v} = 0 \tag{5}$$

(four scalar equations altogether). Here $\rho_0$ is the density and $c_0$ is the speed of sound in still air.

Assume now that the curl of $\mathbf{v}$ is zero; this means that there are no sound "eddies" and the velocity $\mathbf{v}$ is irrotational. It follows that $\rho$ and all three components of $\mathbf{v}$ satisfy the wave equation:

$$\frac{\partial^2 \mathbf{v}}{\partial t^2} = c_0^2 \, \Delta \mathbf{v} \quad \text{and} \quad \frac{\partial^2 \rho}{\partial t^2} = c_0^2 \, \Delta \rho. \tag{6}$$

The interested reader will find a derivation of these equations in Section 13.2.

Now if we are describing sound propagation in a closed, sound-insulated room $D$ with *rigid* walls, say a concert hall, then the air molecules at the wall can only move parallel to the boundary, so that no sound can travel in a normal direction to the boundary. So $\mathbf{v} \cdot \mathbf{n} = 0$ on bdy $D$. Since curl $\mathbf{v} = 0$, there is a standard fact in vector calculus (Exercise 1.3.11) which says that there is a "potential" function $\psi$ such that $\mathbf{v} = -\operatorname{grad} \psi$. The potential also satisfies the wave equation $\partial^2 \psi / \partial t^2 = c_0^2 \, \Delta \psi$, and the boundary condition for it is $-\mathbf{v} \cdot \mathbf{n} = \mathbf{n} \cdot \operatorname{grad} \psi = 0$ or Neumann's condition for $\psi$.

At an *open* window of the room $D$, the atmospheric pressure is a constant and there is no difference of pressure across the window. The pressure $p$ is proportional to the density $\rho$, for small disturbances of the air. Thus $\rho$ is a constant at the window, which means that $\rho$ satisfies the Dirichlet boundary condition $\rho = \rho_0$.

At a *soft* wall, such as an elastic membrane covering an open window, the pressure difference $p - p_0$ across the membrane is proportional to the normal velocity $\mathbf{v} \cdot \mathbf{n}$, namely

$$p - p_0 = Z \, \mathbf{v} \cdot \mathbf{n},$$

where Z is called the acoustic impedance of the wall. (A rigid wall has a very large impedance and an open window has zero impedance.) Now $p - p_0$ is in turn proportional to $\rho - \rho_0$ for small disturbances. Thus the system of four equations (4),(5) satisfies the boundary condition

$$\mathbf{v} \cdot \mathbf{n} = a(\rho - \rho_0),$$

where $a$ is a constant proportional to $1/Z$. (See [MI] for further discussion.)

$\square$

A different kind of boundary condition in the case of the wave equation is

$$\frac{\partial u}{\partial n} + b \frac{\partial u}{\partial t} = 0. \tag{7}$$

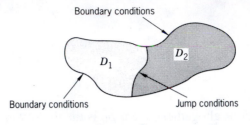

**Figure 2**

This condition means that energy is *radiated to* ($b > 0$) or *absorbed from* ($b < 0$) the exterior through the boundary. For instance, a vibrating string whose ends are immersed in a viscous liquid would satisfy (7) with $b > 0$ since energy is radiated to the liquid.

## CONDITIONS AT INFINITY

In case the domain $D$ is unbounded, the physics usually provides conditions at infinity. These can be tricky. An example is Schrödinger's equation, where the domain $D$ is all of space, and we require that $\int |u|^2 \, d\mathbf{x} = 1$. The finiteness of this integral means, in effect, that $u$ "vanishes at infinity."

A second example is afforded by the scattering of acoustic or electromagnetic waves. If we want to study sound or light waves that are radiating outward (to infinity), the appropriate condition at infinity is "Sommerfeld's outgoing radiation condition"

$$\lim_{r \to \infty} r \left( \frac{\partial u}{\partial r} - \frac{\partial u}{\partial t} \right) = 0, \tag{8}$$

where $r = |\mathbf{x}|$ is the spherical coordinate. (In a given mathematical context this limit would be made more precise.) (See Section 13.3.)

## JUMP CONDITIONS

These occur when the domain $D$ has two parts, $D = D_1 \cup D_2$ (see Figure 2), with different physical properties. An example is heat conduction, where $D_1$ and $D_2$ consist of two different materials (see Exercise 6).

## EXERCISES

1. By trial and error, find a solution of the diffusion equation $u_t = u_{xx}$ with the initial condition $u(x, 0) = x^2$.

2. (a) Show that the temperature of a metal rod, insulated at the end $x = 0$, satisfies the boundary condition $\partial u / \partial x = 0$. (Use Fourier's law.)

   (b) Do the same for the diffusion of gas along a tube that is closed off at the end $x = 0$. (Use Fick's law.)

(c)  Show that the three-dimensional version of (a) (insulated solid) or (b) (impermeable container) leads to the boundary condition $\partial u/\partial n = 0$.

3.  A homogeneous body occupying the solid region $D$ is completely insulated. Its initial temperature is $f(\mathbf{x})$. Find the steady-state temperature that it reaches after a long time. (*Hint:* No heat is gained or lost.)

4.  A rod occupying the interval $0 \le x \le l$ is subject to the heat source $f(x) = 0$ for $0 < x < \frac{l}{2}$, and $f(x) = H$ for $\frac{l}{2} < x < l$ where $H > 0$. The rod has physical constants $c = \rho = \kappa = 1$, and its ends are kept at zero temperature.
    (a)  Find the steady-state temperature of the rod.
    (b)  Which point is the hottest, and what is the temperature there?

5.  In Exercise 1.3.4, find the boundary condition if the particles lie above an impermeable horizontal plane $z = a$.

6.  Two homogeneous rods have the same cross section, specific heat $c$, and density $\rho$ but different heat conductivities $\kappa_1$ and $\kappa_2$ and lengths $L_1$ and $L_2$. Let $k_j = \kappa_j/c\rho$ be their diffusion constants. They are welded together so that the temperature $u$ and the heat flux $\kappa u_x$ at the weld are continuous. The left-hand rod has its left end maintained at temperature zero. The right-hand rod has its right end maintained at temperature $T$ degrees.
    (a)  Find the *equilibrium* temperature distribution in the composite rod.
    (b)  Sketch it as a function of $x$ in case $k_1 = 2$, $k_2 = 1$, $L_1 = 3$, $L_2 = 2$, and $T = 10$. (This exercise requires a lot of elementary algebra, but it's worth it.)

7.  In linearized gas dynamics (sound), verify the following.
    (a)  If curl $\mathbf{v} = \mathbf{0}$ at $t = 0$, then curl $\mathbf{v} = \mathbf{0}$ at all later times.
    (b)  Each component of $\mathbf{v}$ and $\rho$ satisfies the wave equation.

## 1.5  WELL-POSED PROBLEMS

Well-posed problems consist of a PDE in a domain together with a set of initial and/or boundary conditions (or other auxiliary conditions) that enjoy the following fundamental properties:

(i)  *Existence:* There exists at least one solution $u(x, t)$ satisfying all these conditions.

(ii)  *Uniqueness:* There is at most one solution.

(iii)  *Stability:* The unique solution $u(x, t)$ depends in a stable manner on the data of the problem. This means that if the data are changed a little, the corresponding solution changes only a little.

For a physical problem modeled by a PDE, the scientist normally tries to formulate physically realistic auxiliary conditions which all together make a well-posed problem. The mathematician tries to prove that a given problem is or is not well-posed. If too few auxiliary conditions are imposed, then there may be more than one solution (*nonuniqueness*) and the problem is called underdetermined. If, on the other hand, there are too many auxiliary conditions, there may be no solution at all (*nonexistence*) and the problem is called overdetermined.

The stability property (iii) is normally required in models of physical problems. This is because you could never measure the data with mathematical precision but only up to some number of decimal places. You cannot distinguish a set of data from a tiny perturbation of it. The solution ought not be significantly affected by such tiny perturbations, so it should change very little.

Let us take an example. We know that a vibrating string with an external force, whose ends are moved in a specified way, satisfies the problem

$$Tu_{tt} - \rho u_{xx} = f(x, t)$$
$$u(x, 0) = \phi(x) \qquad u_t(x, 0) = \psi(x) \qquad (1)$$
$$u(0, t) = g(t) \qquad u(L, t) = h(t)$$

for $0 < x < L$. The *data* for this problem consist of the five functions $f(x, t)$, $\phi(x)$, $\psi(x)$, $g(t)$, and $h(t)$. Existence and uniqueness would mean that there is exactly one solution $u(x, t)$ for arbitrary (differentiable) functions $f, \phi, \psi, g, h$. Stability would mean that if any of these five functions are slightly perturbed, then $u$ is also changed only slightly. To make this precise requires a definition of the "*nearness*" of functions. Mathematically, this requires the concept of a "distance", "metric", "norm", or "topology" in function space and will be discussed in the context of specific examples (see Sections 2.3, 3.4, or 5.5). Problem (1) is indeed well-posed if we make the appropriate choice of "nearness."

As a second example, consider the diffusion equation. Given an initial condition $u(x, 0) = f(x)$, we expect a unique solution, in fact, well-posedness, for $t > 0$. But consider the backwards problem! Given $f(x)$, find $u(x, t)$ for $t < 0$. What past behavior could have led up to the concentration $f(x)$ at time 0? Any chemist knows that diffusion is a smoothing process since the concentration of a substance tends to flatten out. Going backward ("*antidiffusion*"), the situation becomes more and more chaotic. Therefore, you would *not* expect well-posedness of the backward-in-time problem for the diffusion equation.

As a third example, consider solving a matrix equation instead of a PDE: namely, $Au = b$, where $A$ is an $m \times n$ matrix and $b$ is a given $m$-vector. The "data" of this problem comprise the vector $b$. If $m > n$, there are more rows than columns and the system is overdetermined. This means that no solution can exist for certain vectors $b$; that is, you don't necessarily have existence. If, on the other hand, $n > m$, there are more columns than rows and the system is underdetermined. This means that there are lots of solutions for certain vectors $b$; that is, you can't have uniqueness.

Now suppose that $m = n$ but $A$ is a singular matrix; that is, $\det A = 0$ or $A$ has no inverse. Then the problem is still ill-posed (neither existence nor uniqueness). It is also unstable. To illustrate the instability further, consider a nonsingular matrix $A$ with one very small eigenvalue. The solution is unique but if $b$ is slightly perturbed, then the error will be greatly magnified in the solution $u$. Such a matrix, in the context of scientific computation, is called ill-conditioned. The ill-conditioning comes from the instability of the matrix equation with a singular matrix.

As a fourth example, consider Laplace's equation $u_{xx} + u_{yy} = 0$ in the region $D = \{-\infty < x < \infty, 0 < y < \infty\}$. It is *not* a well-posed problem to specify both $u$ and $u_y$ on the boundary of $D$, for the following reason. It has the solutions

$$u_n(x, y) = \frac{1}{n} e^{-\sqrt{n}} \sin nx \sinh ny. \tag{2}$$

Notice that they have boundary data $u_n(x, 0) = 0$ and $\partial u_n / \partial y(x, 0) = e^{-\sqrt{n}} \sin nx$, which tends to zero as $n \to \infty$. But for $y \neq 0$ the solutions $u_n(x, y)$ do not tend to zero as $n \to \infty$. Thus the stability condition (iii) is violated.

## EXERCISES

1.  Consider the problem

$$\frac{d^2 u}{dx^2} + u = 0$$

$$u(0) = 0 \quad \text{and} \quad u(L) = 0,$$

consisting of an ODE and a pair of boundary conditions. Clearly, the function $u(x) \equiv 0$ is a solution. Is this solution *unique, or not*? Does the answer depend on $L$?

2.  Consider the problem

$$u''(x) + u'(x) = f(x)$$

$$u'(0) = u(0) = \tfrac{1}{2}[u'(l) + u(l)],$$

with $f(x)$ a given function.
(a)  Is the solution *unique*? Explain.
(b)  Does a solution necessarily *exist*, or is there a condition that $f(x)$ must satisfy for existence? Explain.

3.  Solve the boundary problem $u'' = 0$ for $0 < x < 1$ with $u'(0) + ku(0) = 0$ and $u'(1) \pm ku(1) = 0$. Do the $+$ and $-$ cases separately. What is special about the case $k = 2$?

4.  Consider the Neumann problem

$$\Delta u = f(x, y, z) \quad \text{in } D$$

$$\frac{\partial u}{\partial n} = 0 \quad \text{on bdy } D.$$

(a)  What can we surely add to any solution to get another solution? So we don't have uniqueness.

(b)  Use the divergence theorem and the PDE to show that

$$\iiint_D f(x, y, z)\, dx\, dy\, dz = 0$$

is a necessary condition for the Neumann problem to have a solution.

(c)  Can you give a physical interpretation of part (a) and/or (b) for either heat flow or diffusion?

5.  Consider the equation

$$u_x + y u_y = 0$$

with the boundary condition $u(x, 0) = \phi(x)$.

(a)  For $\phi(x) \equiv x$, show that no solution exists.

(b)  For $\phi(x) \equiv 1$, show that there are many solutions.

6.  Solve the equation $u_x + 2xy^2 u_y = 0$ with $u(x, 0) = \phi(x)$.


## 1.6   TYPES OF SECOND-ORDER EQUATIONS

In this section we show how the Laplace, wave, and diffusion equations are in some sense typical among all second-order PDEs. However, these three equations are quite different from each other. It is natural that the Laplace equation $u_{xx} + u_{yy} = 0$ and the wave equation $u_{xx} - u_{yy} = 0$ should have very different properties. After all, the *algebraic* equation $x^2 + y^2 = 1$ represents a circle, whereas the equation $x^2 - y^2 = 1$ represents a hyperbola. The parabola is somehow in between.

In general, let's consider the PDE

$$\boxed{a_{11}u_{xx} + 2a_{12}u_{xy} + a_{22}u_{yy} + a_1 u_x + a_2 u_y + a_0 u = 0.} \tag{1}$$

This is a linear equation of order two in two variables with six real constant coefficients. (The factor 2 is introduced for convenience.)

**Theorem 1.**  By a linear transformation of the independent variables, the equation can be reduced to one of three forms, as follows.

(i)  *Elliptic case:* If $a_{12}^2 < a_{11}a_{22}$, it is reducible to

$$u_{xx} + u_{yy} + \cdots = 0$$

(where $\cdots$ denotes terms of order 1 or 0).

(ii)  *Hyperbolic case:* If $a_{12}^2 > a_{11}a_{22}$, it is reducible to

$$u_{xx} - u_{yy} + \cdots = 0.$$

(iii)  *Parabolic case:* If $a_{12}^2 = a_{11}a_{22}$, it is reducible to

$$u_{xx} + \cdots = 0$$

(unless $a_{11} = a_{12} = a_{22} = 0$).

The proof is easy and is just like the analysis of conic sections in analytic geometry as either ellipses, hyperbolas, or parabolas. For simplicity, let's suppose that $a_{11} = 1$ and $a_1 = a_2 = a_0 = 0$. By completing the square, we can then write (1) as

$$(\partial_x + a_{12}\partial_y)^2 u + \left(a_{22} - a_{12}^2\right)\partial_y^2 u = 0 \tag{2}$$

(where we use the operator notation $\partial_x = \partial/\partial x$, $\partial_y^2 = \partial^2/\partial y^2$, etc.). In the elliptic case, $a_{12}^2 < a_{22}$. Let $b = \left(a_{22} - a_{12}^2\right)^{1/2} > 0$. Introduce the new variables $\xi$ and $\eta$ by

$$x = \xi, \quad y = a_{12}\xi + b\eta. \tag{3}$$

Then $\partial_\xi = 1 \cdot \partial_x + a_{12}\partial_y$, $\partial_\eta = 0 \cdot \partial_x + b\partial_y$, so that the equation becomes

$$\partial_\xi^2 u + \partial_\eta^2 u = 0, \tag{4}$$

which is Laplace's. The procedure is similar in the other cases.   □

## Example 1.

Classify each of the equations
(a)  $u_{xx} - 5u_{xy} = 0$.
(b)  $4u_{xx} - 12u_{xy} + 9u_{yy} + u_y = 0$.
(c)  $4u_{xx} + 6u_{xy} + 9u_{yy} = 0$.

Indeed, we check the sign of the "discriminant" $\mathcal{D} = a_{12}^2 - a_{11}a_{22}$. For (a) we have $\mathcal{D} = (-5/2)^2 - (1)(0) = 25/4 > 0$, so it is hyperbolic. For (b), we have $\mathcal{D} = (-6)^2 - (4)(9) = 36 - 36 = 0$, so it is parabolic. For (c), we have $\mathcal{D} = 3^2 - (4)(9) = 9 - 36 < 0$, so it is elliptic.   □

The same analysis can be done in any number of variables, using a bit of linear algebra. Suppose that there are $n$ variables, denoted $x_1, x_2 \ldots, x_n$, and the equation is

$$\sum_{i,j=1}^{n} a_{ij}u_{x_i x_j} + \sum_{i=1}^{n} a_i u_{x_i} + a_0 u = 0, \tag{5}$$

with real constants $a_{ij}$, $a_i$, and $a_0$. Since the mixed derivatives are equal, we may as well assume that $a_{ij} = a_{ji}$. Let $\mathbf{x} = (x_1, \ldots, x_n)$. Consider any linear change of independent variables:

$$(\xi_1, \ldots, \xi_n) = \boldsymbol{\xi} = B\mathbf{x},$$

where $B$ is an $n \times n$ matrix. That is,

$$\xi_k = \sum_m b_{km}x_m. \tag{6}$$

Convert to the new variables using the chain rule:

$$\frac{\partial}{\partial x_i} = \sum_k \frac{\partial \xi_k}{\partial x_i} \frac{\partial}{\partial \xi_k}$$

and

$$u_{x_i x_j} = \left(\sum_k b_{ki} \frac{\partial}{\partial \xi_k}\right)\left(\sum_l b_{lj} \frac{\partial}{\partial \xi_l}\right) u.$$

Therefore the PDE is converted to

$$\sum_{i,j} a_{ij} u_{x_i x_j} = \sum_{k,l} \left(\sum_{i,j} b_{ki} a_{ij} b_{lj}\right) u_{\xi_k \xi_l}. \tag{7}$$

(Watch out that on the left side $u$ is considered as a function of $\mathbf{x}$, whereas on the right side it is considered as a function of $\boldsymbol{\xi}$.) So you get a second-order equation in the new variables $\boldsymbol{\xi}$, but with the *new coefficient matrix* given within the parentheses. That is, the new matrix is

$$BA^tB,$$

where $A = (a_{ij})$ is the original coefficient matrix, the matrix $B = (b_{ij})$ defines the transformation, and $^tB = (b_{ji})$ is its transpose.

Now a theorem of linear algebra says that for any symmetric real matrix $A$, there is a rotation $B$ (an orthogonal matrix with determinant 1) such that $BA^tB$ is the diagonal matrix

$$BA^tB = D = \begin{pmatrix} d_1 & & & & \\ & d_2 & & & \\ & & \cdot & & \\ & & & \cdot & \\ & & & & \cdot \\ & & & & & d_n \end{pmatrix}. \tag{8}$$

The real numbers $d_1, \ldots, d_n$ are the eigenvalues of $A$. Finally, a change of scale would convert $D$ into a diagonal matrix with each of the $d$'s equal to $+1, -1,$ or $0$. (This is what we did, in effect, early in this section for the case $n = 2$.)

Thus any PDE of the form (5) can be converted by means of a linear change of variables into a PDE with a diagonal coefficient matrix.

**Definition.** The PDE (5) is called *elliptic* if all the eigenvalues $d_1, \ldots, d_n$ are positive or all are negative. [This is equivalent to saying that the original coefficient matrix $A$ (or $-A$) is positive definite.] The PDE is called *hyperbolic* if none of the $d_1, \ldots, d_n$ vanish and one of them has the opposite sign from the $(n - 1)$ others. If none vanish, but at least two of them are positive and at least two are negative, it is called *ultrahyperbolic*. If exactly one of the eigenvalues is zero and all the others have the same sign, the PDE is called *parabolic*.

Ultrahyperbolic equations occur quite rarely in physics and mathematics, so we shall not discuss them further. Just as each of the three conic sections has quite distinct properties (boundedness, shape, asymptotes), so do each of the three main types of PDEs.  □

More generally, if the coefficients are variable, that is, the $a_{ij}$ are functions of **x**, the equation may be elliptic in one region and hyperbolic in another.

## Example 2.

Find the regions in the $xy$ plane where the equation

$$yu_{xx} - 2u_{xy} + xu_{yy} = 0$$

is elliptic, hyperbolic, or parabolic. Indeed, $\mathscr{D} = (-1)^2 - (y)(x) = 1 - xy$. So the equation is parabolic on the hyperbola $(xy = 1)$, elliptic in the two convex regions $(xy > 1)$, and hyperbolic in the connected region $(xy < 1)$.  □

If the equation is nonlinear, the regions of ellipticity (and so on) may depend on which solution we are considering. Sometimes nonlinear transformations, instead of linear transformations such as $B$ above, are important. But this is a complicated subject that is poorly understood.

## EXERCISES

1.  What is the type of each of the following equations?
    (a)  $u_{xx} - u_{xy} + 2u_y + u_{yy} - 3u_{yx} + 4u = 0$.
    (b)  $9u_{xx} + 6u_{xy} + u_{yy} + u_x = 0$.
2.  Find the regions in the $xy$ plane where the equation

    $$(1 + x)u_{xx} + 2xyu_{xy} - y^2u_{yy} = 0$$

    is elliptic, hyperbolic, or parabolic. Sketch them.
3.  Among all the equations of the form (1), show that the only ones that are unchanged under all rotations (*rotationally invariant*) have the form $a(u_{xx} + u_{yy}) + bu = 0$.
4.  What is the *type* of the equation

    $$u_{xx} - 4u_{xy} + 4u_{yy} = 0?$$

    Show by direct substitution that $u(x, y) = f(y + 2x) + xg(y + 2x)$ is a solution for arbitrary functions $f$ and $g$.
5.  Reduce the elliptic equation

    $$u_{xx} + 3u_{yy} - 2u_x + 24u_y + 5u = 0$$

to the form $v_{xx} + v_{yy} + cv = 0$ by a change of dependent variable $u = ve^{\alpha x + \beta y}$ and then a change of scale $y' = \gamma y$.

6. Consider the equation $3u_y + u_{xy} = 0$.
    (a)   What is its type?
    (b)   Find the general solution. (*Hint:* Substitute $v = u_y$.)
    (c)   With the auxiliary conditions $u(x, 0) = e^{-3x}$ and $u_y(x, 0) = 0$, does a solution exist? Is it unique?

# 2

# WAVES AND DIFFUSIONS

In this chapter we study the wave and diffusion equations on the whole real line $-\infty < x < +\infty$. Real physical situations are usually on finite intervals. We are justified in taking $x$ on the whole real line for two reasons. Physically speaking, if you are sitting far away from the boundary, it will take a certain time for the boundary to have a substantial effect on you, and until that time the solutions we obtain in this chapter are valid. Mathematically speaking, the absence of a boundary is a big simplification. The most fundamental properties of the PDEs can be found most easily without the complications of boundary conditions. That is the purpose of this chapter. We begin with the wave equation.

## 2.1 THE WAVE EQUATION

We write the wave equation as

$$u_{tt} = c^2 u_{xx} \qquad \text{for } -\infty < x < +\infty. \tag{1}$$

(Physically, you can imagine a very long string.) This is the simplest second-order equation. The reason is that the operator factors nicely:

$$u_{tt} - c^2 u_{xx} = \left( \frac{\partial}{\partial t} - c \frac{\partial}{\partial x} \right)\left( \frac{\partial}{\partial t} + c \frac{\partial}{\partial x} \right) u = 0. \tag{2}$$

This means that, starting from a function $u(x, t)$, you compute $u_t + cu_x$, call the result $v$, then you compute $v_t - cv_x$, and you ought to get the zero function. The *general solution* is

$$u(x, t) = f(x + ct) + g(x - ct) \tag{3}$$

where $f$ and $g$ are two *arbitrary* (twice differentiable) functions of a single variable.

**Proof.**    Because of (2), if we let $v = u_t + cu_x$, we must have $v_t - cv_x = 0$. Thus we have two first-order equations

$$v_t - cv_x = 0 \qquad (4a)$$

and

$$u_t + cu_x = v. \qquad (4b)$$

These two first-order equations are equivalent to (1) itself. Let's solve them one at a time. As we know from Section 1.2, equation (4a) has the solution $v(x, t) = h(x + ct)$, where $h$ is any function.

So we must solve the other equation, which now takes the form

$$u_t + cu_x = h(x + ct) \qquad (4c)$$

for the unknown function $u(x, t)$. It is easy to check directly by differentiation that one solution is $u(x, t) = f(x + ct)$, where $f'(s) = h(s)/2c$. [A prime (') denotes the derivative of a function of one variable.] To the solution $f(x + ct)$ we can add $g(x - ct)$ to get another solution (since the equation is linear). The most general solution of (4b) in fact turns out to be a particular solution plus any solution of the homogeneous equation; that is,

$$u(x, t) = f(x + ct) + g(x - ct),$$

as asserted by the theorem. The complete justification is left to be worked out in Exercise 4.

A different method to derive the solution formula (3) is to introduce the *characteristic coordinates*

$$\xi = x + ct \qquad \eta = x - ct.$$

By the chain rule, we have $\partial_x = \partial_\xi + \partial_\eta$ and $\partial_t = c\partial_\xi + c\partial_\eta$. Therefore, $\partial_t - c\partial_x = -2c\partial_\eta$ and $\partial_t + c\partial_x = 2c\partial_\xi$. So equation (1) takes the form

$$(\partial_t - c\partial_x)(\partial_t + c\partial_x)u = (-2c\partial_\xi)(2c\partial_\eta)u = 0,$$

which means that $u_{\xi\eta} = 0$ since $c \neq 0$. The solution of this transformed equation is

$$u = f(\xi) + g(\eta)$$

(see Section 1.1), which agrees exactly with the previous answer (3).    □

The wave equation has a nice simple geometry. There are *two* families of characteristic lines, $x \pm ct = $ constant, as indicated in Figure 1. The most general solution is the sum of two functions. One, $g(x - ct)$, is a wave of arbitrary shape traveling to the *right* at speed $c$. The other, $f(x + ct)$, is another shape traveling to the *left* at speed $c$. A "movie" of $g(x - ct)$ is sketched in Figure 1 of Section 1.3.

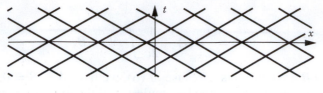

Figure 1

## INITIAL VALUE PROBLEM

The initial-value problem is to solve the wave equation

$$u_{tt} = c^2 u_{xx} \qquad \text{for } -\infty < x < +\infty \tag{1}$$

with the initial conditions

$$u(x, 0) = \phi(x) \qquad u_t(x, 0) = \psi(x), \tag{5}$$

where $\phi$ and $\psi$ are arbitrary functions of $x$. There is one, and only one, solution of this problem. For instance, if $\phi(x) = \sin x$ and $\psi(x) = 0$, then $u(x, t) = \sin x \cos ct$.

The solution of (1),(5) is easily found from the general formula (3). First, setting $t = 0$ in (3), we get

$$\phi(x) = f(x) + g(x). \tag{6}$$

Then, using the chain rule, we differentiate (3) with respect to $t$ and put $t = 0$ to get

$$\psi(x) = cf'(x) - cg'(x). \tag{7}$$

Let's regard (6) and (7) as two equations for the two unknown functions $f$ and $g$. To solve them, it is convenient temporarily to change the name of the variable to some neutral name; we change the name of $x$ to $s$. Now we differentiate (6) and divide (7) by $c$ to get

$$\phi' = f' + g' \qquad \text{and} \qquad \frac{1}{c}\psi = f' - g'.$$

Adding and subtracting the last pair of equations gives us

$$f' = \frac{1}{2}\left(\phi' + \frac{\psi}{c}\right) \qquad \text{and} \qquad g' = \frac{1}{2}\left(\phi' - \frac{\psi}{c}\right).$$

Integrating, we get

$$f(s) = \frac{1}{2}\phi(s) + \frac{1}{2c}\int_0^s \psi + A$$

and

$$g(s) = \frac{1}{2}\phi(s) - \frac{1}{2c}\int_0^s \psi + B,$$

where $A$ and $B$ are constants. Because of (6), we have $A + B = 0$. This tells us what $f$ and $g$ are in the general formula (3). Substituting $s = x + ct$ into the formula for $f$ and $s = x - ct$ into that of $g$, we get

$$u(x, t) = \frac{1}{2}\phi(x + ct) + \frac{1}{2c}\int_0^{x+ct} \psi + \frac{1}{2}\phi(x - ct) - \frac{1}{2c}\int_0^{x-ct} \psi.$$

This simplifies to

$$u(x, t) = \frac{1}{2}[\phi(x + ct) + \phi(x - ct)] + \frac{1}{2c}\int_{x-ct}^{x+ct} \psi(s)\,ds. \qquad (8)$$

   This is the solution formula for the initial-value problem, due to d'Alembert in 1746. Assuming $\phi$ to have a continuous second derivative (written $\phi \in C^2$) and $\psi$ to have a continuous first derivative ($\psi \in C^1$), we see from (8) that $u$ itself has continuous second partial derivatives in $x$ and $t$ ($u \in C^2$). Then (8) is a *bona fide* solution of (1) and (5). You may check this directly by differentiation and by setting $t = 0$.

### Example 1.

   For $\phi(x) \equiv 0$ and $\psi(x) = \cos x$, the solution is $u(x, t) = (1/2c)$ $[\sin(x + ct) - \sin(x - ct)] = (1/c)\cos x \sin ct$. Checking this result directly, we have $u_{tt} = -c\cos x \sin ct$, $u_{xx} = -(1/c)\cos x \sin ct$, so that $u_{tt} = c^2 u_{xx}$. The initial condition is easily checked.   □

### Example 2. The Plucked String

For a vibrating string the speed is $c = \sqrt{T/\rho}$. Consider an infinitely long string with initial position

$$\phi(x) = \begin{cases} b - \dfrac{b|x|}{a} & \text{for } |x| < a \\ 0 & \text{for } |x| > a \end{cases} \qquad (9)$$

and initial velocity $\psi(x) \equiv 0$ for all $x$. This is a "three-finger" pluck, with all three fingers removed at once. A "movie" of this solution $u(x, t) = \frac{1}{2}[\phi(x + ct) + \phi(x - ct)]$ is shown in Figure 2. (Even though this solution is not twice differentiable, it can be shown to be a "weak" solution, as discussed later in Section 12.1.)

   Each of these pictures is the sum of two triangle functions, one moving to the right and one to the left, as is clear graphically. To write down the formulas that correspond to the pictures requires a lot more work. The formulas depend on the relationships among the five numbers $0, \pm a, x \pm ct$. For instance, let $t = a/2c$. Then $x \pm ct = x \pm a/2$. First, if $x < -3a/2$, then $x \pm a/2 < -a$ and $u(x, t) \equiv 0$. Second, if $-3a/2 < x < -a/2$, then

$$u(x, t) = \frac{1}{2}\phi\left(x + \frac{1}{2}a\right) = \frac{1}{2}\left(b - \frac{b|x + \frac{1}{2}a|}{a}\right) = \frac{3b}{4} + \frac{bx}{2a}.$$

Third, if $|x| < a/2$, then

$$u(x, t) = \frac{1}{2}\left[\phi\left(x + \frac{1}{2}a\right) + \phi\left(x - \frac{1}{2}a\right)\right]$$

$$= \frac{1}{2}\left[b - \frac{b\left(x + \frac{1}{2}a\right)}{a} + b - \frac{b\left(\frac{1}{2}a - x\right)}{a}\right]$$

$$= \frac{1}{2}b$$

and so on [see Figure 2].                                                                          □

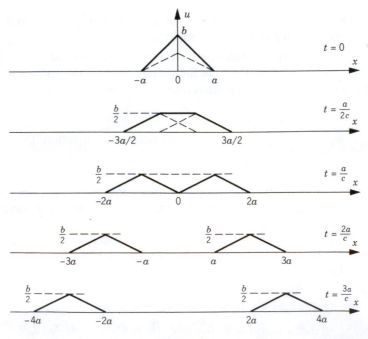

Figure 2

## EXERCISES

1. Solve $u_{tt} = c^2 u_{xx}$, $u(x, 0) = e^x$, $u_t(x, 0) = \sin x$.

2. Solve $u_{tt} = c^2 u_{xx}$, $u(x, 0) = \log(1 + x^2)$, $u_t(x, 0) = 4 + x$.

3. The midpoint of a piano string of tension $T$, density $\rho$, and length $l$ is hit by a hammer whose head diameter is $2a$. A flea is sitting at a distance $l/4$ from one end. (Assume that $a < l/4$; otherwise, poor flea!) How long does it take for the disturbance to reach the flea?

4. Justify the conclusion at the beginning of Section 2.1 that *every* solution of the wave equation has the form $f(x + ct) + g(x - ct)$.

5. (*The hammer blow*) Let $\phi(x) \equiv 0$ and $\psi(x) = 1$ for $|x| < a$ and $\psi(x) = 0$ for $|x| \geq a$. Sketch the string profile ($u$ versus $x$) at each of the successive instants $t = a/2c$, $a/c$, $3a/2c$, $2a/c$, and $5a/c$. [*Hint:* Calculate

$$u(x, t) = \frac{1}{2c} \int_{x-ct}^{x+ct} \psi(s)\, ds = \frac{1}{2c} \{\text{length of } (x - ct, x + ct) \cap (-a, a)\}.$$

Then $u(x, a/2c) = (1/2c)$ {length of $(x - a/2, x + a/2) \cap (-a, a)$}. This takes on different values for $|x| < a/2$, for $a/2 < x < 3a/2$, and for $x > 3a/2$. Continue in this manner for each case.]

6. In Exercise 5, find the greatest displacement, $\max_x u(x, t)$, as a function of $t$.

7. If both $\phi$ and $\psi$ are odd functions of $x$, show that the solution $u(x, t)$ of the wave equation is also odd in $x$ for all $t$.

8. A *spherical wave* is a solution of the three-dimensional wave equation of the form $u(r, t)$, where $r$ is the distance to the origin (the spherical coordinate). The wave equation takes the form

$$u_{tt} = c^2 \left( u_{rr} + \frac{2}{r} u_r \right) \quad \text{(``spherical wave equation'')}.$$

   (a) Change variables $v = ru$ to get the equation for $v$: $v_{tt} = c^2 v_{rr}$.
   (b) Solve for $v$ using (3) and thereby solve the spherical wave equation.
   (c) Use (8) to solve it with initial conditions $u(r, 0) = \phi(r)$, $u_t(r, 0) = \psi(r)$, taking both $\phi(r)$ and $\psi(r)$ to be even functions of $r$.

9. Solve $u_{xx} - 3u_{xt} - 4u_{tt} = 0$, $u(x, 0) = x^2$, $u_t(x, 0) = e^x$. (*Hint:* Factor the operator as we did for the wave equation.)

10. Solve $u_{xx} + u_{xt} - 20u_{tt} = 0$, $u(x, 0) = \phi(x)$, $u_t(x, 0) = \psi(x)$.

11. Find the general solution of $3u_{tt} + 10u_{xt} + 3u_{xx} = \sin(x + t)$.

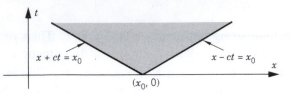

Figure 1

## 2.2 CAUSALITY AND ENERGY

### CAUSALITY

We have just learned that the effect of an initial position $\phi(x)$ is a pair of waves traveling in either direction at speed $c$ and at half the original amplitude. The effect of an initial velocity $\psi$ is a wave spreading out at speed $\leq c$ in both directions (see Exercise 2.1.5 for an example). So part of the wave may lag behind (if there is an initial velocity), but *no part goes faster than speed $c$.* The last assertion is called the *principle of causality*. It can be visualized in the $xt$ plane in Figure 1.

An initial condition (position or velocity or both) at the point $(x_0,\ 0)$ can affect the solution for $t > 0$ only in the shaded sector, which is called the *domain of influence* of the point $(x_0,\ 0)$. As a consequence, if $\phi$ and $\psi$ vanish for $|x| > R$, then $u(x,\ t) = 0$ for $|x| > R + ct$. In words, the domain of influence of an interval $(|x| \leq R)$ is a sector $(|x| \leq R + ct)$.

An "inverse" way to express causality is the following. *Fix a point* $(x,\ t)$ for $t > 0$ (see Figure 2). How is the number $u(x,\ t)$ synthesized from the initial data $\phi$, $\psi$? It depends only on the values of $\phi$ at the two points $x \pm ct$, and it depends only on the values of $\psi$ within the interval $[x - ct,\ x + ct]$. We therefore say that the interval $(x - ct,\ x + ct)$ is the interval of dependence of the point $(x,\ t)$ on $t = 0$. Sometimes we call the entire shaded triangle $\Delta$ the *domain of dependence* or the *past history* of the point $(x,\ t)$. The domain of dependence is bounded by the pair of characteristic lines that pass through $(x,\ t)$.

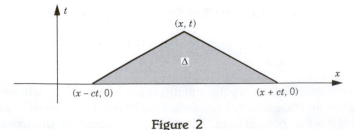

Figure 2

## ENERGY

Imagine an infinite string with constants $\rho$ and $T$. Then $\rho u_{tt} = T u_{xx}$ for $-\infty < x < +\infty$. From physics we know that the kinetic energy is $\frac{1}{2}mv^2$, which in our case takes the form KE $= \frac{1}{2}\rho \int u_t^2\, dx$. This integral, and the following ones, are evaluated from $-\infty$ to $+\infty$. To be sure that the integral converges, we assume that $\phi(x)$ and $\psi(x)$ vanish outside an interval $\{|x| \leq R\}$. As mentioned above, $u(x, t)$ [and therefore $u_t(x, t)$] vanish for $|x| > R + ct$. Differentiating the kinetic energy, we can pass the derivative under the integral sign (see Section A.3) to get

$$\frac{dKE}{dt} = \rho \int u_t u_{tt}\, dx.$$

Then we substitute the PDE $\rho u_{tt} = T u_{xx}$ and integrate by parts to get

$$\frac{dKE}{dt} = T \int u_t u_{xx}\, dx = T u_t u_x - T \int u_{tx} u_x\, dx.$$

The term $T u_t u_x$ is evaluated at $x = \pm\infty$ and so it vanishes. But the final term is a pure derivative since $u_{tx} u_x = \left(\frac{1}{2}u_x^2\right)_t$. Therefore,

$$\frac{dKE}{dt} = -\frac{d}{dt} \int \frac{1}{2} T u_x^2\, dx.$$

Let $PE = \frac{1}{2}T \int u_x^2\, dx$ and let $E = KE + PE$. Then $dKE/dt = -dPE/dt$, or $dE/dt = 0$. Thus

$$\boxed{E = \frac{1}{2} \int_{-\infty}^{+\infty} \left(\rho u_t^2 + T u_x^2\right) dx} \tag{1}$$

is a constant independent of $t$. This is the law of *conservation of energy*.

In physics courses we learn that $PE$ has the interpretation of the potential energy. The only thing we need mathematically is the total energy $E$. The conservation of energy is one of the most basic facts about the wave equation. Sometimes the definition of $E$ is modified by a constant factor, but that does not affect its conservation. Notice that the energy is necessarily positive. The energy can also be used to derive causality (as will be done in Section 9.1).

### Example 1.

The plucked string, Example 2 of Section 2.1, has the energy

$$E = \frac{1}{2}T \int \phi_x^2\, dx = \frac{1}{2}T \left(\frac{b}{a}\right)^2 2a = \frac{Tb^2}{a}. \qquad \square$$

In electromagnetic theory the equations are Maxwell's. Each component of the electric and magnetic fields satisfies the (three-dimensional) wave equation, where $c$ is the speed of light. The principle of causality, discussed above, is the cornerstone of the theory of relativity. It means that a signal located at the position $x_0$ at the instant $t_0$ cannot move faster than the speed of light. The domain of influence of $(x_0, t_0)$ consists of all the points that can be reached by a signal of speed $c$ starting from the point $x_0$ at the time $t_0$. It turns out that the solutions of the *three*-dimensional wave equation always travel at speeds exactly equal to $c$ and never slower. Therefore, the causality principle is sharper in three dimensions than in one. This sharp form is called *Huygens's principle* (see Chapter 9).

Flatland is an imaginary two-dimensional world. You can think of yourself as a waterbug confined to the surface of a pond. You wouldn't want to live there because Huygens's principle is not valid in two dimensions (see Section 9.2). Each sound you make would automatically mix with the "echoes" of your previous sounds. And each view would be mixed fuzzily with the previous views. Three is the best of all possible dimensions.

## EXERCISES

1.  Use the energy conservation of the wave equation to prove that the only solution with $\phi \equiv 0$ and $\psi \equiv 0$ is $u \equiv 0$. (*Hint:* Use the first vanishing theorem in Section A.1.)

2.  For a solution $u(x, t)$ of the wave equation with $\rho = T = c = 1$, the energy density is defined as $e = \frac{1}{2}(u_t^2 + u_x^2)$ and the momentum density as $p = u_t u_x$.
    (a)  Show that $\partial e/\partial t = \partial p/\partial x$ and $\partial p/\partial t = \partial e/\partial x$.
    (b)  Show that both $e(x, t)$ and $p(x, t)$ also satisfy the wave equation.

3.  Show that the wave equation has the following invariance properties.
    (a)  Any translate $u(x - y, t)$, where $y$ is fixed, is also a solution.
    (b)  Any derivative, say $u_x$, of a solution is also a solution.
    (c)  The dilated function $u(ax, at)$ is also a solution, for any constant $a$.

4.  If $u(x, t)$ satisfies the wave equation $u_{tt} = u_{xx}$, prove the identity

    $$u(x + h, t + k) + u(x - h, t - k) = u(x + k, t + h) + u(x - k, t - h)$$

    for all $x$, $t$, $h$, and $k$. Sketch the quadrilateral $Q$ whose vertices are the arguments in the identity.

5.  For the *damped* string, equation (1.3.3), show that the energy decreases.

6.  Prove that, among all possible dimensions, only in three dimensions can one have distortionless spherical wave propagation with attenuation. This means the following. A spherical wave in $n$-dimensional space satisfies the PDE

    $$u_{tt} = c^2 \left( u_{rr} + \frac{n-1}{r} u_r \right),$$

    where $r$ is the spherical coordinate. Consider such a wave that has the special form $u(r, t) = \alpha(r) f(t - \beta(r))$, where $\alpha(r)$ is called the attenuation and $\beta(r)$ the delay. The question is whether such solutions exist for "arbitrary" functions $f$.

(a) Plug the special form into the PDE to get an ODE for $f$.
(b) Set the coefficients of $f''$, $f'$, and $f$ equal to zero.
(c) Solve the ODEs to see that $n = 1$ or $n = 3$ (unless $u \equiv 0$).
(d) If $n = 1$, show that $\alpha(r)$ is a constant (so that "there is no attenuation").

(T. Morley, *American Mathematical Monthly*, Vol. 27, pp. 69–71, 1985)

## 2.3 THE DIFFUSION EQUATION

In this section we begin a study of the one-dimensional diffusion equation

$$u_t = ku_{xx}. \tag{1}$$

Diffusions are very different from waves, and this is reflected in the mathematical properties of the equations. Because (1) is harder to solve than the wave equation, we begin this section with a general discussion of some of the properties of diffusions. We begin with the maximum principle, from which we'll deduce the uniqueness of an initial-boundary problem. We postpone until the next section the derivation of the solution formula for (1) on the whole real line.

**Maximum Principle.** If $u(x, t)$ satisfies the diffusion equation in a rectangle (say, $0 \leq x \leq l$, $0 \leq t \leq T$) in space-time, then the maximum value of $u(x, t)$ is assumed either initially ($t = 0$) or on the lateral sides ($x = 0$ or $x = l$) (see Figure 1).

In fact, there is a *stronger version* of the maximum principle which asserts that the maximum cannot be assumed anywhere inside the rectangle but *only on the bottom or the lateral sides* (unless $u$ is a constant). The corners are allowed.

The minimum value has the same property; it too can be attained only on the bottom or the lateral sides. To prove the minimum principle, just apply the maximum principle to $[-u(x, t)]$.

These principles have a natural interpretation in terms of diffusion or heat flow. If you have a rod with no internal heat source, the hottest spot and the

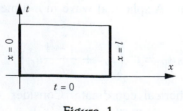

Figure 1

coldest spot can occur only initially or at one of the two ends of the rod. Thus a hot spot at time zero will cool off (unless heat is fed into the rod at an end). You can burn one of its ends but the maximum temperature will always be at the hot end, so that it will be cooler away from that end. Similarly, if you have a substance diffusing along a tube, its highest concentration can occur only initially or at one of the ends of the tube.

If we draw a "movie" of the solution, the maximum drops down while the minimum comes up. So the differential equation tends to smooth the solution out. (This is very different from the behavior of the wave equation!)

**Proof of the Maximum Principle.**   We'll prove only the weaker version. (Surprisingly, its strong form is much more difficult to prove.) For the strong version, see [PW]. The idea of the proof is to use the fact, from calculus, that at an interior maximum the first derivatives vanish and the second derivatives satisfy inequalities such as $u_{xx} \leq 0$. If we knew that $u_{xx} \neq 0$ at the maximum (which we do not), then we'd have $u_{xx} < 0$ as well as $u_t = 0$, so that $u_t \neq ku_{xx}$. This contradiction would show that the maximum could only be somewhere on the boundary of the rectangle. However, because $u_{xx}$ could in fact be equal to zero, we need to play a mathematical game to make the argument work.

So let $M$ denote the maximum value of $u(x, t)$ on the three sides $t = 0$, $x = 0$, and $x = l$. (Recall that any continuous function on any bounded closed set is bounded and assumes its maximum on that set.) We must show that $u(x, t) \leq M$ throughout the rectangle $R$.

Let $\epsilon$ be a positive constant and let $v(x, t) = u(x, t) + \epsilon x^2$. Our goal is to show that $v(x, t) \leq M + \epsilon l^2$ throughout $R$. Once this is accomplished, we'll have $u(x, t) \leq M + \epsilon(l^2 - x^2)$. This conclusion is true for any $\epsilon > 0$. Therefore, $u(x, t) \leq M$ throughout $R$, which is what we are trying to prove.

Now from the definition of $v$, it is clear that $v(x, t) \leq M + \epsilon l^2$ on $t = 0$, on $x = 0$, and on $x = l$. This function $v$ satisfies

$$v_t - kv_{xx} = u_t - k(u + \epsilon x^2)_{xx} = u_t - ku_{xx} - 2\epsilon k = -2\epsilon k < 0, \qquad (2)$$

which is the "diffusion inequality." Now suppose that $v(x, t)$ attains its maximum at an *interior* point $(x_0, t_0)$. That is, $0 < x_0 < l, 0 < t_0 < T$. By ordinary calculus, we know that $v_t = 0$ and $v_{xx} \leq 0$ at $(x_0, t_0)$. This contradicts the diffusion inequality (2). So there can't be an interior maximum. Suppose now that $v(x, t)$ has a maximum (in the closed rectangle) at a point on the *top* edge $\{t_0 = T$ and $0 < x < l\}$. Then $v_x(x_0, t_0) = 0$ and $v_{xx}(x_0, t_0) \leq 0$, as before. Furthermore, because $v(x_0, t_0)$ is bigger than $v(x_0, t_0 - \delta)$, we have

$$v_t(x_0, t_0) = \lim \frac{v(x_0, t_0) - v(x_0, t_0 - \delta)}{\delta} \geq 0$$

as $\delta \to 0$ through positive values. (This is not an equality because the maximum is only "one-sided" in the variable $t$.) We again reach a contradiction to the diffusion inequality.

But $v(x, t)$ does have a maximum *somewhere* in the closed rectangle $0 \le x \le l$, $0 \le t \le T$. This maximum must be on the bottom or sides. Therefore $v(x, t) \le M + \epsilon l^2$ throughout $R$. This proves the maximum principle (in its weaker version).

## UNIQUENESS

The maximum principle can be used to give a proof of *uniqueness for the Dirichlet problem for the diffusion equation.* That is, there is at most one solution of

$$u_t - ku_{xx} = f(x, t) \quad \text{for } 0 < x < l \text{ and } t > 0$$
$$u(x, 0) = \phi(x) \tag{3}$$
$$u(0, t) = g(t) \qquad u(l, t) = h(t)$$

for four given functions $f$, $\phi$, $g$, and $h$. Uniqueness means that any solution is determined completely by its initial and boundary conditions. Indeed, let $u_1(x, t)$ and $u_2(x, t)$ be two solutions of (3). Let $w = u_1 - u_2$ be their difference. Then $w_t - kw_{xx} = 0$, $w(x, 0) = 0$, $w(0, t) = 0$, $w(l, t) = 0$. Let $T > 0$. By the maximum principle, $w(x, t)$ has its maximum for the rectangle on its bottom or sides—exactly where it vanishes. So $w(x, t) \le 0$. The same type of argument for the minimum shows that $w(x, t) \ge 0$. Therefore, $w(x, t) \equiv 0$, so that $u_1(x, t) \equiv u_2(x, t)$ for all $t \ge 0$.

Here is a second proof of uniqueness for problem (3), by a very different technique, the *energy method.* Multiplying the equation for $w = u_1 - u_2$ by $w$ itself, we can write

$$0 = 0 \cdot w = (w_t - kw_{xx})(w) = \left(\tfrac{1}{2}w^2\right)_t + (-kw_x w)_x + kw_x^2.$$

(Verify this by carrying out the derivatives on the right side.) Upon integrating over the interval $0 < x < l$, we get

$$0 = \int_0^l \left(\tfrac{1}{2}w^2\right)_t dx - kw_x w \Big|_{x=0}^{x=l} + k \int_0^l w_x^2 dx.$$

Because of the boundary conditions ($w = 0$ at $x = 0, l$),

$$\frac{d}{dt} \int_0^l \frac{1}{2}[w(x, t)]^2 dx = -k \int_0^l [w_x(x, t)]^2 dx \le 0,$$

where the time derivative has been pulled out of the $x$ integral (see Section A.3). Therefore, $\int w^2 dx$ is decreasing, so

$$\int_0^l [w(x, t)]^2 dx \le \int_0^l [w(x, 0)]^2 dx \tag{4}$$

for $t \ge 0$. The right side of (4) vanishes because the initial conditions of $u$ and $v$ are the same, so that $\int [w(x, t)]^2 dx = 0$ for all $t > 0$. So $w \equiv 0$ and $u_1 \equiv u_2$ for all $t \ge 0$.

## STABILITY

This is the third ingredient of well-posedness (see Section 1.5). It means that the initial and boundary conditions are correctly formulated. The energy method leads to the following form of stability of problem (3), in case $h = g = f = 0$. Let $u_1(x, 0) = \phi_1(x)$ and $u_2(x, 0) = \phi_2(x)$. Then $w = u_1 - u_2$ is the solution with the initial datum $\phi_1 - \phi_2$. So from (4) we have

$$\int_0^l [u_1(x, t) - u_2(x, t)]^2\, dx \leq \int_0^l [\phi_1(x) - \phi_2(x)]^2\, dx. \qquad (5)$$

On the right side is a quantity that measures the nearness of the initial data for two solutions, and on the left we measure the nearness of the solutions at any later time. Thus, if we start nearby (at $t = 0$), we stay nearby. This is exactly the meaning of stability in the "square integral" sense (see Sections 1.5 and 5.4).

The maximum principle also proves the stability, but with a different way to measure nearness. Consider two solutions of (3) in a rectangle. We then have $w \equiv u_1 - u_2 = 0$ on the lateral sides of the rectangle and $w = \phi_1 - \phi_2$ on the bottom. The maximum principle asserts that throughout the rectangle

$$u_1(x, t) - u_2(x, t) \leq \max|\phi_1 - \phi_2|.$$

The "minimum" principle says that

$$u_1(x, t) - u_2(x, t) \geq -\max|\phi_1 - \phi_2|.$$

Therefore,

$$\max_{0 \leq x \leq l} |u_1(x, t) - u_2(x, t)| \leq \max_{0 \leq x \leq l} |\phi_1(x) - \phi_2(x)|, \qquad (6)$$

valid for all $t > 0$. Equation (6) is in the same spirit as (5), but with a quite different method of measuring the nearness of functions. It is called stability in the "uniform" sense.

## EXERCISES

1.  Consider the solution $1 - x^2 - 2kt$ of the diffusion equation. Find the locations of its maximum and its minimum in the closed rectangle $\{0 \leq x \leq 1, 0 \leq t \leq T\}$.

2.  Consider a solution of the diffusion equation $u_t = u_{xx}$ in $\{0 \leq x \leq l, 0 \leq t < \infty\}$.
    (a)  Let $M(T) =$ the maximum of $u(x, t)$ in the closed rectangle $\{0 \leq x \leq l, 0 \leq t \leq T\}$. Does $M(T)$ increase or decrease as a function of $T$?
    (b)  Let $m(T) =$ the minimum of $u(x, t)$ in the closed rectangle $\{0 \leq x \leq l, 0 \leq t \leq T\}$. Does $m(T)$ increase or decrease as a function of $T$?

3.  Consider the diffusion equation $u_t = u_{xx}$ in the interval $(0, 1)$ with $u(0, t) = u(1, t) = 0$ and $u(x, 0) = 1 - x^2$. Note that this initial function does not satisfy the boundary condition at the left end, but that the solution will satisfy it for all $t > 0$.

(a)   Show that $u(x, t) > 0$ at all interior points $0 < x < 1$, $0 < t < \infty$.

(b)   For each $t > 0$, let $\mu(t) =$ the maximum of $u(x, t)$ over $0 \le x \le 1$. Show that $\mu(t)$ is a decreasing (i.e., nonincreasing) function of $t$. (*Hint:* Let the maximum occur at the point $X(t)$, so that $\mu(t) = u(X(t), t)$. Differentiate $\mu(t)$, assuming that $X(t)$ is differentiable.)

(c)   Draw a rough sketch of what you think the solution looks like ($u$ versus $x$) at a few times. (If you have appropriate software available, compute it.)

4.   Consider the diffusion equation $u_t = u_{xx}$ in $\{0 < x < 1, \ 0 < t < \infty\}$ with $u(0, t) = u(1, t) = 0$ and $u(x, 0) = 4x(1 - x)$.

(a)   Show that $0 < u(x, t) < 1$ for all $t > 0$ and $0 < x < 1$.

(b)   Show that $u(x, t) = u(1 - x, t)$ for all $t \ge 0$ and $0 \le x \le 1$.

(c)   Use the energy method to show that $\int_0^1 u^2 \, dx$ is a strictly decreasing function of $t$.

5.   The purpose of this exercise is to show that the maximum principle is not true for the equation $u_t = xu_{xx}$, which has a variable coefficient.

(a)   Verify that $u = -2xt - x^2$ is a solution. Find the location of its maximum in the closed rectangle $\{-2 \le x \le 2, \ 0 \le t \le 1\}$.

(b)   Where precisely does our proof of the maximum principle break down for this equation?

6.   Prove the *comparison principle* for the diffusion equation: If $u$ and $v$ are two solutions, and if $u \le v$ for $t = 0$, for $x = 0$, and for $x = l$, then $u \le v$ for $0 \le t < \infty$, $0 \le x \le l$.

7.   (a)   More generally, if $u_t - ku_{xx} = f$, $v_t - kv_{xx} = g$, $f \le g$, and $u \le v$ at $x = 0$, $x = l$ and $t = 0$, prove that $u \le v$ for $0 \le x \le l$, $0 \le t < \infty$.

(b)   If $v_t - v_{xx} \ge \sin x$ for $0 \le x \le \pi$, $0 < t < \infty$, and if $v(0, t) \ge 0$, $v(\pi, t) \ge 0$ and $v(x, 0) \ge \sin x$, use part (a) to show that $v(x, t) \ge (1 - e^{-t}) \sin x$.

8.   Consider the diffusion equation on $(0, l)$ with the Robin boundary conditions $u_x(0, t) - a_0 u(0, t) = 0$ and $u_x(l, t) + a_l u(l, t) = 0$. If $a_0 > 0$ and $a_l > 0$, use the energy method to show that the endpoints contribute to the decrease of $\int_0^l u^2(x, t) \, dx$. (This is interpreted to mean that part of the "energy" is lost at the boundary, so we call the boundary conditions "*radiating*" or "*dissipative*.")

## 2.4   DIFFUSION ON THE WHOLE LINE

Our purpose in this section is to solve the problem

$$u_t = ku_{xx} \quad (-\infty < x < \infty, \ 0 < t < \infty) \tag{1}$$
$$u(x, 0) = \phi(x). \tag{2}$$

As with the wave equation, the problem on the infinite line has a certain "purity", which makes it easier to solve than the finite-interval problem. (The effects of boundaries will be discussed in the next several chapters.) Also as with the wave equation, we will end up with an explicit formula. But it will be derived by a method *very different* from the methods used before. (The characteristics for the diffusion equation are just the lines $t = $ constant and play no major role in the analysis.) Because the solution of (1) is not easy to derive, we first set the stage by making some general comments.

Our method is to solve it for a *particular* $\phi(x)$ and then build the general solution from this particular one. We'll use five basic *invariance properties* of the diffusion equation (1).

(a)   The *translate* $u(x - y, t)$ of any solution $u(x, t)$ is another solution, for any fixed $y$.

(b)   Any *derivative* ($u_x$ or $u_t$ or $u_{xx}$, etc.) of a solution is again a solution.

(c)   A *linear combination* of solutions of (1) is again a solution of (1). (This is just linearity.)

(d)   An *integral* of solutions is again a solution. Thus if $S(x, t)$ is a solution of (1), then so is $S(x - y, t)$ and so is

$$v(x, t) = \int_{-\infty}^{\infty} S(x - y, t)g(y)\,dy$$

for any function $g(y)$, as long as this improper integral converges appropriately. (We'll worry about convergence later.) In fact, (d) is just a limiting form of (c).

(e)   If $u(x, t)$ is a solution of (1), so is the *dilated* function $u(\sqrt{a}x, at)$, for any $a > 0$. Prove this by the chain rule: Let $v(x, t) = u(\sqrt{a}x, at)$. Then $v_t = [\partial(at)/\partial t]u_t = au_t$ and $v_x = [\partial(\sqrt{a}x)/\partial x]u_x = \sqrt{a}u_x$ and $v_{xx} = \sqrt{a} \cdot \sqrt{a}u_{xx} = au_{xx}$.

Our goal is to find a particular solution of (1) and then to construct all the other solutions using property (d). The particular solution we will look for is the one, denoted $Q(x, t)$, which satisfies the *special initial condition*

$$\boxed{Q(x, 0) = 1 \quad \text{for } x > 0 \quad Q(x, 0) = 0 \quad \text{for } x < 0.} \tag{3}$$

The reason for this choice is that this initial condition does not change under dilation. We'll find $Q$ in three steps.

**Step 1**   We'll look for $Q(x, t)$ of the special form

$$Q(x, t) = g(p) \quad \text{where } p = \frac{x}{\sqrt{4kt}} \tag{4}$$

and $g$ is a function of only one variable (to be determined). (The $\sqrt{4k}$ factor is included only to simplify a later formula.)

*Why* do we expect $Q$ to have this special form? Because property (e) says that equation (1) doesn't "see" the dilation $x \to \sqrt{a}x, t \to at$. Clearly, (3) doesn't change at all under the dilation. So $Q(x, t)$, which is defined by conditions (1) and (3), ought not see the dilation either. How could that happen? In only one way: if $Q$ depends on $x$ and $t$ solely through the combination $x/\sqrt{t}$. For the dilation takes $x/\sqrt{t}$ into $\sqrt{a}x/\sqrt{at} = x/\sqrt{t}$. Thus let $p = x/\sqrt{4kt}$ and look for $Q$ which satisfies (1) and (3) and has the form (4).

**Step 2**   Using (4), we convert (1) into an ODE for $g$ by use of the chain rule:

$$Q_t = \frac{dg}{dp}\frac{\partial p}{\partial t} = -\frac{1}{2t}\frac{x}{\sqrt{4kt}}g'(p)$$

$$Q_x = \frac{dg}{dp}\frac{\partial p}{\partial x} = \frac{1}{\sqrt{4kt}}g'(p)$$

$$Q_{xx} = \frac{dQ_x}{dp}\frac{\partial p}{\partial x} = \frac{1}{4kt}g''(p)$$

$$0 = Q_t - kQ_{xx} = \frac{1}{t}\left[-\frac{1}{2}pg'(p) - \frac{1}{4}g''(p)\right].$$

Thus

$$g'' + 2pg' = 0.$$

This ODE is easily solved using the integrating factor $\exp \int 2p\, dp = \exp(p^2)$. We get $g'(p) = c_1 \exp(-p^2)$ and

$$Q(x, t) = g(p) = c_1 \int e^{-p^2}\, dp + c_2.$$

**Step 3**   We find a completely explicit formula for $Q$. We've just shown that

$$Q(x, t) = c_1 \int_0^{x/\sqrt{4kt}} e^{-p^2}\, dp + c_2.$$

This formula is valid only for $t > 0$. Now use (3), expressed as a limit as follows.

If $x > 0$, $\quad 1 = \lim_{t \searrow 0} Q = c_1 \int_0^{+\infty} e^{-p^2}\, dp + c_2 = c_1\frac{\sqrt{\pi}}{2} + c_2.$

If $x < 0$, $\quad 0 = \lim_{t \searrow 0} Q = c_1 \int_0^{-\infty} e^{-p^2}\, dp + c_2 = -c_1\frac{\sqrt{\pi}}{2} + c_2.$

See Exercise 6. Here $\lim_{t \searrow 0}$ means limit from the right. This determines the coefficients $c_1 = 1/\sqrt{\pi}$ and $c_2 = \frac{1}{2}$. Therefore, $Q$ is the function

$$Q(x, t) = \frac{1}{2} + \frac{1}{\sqrt{\pi}} \int_0^{x/\sqrt{4kt}} e^{-p^2} dp \tag{5}$$

for $t > 0$. Notice that it does indeed satisfy (1), (3), and (4).

**Step 4**  Having found $Q$, we now *define* $S = \partial Q/\partial x$. (The explicit formula for $S$ will be written below.) By property (b), $S$ is also a solution of (1). Given any function $\phi$, we also *define*

$$u(x, t) = \int_{-\infty}^{\infty} S(x - y, t)\phi(y) \, dy \quad \text{for } t > 0. \tag{6}$$

By property (d), $u$ is another solution of (1). *We claim that $u$ is the unique solution of* (1), (2). To verify the validity of (2), we write

$$u(x, t) = \int_{-\infty}^{\infty} \frac{\partial Q}{\partial x}(x - y, t)\phi(y) \, dy$$

$$= -\int_{-\infty}^{\infty} \frac{\partial}{\partial y}[Q(x - y, t)]\phi(y) \, dy$$

$$= +\int_{-\infty}^{\infty} Q(x - y, t)\phi'(y) \, dy - Q(x - y, t)\phi(y)\Big|_{y=-\infty}^{y=+\infty}$$

upon integrating by parts. We assume these limits vanish. In particular, let's temporarily assume that $\phi(y)$ itself equals zero for $|y|$ large. Therefore,

$$u(x, 0) = \int_{-\infty}^{\infty} Q(x - y, 0)\phi'(y) \, dy$$

$$= \int_{-\infty}^{x} \phi'(y) \, dy = \phi\Big|_{-\infty}^{x} = \phi(x)$$

because of the initial condition for $Q$ and the assumption that $\phi(-\infty) = 0$. This is the initial condition (2). We conclude that (6) is our solution formula, where

$$S = \frac{\partial Q}{\partial x} = \frac{1}{2\sqrt{\pi kt}} e^{-x^2/4kt} \quad \text{for } t > 0. \tag{7}$$

That is,

$$u(x, t) = \frac{1}{\sqrt{4\pi kt}} \int_{-\infty}^{\infty} e^{-(x-y)^2/4kt} \phi(y) \, dy. \tag{8}$$

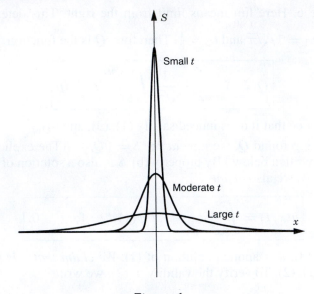

**Figure 1**

$S(x, t)$ is known as the *source function, Green's function, fundamental solution, gaussian,* or *propagator* of the diffusion equation, or simply the *diffusion kernel*. It gives the solution of (1),(2) with any initial datum $\phi$. The formula only gives the solution for $t > 0$. When $t = 0$ it makes no sense.    □

The *source function* $S(x, t)$ is defined for all real $x$ and for all $t > 0$. $S(x, t)$ is positive and is even in $x$ [$S(-x, t) = S(x, t)$]. It looks like Figure 1 for various values of $t$. For large $t$, it is very spread out. For small $t$, it is a very tall thin spike (a "*delta function*") of height $(4\pi kt)^{-1/2}$. The area under its graph is

$$\int_{-\infty}^{\infty} S(x, t)\, dx = \frac{1}{\sqrt{\pi}} \int_{-\infty}^{\infty} e^{-q^2}\, dq = 1$$

by substituting $q = x/\sqrt{4kt}$, $dq = (dx)/\sqrt{4kt}$ (see Exercise 7). Now look more carefully at the sketch of $S(x, t)$ for a very small $t$. If we cut out the tall spike, the rest of $S(x, t)$ is very small. Thus

$$\max_{|x|>\delta} S(x, t) \to 0 \qquad \text{as} \quad t \to 0 \tag{9}$$

Notice that the value of the solution $u(x, t)$ given by (6) is a kind of weighted *average* of the initial values around the point $x$. Indeed, we can write

$$u(x, t) = \int_{-\infty}^{\infty} S(x - y, t)\phi(y)\, dy \simeq \sum_{t} S(x - y_i, t)\phi(y_i)\Delta y_i$$

approximately. This is the average of the solutions $S(x - y_i, t)$ with the weights $\phi(y_i)$. For very small $t$, the source function is a spike so that the formula exaggerates the values of $\phi$ near $x$. For *any* $t > 0$ the solution is a spread-out version of the initial values at $t = 0$.

Here's the physical interpretation. Consider diffusion. $S(x - y, t)$ represents the result of a unit mass (say, 1 gram) of substance located at time zero exactly at the position $y$ which is diffusing (spreading out) as time advances. For any initial distribution of concentration, the amount of substance initially in the interval $\Delta y$ spreads out in time and contributes approximately the term $S(x - y_i, t)\phi(y_i)\Delta y_i$. All these contributions are added up to get the whole distribution of matter. Now consider heat flow. $S(x - y, t)$ represents the result of a "hot spot" at $y$ at time 0. The hot spot is cooling off and spreading its heat along the rod.

Another physical interpretation is brownian motion, where particles move randomly in space. For simplicity, we assume that the motion is one-dimensional; that is, the particles move along a tube. Then the probability that a particle which begins at position $x$ ends up in the interval $(a, b)$ at time $t$ is precisely $\int_a^b S(x - y, t)\,dy$ for some constant $k$, where $S$ is defined in (7). In other words, if we let $u(x, t)$ be the probability density (probability per unit length) and if the initial probability density is $\phi(x)$, then the probability at all later times is given by formula (6). That is, $u(x, t)$ satisfies the diffusion equation.

It is usually impossible to evaluate integral (8) completely in terms of elementary functions. Answers to particular problems, that is, to particular initial data $\phi(x)$, are sometimes expressible in terms of the *error function* of statistics,

$$\mathscr{E}\mathrm{rf}(x) = \frac{2}{\sqrt{\pi}} \int_0^x e^{-p^2}\,dp. \tag{10}$$

Notice that $\mathscr{E}\mathrm{rf}(0) = 0$. By Exercise 6, $\lim\limits_{x \to +\infty} \mathscr{E}\mathrm{rf}(x) = 1$.

## Example 1.

From (5) we can write $Q(x, t)$ in terms of $\mathscr{E}\mathrm{rf}$ as

$$Q(x, t) = \frac{1}{2} + \frac{1}{2}\mathscr{E}\mathrm{rf}\left(\frac{x}{\sqrt{4kt}}\right). \qquad \square$$

## Example 2.

Solve the diffusion equation with the initial condition $u(x, 0) = e^{-x}$. To do so, we simply plug this into the general formula (8):

$$u(x, t) = \frac{1}{\sqrt{4\pi kt}} \int_{-\infty}^{\infty} e^{-(x-y)^2/4kt}e^{-y}\,dy.$$

This is one of the few fortunate examples that can be integrated. The exponent is

$$-\frac{x^2 - 2xy + y^2 + 4kty}{4kt}.$$

Completing the square in the $y$ variable, it is

$$-\frac{(y + 2kt - x)^2}{4kt} + kt - x.$$

We let $p = (y + 2kt - x)/\sqrt{4kt}$ so that $dp = dy/\sqrt{4kt}$. Then

$$u(x, t) = e^{kt-x} \int_{-\infty}^{\infty} e^{-p^2} \frac{dp}{\sqrt{\pi}} = e^{kt-x}.$$

By the maximum principle, a solution in a bounded interval cannot grow in time. However, this particular solution grows, rather than decays, in time. The reason is that the left side of the rod is initially very hot [$u(x, 0) \to +\infty$ as $x \to -\infty$] and the heat gradually diffuses throughout the rod.    □

## EXERCISES

1.  Solve the diffusion equation with the initial condition

    $$\phi(x) = 1 \quad \text{for } |x| < l \quad \text{and} \quad \phi(x) = 0 \quad \text{for } |x| > l.$$

    Write your answer in terms of $\mathscr{E}rf(x)$.

2.  Do the same for $\phi(x) = 1$ for $x > 0$ and $\phi(x) = 3$ for $x < 0$.

3.  Use (8) to solve the diffusion equation if $\phi(x) = e^{3x}$. (You may also use Exercises 6 and 7 below.)

4.  Solve the diffusion equation if $\phi(x) = e^{-x}$ for $x > 0$ and $\phi(x) = 0$ for $x < 0$.

5.  Prove properties (a) to (e) of the diffusion equation (1).

6.  Compute $\int_0^\infty e^{-x^2} dx$. (*Hint:* This is a function that *cannot* be integrated by formula. So use the following trick. Transform the double integral $\int_0^\infty e^{-x^2} dx \cdot \int_0^\infty e^{-y^2} dy$ into polar coordinates and you'll end up with a function that can be integrated easily.)

7.  Use Exercise 6 to show that $\int_{-\infty}^\infty e^{-p^2} dp = \sqrt{\pi}$. Then substitute $p = x/\sqrt{4kt}$ to show that

    $$\int_{-\infty}^{\infty} S(x, t) \, dx = 1.$$

8.  Show that for any fixed $\delta > 0$ (no matter how small),

    $$\max_{\delta \leq |x| < \infty} S(x, t) \to 0 \qquad \text{as } t \to 0.$$

[This means that the tail of $S(x, t)$ is "uniformly small".]

9. Solve the diffusion equation $u_t = ku_{xx}$ with the initial condition $u(x, 0) = x^2$ by the following special method. First show that $u_{xxx}$ satisfies the diffusion equation with *zero* initial condition. Therefore, by uniqueness, $u_{xxx} \equiv 0$. Integrating this result thrice, obtain $u(x, t) = A(t)x^2 + B(t)x + C(t)$. Finally, it's easy to solve for $A$, $B$, and $C$ by plugging into the original problem.

10. (a) Solve Exercise 9 using the general formula discussed in the text. This expresses $u(x, t)$ as a certain integral. Substitute $p = (x - y)/\sqrt{4kt}$ in this integral.
    (b) Since the solution is unique, the resulting formula must agree with the answer to Exercise 9. Deduce the value of

$$\int_{-\infty}^{\infty} p^2 e^{-p^2} dp.$$

11. (a) Consider the diffusion equation on the whole line with the usual initial condition $u(x, 0) = \phi(x)$. If $\phi(x)$ is an *odd* function, show that the solution $u(x, t)$ is also an *odd* function of $x$. (*Hint:* Consider $u(-x, t) + u(x, t)$ and use the uniqueness.)
    (b) Show that the same is true if "odd" is replaced by "even."
    (c) Show that the analogous statements are true for the wave equation.

12. The purpose of this exercise is to calculate $Q(x, t)$ approximately for large $t$. Recall that $Q(x, t)$ is the temperature of an infinite rod that is initially at temperature 1 for $x > 0$, and 0 for $x < 0$.
    (a) Express $Q(x, t)$ in terms of $\mathscr{E}$rf.
    (b) Find the Taylor series of $\mathscr{E}$rf$(x)$ around $x = 0$. (*Hint:* Expand $e^z$, substitute $z = -y^2$, and integrate term by term.)
    (c) Use the first two nonzero terms in this Taylor expansion to find an approximate formula for $Q(x, t)$.
    (d) *Why* is this formula a good approximation for $x$ fixed and $t$ large?

13. Prove from first principles that $Q(x, t)$ *must* have the form (4), as follows.
    (a) Assuming uniqueness show that $Q(x, t) = Q(\sqrt{a} x, at)$. This identity is valid for all $a > 0$, all $t > 0$, and all $x$.
    (b) Choose $a = 1/(4kt)$.

14. Let $\phi(x)$ be a continuous function such that $|\phi(x)| \le Ce^{ax^2}$. Show that formula (8) for the solution of the diffusion equation makes sense for $0 < t < 1/(4ak)$, but not necessarily for larger $t$.

15. Prove the uniqueness of the diffusion problem with Neumann boundary conditions:

$$u_t - ku_{xx} = f(x, t) \quad \text{for } 0 < x < l, t > 0 \quad u(x, 0) = \phi(x)$$
$$u_x(0, t) = g(t) \quad u_x(l, t) = h(t)$$

by the energy method.

16. Solve the diffusion equation with constant dissipation:

$$u_t - ku_{xx} + bu = 0 \qquad \text{for } -\infty < x < \infty \qquad \text{with } u(x, 0) = \phi(x),$$

where $b > 0$ is a constant. (*Hint:* Make the change of variables $u(x, t) = e^{-bt}v(x, t)$.)

17. Solve the diffusion equation with variable dissipation:

$$u_t - ku_{xx} + bt^2 u = 0 \qquad \text{for } -\infty < x < \infty \qquad \text{with } u(x, 0) = \phi(x),$$

where $b > 0$ is a constant. (*Hint:* The solutions of the ODE $w_t + bt^2 w = 0$ are $Ce^{-bt^3/3}$. So make the change of variables $u(x, t) = e^{-bt^3/3}v(x, t)$ and derive an equation for $v$.)

18. Solve the heat equation with convection:

$$u_t - ku_{xx} + Vu_x = 0 \qquad \text{for } -\infty < x < \infty \qquad \text{with } u(x, 0) = \phi(x),$$

where $V$ is a constant. (*Hint:* Go to a moving frame of reference by substituting $y = x - Vt$.)

19. (a) Show that $S_2(x, y, t) = S(x, t)S(y, t)$ satisfies the diffusion equation $S_t = k(S_{xx} + S_{yy})$.
    (b) Deduce that $S_2(x, y, t)$ is the source function for two-dimensional diffusions.

## 2.5 COMPARISON OF WAVES AND DIFFUSIONS

We have seen that the basic property of waves is that information gets transported in both directions at a finite speed. The basic property of diffusions is that the initial disturbance gets spread out in a smooth fashion and gradually disappears. The fundamental properties of these two equations can be summarized in the following table.

| Property | | Waves | Diffusions |
|---|---|---|---|
| (i) | Speed of propagation? | Finite ($\leq c$) | Infinite |
| (ii) | Singularities for $t > 0$? | Transported along characteristics (speed $= c$) | Lost immediately |
| (iii) | Well-posed for $t > 0$? | Yes | Yes (at least for bounded solutions) |
| (iv) | Well-posed for $t < 0$? | Yes | No |
| (v) | Maximum principle | No | Yes |
| (vi) | Behavior as $t \to +\infty$? | Energy is constant so does not decay | Decays to zero (if $\phi$ integrable) |
| (vii) | Information | Transported | Lost gradually |

For the wave equation we have seen most of these properties already. That there is no maximum principle is easy to see. Generally speaking, the wave equation just moves information along the characteristic lines. In more than one dimension we'll see that it spreads information in expanding circles or spheres.

For the diffusion equation we discuss property (ii), that singularities are immediately lost, in Section 3.5. The solution is differentiable to all orders even if the initial data are not. Properties (iii), (v), and (vi) have been shown already. The fact that information is gradually lost [property (vii)] is clear from the graph of a typical solution, for instance, from $S(x, t)$.

As for property (i) for the diffusion equation, notice from formula (2.4.8) that the value of $u(x, t)$ depends on the values of the initial datum $\phi(y)$ for *all y*, where $-\infty < y < \infty$. Conversely, the value of $\phi$ at a point $x_0$ has an *immediate effect everywhere* (for $t > 0$), even though most of its effect is only for a short time near $x_0$. Therefore, the *speed of propagation is infinite*. Exercise 2(b) shows that solutions of the diffusion equation can travel at any speed. This is in stark contrast to the wave equation (and all hyperbolic equations).

As for (iv), there are several ways to see that *the diffusion equation is not well-posed for t < 0* ("backward in time"). One way is the following. Let

$$u_n(x, t) = \frac{1}{n} \sin nx \, e^{-n^2 kt}. \tag{1}$$

You can check that this satisfies the diffusion equation for all $x$, $t$. Also, $u_n(x, 0) = n^{-1} \sin nx \to 0$ uniformly as $n \to \infty$. But consider any $t < 0$, say $t = -1$. Then $u_n(x, -1) = n^{-1} \sin nx \, e^{+kn^2} \to \pm\infty$ uniformly as $n \to \infty$ except for a few $x$. Thus $u_n$ is close to the zero solution at time $t = 0$ but not at time $t = -1$. This violates the stability, in the uniform sense at least.

Another way is to let $u(x, t) = S(x, t + 1)$. This is a solution of the diffusion equation $u_t = ku_{xx}$ for $t > -1$, $-\infty < x < \infty$. But $u(0, t) \to \infty$ as $t \searrow -1$, as we saw above. So we cannot solve backwards in time with the perfectly nice-looking initial data $e^{-x^2/4k}$.

Besides, any physicist knows that heat flow, brownian motion, and so on, are irreversible processes. Going backward leads to chaos.

## EXERCISES

1. Show that there is no maximum principle for the wave equation.

2. Consider a traveling wave $u(x, t) = f(x - at)$ where $f$ is a given function of one variable.
   (a) If it is a solution of the wave equation, show that the speed must be $a = \pm c$ (unless $f$ is a linear function).
   (b) If it is a solution of the diffusion equation, find $f$ and show that the speed $a$ is arbitrary.

3.  Let $u$ satisfy the diffusion equation $u_t = \frac{1}{2}u_{xx}$. Let

$$v(x, t) = \frac{1}{\sqrt{t}} e^{x^2/2t} v\left(\frac{x}{t}, \frac{1}{t}\right).$$

Show that $v$ satisfies the "backward" diffusion equation $v_t = -\frac{1}{2}v_{xx}$ for $t > 0$.

4.  Here is a direct relationship between the wave and diffusion equations. Let $u(x, t)$ solve the wave equation on the whole line with bounded second derivatives. Let

$$v(x, t) = \frac{c}{\sqrt{4\pi kt}} \int_{-\infty}^{\infty} e^{-s^2c^2/4kt} u(x, s)\, ds.$$

(a)  Show that $v(x, t)$ solves the diffusion equation!
(b)  Show that $\lim_{t \to 0} v(x, t) = u(x, 0)$.

(*Hint:* (a) Write the formula as $v(x, t) = \int_{-\infty}^{\infty} H(s, t)u(x, s)\, ds$, where $H(x, t)$ solves the diffusion equation with constant $k/c^2$ for $t > 0$. Then differentiate $v(x, t)$ using Section A.3. (b) Use the fact that $H(s, t)$ is essentially the source function of the diffusion equation with the spatial variable $s$.)

# 3

# REFLECTIONS AND SOURCES

In this chapter we solve the simplest reflection problems, when there is only a single point of reflection at one end of a semi-infinite line. In Chapter 4 we shall begin a systematic study of more complicated reflection problems. In Sections 3.3 and 3.4 we solve problems with sources: that is, the inhomogeneous wave and diffusion equations. Finally, in Section 3.5 we analyze the solution of the diffusion equation more carefully.

## 3.1 DIFFUSION ON THE HALF-LINE

Let's take the domain to be $D$ = the half-line $(0, \infty)$ and take the *Dirichlet boundary condition* at the single endpoint $x = 0$. So the problem is

$$
\begin{array}{ll}
v_t - k v_{xx} = 0 & \text{in } \{0 < x < \infty, \quad 0 < t < \infty\}, \\
v(x, 0) = \phi(x) & \text{for } t = 0 \\
v(0, t) = 0 & \text{for } x = 0
\end{array}
\tag{1}
$$

The PDE is supposed to be satisfied in the open region $\{0 < x < \infty,\ 0 < t < \infty\}$. If it exists, we know that the solution $v(x, t)$ of this problem is unique because of our discussion in Section 2.3. It can be interpreted, for instance, as the temperature in a very long rod with one end immersed in a reservoir of temperature zero and with insulated sides.

We are looking for a solution formula analogous to (2.4.8). In fact, we shall reduce our new problem to our old one. Our method uses the idea of an *odd function*. Any function $\psi(x)$ that satisfies $\psi(-x) \equiv -\psi(+x)$ is called an odd function. Its graph $y = \psi(x)$ is symmetric with respect to the origin

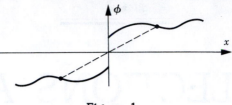

**Figure 1**

(see Figure 1). Automatically (by putting $x = 0$ in the definition), $\psi(0) = 0$. For a detailed discussion of odd and even functions, see Section 5.2.

Now the initial datum $\phi(x)$ of our problem is defined only for $x \geq 0$. Let $\phi_{\text{odd}}$ be the unique *odd extension* of $\phi$ to the whole line. That is,

$$\phi_{\text{odd}}(x) = \begin{cases} \phi(x) & \text{for } x > 0 \\ -\phi(-x) & \text{for } x < 0 \\ 0 & \text{for } x = 0. \end{cases} \tag{2}$$

The extension concept too is discussed in Section 5.2.

Let $u(x, t)$ be the solution of

$$\begin{aligned} u_t - k u_{xx} &= 0 \\ u(x, 0) &= \phi_{\text{odd}}(x) \end{aligned} \tag{3}$$

for the *whole line* $-\infty < x < \infty, 0 < t < \infty$. According to Section 2.3, it is given by the formula

$$u(x, t) = \int_{-\infty}^{\infty} S(x - y, \, t)\phi_{\text{odd}}(y)\,dy. \tag{4}$$

Its "restriction,"

$$v(x, t) = u(x, t) \quad \text{for } x > 0, \tag{5}$$

will be the unique solution of our new problem (1). There is no difference at all between $v$ and $u$ except that the negative values of $x$ are not considered when discussing $v$.

Why is $v(x, t)$ the solution of (1)? Notice first that $u(x, t)$ must also be an odd function of $x$ (see Exercise 2.4.11). That is, $u(-x, t) = -u(x, t)$. Putting $x = 0$, it is clear that $u(0, t) = 0$. So the boundary condition $v(0, t) = 0$ is *automatically* satisfied! Furthermore, $v$ solves the PDE as well as the initial condition for $x > 0$, simply because it is equal to $u$ for $x > 0$ and $u$ satisfies the same PDE for all $x$ and the same initial condition for $x > 0$.

The explicit formula for $v(x, t)$ is easily deduced from (4) and (5). From (4) and (2) we have

$$u(x, \, t) = \int_0^{\infty} S(x - y, \, t)\phi(y)\,dy - \int_{-\infty}^0 S(x - y, \, t)\phi(-y)\,dy.$$

Changing the variable $-y$ to $+y$ in the second integral, we get

$$u(x, t) = \int_0^\infty [S(x - y, \, t) - S(x + y, \, t)] \, \phi(y) \, dy.$$

(Notice the change in the limits of integration.) Hence for $0 < x < \infty$, $0 < t < \infty$, we have

$$v(x, \, t) = \frac{1}{\sqrt{4\pi kt}} \int_0^\infty [e^{-(x-y)^2/4kt} - e^{-(x+y)^2/4kt}] \, \phi(y) \, dy. \qquad (6)$$

This is the complete solution formula for (1).

We have just carried out the *method of odd extensions* or *reflection method*, so called because the graph of $\phi_{\text{odd}}(x)$ is the reflection of the graph of $\phi(x)$ across the origin.

## Example 1.

Solve (1) with $\phi(x) \equiv 1$. The solution is given by formula (6). This case can be simplified as follows. Let $p = (x - y)/\sqrt{4kt}$ in the first integral and $q = (x + y)/\sqrt{4kt}$ in the second integral. Then

$$u(x, t) = \int_{-\infty}^{x/\sqrt{4kt}} e^{-p^2} dp/\sqrt{\pi} - \int_{x/\sqrt{4kt}}^{+\infty} e^{-q^2} dq/\sqrt{\pi}$$

$$= \left[\frac{1}{2} + \frac{1}{2}\mathscr{E}\text{rf}\left(\frac{x}{\sqrt{4kt}}\right)\right] - \left[\frac{1}{2} - \frac{1}{2}\mathscr{E}\text{rf}\left(\frac{x}{\sqrt{4kt}}\right)\right]$$

$$= \mathscr{E}\text{rf}\left(\frac{x}{\sqrt{4kt}}\right). \qquad \square$$

Now let's play the same game with the *Neumann problem*

$$\boxed{\begin{aligned} w_t - kw_{xx} &= 0 \quad \text{for } 0 < x < \infty, \, 0 < t < \infty \\ w(x, \, 0) &= \phi(x) \\ w_x(0, \, t) &= 0. \end{aligned}} \qquad (7)$$

In this case the reflection method is to use *even*, rather than odd, extensions. An even function is a function $\psi$ such that $\psi(-x) = +\psi(x)$. If $\psi$ is an even function, then differentiation shows that its derivative is an odd function. So automatically its slope at the origin is zero: $\psi'(0) = 0$. If $\phi(x)$ is defined only on the half-line, its *even extension* is defined to be

$$\phi_{\text{even}}(x) = \begin{cases} \phi(x) & \text{for } x \geq 0 \\ +\phi(-x) & \text{for } x \leq 0 \end{cases} \qquad (8)$$

By the same reasoning as we used above, we end up with an explicit formula for $w(x, t)$. It is

$$w(x, t) = \frac{1}{\sqrt{4\pi kt}} \int_0^\infty [e^{-(x-y)^2/4kt} + e^{-(x+y)^2/4kt}] \phi(y) \, dy. \qquad (9)$$

This is carried out in Exercise 3. Notice that the only difference between (6) and (9) is a single minus sign!

### Example 2.

Solve (7) with $\phi(x) = 1$. This is the same as Example 1 except for the single sign. So we can copy from that example:

$$u(x, t) = \left[\frac{1}{2} + \frac{1}{2}\mathscr{E}\mathrm{rf}\left(\frac{x}{4kt}\right)\right] + \left[\frac{1}{2} - \frac{1}{2}\mathscr{E}\mathrm{rf}\left(\frac{x}{4kt}\right)\right] = 1.$$

(That was stupid: We could have guessed it!)   □

### EXERCISES

1.  Solve $u_t = ku_{xx}$; $u(x, 0) = e^{-x}$; $u(0, t) = 0$ on the half-line $0 < x < \infty$.
2.  Solve $u_t = ku_{xx}$; $u(x, 0) = 0$; $u(0, t) = 1$ on the half-line $0 < x < \infty$.
3.  Derive the solution formula for the half-line Neumann problem $w_t - kw_{xx} = 0$ for $0 < x < \infty$, $0 < t < \infty$; $w_x(0, t) = 0$; $w(x, 0) = \phi(x)$.
4.  Consider the following problem with a Robin boundary condition:

    DE:   $u_t = ku_{xx}$          on the half-line $0 < x < \infty$
                              (and $0 < t < \infty$)

    IC:   $u(x, 0) = x$          for $t = 0$ and $0 < x < \infty$          (*)

    BC:   $u_x(0, t) - 2u(0, t) = 0$   for $x = 0$.

    The purpose of this exercise is to verify the solution formula for (*). Let $f(x) = x$ for $x > 0$, let $f(x) = x + 1 - e^{2x}$ for $x < 0$, and let

    $$v(x, t) = \frac{1}{\sqrt{4\pi kt}} \int_{-\infty}^\infty e^{-(x-y)^2/4kt} f(y) \, dy.$$

    (a)   What PDE and initial condition does $v(x, t)$ satisfy for $-\infty < x < \infty$?
    (b)   Let $w = v_x - 2v$. What PDE and initial condition does $w(x, t)$ satisfy for $-\infty < x < \infty$?
    (c)   Show that $f'(x) - 2f(x)$ is an odd function (for $x \neq 0$).
    (d)   Use Exercise 2.4.11 to show that $w$ is an odd function of $x$.

(e)  Deduce that $v(x, t)$ satisfies (*) for $x > 0$. Assuming uniqueness, deduce that the solution of (*) is given by

$$u(x, t) = \frac{1}{\sqrt{4\pi kt}} \int_{-\infty}^{\infty} e^{-(x-y)^2/4kt} f(y)\, dy.$$

5. (a)  Use the method of Exercise 4 to solve the Robin problem:

    DE:   $u_t = k u_{xx}$                  on the half-line $0 < x < \infty$
                                             (and $0 < t < \infty$)

    IC:   $u(x, 0) = x$                 for $t = 0$ and $0 < x < \infty$
    BC:   $u_x(0, t) - hu(0, t) = 0$    for $x = 0$,

    where $h$ is a constant.

(b)  Generalize the method to the case of general initial data $\phi(x)$.

## 3.2   REFLECTIONS OF WAVES

Now we try the same kind of problem for the wave equation as we did in Section 3.1 for the diffusion equation. We again begin with the *Dirichlet problem* on the half-line $(0, \infty)$. Thus the problem is

$$
\boxed{
\begin{array}{lll}
\text{DE}: & v_{tt} - c^2 v_{xx} = 0 & \text{for } 0 < x < \infty \\
& & \text{and } -\infty < t < \infty \\
\text{IC}: & v(x, 0) = \phi(x), \quad v_t(x, 0) = \psi(x) & \text{for } t = 0 \\
& & \text{and } 0 < x < \infty \\
\text{BC}: & v(0, t) = 0 & \text{for } x = 0 \\
& & \text{and } -\infty < t < \infty.
\end{array}
}
\tag{1}
$$

The reflection method is carried out in the same way as in Section 3.1. Consider the *odd* extensions of both of the initial functions to the whole line, $\phi_{odd}(x)$ and $\psi_{odd}(x)$. Let $u(x, t)$ be the solution of the initial-value problem on $(-\infty, \infty)$ with the initial data $\phi_{odd}$ and $\psi_{odd}$. Then $u(x, t)$ is once again an odd function of $x$ (see Exercise 2.1.7). Therefore, $u(0, t) = 0$, so that the boundary condition is satisfied automatically. Define $v(x, t) = u(x, t)$ for $0 < x < \infty$ [the restriction of $u$ to the half-line]. Then $v(x, t)$ is precisely the solution we are looking for. From the formula in Section 2.1, we have for $x \geq 0$,

$$v(x, t) = u(x, t) = \frac{1}{2}[\phi_{odd}(x + ct) + \phi_{odd}(x - ct)] + \frac{1}{2c} \int_{x-ct}^{x+ct} \psi_{odd}(y)\, dy.$$

Let's "unwind" this formula, recalling the meaning of the odd extensions. First we notice that for $x > c|t|$ only positive arguments occur in the formula,

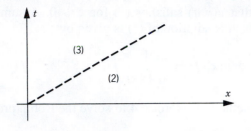

**Figure 1**

so that $u(x, t)$ is given by the *usual* formula:

$$v(x, t) = \frac{1}{2}[\phi(x + ct) + \phi(x - ct)] + \frac{1}{2c}\int_{x-ct}^{x+ct} \psi(y)\,dy$$

$$\text{for } x > c|t|.$$

(2)

But in the *other* region $0 < x < c|t|$, we have $\phi_{\text{odd}}(x - ct) = -\phi(ct - x)$, and so on, so that

$$v(x, t) = \frac{1}{2}[\phi(x + ct) - \phi(ct - x)] + \frac{1}{2c}\int_0^{x+ct} \psi(y)\,dy + \frac{1}{2c}\int_{x-ct}^0 [-\psi(-y)]\,dy.$$

Notice the switch in signs! In the last term we change variables $y \to -y$ to get $1/2c \int_{ct-x}^{ct+x} \psi(y)\,dy$. Therefore,

$$v(x, t) = \frac{1}{2}[\phi(ct + x) - \phi(ct - x)] + \frac{1}{2c}\int_{ct-x}^{ct+x} \psi(y)\,dy$$

(3)

for $0 < x < c|t|$. The complete solution is given by the pair of formulas (2) and (3). The two regions are sketched in Figure 1 for $t > 0$.

Graphically, the result can be interpreted as follows. Draw the backward characteristics from the point $(x, t)$. In case $(x, t)$ is in the region $x < ct$, one of the characteristics hits the $t$ axis ($x = 0$) before it hits the $x$ axis, as indicated in Figure 2. The formula (3) shows that *the reflection induces a change of*

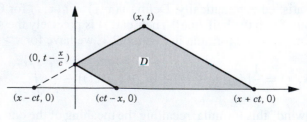

**Figure 2**

*sign.* The value of $v(x, t)$ now depends on the values of $\phi$ at the pair of points $ct \pm x$ and on the values of $\psi$ in the short interval between these points. Note that the other values of $\psi$ have canceled out. The shaded area $D$ in Figure 2 is called the *domain of dependence of the point* $(x, t)$.

The case of the Neumann problem is left as an exercise.

## THE FINITE INTERVAL

Now let's consider the guitar string with fixed ends:

$$v_{tt} = c^2 v_{xx} \quad v(x, 0) = \phi(x) \quad v_t(x, 0) = \psi(x) \quad \text{for } 0 < x < l,$$
$$v(0, t) = v(l, t) = 0. \tag{4}$$

This problem is much more difficult because a typical wave will bounce back and forth an infinite number of times. Nevertheless, let's use the method of reflection. This is a bit tricky, so you are invited to skip the rest of this section if you wish.

The initial data $\phi(x)$ and $\psi(x)$ are now given only for $0 < x < l$. We extend them to the whole line to be "*odd*" with respect to *both* $x = 0$ *and* $x = l$:

$$\phi_{ext}(-x) = -\phi_{ext}(x) \quad \text{and} \quad \phi_{ext}(2l - x) = -\phi_{ext}(x).$$

The simplest way to do this is to define

$$\phi_{ext}(x) = \begin{cases} \phi(x) & \text{for} & 0 < x < l \\ -\phi(-x) & \text{for} & -l < x < 0 \\ \text{extended to be of period } 2l. \end{cases}$$

See Figure 3 for an example. And see Section 5.2 for further discussion. "Period $2l$" means that $\phi_{ext}(x + 2l) = \phi_{ext}(x)$ for all $x$. We do exactly the same for $\psi(x)$ (defined for $0 < x < l$) to get $\psi_{ext}(x)$ defined for $-\infty < x < \infty$.

Now let $u(x, t)$ be the solution of the infinite line problem with the extended initial data. Let $v$ be the restriction of $u$ to the interval $(0, l)$. Thus $v(x, t)$ is

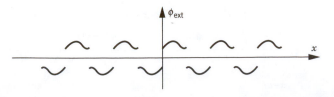

Figure 3

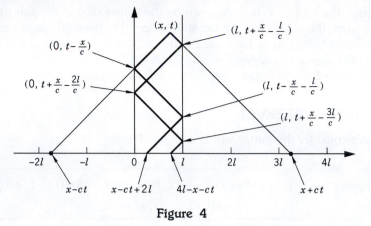

**Figure 4**

given by the formula

$$v(x, t) = \frac{1}{2}\phi_{\text{ext}}(x + ct) + \frac{1}{2}\phi_{\text{ext}}(x - ct) + \frac{1}{2c}\int_{x-ct}^{x+ct} \psi_{\text{ext}}(s)\,ds \qquad (5)$$

for $0 \le x \le l$. This simple formula contains all the information we need. But to see it explicitly we must unwind the definitions of $\phi_{\text{ext}}$ and $\psi_{\text{ext}}$. This will give a resulting formula which appears quite complicated because it includes a precise description of *all* the reflections of the wave at both of the boundary points $x = 0$ and $x = l$.

The way to understand the explicit result we are about to get is by drawing a space-time diagram (Figure 4). From the point $(x, t)$, we draw the two characteristic lines and reflect them each time they hit the boundary. We keep track of the change of sign at each reflection. We illustrate the result in Figure 4 for the case of a typical point $(x, t)$. We also illustrate in Figure 5 the definition of the extended function $\phi_{\text{ext}}(x)$. (The same picture is valid for $\psi_{\text{ext}}$.) For instance, for the point $(x, t)$ as drawn in Figures 4 and 5, we have

$$\phi_{\text{ext}}(x + ct) = -\phi(4l - x - ct) \quad \text{and} \quad \phi_{\text{ext}}(x - ct) = +\phi(x - ct + 2l).$$

The minus coefficient on $-\phi(-x - ct + 4l)$ comes from the odd number of reflections ($= 3$). The plus coefficient on $\phi(x - ct + 2l)$ comes from the even

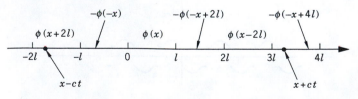

**Figure 5**

number of reflections ($= 2$). Therefore, the general formula (5) reduces to

$$v(x, t) = \frac{1}{2}\phi(x - ct + 2l) - \frac{1}{2}\phi(4l - x - ct)$$

$$+ \frac{1}{2c}\left[\int_{x-ct}^{-l} \psi(y + 2l)\,dy + \int_{-l}^{0} -\psi(-y)\,dy\right.$$

$$+ \int_{0}^{1} \psi(y)\,dy + \int_{l}^{2l} -\psi(-y + 2l)\,dy$$

$$+ \left.\int_{2l}^{3l} \psi(y - 2l)\,dy + \int_{3l}^{x+ct} -\psi(-y + 4l)\,dy\right]$$

But notice that there is an exact cancellation of the four middle integrals, as we see by changing $y \to -y$ and $y - 2l \to -y + 2l$. So, changing variables in the two remaining integrals, the formula simplifies to

$$v(x, t) = \frac{1}{2}\phi(x - ct + 2l) - \frac{1}{2}\phi(4l - x - ct)$$

$$+ \frac{1}{2c}\int_{x-ct+2l}^{l} \psi(s)\,ds + \frac{1}{2c}\int_{l}^{4l-x-ct} \psi(s)\,ds.$$

Therefore, we end up with the formula

$$v(x, t) = \frac{1}{2}\phi(x - ct + 2l) - \frac{1}{2}\phi(4l - x - ct) + \int_{x-ct+2l}^{4l-x-ct} \psi(s)\frac{ds}{2c} \qquad (6)$$

*at the point* $(x, t)$ *illustrated*, which has three reflections on one end and two on the other. Formula (6) is valid only for such points.

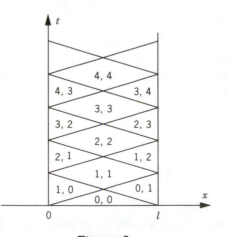

Figure 6

The solution formula at any other point $(x, t)$ is characterized by the number of reflections at each end ($x = 0, l$). This divides the space-time picture into diamond-shaped regions as illustrated in Figure 6. *Within each diamond the solution $v(x, t)$ is given by a different formula.* Further examples may be found in the exercises.

The formulas explain in detail how the solution looks. However, the method is impossible to generalize to two- or three-dimensional problems, nor does it work for the diffusion equation at all. Also, it is very complicated! Therefore, in Chapter 4 we shall introduce a completely different method (Fourier's) for solving problems on a finite interval.

## EXERCISES

1. Solve the Neumann problem for the wave equation on the half-line $0 < x < \infty$.

2. The longitudinal vibrations of a semi-infinite flexible rod satisfy the wave equation $u_{tt} = c^2 u_{xx}$ for $x > 0$. Assume that the end $x = 0$ is free ($u_x = 0$); it is initially at rest but has a constant initial velocity $V$ for $a < x < 2a$ and has zero initial velocity elsewhere. Plot $u$ versus $x$ at the times $t = 0, a/c, 3a/2c, 2a/c$, and $3a/c$.

3. A wave $f(x + ct)$ travels along a semi-infinite string ($0 < x < \infty$) for $t < 0$. Find the vibrations $u(x, t)$ of the string for $t > 0$ if the end $x = 0$ is fixed.

4. Repeat Exercise 3 if the end is free.

5. Solve $u_{tt} = 4u_{xx}$ for $0 < x < \infty, u(0, t) = 0, u(x, 0) \equiv 1, u_t(x, 0) \equiv 0$ using the reflection method. This solution has a singularity; find its location.

6. Solve $u_{tt} = c^2 u_{xx}$ in $0 < x < \infty, 0 \le t < \infty, u(x, 0) = 0, u_t(x, 0) = V$,

$$u_t(0, t) + au_x(0, t) = 0,$$

where $V$, $a$, and $c$ are positive constants and $a > c$.

7. (a) Show that $\phi_{odd}(x) = (\text{sign } x)\phi(|x|)$.
   (b) Show that $\phi_{ext}(x) = \phi_{odd}(x - 2l[x/2l])$, where $[\cdot]$ denotes the greatest integer function.
   (c) Show that

$$
\phi_{ext}(x) = \begin{cases} \phi\left(x - \left[\dfrac{x}{l}\right]l\right) & \text{if } \left[\dfrac{x}{l}\right] \text{ even} \\[2mm] -\phi\left(-x - \left[\dfrac{x}{l}\right]l - l\right) & \text{if } \left[\dfrac{x}{l}\right] \text{ odd.} \end{cases}
$$

8. For the wave equation in a finite interval $(0, l)$ with Dirichlet conditions, explain the solution formula within each diamond-shaped region.

9. (a) Find $u(\frac{2}{3}, 2)$ if $u_{tt} = u_{xx}$ in $0 < x < 1, u(x, 0) = x^2(1 - x)$,
$u_t(x, 0) = (1 - x)^2, u(0, t) = u(1, t) = 0$.
(b) Find $u(\frac{1}{4}, \frac{7}{2})$.

10. Solve $u_{tt} = 9u_{xx}$ in $0 < x < \pi/2, u(x, 0) = \cos x, u_t(x, 0) = 0$,
$u_x(0, t) = 0, u(\pi/2, t) = 0$.

11. Solve $u_{tt} = c^2 u_{xx}$ in $0 < x < l, u(x, 0) = 0, u_t(x, 0) = x, u(0, t) = u(l, t) = 0$.

## 3.3 DIFFUSION WITH A SOURCE

In this section we solve the *inhomogeneous* diffusion equation on the whole line,

$$
\boxed{
\begin{aligned}
&u_t - ku_{xx} = f(x, t) \qquad (-\infty < x < \infty, \quad 0 < t < \infty) \\
&u(x, 0) = \phi(x)
\end{aligned}
}
\tag{1}
$$

with $f(x, t)$ and $\phi(x)$ arbitrary given functions. For instance, if $u(x, t)$ represents the temperature of a rod, then $\phi(x)$ is the initial temperature distribution and $f(x, t)$ is a source (or sink) of heat provided to the rod at later times.

We will show that the solution of (1) is

$$
\boxed{
\begin{aligned}
u(x, t) &= \int_{-\infty}^{\infty} S(x - y, t)\phi(y)\, dy \\
&\quad + \int_0^t \int_{-\infty}^{\infty} S(x - y, t - s)f(y, s)\, dy\, ds.
\end{aligned}
}
\tag{2}
$$

Notice that there is the usual term involving the initial data $\phi$ and another term involving the source $f$. Both terms involve the source function $S$.

Let's begin by explaining where (2) comes from. Later we will actually prove the validity of the formula. (If a strictly mathematical proof is satisfactory to you, this paragraph and the next two can be skipped.) Our explanation is an analogy. The simplest analogy is the ODE

$$
\frac{du}{dt} + Au(t) = f(t), \qquad u(0) = \phi,
\tag{3}
$$

where $A$ is a constant. Using the integrating factor $e^{tA}$, the solution is

$$
u(t) = e^{-tA}\phi + \int_0^t e^{(s-t)A} f(s)\, ds.
\tag{4}
$$

A more elaborate analogy is the following. Let's suppose that $\phi$ is an $n$-vector, $u(t)$ is an $n$-vector function of time, and $A$ is a fixed $n \times n$ matrix.

Then (3) is a coupled system of $n$ linear ODEs. In case $f(t) \equiv 0$, the solution of (3) is given as $u(t) = S(t)\phi$, where $S(t)$ is the matrix $S(t) = e^{-tA}$. So in case $f(t) \neq 0$, an integrating factor for (3) is $S(-t) = e^{tA}$. Now we multiply (3) on the left by this integrating factor to get

$$\frac{d}{dt}[S(-t)u(t)] = S(-t)\frac{du}{dt} + S(-t)Au(t) = S(-t)f(t).$$

Integrating from 0 to $t$, we get

$$S(-t)u(t) - \phi = \int_0^t S(-s)f(s)\,ds.$$

Multiplying this by $S(t)$, we end up with the solution formula

$$u(t) = S(t)\phi + \int_0^t S(t-s)f(s)\,ds. \tag{5}$$

The first term in (5) represents the solution of the homogeneous equation, the second the effect of the source $f(t)$. For a single equation, of course, (5) reduces to (4).    □

Now let's return to the original diffusion problem (1). There is an analogy between (2) and (5) which we now explain. The solution of (1) will have two terms. The first one will be the solution of the homogeneous problem, already solved in Section 2.4, namely

$$\int_{-\infty}^{\infty} S(x - y, t)\phi(y)\,dy = (\mathcal{S}(t)\phi)(x). \tag{6}$$

$S(x - y, t)$ is the source function given by the formula (2.4.7). Here we are using $\mathcal{S}(t)$ to denote the *source operator*, which transforms any function $\phi$ to the new function given by the integral in (6). (Remember: Operators transform functions into functions.) We can now *guess* what the whole solution to (1) must be. In analogy to formula (5), we guess that the solution of (1) is

$$u(t) = \mathcal{S}(t)\phi + \int_0^t \mathcal{S}(t-s)f(s)\,ds. \tag{7}$$

Formula (7) is exactly the same as (2):

$$u(x, t) = \int_{-\infty}^{\infty} S(x - y, t)\phi(y)\,dy$$

$$+ \int_0^t \int_{-\infty}^{\infty} S(x - y, t - s)f(y, s)\,dy\,ds. \tag{2}$$

The method we have just used to find formula (2) is the operator method.

**Proof of (2).**  All we have to do is verify that the function $u(x, t)$, which is *defined* by (2), in fact satisfies the PDE and IC (1). Since the solution of

(1) is unique, we would then know that $u(x, t)$ is that unique solution. For simplicity, we may as well let $\phi \equiv 0$, since we understand the $\phi$ term already.

We first verify the PDE. Differentiating (2), assuming $\phi \equiv 0$ and using the rule for differentiating integrals in Section A.3, we have

$$\frac{\partial u}{\partial t} = \frac{\partial}{\partial t} \int_0^t \int_{-\infty}^{\infty} S(x - y, t - s) f(y, s) \, dy \, ds$$

$$= \int_0^t \int_{-\infty}^{\infty} \frac{\partial S}{\partial t}(x - y, t - s) f(y, s) \, dy \, ds$$

$$+ \lim_{s \to t} \int_{-\infty}^{\infty} S(x - y, t - s) f(y, s) \, dy,$$

taking special care due to the singularity of $S(x - y, t - s)$ at $t - s = 0$. Using the fact that $S(x - y, t - s)$ satisfies the diffusion equation, we get

$$\frac{\partial u}{\partial t} = \int_0^t \int_{-\infty}^{\infty} k \frac{\partial^2 S}{\partial x^2}(x - y, t - s) f(y, s) \, dy \, ds$$

$$+ \lim_{\epsilon \to 0} \int_{-\infty}^{\infty} S(x - y, \epsilon) f(y, t) \, dy.$$

Pulling the spatial derivative outside the integral and using the initial condition satisfied by $S$, we get

$$\frac{\partial u}{\partial t} = k \frac{\partial^2}{\partial x^2} \int_0^t \int_{-\infty}^{\infty} S(x - y, t - s) f(y, s) \, dy \, ds + f(x, t)$$

$$= k \frac{\partial^2 u}{\partial x^2} + f(x, t).$$

This identity is exactly the PDE (1). Second, we verify the initial condition. Letting $t \to 0$, the first term in (2) tends to $\phi(x)$ because of the initial condition of $S$. The second term is an integral from 0 to 0. Therefore,

$$\lim_{t \to 0} u(x, t) = \phi(x) + \int_0^0 \cdots = \phi(x).$$

This proves that (2) is the unique solution.    □

Remembering that $S(x, t)$ is the gaussian distribution (2.4.7), the formula (2) takes the explicit form

$$u(x, t) = \int_0^t \int_{-\infty}^{\infty} S(x - y, t - s) f(y, s) \, dy \, ds$$

$$= \int_0^t \int_{-\infty}^{\infty} \frac{1}{\sqrt{4\pi k(t - s)}} e^{-(x-y)^2/4k(t-s)} f(y, s) \, dy \, ds. \qquad (8)$$

in the case that $\phi \equiv 0$.

## SOURCE ON A HALF-LINE

For inhomogeneous diffusion on the half-line we can use the method of reflection just as in Section 3.1 (see Exercise 1).

Now consider the more complicated problem of a *boundary source* $h(t)$ on the half-line; that is,

$$v_t - kv_{xx} = f(x, t) \quad \text{for } 0 < x < \infty, \quad 0 < t < \infty$$
$$\boldsymbol{v(0, t) = h(t)} \tag{9}$$
$$v(x, 0) = \phi(x).$$

We may use the following subtraction device to reduce (9) to a simpler problem. Let $V(x, t) = v(x, t) - h(t)$. Then $V(x, t)$ will satisfy

$$V_t - kV_{xx} = f(x, t) - h'(t) \quad \text{for } 0 < x < \infty, \quad 0 < t < \infty$$
$$V(0, t) = 0 \tag{10}$$
$$V(x, 0) = \phi(x) - h(0).$$

To verify (10), just subtract! This new problem has a homogeneous boundary condition to which we can apply the method of reflection. Once we find $V$, we recover $v$ by $v(x, t) = V(x, t) + h(t)$. This simple subtraction device is often used to reduce one linear problem to another.

The domain of independent variables $(x, t)$ in this case is a quarter-plane with specified conditions on both of its half-lines. If they do not agree at the corner [i.e., if $\phi(0) \neq h(0)$], then the solution is discontinuous there (but continuous everywhere else). This is physically sensible. Think for instance, of suddenly at $t = 0$ sticking a hot iron bar into a cold bath.

For the inhomogeneous *Neumann* problem on the half-line,

$$w_t - kw_{xx} = f(x, t) \quad \text{for } 0 < x < \infty, \quad 0 < t < \infty$$
$$\boldsymbol{w_x(0, t) = h(t)} \tag{11}$$
$$w(x, 0) = \phi(x),$$

we would subtract off the function $xh(t)$. That is, $W(x, t) = w(x, t) - xh(t)$. Differentiation implies that $W_x(0, t) = 0$. Some of these problems are worked out in the exercises.

## EXERCISES

1. Solve the inhomogeneous diffusion equation on the half-line with Dirichlet boundary condition:

$$u_t - ku_{xx} = f(x, t) \quad (0 < x < \infty, \quad 0 < t < \infty)$$
$$u(0, t) = 0 \quad u(x, 0) = \phi(x)$$

using the method of reflection.

2.  Solve the completely inhomogeneous diffusion problem on the half-line

$$v_t - kv_{xx} = f(x, t) \qquad \text{for } 0 < x < \infty, \quad 0 < t < \infty$$
$$v(0, t) = h(t) \qquad v(x, 0) = \phi(x),$$

by carrying out the subtraction method begun in the text.

3.  Solve the inhomogeneous Neumann diffusion problem on the half-line

$$w_t - kw_{xx} = 0 \qquad \text{for } 0 < x < \infty, \quad 0 < t < \infty$$
$$w_x(0, t) = h(t) \qquad w(x, 0) = \phi(x),$$

by the subtraction method indicated in the text.

## 3.4  WAVES WITH A SOURCE

The purpose of this section is to solve

$$u_{tt} - c^2 u_{xx} = f(x, t) \tag{1}$$

on the whole line, together with the usual initial conditions

$$\boxed{\begin{aligned} u(x, 0) &= \phi(x) \\ u_t(x, 0) &= \psi(x) \end{aligned}} \tag{2}$$

where $f(x, t)$ is a given function. For instance, $f(x, t)$ could be interpreted as an external force acting on an infinitely long vibrating string.

Because $L = \partial_t^2 - c^2 \partial_x^2$ is a linear operator, the solution will be the *sum of three terms*, one for $\phi$, one for $\psi$, and one for $f$. The first two terms are given already in Section 2.1 and we must find the third term. We'll derive the following formula.

**Theorem 1.**   The unique solution of (1),(2) is

$$\boxed{u(x, t) = \frac{1}{2}[\phi(x + ct) + \phi(x - ct)] + \frac{1}{2c}\int_{x-ct}^{x+ct} \psi + \frac{1}{2c}\iint_\Delta f} \tag{3}$$

where $\Delta$ is the characteristic triangle (see Figure 1).

The double integral in (3) is equal to the iterated integral

$$\int_0^t \int_{x-c(t-s)}^{x+c(t-s)} f(y, s) \, dy \, ds.$$

We will give three different derivations of this formula! But first, let's note what the formula says. It says that the effect of a force $f$ on $u(x, t)$ is obtained

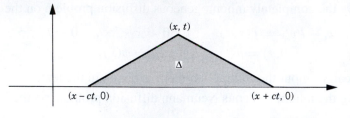

**Figure 1**

by simply integrating $f$ over the past history of the point $(x, t)$ back to the initial time $t = 0$. This is yet another example of the causality principle.

## WELL-POSEDNESS

We first show that the problem (1),(2) is well-posed in the sense of Section 1.5. The well-posedness has three ingredients, as follows. Existence is clear, given that the formula (3) itself is an explicit solution. If $\phi$ has a continuous second derivative, $\psi$ has a continuous first derivative, and $f$ is continuous, then the formula (3) yields a function $u$ with continuous second partials that satisfies the equation. Uniqueness means that there are no other solutions of (1),(2). This will follow from any one of the derivations given below.

Third, we claim that the problem (1),(2) is stable in the sense of Section 1.5. This means that if the data $(\phi, \psi, f)$ change a little, then $u$ also changes only a little. To make this precise, we need a way to measure the "nearness" of functions, that is, a *metric* or *norm* on function spaces. We will illustrate this concept using the *uniform norms:*

$$\|w\| = \max_{-\infty < x < \infty} |w(x)|$$

and

$$\|w\|_T = \max_{-\infty < x < \infty,\ 0 \le t \le T} |w(x, t)|.$$

Here $T$ is fixed. Suppose that $u_1(x, t)$ is the solution with data $(\phi_1(x), \psi_1(x), f_1(x, t))$ and $u_2(x, t)$ is the solution with data $(\phi_2(x), \psi_2(x), f_2(x, t))$ (six given functions). We have the same formula (3) satisfied by $u_1$ and by $u_2$ except for the different data. We subtract the two formulas. We let $u = u_1 - u_2$. Since the area of $\Delta$ equals $ct^2$, we have from (3) the inequality

$$|u(x, t)| \le \max|\phi| + \frac{1}{2c} \cdot \max|\psi| \cdot 2ct + \frac{1}{2c} \cdot \max|f| \cdot ct^2$$

$$= \max|\phi| + t \cdot \max|\psi| + \frac{t^2}{2} \cdot \max|f|.$$

Therefore,

$$\|u_1 - u_2\|_T \le \|\phi_1 - \phi_2\| + T\|\psi_1 - \psi_2\| + \frac{T^2}{2}\|f_1 - f_2\|_T. \qquad (4)$$

So if $\|\phi_1 - \phi_2\| < \delta$, $\|\psi_1 - \psi_2\| < \delta$, and $\|f_1 - f_2\|_T < \delta$, where $\delta$ is small, then

$$\|u_1 - u_2\|_T < \delta(1 + T + T^2) \le \epsilon$$

provided that $\delta \le \epsilon/(1 + T + T^2)$. Since $\epsilon$ is arbitrarily small, this argument proves the well-posedness of the problem (1),(2) with respect to the uniform norm.

## PROOF OF THEOREM 1

**Method of Characteristic Coordinates**    We introduce the usual characteristic coordinates $\xi = x + ct$, $\eta = x - ct$, (see Figure 2). As in Section 2.1, we have

$$Lu \equiv u_{tt} - c^2 u_{xx} = -4c^2 u_{\xi\eta} = f\left(\frac{\xi + \eta}{2}, \frac{\xi - \eta}{2c}\right).$$

We integrate this equation with respect to $\eta$, leaving $\xi$ as a constant. Thus $u_\xi = -(1/4c^2)\int^\eta f\,d\eta$. Then we integrate with respect to $\xi$ to get

$$u = -\frac{1}{4c^2}\int^\xi \int^\eta f\,d\eta\,d\xi \qquad (5)$$

The lower limits of integration here are arbitrary: They correspond to constants of integration. The calculation is much easier to understand if we fix a point $P_0$ with coordinates $x_0$, $t_0$ and

$$\xi_0 = x_0 + ct_0 \qquad \eta_0 = x_0 - ct_0.$$

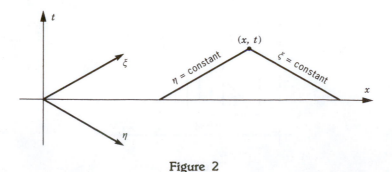

Figure 2

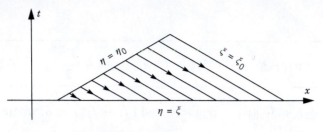

**Figure 3**

We evaluate (5) at $P_0$ *and make a particular choice of the lower limits.* Thus

$$u(P_0) = -\frac{1}{4c^2} \int_{\eta_0}^{\xi_0} \int_{\xi}^{\eta_0} f\left(\frac{\xi + \eta}{2}, \frac{\xi - \eta}{2c}\right) d\eta \, d\xi$$

$$= +\frac{1}{4c^2} \int_{\eta_0}^{\xi_0} \int_{\eta_0}^{\xi} f\left(\frac{\xi + \eta}{2}, \frac{\xi - \eta}{2c}\right) d\eta \, d\xi$$

(6)

is a particular solution. As Figure 3 indicates, $\eta$ now represents a variable going along a line segment to the base $\eta = \xi$ of the triangle $\Delta$ from the left-hand edge $\eta = \eta_0$, while $\xi$ runs from the left-hand corner to the right-hand edge. Thus we have integrated over the whole triangle $\Delta$.

The iterated integral, however, is not exactly the double integral over $\Delta$ because the coordinate axes are not orthogonal. The original axes ($x$ and $t$) are orthogonal, so we make a change of variables back to $x$ and $t$. This amounts to substituting back

$$x = \frac{\xi + \eta}{2} \qquad t = \frac{\xi - \eta}{2c}.$$

(7)

A little square in Figure 4 goes into a parallelogram in Figure 5. The change in its area is measured by the jacobian determinant $J$ (see Section A.1). Since

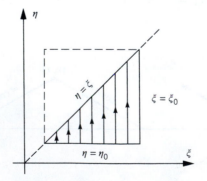

**Figure 4**

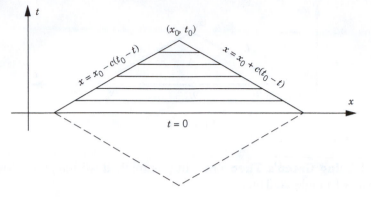

**Figure 5**

our change of variable is a linear transformation, the jacobian is just the determinant of its coefficient matrix:

$$J = \left| \det \begin{pmatrix} \dfrac{\partial \xi}{\partial x} & \dfrac{\partial \xi}{\partial t} \\ \dfrac{\partial \eta}{\partial x} & \dfrac{\partial \eta}{\partial t} \end{pmatrix} \right| = \left| \det \begin{pmatrix} 1 & c \\ 1 & -c \end{pmatrix} \right| = 2c.$$

Thus $d\eta\, d\xi = J\, dx\, dt = 2c\, dx\, dt$. Therefore, the rule for changing variables in a multiple integral (the jacobian theorem) then gives

$$u(P_0) = \frac{1}{4c^2} \iint_\Delta f(x, t) J\, dx\, dt. \tag{8}$$

This is precisely Theorem 1. The formula can also be written as the iterated integral in $x$ and $t$:

$$u(x_0, t_0) = \frac{1}{2c} \int_0^{t_0} \int_{x_0 - c(t_0 - t)}^{x_0 + c(t_0 - t)} f(x, t)\, dx\, dt, \tag{9}$$

integrating first over the horizontal line segments in Figure 5 and then vertically.

A variant of the method of characteristic coordinates is to write (1) as the system of two equations

$$u_t + c u_x = v \qquad v_t - c v_x = f,$$

the first equation being the definition of $v$, as in Section 2.1. If we first solve the second equation, then $v$ is a line integral of $f$ over a characteristic line segment $x + ct = \text{constant}$. The first equation then gives $u(x, t)$ by sweeping out these line segments over the characteristic triangle $\Delta$. To carry out this variant is a little tricky, however, and we leave it as an exercise.

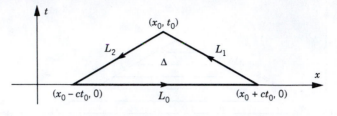

**Figure 6**

**Method Using Green's Theorem**    In this method we integrate $f$ over the past history triangle $\Delta$. Thus

$$\iint_{\Delta} f \, dx \, dt = \iint_{\Delta} (u_{tt} - c^2 u_{xx}) \, dx \, dt. \tag{10}$$

But Green's theorem says that

$$\iint_{\Delta} (P_x - Q_t) \, dx \, dt = \int_{\text{bdy}} P \, dt + Q \, dx$$

for any functions $P$ and $Q$, where the line integral on the boundary is taken counterclockwise (see Section A.3). Thus we get

$$\iint_{\Delta} f \, dx \, dt = \int_{L_0 + L_1 + L_2} (-c^2 u_x \, dt - u_t \, dx). \tag{11}$$

This is the sum of three line integrals over straight line segments (see Figure 6). We evaluate each piece separately. On $L_0$, $dt = 0$ and $u_t(x, 0) = \psi(x)$, so that

$$\int_{L_0} = -\int_{x_0 - ct_0}^{x_0 + ct_0} \psi(x) \, dx.$$

On $L_1$, $x + ct = x_0 + ct_0$, so that $dx + c \, dt = 0$, whence $-c^2 u_x \, dt - u_t \, dx = cu_x \, dx + cu_t \, dt = c \, du$. (We're in luck!) Thus

$$\int_{L_1} = c \int_{L_1} du = cu(x_0, t_0) - c\phi(x_0 + ct_0).$$

In the same way,

$$\int_{L_2} = -c \int_{L_2} du = -c\phi(x_0 - ct_0) + cu(x_0, t_0).$$

Adding these three results, we get

$$\iint_{\Delta} f \, dx \, dt = 2cu(x_0, t_0) - c[\phi(x_0 + ct_0) + \phi(x_0 - ct_0)] - \int_{x_0 - ct_0}^{x_0 + ct_0} \psi(x) \, dx.$$

Thus

$$
u(x_0, t_0) = \frac{1}{2c} \iint_\Delta f \, dx \, dt + \frac{1}{2}[\phi(x_0 + ct_0) + \phi(x_0 - ct_0)]
$$
$$
+ \frac{1}{2c} \int_{x_0-ct_0}^{x_0+ct_0} \psi(x) \, dx, \tag{12}
$$

which is the same as before.

**Operator Method**   This is how we solved the diffusion equation with a source. Let's try it out on the wave equation. The ODE analog is the equation,

$$
\frac{d^2u}{dt^2} + A^2u(t) = f(t), \quad u(0) = \phi, \quad \frac{du}{dt}(0) = \psi. \tag{13}
$$

We could think of $A^2$ as a positive constant (or even a positive square matrix.) The solution of (13) is

$$
u(t) = S'(t)\phi + S(t)\psi + \int_0^t S(t - s)f(s) \, ds, \tag{14}
$$

where

$$
S(t) = A^{-1} \sin tA \quad \text{and} \quad S'(t) = \cos tA. \tag{15}
$$

The key to understanding formula (14) is that $S(t)\psi$ is the solution of problem (13) in the case that $\phi = 0$ and $f = 0$.

Let's return to the PDE

$$
u_{tt} - c^2u_{xx} = f(x, t) \quad u(x, 0) = \phi(x) \quad u_t(x, 0) = \psi(x). \tag{16}
$$

The basic operator ought to be given by the $\psi$ term. That is,

$$
\mathcal{S}(t)\psi = \frac{1}{2c} \int_{x-ct}^{x+ct} \psi(y) \, dy = v(x, t), \tag{17}
$$

where $v(x, t)$ solves $v_{tt} - c^2v_{xx} = 0$, $v(x, 0) = 0$, $v_t(x, 0) = \psi(x)$. $\mathcal{S}(t)$ is the *source operator*. By (14) we would expect the $\phi$ term to be $(\partial/\partial t)\mathcal{S}(t)\phi$. In fact,

$$
\frac{\partial}{\partial t}\mathcal{S}(t)\phi = \frac{\partial}{\partial t}\frac{1}{2c} \int_{x-ct}^{x+ct} \phi(y) \, dy
$$
$$
= \frac{1}{2c} [c\phi(x + ct) - (-c)\phi(x - ct)],
$$

in agreement with our old formula (2.1.8)! So we must be on the right track.

Let's now take the $f$ term; that is, $\phi = \psi = 0$. By analogy with the last term in (14), the solution *ought* to be

$$
u(t) = \int_0^t \mathcal{S}(t - s)f(s) \, ds.
$$

That is, using (17),

$$u(x, t) = \int_0^t \left[ \frac{1}{2c} \int_{x-c(t-s)}^{x+c(t-s)} f(y, s)\, dy \right] ds = \frac{1}{2c} \iint_\Delta f\, dx\, dt.$$

This is once again the same result!

*The moral of the operator method is that if you can solve the homogeneous equation, you can also solve the inhomogeneous equation.* This is sometimes known as *Duhamel's principle.*

## SOURCE ON A HALF-LINE

The solution of the general inhomogeneous problem on a half-line

$$\begin{aligned}
\text{DE:} \quad & v_{tt} - c^2 v_{xx} = f(x, t) \quad \text{in} \quad 0 < x < \infty \\
\text{IC:} \quad & v(x, 0) = \phi(x) \quad v_t(x, 0) = \psi(x) \qquad\qquad (18) \\
\text{BC:} \quad & v(0, t) = h(t)
\end{aligned}$$

is the sum of four terms, one for each data function $\phi, \psi, f$, and $h$. For $x > ct > 0$, the solution has precisely the same form as in (3), with the backward triangle $\Delta$ as the domain of dependence. For $0 < x < ct$, however, it is given by

$$v(x, t) = \phi \text{ term} + \psi \text{ term} + h\left(t - \frac{x}{c}\right) + \frac{1}{2c} \iint_D f \qquad (19)$$

where $t - x/c$ is the reflection point and $D$ is the shaded region in Figure 3.2.2. The only caveat is that the given conditions had better coincide at the origin. That is, we require that $\phi(0) = h(0)$ and $\psi(0) = h'(0)$. If this were not assumed, there would be a singularity on the characteristic line emanating from the corner.

Let's derive the boundary term $h(t - x/c)$ for $x < ct$. To accomplish this, it is convenient to assume that $\phi = \psi = f = 0$. We shall derive the solution from scratch using the fact that $v(x, t)$ must take the form $v(x, t) = j(x + ct) + g(x - ct)$. From the initial conditions ($\phi = \psi = 0$), we find that $j(s) = g(s) = 0$ for $s > 0$. From the boundary condition we have $h(t) = v(0, t) = g(-ct)$ for $t > 0$. Thus $g(s) = h(-s/c)$ for $s < 0$. Therefore, if $x < ct, t > 0$, we have $v(x, t) = 0 + h(-[x - ct]/c) = h(t - x/c)$.

## FINITE INTERVAL

For a finite interval $(0, l)$ with inhomogeneous boundary conditions $v(0, t) = h(t), v(l, t) = k(t)$, we get the whole series of terms

$$\begin{aligned}
v(x, t) = h\left(t - \frac{x}{c}\right) - h\left(t + \frac{x - 2l}{c}\right) + h\left(t - \frac{x + 2l}{c}\right) + \cdots \\
+ k\left(t + \frac{x - l}{c}\right) - k\left(t - \frac{x + l}{c}\right) + k\left(t + \frac{x - 3l}{c}\right) + \cdots
\end{aligned}$$

(see Exercise 15 and Figure 3.2.4).

## EXERCISES

1. Solve $u_{tt} = c^2 u_{xx} + xt$,     $u(x, 0) = 0$,     $u_t(x, 0) = 0$.
2. Solve $u_{tt} = c^2 u_{xx} + e^{ax}$,     $u(x, 0) = 0$,     $u_t(x, 0) = 0$.
3. Solve $u_{tt} = c^2 u_{xx} + \cos x$,     $u(x, 0) = \sin x$,     $u_t(x, 0) = 1 + x$.
4. Show that the solution of the inhomogeneous wave equation

$$u_{tt} = c^2 u_{xx} + f, \quad u(x, 0) = \phi(x), \quad u_t(x, 0) = \psi(x),$$

is the sum of three terms, one each for $f$, $\phi$, and $\psi$.

5. Let $f(x, t)$ be any function and let $u(x, t) = (1/2c)\iint_\Delta f$, where $\Delta$ is the triangle of dependence. Verify directly by differentiation that

$$u_{tt} = c^2 u_{xx} + f \quad \text{and} \quad u(x, 0) \equiv u_t(x, 0) \equiv 0.$$

(*Hint:* Begin by writing the formula as the *iterated* integral

$$u(x, t) = \frac{1}{2c} \int_0^t \int_{x-ct+cs}^{x+ct-cs} f(y, s)\, dy\, ds$$

and differentiate with care using the rule in the Appendix. This exercise is not easy.)

6. Derive the formula for the inhomogeneous wave equation in yet another way.
   (a)   Write it as the system

$$u_t + cu_x = v, \quad v_t - cv_x = f.$$

   (b)   Solve the first equation for $u$ in terms of $v$ as

$$u(x, t) = \int_0^t v(x - ct + cs, s)\, ds.$$

   (c)   Similarly, solve the second equation for $v$ in terms of $f$.
   (d)   Substitute part (c) into part (b) and write as an iterated integral.

7. Let $A$ be a positive-definite $n \times n$ matrix. Let

$$S(t) = \sum_{m=0}^{\infty} \frac{(-1)^m A^{2m} t^{2m+1}}{(2m + 1)!}.$$

   (a)   Show that this series of matrices converges uniformly for bounded $t$ and its sum $S(t)$ solves the problem $S''(t) + A^2 S(t) = 0$, $S(0) = 0$, $S'(0) = I$, where $I$ is the identity matrix. Therefore, it makes sense to denote $S(t)$ as $A^{-1} \sin tA$ and to denote its derivative $S'(t)$ as $\cos(tA)$.
   (b)   Show that the solution of (13) is (14).

8. Show that the source operator for the wave equation solves the problem

$$\mathcal{S}_{tt} - c^2 \mathcal{S}_{xx} = 0, \quad \mathcal{S}(0) = 0, \quad \mathcal{S}_t(0) = I,$$

where $I$ is the identity operator.

9.  Let $u(t) = \int_0^t \mathcal{S}(t - s) f(s) \, ds$. Using *only* Exercise 8, show that $u$ solves the inhomogeneous wave equation with zero initial data.

10. Use any method to show that $u = 1/(2c) \iint_D f$ solves the inhomogeneous wave equation on the half-line with zero initial and boundary data, where $D$ is the domain of dependence for the half-line.

11. Show by direct substitution that $u(x, t) = h(t - x/c)$ for $x < ct$ and $u(x, t) = 0$ for $x \geq ct$ solves the wave equation on the half-line $(0, \infty)$ with zero initial data and boundary condition $u(0, t) = h(t)$.

12. Derive the solution of the fully inhomogeneous wave equation on the half-line

$$v_{tt} - c^2 v_{xx} = f(x, t) \quad \text{in } 0 < x < \infty$$
$$v(x, 0) = \phi(x), \quad v_t(x, 0) = \psi(x)$$
$$v(0, t) = h(t),$$

by means of the method using Green's theorem. (*Hint:* Integrate over the domain of dependence.)

13. Solve $u_{tt} = c^2 u_{xx}$ for $0 < x < \infty$,
$u(0, t) = t^2, \quad u(x, 0) = x, \quad u_t(x, 0) = 0$.

14. Solve the homogeneous wave equation on the half-line $(0, \infty)$ with zero initial data and with the Neumann boundary condition $u_x(0, t) = k(t)$. Use any method you wish.

15. Derive the solution of the wave equation in a finite interval with inhomogeneous boundary conditions $v(0, t) = h(t)$, $v(l, t) = k(t)$, and with $\phi = \psi = f = 0$.

## 3.5   DIFFUSION REVISITED

In this section we make a careful mathematical analysis of the solution of the diffusion equation that we found in Section 2.4. (On the other hand, the formula for the solution of the wave equation is so much simpler that it doesn't require a special justification.)

The solution formula for the diffusion equation is an example of a *convolution*, the convolution of $\phi$ with $S$ (at a fixed $t$). It is

$$u(x, t) = \int_{-\infty}^{\infty} S(x - y, t)\, \phi(y)\, dy = \int_{-\infty}^{\infty} S(z, t)\, \phi(x - z)\, dz, \qquad (1)$$

where $S(z, t) = 1/\sqrt{4\pi kt}\, e^{-z^2/4kt}$. If we introduce the variable $p = z/\sqrt{kt}$, it takes the equivalent form

$$u(x, t) = \frac{1}{\sqrt{4\pi}} \int_{-\infty}^{\infty} e^{-p^2/4} \phi(x - p\sqrt{kt})\, dp. \qquad (2)$$

Now we are prepared to state a precise theorem.

**Theorem 1.**   Let $\phi(x)$ be a bounded continuous function for $-\infty < x < \infty$. Then the formula (2) defines an infinitely differentiable function $u(x, t)$ for $-\infty < x < \infty$, $0 < t < \infty$, which satisfies the equation $u_t = k u_{xx}$ and $\lim_{t \searrow 0} u(x, t) = \phi(x)$ for each $x$.

**Proof.**   The integral converges easily because

$$|u(x, t)| \le \frac{1}{\sqrt{4\pi}} (\max |\phi|) \int_{-\infty}^{\infty} e^{-p^2/4} \, dp = \max |\phi|.$$

(This inequality is related to the maximum principle.) Thus the integral converges uniformly and absolutely. Let us show that $\partial u / \partial x$ exists. It equals $\int (\partial S/\partial x)(x - y, t)\phi(y) \, dy$ provided that this new integral also converges absolutely. Now

$$\int_{-\infty}^{\infty} \frac{\partial S}{\partial x}(x - y, t)\phi(y) \, dy = -\frac{1}{\sqrt{4\pi kt}} \int_{-\infty}^{\infty} \frac{x - y}{2kt} e^{-(x-y)^2/4kt} \phi(y) \, dy$$

$$= \frac{c}{\sqrt{t}} \int_{-\infty}^{\infty} p e^{-p^2/4} \phi(x - p\sqrt{kt}) \, dp$$

$$\le \frac{c}{\sqrt{t}} (\max |\phi|) \int_{-\infty}^{\infty} |p| \, e^{-p^2/4} \, dp,$$

where $c$ is a constant. The last integral is finite. So this integral also converges uniformly and absolutely. Therefore, $u_x = \partial u / \partial x$ exists and is given by this formula. All derivatives of all orders $(u_t, u_{xt}, u_{xx}, u_{tt}, \ldots)$ work the same way because each differentiation brings down a power of $p$ so that we end up with convergent integrals like $\int p^n e^{-p^2/4} \, dp$. So $u(x, t)$ is differentiable to all orders. Since $S(x, t)$ satisfies the diffusion equation for $t > 0$, so does $u(x, t)$.

It remains to prove the initial condition. It has to be understood in a limiting sense because the formula itself has meaning only for $t > 0$. Because the integral of $S$ is 1, we have

$$u(x, t) - \phi(x) = \int_{-\infty}^{\infty} S(x - y, t) \, [\phi(y) - \phi(x)] \, dy$$

$$= \frac{1}{\sqrt{4\pi}} \int_{-\infty}^{\infty} e^{-p^2/4} [\phi(x - p\sqrt{kt}) - \phi(x)] \, dp.$$

For fixed $x$ we must show that this tends to zero as $t \to 0$. The idea is that for $p\sqrt{t}$ small, the continuity of $\phi$ makes the integral small; while for $p\sqrt{t}$ not small, $p$ is large and the exponential factor is small.

To carry out this idea, let $\epsilon > 0$. Let $\delta > 0$ be so small that

$$\max_{|y - x| \le \delta} |\phi(y) - \phi(x)| < \frac{\epsilon}{2}.$$

This can be done because $\phi$ is continuous at $x$. We break up the integral into the part where $|p| < \delta/\sqrt{kt}$ and the part where $|p| \geq \delta/\sqrt{kt}$. The first part is

$$\left| \int_{|p|<\delta/\sqrt{kt}} \right| \leq \left( \frac{1}{\sqrt{4\pi}} \int e^{-p^2/4} dp \right) \cdot \max_{|y-x|\leq\delta} |\phi(y) - \phi(x)|$$

$$< 1 \cdot \frac{\epsilon}{2} = \frac{\epsilon}{2}.$$

The second part is

$$\left| \int_{|p|\geq\delta/\sqrt{kt}} \right| \leq \frac{1}{\sqrt{4\pi}} \cdot 2(\max |\phi|) \cdot \int_{|p|\geq\delta/\sqrt{kt}} e^{-p^2/4} dp < \frac{\epsilon}{2}$$

by choosing $t$ sufficiently small, since the integral $\int_{-\infty}^{\infty} e^{-p^2/4} \, dp$ converges and $\delta$ is fixed. (That is, the "tails" $\int_{|p|\geq N} e^{-p^2/4} \, dp$ are as small as we wish if $N = \delta/\sqrt{kt}$ is large enough.) Therefore,

$$|u(x, t) - \phi(x)| < \tfrac{1}{2}\epsilon + \tfrac{1}{2}\epsilon = \epsilon$$

provided that $t$ is small enough. This means exactly that $u(x, t) \to \phi(x)$ as $t \to 0$.    $\square$

**Corollary.**    The solution has all derivatives of all orders for $t > 0$, even if $\phi$ is not differentiable. We can say therefore that all solutions become smooth as soon as diffusion takes effect. There are no singularities, in sharp contrast to the wave equation.

**Proof.**    We use formula (1)

$$u(x, t) = \int_{-\infty}^{\infty} S(x - y, t)\phi(y) \, dy$$

together with the rule for differentiation under an integral sign, Theorem 2 in Section A.3.

**Piecewise Continuous Initial Data.**    Notice that the continuity of $\phi(x)$ was used in only one part of the proof. With an appropriate change we can allow $\phi(x)$ to have a jump discontinuity. [Consider, for instance, the initial data for $Q(x, t)$.]

A function $\phi(x)$ is said to have a *jump* at $x_0$ if both the limit of $\phi(x)$ as $x \to x_0$ from the right exists [denoted $\phi(x_0+)$] and the limit from the left [denoted $\phi(x_0-)$] exists but these two limits are not equal. A function is called *piecewise continuous* if in each finite interval it has only a finite number of jumps and it is continuous at all other points. This concept is discussed in more detail in Section 5.2.

**Theorem 2.**   Let $\phi(x)$ be a bounded function that is piecewise continuous. Then (1) is an infinitely differentiable solution for $t > 0$ and

$$\lim_{t \searrow 0} u(x, t) = \tfrac{1}{2}[\phi(x+) + \phi(x-)]$$

for all $x$. At every point of continuity this limit equals $\phi(x)$.

**Proof.**   The idea is the same as before. The only difference is to split the integrals into $p > 0$ and $p < 0$. We need to show that

$$\frac{1}{\sqrt{4\pi}} \int_0^{\pm\infty} e^{-p^2/4} \phi(x + \sqrt{kt}\, p)\, dp \to \pm\frac{1}{2}\phi(x\pm).$$

The details are left as an exercise.   $\square$

## EXERCISES

1.   Prove that if $\phi$ is any piecewise continuous function, then

$$\frac{1}{\sqrt{4\pi}} \int_0^{\pm\infty} e^{-p^2/4} \phi(x + \sqrt{kt}\, p)\, dp \to \pm\frac{1}{2}\phi(x\pm) \quad \text{as } t \searrow 0.$$

2.   Use Exercise 1 to prove Theorem 2.

# 4

# BOUNDARY PROBLEMS

In this chapter we finally come to the physically realistic case of a finite interval $0 < x < l$. The methods we introduce will frequently be used in the rest of this book.

## 4.1 SEPARATION OF VARIABLES, THE DIRICHLET CONDITION

We first consider the homogeneous Dirichlet conditions for the wave equation:

$$u_{tt} = c^2 u_{xx} \quad \text{for } 0 < x < l \tag{1}$$
$$u(0, t) = 0 = u(l, t) \tag{2}$$

with some initial conditions

$$u(x, 0) = \phi(x) \quad u_t(x, 0) = \psi(x). \tag{3}$$

The method we shall use consists of building up the general solution as a linear combination of special ones that are easy to find. (Once before, in Section 2.4, we followed this program, but with different building blocks.)

A *separated solution* is a solution of (1) and (2) of the form

$$u(x, t) = X(x)T(t). \tag{4}$$

(It is important to distinguish between the independent variable written as a lowercase letter and the function written as a capital letter.) Our first goal is to look for as many separated solutions as possible.

Plugging the form (4) into the wave equation (1), we get

$$X(x)T''(t) = c^2 X''(x)T(t)$$

or, dividing by $-c^2 XT$,

$$-\frac{T''}{c^2 T} = -\frac{X''}{X} = \lambda.$$

This defines a quantity $\lambda$, *which must be a constant*. (*Proof:* $\partial\lambda/\partial x = 0$ and $\partial\lambda/\partial t = 0$, so $\lambda$ is a constant. Alternatively, we can argue that $\lambda$ doesn't depend on $x$ because of the first expression and doesn't depend on $t$ because of the second expression, so that it doesn't depend on any variable.) We will show at the end of this section that $\lambda > 0$. (This is the reason for introducing the minus signs the way we did.)

So let $\lambda = \beta^2$, where $\beta > 0$. Then the equations above are a pair of *separate* (!) ordinary differential equations for $X(x)$ and $T(t)$:

$$X'' + \beta^2 X = 0 \quad \text{and} \quad T'' + c^2\beta^2 T = 0. \tag{5}$$

These ODEs are easy to solve. The solutions have the form

$$X(x) = C \cos \beta x + D \sin \beta x \tag{6}$$

$$T(t) = A \cos \beta ct + B \sin \beta ct, \tag{7}$$

where $A, B, C$, and $D$ are constants.

The second step is to impose the boundary conditions (2) on the separated solution. They simply require that $X(0) = 0 = X(l)$. Thus

$$0 = X(0) = C \quad \text{and} \quad 0 = X(l) = D \sin \beta l.$$

Surely we are not interested in the obvious solution $C = D = 0$. So we must have $\beta l = n\pi$, a root of the sine function. That is,

$$\lambda_n = \left(\frac{n\pi}{l}\right)^2, \quad X_n(x) = \sin\frac{n\pi x}{l} \quad (n = 1, 2, 3, \ldots) \tag{8}$$

are distinct solutions. Each sine function may be multiplied by an arbitrary constant.

Therefore, there are an *infinite* (!) number of separated solutions of (1) and (2), one for each $n$. They are

$$u_n(x, t) = \left(A_n \cos\frac{n\pi ct}{l} + B_n \sin\frac{n\pi ct}{l}\right) \sin\frac{n\pi x}{l}$$

$(n = 1, 2, 3, \ldots)$, where $A_n$ and $B_n$ are arbitrary constants. The sum of solutions is again a solution, so *any finite sum*

$$u(x, t) = \sum_n \left(A_n \cos\frac{n\pi ct}{l} + B_n \sin\frac{n\pi ct}{l}\right) \sin\frac{n\pi x}{l} \tag{9}$$

is also a solution of (1) and (2).

Formula (9) solves (3) as well as (1) and (2), provided that

$$\phi(x) = \sum_n A_n \sin \frac{n\pi x}{l} \tag{10}$$

and

$$\psi(x) = \sum_n \frac{n\pi c}{l} B_n \sin \frac{n\pi x}{l}. \tag{11}$$

Thus for any initial data of this form, the problem (1), (2), and (3) has a simple explicit solution.

But such data (10) and (11) clearly are very special. So let's try (following Fourier in 1827) to take *infinite sums*. Then we ask what kind of data pairs $\phi(x)$, $\psi(x)$ can be expanded as in (10), (11) for some choice of coefficients $A_n$, $B_n$? This question was the source of great disputes for half a century around 1800, but the final result of the disputes was very simple: *Practically any* (!) *function $\phi(x)$ on the interval $(0, l)$ can be expanded in an infinite series* (10). We will show this in Chapter 5. It will have to involve technical questions of convergence and differentiability of infinite series like (9). The series in (10) is called a *Fourier sine series* on $(0, l)$. But for the time being let's not worry about these mathematical points. Let's just forge ahead to see what their implications are.

First of all, (11) is the same kind of series for $\psi(x)$ as (10) is for $\phi(x)$. What we've shown is simply that *if* (10), (11) *are true, then the infinite series* (9) *ought to be the solution of the whole problem* (1), (2), (3).

A sketch of the first few functions $\sin(\pi x/l)$, $\sin(2\pi x/l)$, ... is shown in Figure 1. The functions $\cos(n\pi ct/l)$ and $\sin(n\pi ct/l)$ which describe the

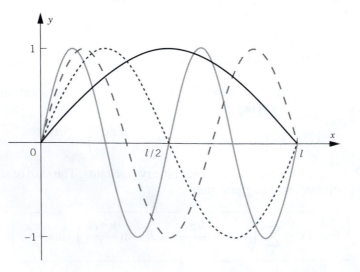

Figure 1

behavior in *time* have a similar form. The coefficients of $t$ inside the sines and cosines, namely $n\pi c/l$, are called the *frequencies*. (In some texts, the frequency is defined as $nc/2l$.)

If we return to the violin string that originally led us to the problem (1), (2), (3), we find that the frequencies are

$$\frac{n\pi\sqrt{T}}{l\sqrt{\rho}} \quad \text{for } n = 1, 2, 3, \ldots \tag{12}$$

The "fundamental" note of the string is the smallest of these, $\pi\sqrt{T}/(l\sqrt{\rho})$. The "overtones" are *exactly* the double, the triple, and so on, of the fundamental! The discovery by Euler in 1749 that the musical notes have such a simple mathematical description created a sensation. It took over half a century to resolve the ensuing controversy over the relationship between the infinite series (9) and d'Alembert's solution in Section 2.1. ☐

The analogous problem for *diffusion* is

$$
\begin{array}{lll}
\text{DE:} & u_t = ku_{xx} \quad (0 < x < l, 0 < t < \infty) & (13) \\
\text{BC:} & u(0, t) = u(l, t) = 0 & (14) \\
\text{IC:} & u(x, 0) = \phi(x). & (15)
\end{array}
$$

To solve it, we separate the variables $u = T(t)X(x)$ as before. This time we get

$$\frac{T'}{kT} = \frac{X''}{X} = -\lambda = \text{constant.}$$

Therefore, $T(t)$ satisfies the equation $T' = -\lambda kT$, whose solution is $T(t) = Ae^{-\lambda kt}$. Furthermore,

$$-X'' = \lambda X \quad \text{in } 0 < x < l \quad \text{with} \quad X(0) = X(l) = 0. \tag{16}$$

This is *precisely the same problem for $X(x)$ as before* and so has the same solutions. Because of the form of $T(t)$,

$$u(x, t) = \sum_{n=1}^{\infty} A_n e^{-(n\pi/l)^2 kt} \sin\frac{n\pi x}{l} \tag{17}$$

is the solution of (13)–(15) provided that

$$\phi(x) = \sum_{n=1}^{\infty} A_n \sin\frac{n\pi x}{l}. \tag{18}$$

Once again, our solution is expressible for each $t$ as a Fourier sine series in $x$ provided that the initial data are.

For example, consider the diffusion of a substance in a tube of length $l$. Each end of the tube opens up into a very large empty vessel. So the concentration $u(x, t)$ at each end is essentially zero. Given an initial concentration $\phi(x)$ in the tube, the concentration at all later times is given by formula (17). Notice that as $t \to \infty$, each term in (17) goes to zero. Thus the substance gradually empties out into the two vessels and less and less remains in the tube. $\qquad\square$

The numbers $\lambda_n = (n\pi/l)^2$ are called *eigenvalues* and the functions $X_n(x) = \sin(n\pi x/l)$ are called *eigenfunctions*. The reason for this terminology is as follows. They satisfy the conditions

$$-\frac{d^2}{dx^2}X = \lambda X, \quad X(0) = X(l) = 0. \tag{19}$$

This is an ODE with conditions at two points. Let $A$ denote the operator $-d^2/dx^2$, which acts on the functions that satisfy the Dirichlet boundary conditions. The differential equation has the form $AX = \lambda X$. An eigenfunction is a solution $X \not\equiv 0$ of this equation and an eigenvalue is a number $\lambda$ for which there exists a solution $X \not\equiv 0$.

This situation is analogous to the more familiar case of an $N \times N$ matrix $A$. A vector $X$ that satisfies $AX = \lambda X$ with $X \neq 0$ is called an eigenvector and $\lambda$ is called an eigenvalue. For an $N \times N$ matrix there are at most $N$ eigenvalues. But for the differential operator that we are interested in, there are an *infinite number of eigenvalues* $\pi^2/l^2$, $4\pi^2/l^2$, $9\pi^2/l^2, \ldots$. Thus you might say that we are dealing with *infinite-dimensional linear algebra!*

In physics and engineering the eigenfunctions are called *normal modes* because they are the natural shapes of solutions that persist for all time.

*Why are all the eigenvalues of this problem positive?* We assumed this in the discussion above, but now let's *prove* it. First, could $\lambda = 0$ be an eigenvalue? This would mean that $X'' = 0$, so that $X(x) = C + Dx$. But $X(0) = X(l) = 0$ implies that $C = D = 0$, so that $X(x) \equiv 0$. Therefore, zero is *not* an eigenvalue.

Next, could there be *negative* eigenvalues? If $\lambda < 0$, let's write it as $\lambda = -\gamma^2$. Then $X'' = \gamma^2 X$, so that $X(x) = C \cosh \gamma x + D \sinh \gamma x$. Then $0 = X(0) = C$ and $0 = X(l) = D \sinh \gamma l$. Hence $D = 0$ since $\sinh \gamma l \neq 0$.

Finally, let $\lambda$ be any *complex* number. Let $\gamma$ be either one of the two square roots of $-\lambda$; the other one is $-\gamma$. Then

$$X(x) = Ce^{\gamma x} + De^{-\gamma x},$$

where we are using the complex exponential function (see Section 5.2). The boundary conditions yield $0 = X(0) = C + D$ and $0 = Ce^{\gamma l} + De^{-\gamma l}$. Therefore $e^{2\gamma l} = 1$. By a well-known property of the complex exponential function, this implies that $\text{Re}(\gamma) = 0$ and $2l \, \text{Im}(\gamma) = 2\pi n$ for some integer $n$. Hence $\gamma = n\pi i/l$ and $\lambda = -\gamma^2 = n^2\pi^2/l^2$, which is real and positive. Thus the only eigenvalues $\lambda$ of our problem (16) are positive numbers; in fact, they are $(\pi/l)^2$, $(2\pi/l)^2, \ldots$.

## EXERCISES

1. (a) Use the Fourier expansion to explain why the note produced by a violin string rises sharply by one octave when the string is clamped exactly at its midpoint.
   (b) Explain why the note rises when the string is tightened.

2. Consider a metal rod ($0 < x < l$), insulated along its sides but not at its ends, which is initially at temperature $= 1$. Suddenly both ends are plunged into a bath of temperature $= 0$. Write the differential equation, boundary conditions, and initial condition. Write the formula for the temperature $u(x, t)$ at later times. In this problem, *assume* the infinite series expansion

$$1 = \frac{4}{\pi} \left( \sin \frac{\pi x}{l} + \frac{1}{3} \sin \frac{3\pi x}{l} + \frac{1}{5} \sin \frac{5\pi x}{l} + \cdots \right)$$

3. A quantum-mechanical particle on the line with an infinite potential outside the interval $(0, l)$ ("particle in a box") is given by Schrödinger's equation $u_t = iu_{xx}$ on $(0, l)$ with Dirichlet conditions at the ends. Separate the variables and use (8) to find its representation as a series.

4. Consider waves in a resistant medium that satisfy the problem

$$u_{tt} = c^2 u_{xx} - r u_t \quad \text{for } 0 < x < l$$
$$u = 0 \quad \text{at both ends}$$
$$u(x, 0) = \phi(x) \quad u_t(x, 0) = \psi(x),$$

   where $r$ is a constant, $0 < r < 2\pi c/l$. Write down the series expansion of the solution.

5. Do the same for $2\pi c/l < r < 4\pi c/l$.

6. Separate the variables for the equation $tu_t = u_{xx} + 2u$ with the boundary conditions $u(0, t) = u(\pi, t) = 0$. Show that there are an infinite number of solutions that satisfy the initial condition $u(x, 0) = 0$. So uniqueness is false for this equation!

## 4.2  THE NEUMANN CONDITION

The same method works for both the Neumann and Robin boundary conditions (BCs). In the former case, (4.1.2) is replaced by $u_x(0, t) = u_x(l, t) = 0$. Then the eigenfunctions are the solutions $X(x)$ of

$$\boxed{-X'' = \lambda X, \quad X'(0) = X'(l) = 0,} \tag{1}$$

other than the trivial solution $X(x) \equiv 0$.

   As before, let's first search for the positive eigenvalues $\lambda = \beta^2 > 0$. As in (4.1.6), $X(x) = C \cos \beta x + D \sin \beta x$, so that

$$X'(x) = -C\beta \sin \beta x + D\beta \cos \beta x.$$

The boundary conditions (1) mean first that $0 = X'(0) = D\beta$, so that $D = 0$, and second that

$$0 = X'(l) = -C\beta \sin \beta l.$$

Since we don't want $C = 0$, we must have $\sin \beta l = 0$. Thus $\beta = \pi/l$, $2\pi/l$, $3\pi/l$, .... Therefore, we have the

Eigenvalues:  $\left(\dfrac{\pi}{l}\right)^2$, $\left(\dfrac{2\pi}{l}\right)^2$, ...  (2)

Eigenfunctions:  $X_n(x) = \cos \dfrac{n\pi x}{l}$   $(n = 1, 2, ...)$  (3)

Next let's check whether zero is an eigenvalue. Set $\lambda = 0$ in the ODE (1). Then $X'' = 0$, so that $X(x) = C + Dx$ and $X'(x) \equiv D$. The Neumann boundary conditions are both satisfied if $D = 0$. $C$ can be any number. Therefore, $\lambda = 0$ *is an eigenvalue*, and any constant function is its eigenfunction.

If $\lambda < 0$ or if $\lambda$ is complex (nonreal), it can be shown directly, as in the Dirichlet case, that there is no eigenfunction. (Another proof will be given in Section 5.3.) Therefore, the list of all the eigenvalues is

$$\lambda_n = \left(\frac{n\pi}{l}\right)^2 \quad \text{for } n = 0, 1, 2, 3, .... \quad (4)$$

*Note that $n = 0$ is included among them!*

So, for instance, the *diffusion* equation with the Neumann BCs has the solution

$$u(x, t) = \frac{1}{2} A_0 + \sum_{n=1}^{\infty} A_n e^{-(n\pi/l)^2 kt} \cos \frac{n\pi x}{l}. \quad (5)$$

This solution requires the initial data to have the "Fourier cosine expansion"

$$\phi(x) = \frac{1}{2} A_0 + \sum_{n=1}^{\infty} A_n \cos \frac{n\pi x}{l}. \quad (6)$$

All the coefficients $A_0, A_1, A_2, ...$ are just constants. The first term in (5) and (6), which comes from the eigenvalue $\lambda = 0$, is written separately in the form $\frac{1}{2} A_0$ just for later convenience. (The reader is asked to bear with this ridiculous factor $\frac{1}{2}$ until Section 5.1 when its convenience will become apparent.)

What is the behavior of $u(x, t)$ as $t \to +\infty$? Since all but the first term in (5) contains an exponentially decaying factor, the solution decays quite fast to the first term $\frac{1}{2}A_0$, which is just a constant. Since these boundary conditions correspond to insulation at both ends, this agrees perfectly with our intuition of Section 2.5 that the solution "spreads out." This is the eventual behavior if we wait long enough. (To actually *prove* that the limit as $t \to \infty$ is given term by term in (5) requires the use of one of the convergence theorems in Section A.2. We omit this verification here.)

Consider now the *wave* equation with the Neumann BCs. The eigenvalue $\lambda = 0$ then leads to $X(x) = $ constant and to the differential equation $T''(t) = \lambda c^2 T(t) = 0$, which has the solution $T(t) = A + Bt$. Therefore, the wave equation with Neumann BCs has the solutions

$$
\boxed{\begin{aligned}
u(x, t) = {}& \frac{1}{2}A_0 + \frac{1}{2}B_0 t \\
& + \sum_{n=1}^{\infty}\left( A_n \cos \frac{n\pi ct}{l} + B_n \sin \frac{n\pi ct}{l} \right)\cos\frac{n\pi x}{l}.
\end{aligned}}
\tag{7}
$$

(Again, the factor $\frac{1}{2}$ will be justified later.) Then the initial data must satisfy

$$
\phi(x) = \frac{1}{2}A_0 + \sum_{n=1}^{\infty} A_n \cos \frac{n\pi x}{l}
\tag{8}
$$

and

$$
\psi(x) = \frac{1}{2}B_0 + \sum_{n=1}^{\infty} \frac{n\pi c}{l} B_n \cos \frac{n\pi x}{l}.
\tag{9}
$$

Equation (9) comes from first differentiating (7) with respect to $t$ and then setting $t = 0$.   □

A "mixed" boundary condition would be Dirichlet at one end and Neumann at the other. For instance, in case the BCs are $u(0, t) = u_x(l, t) = 0$, the eigenvalue problem is

$$
\boxed{-X'' = \lambda X \qquad X(0) = X'(l) = 0.}
\tag{10}
$$

The eigenvalues then turn out to be $(n + \frac{1}{2})^2 \pi^2 / l^2$ and the eigenfunctions $\sin[(n + \frac{1}{2})\pi x / l]$ for $n = 0, 1, 2, \ldots$ (see Exercises 1 and 2). For a discussion of boundary conditions in the context of musical instruments, see [HJ].

For another example, consider the *Schrödinger* equation $u_t = iu_{xx}$ in $(0, l)$ with the Neumann BCs $u_x(0, t) = u_x(l, t) = 0$ and initial condition $u(x, 0) = \phi(x)$. Separation of variables leads to the equation

$$
\frac{T'}{iT} = \frac{X''}{X} = -\lambda = \text{constant},
$$

so that $T(t) = e^{-i\lambda t}$ and $X(x)$ satisfies exactly the same problem (1) as before. Therefore, the solution is

$$u(x, t) = \frac{1}{2}A_0 + \sum_{n=1}^{\infty} A_n e^{-i(n\pi/l)^2 t} \cos \frac{n\pi x}{l}.$$

The initial condition requires the cosine expansion (6).

## EXERCISES

1.  Solve the diffusion problem $u_t = k u_{xx}$ in $0 < x < l$, with the mixed boundary conditions $u(0, t) = u_x(l, t) = 0$.
2.  Consider the equation $u_{tt} = c^2 u_{xx}$ for $0 < x < l$, with the boundary conditions $u_x(0, t) = 0$, $u(l, t) = 0$ (Neumann at the left, Dirichlet at the right).
    (a) Show that the eigenfunctions are $\cos[(n + \frac{1}{2})\pi x/l]$.
    (b) Write the series expansion for a solution $u(x, t)$.
3.  Solve the Schrödinger equation $u_t = i k u_{xx}$ for real $k$ in the interval $0 < x < l$ with the boundary conditions $u_x(0, t) = 0$, $u(l, t) = 0$.
4.  Consider diffusion inside an enclosed circular tube. Let its length (circumference) be $2l$. Let $x$ denote the arc length parameter where $-l \le x \le l$. Then the concentration of the diffusing substance satisfies

    $$u_t = k u_{xx} \quad \text{for} \ -l \le x \le l$$

    $$u(-l, t) = u(l, t) \quad \text{and} \quad u_x(-l, t) = u_x(l, t).$$

    These are called *periodic boundary conditions*.
    (a) Show that the eigenvalues are $\lambda = (n\pi/l)^2$ for $n = 0, 1, 2, 3, \ldots$.
    (b) Show that the concentration is

    $$u(x, t) = \frac{1}{2}A_0 + \sum_{n=1}^{\infty} \left( A_n \cos \frac{n\pi x}{l} + B_n \sin \frac{n\pi x}{l} \right) e^{-n^2\pi^2 kt/l^2}.$$

## 4.3 THE ROBIN CONDITION

We continue the method of separation of variables for the case of the Robin condition. The Robin condition means that we are solving $-X'' = \lambda X$ with the boundary conditions

$$
\begin{array}{lll}
X' - a_0 X = 0 & \text{at } x = 0 & (1) \\
X' + a_l X = 0 & \text{at } x = l. & (2)
\end{array}
$$

The two constants $a_0$ and $a_l$ should be considered as given.

The physical reason they are written with opposite signs is that they correspond to *radiation* of energy if $a_0$ and $a_l$ are positive, *absorption* of energy if $a_0$ and $a_l$ are negative, and *insulation* if $a_0 = a_l = 0$. This is the interpretation for a heat problem: See the discussion in Section 1.4 or Exercise 2.3.8. For the case of the vibrating string, the interpretation is that the string shares its energy with the endpoints if $a_0$ and $a_l$ are positive, whereas the string gains some energy from the endpoints if $a_0$ and $a_l$ are negative: See Exercise 11.

The mathematical reason for writing the constants in this way is that the unit *outward* normal **n** for the interval $0 \le x \le l$ points to the *left* at $x = 0$ (**n** $= -1$) and to the *right* at $x = l$ (**n** $= +1$). Therefore, we expect that the nature of the eigenfunctions might depend on the signs of the two constants in opposite ways.

## POSITIVE EIGENVALUES

Our task now is to solve the ODE $-X'' = \lambda X$ with the boundary conditions (1), (2). First let's look for the *positive eigenvalues*

$$\lambda = \beta^2 > 0.$$

As usual, the solution of the ODE is

$$X(x) = C \cos \beta x + D \sin \beta x \tag{3}$$

so that

$$X'(x) \pm a X(x) = (\beta D \pm aC) \cos \beta x + (-\beta C \pm aD) \sin \beta x.$$

At the left end $x = 0$ we require that

$$0 = X'(0) - a_0 X(0) = \beta D - a_0 C. \tag{4}$$

So we can solve for $D$ in terms of $C$. At the right end $x = l$ we require that

$$0 = (\beta D + a_l C) \cos \beta l + (-\beta C + a_l D) \sin \beta l. \tag{5}$$

Messy as they may look, equations (4) and (5) are easily solved since they are equivalent to the matrix equation

$$\begin{pmatrix} -a_0 & \beta \\ a_l \cos \beta l - \beta \sin \beta l & \beta \cos \beta l + a_l \sin \beta l \end{pmatrix} \begin{pmatrix} C \\ D \end{pmatrix} = \begin{pmatrix} 0 \\ 0 \end{pmatrix}. \tag{6}$$

Therefore, substituting for $D$, we have

$$0 = (a_0 C + a_l C) \cos \beta l + \left( -\beta C + \frac{a_l a_0 C}{\beta} \right) \sin \beta l. \tag{7}$$

We don't want the trivial solution $C = 0$. We divide by $C \cos \beta l$ and multiply by $\beta$ to get

$$\boxed{(\beta^2 - a_0 a_l) \tan \beta l = (a_0 + a_l)\beta.} \tag{8}$$

Any root $\beta > 0$ of this "algebraic" equation would give us an eigenvalue $\lambda = \beta^2$.

What would be the corresponding eigenfunction? It would be the above $X(x)$ with the required relation between $C$ and $D$, namely,

$$X(x) = C \left( \cos \beta x + \frac{a_0}{\beta} \sin \beta x \right) \qquad (9)$$

for any $C \neq 0$. By the way, because we divided by $\cos \beta l$, there is the exceptional case when $\cos \beta l = 0$; it would mean by (7) that $\beta = \sqrt{a_0 a_l}$.

Our next task is to solve (8) for $\beta$. This is not so easy, as there is no simple formula. One way is to calculate the roots numerically, say by Newton's method. Another way is by graphical analysis, which, instead of precise numerical values, will provide a lot of qualitative information. This is what we'll do. It's here where the nature of $a_0$ and $a_l$ come into play. Let us rewrite the eigenvalue equation (8) as

$$\tan \beta l = \frac{(a_0 + a_l)\beta}{\beta^2 - a_0 a_l}. \qquad (10)$$

Our method is to sketch the graphs of the tangent function $y = \tan \beta l$ and the rational function $y = (a_0 + a_l)\beta/(\beta^2 - a_0 a_l)$ as functions of $\beta > 0$ and to find their points of intersection. What the rational function looks like depends on the constants $a_0$ and $a_l$.

**Case 1**   In Figure 1 is pictured the case of *radiation at both ends*: $a_0 > 0$ and $a_l > 0$. Each of the points of intersection (for $\beta > 0$) provides an eigenvalue $\lambda_n = \beta_n^2$. The results depend very much on the $a_0$ and $a_l$. The exceptional situation mentioned above, when $\cos \beta l = 0$ and $\beta = \sqrt{a_0 a_l}$, will occur when the graphs of the tangent function and the rational function "intersect at infinity."

No matter what they are, as long as they are both positive, the graph clearly shows that

$$n^2 \frac{\pi^2}{l^2} < \lambda_n < (n+1)^2 \frac{\pi^2}{l^2} \qquad (n = 0, 1, 2, 3, \ldots). \qquad (11)$$

Furthermore,

$$\lim_{n \to \infty} \beta_n - n\frac{\pi}{l} = 0, \qquad (12)$$

which means that the larger eigenvalues get relatively closer and closer to $n^2\pi^2/l^2$ (see Exercise 19). You may compare this to the case $a_0 = a_l = 0$, the Neumann problem, where they are all *exactly* equal to $n^2\pi^2/l^2$.

**Case 2**   The case of absorption at $x = 0$ and radiation at $x = l$, but *more radiation than absorption*, is given by the conditions

$$a_0 < 0, \quad a_l > 0, \quad a_0 + a_l > 0. \qquad (13)$$

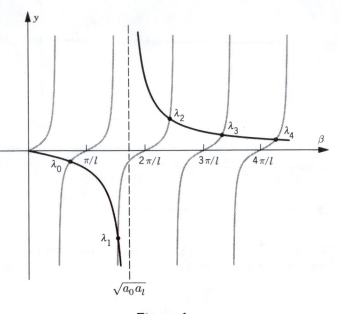

**Figure 1**

Then the graph looks like Figure 2 or 3, depending on the relative sizes of $a_0$ and $a_l$. Once again we see that (11) and (12) hold, except that in Figure 2 there is no eigenvalue $\lambda_0$ in the interval $(0, \pi^2/l^2)$.

There is an eigenvalue in the interval $(0, \pi^2/l^2)$ only if the rational curve crosses the *first* branch of the tangent curve. Since the rational curve has only a single maximum, this crossing can happen only if the slope of the rational curve is greater than the slope of the tangent curve at the origin. Let's

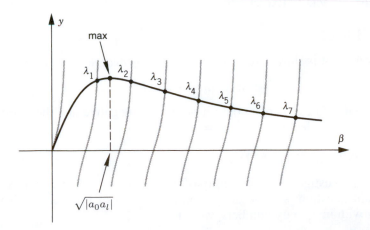

**Figure 2**

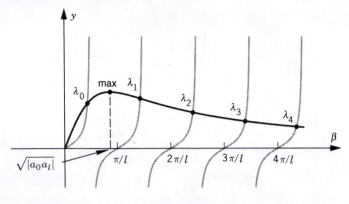

**Figure 3**

calculate these two slopes. A direct calculation shows that the slope $dy/d\beta$ of the rational curve at the origin is

$$\frac{a_0 + a_l}{-a_0 a_l} = \frac{a_l - |a_0|}{a_l\,|a_0|} > 0$$

because of (13). On the other hand, the slope of the tangent curve $y = \tan l\beta$ at the origin is $l\sec^2(l0) = l$. Thus we reach the following conclusion. In case

$$a_0 + a_l > -a_0 a_l l \tag{14}$$

(which means "much more radiation than absorption"), the rational curve will start out at the origin with a greater slope than the tangent curve and the two graphs must intersect at a point in the interval $(0,\ \pi/2l)$. Therefore, we conclude that *in Case 2 there is an eigenvalue $0 < \lambda_0 < (\pi/2l)^2$ if and only if* (14) *holds.*

Other cases, for instance absorption at both ends, may be found in the exercises, especially Exercise 8.

## ZERO EIGENVALUE

In Exercise 2 it is shown that *there is a zero eigenvalue if and only if*

$$a_0 + a_l = -a_0 a_l l. \tag{15}$$

Notice that (15) can happen only if $a_0$ or $a_l$ is negative and the interval has exactly a certain length or else $a_0 = a_l = 0$.

## NEGATIVE EIGENVALUE

Now let's investigate the possibility of a negative eigenvalue. This is a very important question; see the discussion at the end of this section. To avoid dealing with imaginary numbers, we set

$$\lambda = -\gamma^2 < 0$$

and write the solution of the differential equation as

$$X(x) = C \cosh \gamma x + D \sinh \gamma x.$$

(An alternative form, which we used at the end of Section 4.1, is $Ae^{\gamma x} + Be^{-\gamma x}$.) The boundary conditions, much as before, lead to the eigenvalue equation

$$\tanh \gamma l = -\frac{(a_0 + a_l)\gamma}{\gamma^2 + a_0 a_l}. \tag{16}$$

(Verify it!) So we look for intersections of these two graphs [on the two sides of (16)] for $\gamma > 0$. Any such point of intersection would provide a negative eigenvalue $\lambda = -\gamma^2$ and a corresponding eigenfunction

$$X(x) = \cosh \gamma x + \frac{a_0}{\gamma} \sinh \gamma x. \tag{17}$$

Several different cases are illustrated in Figure 4. Thus in Case 1, of radiation at both ends, when $a_0$ and $a_l$ are both positive, there is no intersection and so no negative eigenvalue.

Case 2, the situation with more radiation than absorption ($a_0 < 0$, $a_l > 0$, $a_0 + a_l > 0$), is illustrated by the two solid (14) and dashed (18) curves. There is either one intersection or none, depending on the slopes at the origin. The slope of the tanh curve is $l$, while the slope of the rational curve is

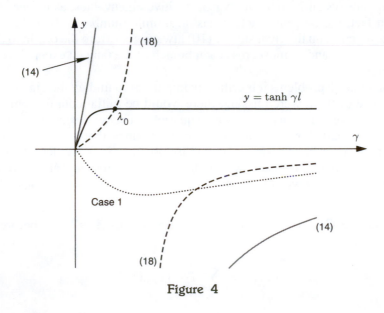

Figure 4

$-(a_0 + a_l)/(a_0 a_l) > 0$. If the last expression is smaller than $l$, there is an intersection; otherwise, there isn't. So our conclusion in Case 2 is as follows.

*Let $a_0 < 0$ and $a_l > -a_0$. If*

$$a_0 + a_l < -a_0 a_l l, \tag{18}$$

*then there exists exactly one negative eigenvalue, which we'll call $\lambda_0 < 0$. If* (14) *holds, then there is no negative eigenvalue.* Notice how the "missing" positive eigenvalue $\lambda_0$ in case (18) now makes its appearance as a negative eigenvalue! Furthermore, the zero eigenvalue is the borderline case (15); therefore, we use the notation $\lambda_0 = 0$ in the case of (15).

## SUMMARY

We summarize the various cases as follows:

*Case 1:* Only positive eigenvalues.

*Case 2 with (14):* Only positive eigenvalues.

*Case 2 with (15):* Zero is an eigenvalue, all the rest are positive.

*Case 2 with (18):* One negative eigenvalue, all the rest are positive.

Exercise 8 provides a complete summary of all the other cases.

In any case, that is, for any values for $a_0$ and $a_l$, there are no complex, nonreal, eigenvalues. This fact can be shown directly as before but will also be shown by a general, more satisfying, argument in Section 5.3. Furthermore, there are always an infinite number of positive eigenvalues, as is clear from (10). In fact, the tangent function has an infinite number of branches. The rational function on the right side of (10) always goes from the origin to the $\beta$ axis as $\beta \to \infty$ and so must cross each branch of the tangent except possibly the first one.

For all these problems it is critically important to find *all* the eigenvalues. If even one of them were missing, there would be initial data for which we could not solve the diffusion or wave equations. This will become clearer in Chapter 5. Exactly how we enumerate the eigenvalues, that is, whether we call the first one $\lambda_0$ or $\lambda_1$ or $\lambda_5$ or $\lambda_{-2}$, is not important. It is convenient, however, to number them in a consistent way. In the examples presented above we have numbered them in a way that neatly exhibits their dependence on $a_0$ and $a_l$.

## What Is the Grand Conclusion for the Robin BCs?   As before, we have an expansion

$$u(x, t) = \sum_n T_n(t) X_n(x), \tag{19}$$

where $X_n(x)$ are the eigenfunctions and where

$$T_n(t) = \begin{cases} A_n e^{-\lambda_n kt} & \text{for diffusions} \\ A_n \cos(\sqrt{\lambda_n}\, ct) + B_n \sin(\sqrt{\lambda_n}\, ct) & \text{for waves.} \end{cases} \qquad (20)$$

## Example 1.

Let $a_0 < 0 < a_0 + a_l < -a_0 a_l l$, which is Case 2 with (18). Then the grand conclusion takes the following explicit form. As we showed above, in this case there is exactly one negative eigenvalue $\lambda_0 = -\gamma_0^2 < 0$ as well as a sequence of positive ones $\lambda_n = +\beta_n^2 > 0$ for $n = 1, 2, 3, \ldots$. The complete solution of the diffusion problem

$$u_t = ku_{xx} \quad \text{for } 0 < x < l, \quad 0 < t < \infty$$
$$u_x - a_0 u = 0 \quad \text{for } x = 0, \quad u_x + a_l u = 0 \quad \text{for } x = l$$
$$u = \phi \quad \text{for } t = 0$$

therefore is

$$u(x, t) = A_0 e^{+\gamma_0^2 kt} \left( \cosh \gamma_0 x + \frac{a_0}{\gamma_0} \sinh \gamma_0 x \right)$$
$$+ \sum_{n=1}^{\infty} A_n e^{-\beta_n^2 kt} \left( \cos \beta_n x + \frac{a_0}{\beta_n} \sin \beta_n x \right). \qquad (21)$$

This conclusion (21) has the following physical interpretation if, say, $u(x, t)$ is the *temperature* in a rod of length $l$. We have taken the case when energy is supplied at $x = 0$ (absorption of energy by the rod, heat flux goes *into* the rod at its left end) and when energy is radiated from the right end (the heat flux goes *out*). For a given length $l$ and a given radiation $a_l > 0$, there is a negative eigenvalue ($\lambda_0 = -\gamma_0^2$) if and only if the absorption is great enough [$|a_0| > a_l/(1 + a_l l)$]. Such a large absorption coefficient allows the temperature to build up to large values, as we see from the expansion (21). In fact, all the terms get smaller as time goes on, except the first one, which *grows* exponentially due to the factor $e^{+\gamma_0^2 kt}$. So the rod gets hotter and hotter (unless $A_0 = 0$, which could only happen for very special initial data).

If, on the other hand, the absorption is relatively small [that is, $|a_0| < a_l/(1 + a_l l)$], then all the eigenvalues are positive and the temperature will remain bounded and will eventually decay to zero. Other interpretations of this sort are left for the exercises. $\qquad \square$

For the *wave equation*, a negative eigenvalue $\lambda_0 = -\gamma_0^2$ would also lead to exponential growth because the expansion for $u(x, t)$ would contain the term

$$(A_0 e^{\gamma_0 ct} + B_0 e^{-\gamma_0 ct}) X_0(x).$$

This term comes from the usual equation $-T'' = \lambda c^2 T = -(\gamma_0 c)^2 T$ for the temporal part of a separated solution (see Exercise 10).

## EXERCISES

1.  Find the eigenvalues graphically for the boundary conditions

    $$X(0) = 0, \qquad X'(l) + aX(l) = 0.$$

    Assume that $a \neq 0$.

2.  Consider the eigenvalue problem with Robin BCs at both ends:

    $$-X'' = \lambda X$$
    $$X'(0) - a_0 X(0) = 0, \quad X'(l) + a_l X(l) = 0.$$

    (a)  Show that $\lambda = 0$ is an eigenvalue if and only if $a_0 + a_l = -a_0 a_l l$.
    (b)  Find the eigenfunctions corresponding to the zero eigenvalue. (*Hint:* First solve the ODE for $X(x)$. The solutions are not sines or cosines.)

3.  Derive the eigenvalue equation (16) for the negative eigenvalues $\lambda = -\gamma^2$ and the formula (17) for the eigenfunctions.

4.  Consider the Robin eigenvalue problem. If

    $$a_0 < 0, \quad a_l < 0 \quad \text{and} \quad -a_0 - a_l < a_0 a_l l,$$

    show that there are *two* negative eigenvalues. This case may be called "substantial absorption at both ends." (*Hint:* Show that the rational curve $y = -(a_0 + a_l)\gamma / (\gamma^2 + a_0 a_l)$ has a single maximum and crosses the line $y = 1$ in two places. Deduce that it crosses the tanh curve in two places.)

5.  In Exercise 4 (substantial absorption at both ends) show graphically that there are an infinite number of positive eigenvalues. Show graphically that they satisfy (11) and (12).

6.  If $a_0 = a_l = a$ in the Robin problem, show that:
    (a)  There are *no* negative eigenvalues if $a \geq 0$, there is *one* if $-2/l < a < 0$, and there are *two* if $a < -2/l$.
    (b)  Zero is an eigenvalue if and only if $a = 0$ or $a = -2/l$.

7.  If $a_0 = a_l = a$, show that as $a \to +\infty$, the eigenvalues tend to the eigenvalues of the Dirichlet problem. That is,

    $$\lim_{a \to \infty} \left\{ \beta_n(a) - \frac{(n+1)\pi}{l} \right\} = 0,$$

    where $\lambda_n(a) = [\beta_n(a)]^2$ is the $(n+1)$st eigenvalue.

8.  Consider again Robin BCs at both ends for arbitrary $a_0$ and $a_l$.
    (a)  In the $a_0 a_l$ plane sketch the hyperbola $a_0 + a_l = -a_0 a_l l$. Indicate the asymptotes. For $(a_0, a_l)$ on this hyperbola, zero is an eigenvalue, according to Exercise 2(a).
    (b)  Show that the hyperbola separates the whole plane into three regions, depending on whether there are two, one, or no negative eigenvalues.
    (c)  Label the directions of increasing absorption and radiation on each axis. Label the point corresponding to Neumann BCs.
    (d)  Where in the plane do the Dirichlet BCs belong?

9.  On the interval $0 \le x \le 1$ of length one, consider the eigenvalue problem

    $$-X'' = \lambda X$$
    $$X'(0) + X(0) = 0 \qquad \text{and} \qquad X(1) = 0$$

    (absorption at one end and zero at the other).
    (a)  Find an eigenfunction with eigenvalue zero. Call it $X_0(x)$.
    (b)  Find an equation for the positive eigenvalues $\lambda = \beta^2$.
    (c)  Show graphically from part (b) that there are an infinite number of positive eigenvalues.
    (d)  Is there a negative eigenvalue?

10.  Solve the wave equation with Robin boundary conditions under the assumption that (18) holds.

11.  (a)  Prove that the (total) energy is conserved for the wave equation with Dirichlet BCs, where the energy is defined to be

    $$E = \tfrac{1}{2} \int_0^l \left( c^{-2} u_t^2 + u_x^2 \right) dx.$$

    (Compare this definition with Section 2.2.)
    (b)  Do the same for the Neumann BCs.
    (c)  For the Robin BCs, show that

    $$E_R = \tfrac{1}{2} \int_0^l \left( c^{-2} u_t^2 + u_x^2 \right) dx + \tfrac{1}{2} a_l [u(l, t)]^2 + \tfrac{1}{2} a_0 [u(0, t)]^2$$

    is conserved. Thus, while the total energy $E_R$ is still a constant, some of the internal energy is "lost" to the boundary if $a_0$ and $a_l$ are positive and "gained" from the boundary if $a_0$ and $a_l$ are negative.

12.  Consider the unusual eigenvalue problem

    $$-v_{xx} = \lambda v \qquad \text{for } 0 < x < l$$

    $$v_x(0) = v_x(l) = \frac{v(l) - v(0)}{l}.$$

    (a)  Show that $\lambda = 0$ is a double eigenvalue.
    (b)  Get an equation for the positive eigenvalues $\lambda > 0$.

(c)   Letting $\gamma = \frac{1}{2}l\sqrt{\lambda}$, reduce the equation in part (b) to the equation

$$\gamma \sin \gamma \cos \gamma = \sin^2 \gamma.$$

(d)   Use part (c) to find half of the eigenvalues explicitly and half of them graphically.

(e)   Assuming that all the eigenvalues are nonnegative, make a list of all the eigenfunctions.

(f)   Solve the problem $u_t = ku_{xx}$ for $0 < x < l$, with the BCs given above, and with $u(x, 0) = \phi(x)$.

(g)   Show that, as $t \to \infty$, $\lim u(x, t) = A + Bx$ for some constants $A, B$, assuming that you can take limits term by term.

13.   Consider a string that is fixed at the end $x = 0$ and is free at the end $x = l$ except that a load (weight) of given mass is attached to the right end.

(a)   Show that it satisfies the problem

$$u_{tt} = c^2 u_{xx} \qquad \text{for } 0 < x < l$$

$$u(0, t) = 0 \qquad u_{tt}(l, t) = -ku_x(l, t)$$

for some constant $k$.

(b)   What is the eigenvalue problem in this case?

(c)   Find the equation for the positive eigenvalues and find the eigenfunctions.

14.   Solve the eigenvalue problem $x^2 u'' + 3xu' + \lambda u = 0$ for $1 < x < e$, with $u(1) = u(e) = 0$. Assume that $\lambda > 1$. (*Hint:* Look for solutions of the form $u = x^m$ for complex $m$.)

15.   Find the equation for the eigenvalues $\lambda$ of the problem

$$(\kappa(x)X')' + \lambda\rho(x)X = 0 \qquad \text{for } 0 < x < l \text{ with } X(0) = X(l) = 0,$$

where $\kappa(x) = \kappa_1^2$ for $x < a$, $\kappa(x) = \kappa_2^2$ for $x > a$, $\rho(x) = \rho_1^2$ for $x < a$, and $\rho(x) = \rho_2^2$ for $x > a$. All these constants are positive and $0 < a < l$.

16.   Find the positive eigenvalues and the corresponding eigenfunctions of the fourth-order operator $+d^4/dx^4$ with the four boundary conditions

$$X(0) = X(l) = X''(0) = X''(l) = 0.$$

17.   Solve the fourth-order eigenvalue problem $X'''' = \lambda X$ in $0 < x < l$, with the four boundary conditions

$$X(0) = X'(0) = X(l) = X'(l) = 0,$$

where $\lambda > 0$. (*Hint:* First solve the fourth-order ODE.)

18.   A tuning fork may be regarded as a pair of vibrating flexible bars with a certain degree of stiffness. Each such bar is clamped at one end and is approximately modeled by the fourth-order PDE $u_{tt} + c^2 u_{xxxx} = 0$. It has initial conditions as for the wave equation. Let's say that on the end $x = 0$ it is clamped (fixed), meaning that it satisfies

$u(0, t) = u_x(0, t) = 0$. On the other end $x = l$ it is free, meaning that it satisfies $u_{xx}(l, t) = u_{xxx}(l, t) = 0$. Thus there are a total of four boundary conditions, two at each end.

(a)   Separate the time and space variables to get the eigenvalue problem $X'''' = \lambda X$.

(b)   Show that zero is not an eigenvalue.

(c)   Assuming that all the eigenvalues are positive, write them as $\lambda = \beta^4$ and find the equation for $\beta$.

(d)   Find the frequencies of vibration.

(e)   Compare your answer in part (d) with the overtones of the vibrating string by looking at the ratio $\beta_2^2 / \beta_1^2$. Explain why you hear an almost pure tone when you listen to a tuning fork.

19.   Show that in Case 1 (radiation at both ends)

$$\lim_{n \to \infty} \left[ \lambda_n - \frac{n^2 \pi^2}{l^2} \right] = \frac{2}{l} (a_0 + a_l).$$

# 5

# FOURIER SERIES

Our first goal in this key chapter is to find the coefficients in a Fourier series. In Section 5.3 we introduce the idea of orthogonality of functions and we show how the different varieties of Fourier series can be treated in a unified fashion. In Section 5.4 we state the basic completeness (convergence) theorems. Proofs are given in Section 5.5. The final section is devoted to the treatment of inhomogeneous boundary conditions. Joseph Fourier developed his ideas on the convergence of trigonometric series while studying heat flow. His 1807 paper was rejected by other scientists as too imprecise and was not published until 1822.

## 5.1 THE COEFFICIENTS

In Chapter 4 we have found Fourier series of several types. How do we find the coefficients? Luckily, there is a very beautiful, conceptual formula for them.

Let us begin with the *Fourier sine series*

$$\phi(x) = \sum_{n=1}^{\infty} A_n \sin \frac{n\pi x}{l} \tag{1}$$

in the interval $(0, l)$. [It turns out that this infinite series converges to $\phi(x)$ for $0 < x < l$, but let's postpone further discussion of the delicate question of convergence for the time being.] The first problem we tackle is to try to find the coefficients $A_n$ if $\phi(x)$ is a given function. The key observation is that the sine functions have the wonderful property that

$$\int_0^l \sin \frac{n\pi x}{l} \sin \frac{m\pi x}{l} \, dx = 0 \quad \text{if } m \neq n, \tag{2}$$

$m$ and $n$ being positive integers. This can be verified directly by integration. [Historically, (1) was first discovered by a horrible expansion in Taylor series!]

**Proof of (2).**  We use the trig identity

$$\sin a \sin b = \tfrac{1}{2}\cos(a - b) - \tfrac{1}{2}\cos(a + b).$$

Therefore, the integral equals

$$\frac{l}{2(m - n)\pi}\sin\frac{(m - n)\pi x}{l}\bigg|_0^l - [\text{same with } (m + n)]$$

if $m \neq n$. This is a linear combination of $\sin(m \pm n)\pi$ and $\sin 0$, and so it vanishes.  □

The far-reaching implications of this observation are astounding. Let's *fix* $m$, multiply (1) by $\sin(m\pi x/l)$, and integrate the series (1) term by term to get

$$\int_0^l \phi(x)\sin\frac{m\pi x}{l}\,dx = \int_0^l \sum_{n=1}^{\infty} A_n \sin\frac{n\pi x}{l}\sin\frac{m\pi x}{l}\,dx$$

$$= \sum_{n=1}^{\infty} A_n \int_0^l \sin\frac{n\pi x}{l}\sin\frac{m\pi x}{l}\,dx.$$

*All but one term in this sum vanishes*, namely the one with $n = m$ ($n$ just being a "dummy" index that takes on all integer values $\geq 1$). Therefore, we are left with the single term

$$A_m \int_0^l \sin^2\frac{m\pi x}{l}\,dx, \tag{3}$$

which equals $\tfrac{1}{2}lA_m$ by explicit integration. Therefore,

$$\boxed{A_m = \frac{2}{l}\int_0^l \phi(x)\sin\frac{m\pi x}{l}\,dx.} \tag{4}$$

This is the famous *formula for the Fourier coefficients* in the series (1). That is, *if $\phi(x)$ has an expansion (1), then* the coefficients must be given by (4).

These are the only possible coefficients in (1). However, the basic question still remains whether (1) is in fact valid with these values of the coefficients. This is a question of convergence, and we postpone it until Section 5.4.

## APPLICATION TO DIFFUSIONS AND WAVES

Going back to the diffusion equation with Dirichlet boundary conditions, the formula (4) provides the final ingredient in the solution formula for arbitrary initial data $\phi(x)$.

As for the wave equation with Dirichlet conditions, the initial data consist of a pair of functions $\phi(x)$ and $\psi(x)$ with expansions (4.1.10) and (4.1.11). The coefficients $A_m$ in (4.1.9) are given by (4), while for the same reason the coefficients $B_m$ are given by the similar formula

$$\frac{m\pi c}{l} B_m = \frac{2}{l} \int_0^l \psi(x) \sin \frac{m\pi x}{l} \, dx. \tag{5}$$

## FOURIER COSINE SERIES

Next let's take the case of the cosine series, which corresponds to the Neumann boundary conditions on $(0, l)$. We write it as

$$\phi(x) = \frac{1}{2} A_0 + \sum_{n=1}^{\infty} A_n \cos \frac{n\pi x}{l}. \tag{6}$$

Again we can verify the magical fact that

$$\int_0^l \cos \frac{n\pi x}{l} \cos \frac{m\pi x}{l} \, dx = 0 \qquad \text{if } m \neq n$$

where $m$ and $n$ are nonnegative integers. (Verify it!) By exactly the same method as above, but with sines replaced by cosines, we get

$$\int_0^l \phi(x) \cos \frac{m\pi x}{l} \, dx = A_m \int_0^l \cos^2 \frac{m\pi x}{l} \, dx = \frac{1}{2} l A_m$$

if $m \neq 0$. For the case $m = 0$, we have

$$\int_0^l \phi(x) \cdot 1 \, dx = \frac{1}{2} A_0 \int_0^l 1^2 \, dx = \frac{1}{2} l A_0.$$

Therefore, for all *nonnegative* integers $m$, we have the formula for the coefficients of the cosine series

$$A_m = \frac{2}{l} \int_0^l \phi(x) \cos \frac{m\pi x}{l} \, dx. \tag{7}$$

[This is the reason for putting the $\frac{1}{2}$ in front of the constant term in (6).]

## FULL FOURIER SERIES

The full Fourier series, or simply the Fourier series, of $\phi(x)$ on the interval $-l < x < l$, is defined as

$$\phi(x) = \frac{1}{2}A_0 + \sum_{n=1}^{\infty}\left(A_n \cos\frac{n\pi x}{l} + B_n \sin\frac{n\pi x}{l}\right). \tag{8}$$

*Watch out:* The interval is twice as long! The eigenfunctions now are all the functions $\{1, \cos(n\pi x/l), \sin(n\pi x/l)\}$, where $n = 1, 2, 3, \ldots$. Again we have the same wonderful coincidence: Multiply any two different eigenfunctions and integrate over the interval and you get zero! That is,

$$\int_{-l}^{l} \cos\frac{n\pi x}{l}\sin\frac{m\pi x}{l}\,dx = 0 \qquad \text{for all } n, m$$

$$\int_{-l}^{l} \cos\frac{n\pi x}{l}\cos\frac{m\pi x}{l}\,dx = 0 \qquad \text{for } n \neq m$$

$$\int_{-l}^{l} \sin\frac{n\pi x}{l}\sin\frac{m\pi x}{l}\,dx = 0 \qquad \text{for } n \neq m$$

$$\int_{-l}^{l} 1\cdot\cos\frac{n\pi x}{l}\,dx = 0 = \int_{-l}^{l} 1\cdot\sin\frac{m\pi x}{l}\,dx.$$

Therefore, the same procedure will work to find the coefficients. We also calculate the integrals of the squares

$$\int_{-l}^{l} \cos^2\frac{n\pi x}{l}\,dx = l = \int_{-l}^{l} \sin^2\frac{n\pi x}{l}\,dx \quad \text{and} \quad \int_{-l}^{l} 1^2\,dx = 2l.$$

(Verify these integrals too!) Then we end up with the formulas

$$A_n = \frac{1}{l}\int_{-l}^{l} \phi(x)\cos\frac{n\pi x}{l}\,dx \quad (n = 0, 1, 2, \ldots) \tag{9}$$

$$B_n = \frac{1}{l}\int_{-l}^{l} \phi(x)\sin\frac{n\pi x}{l}\,dx \quad (n = 1, 2, 3, \ldots) \tag{10}$$

for the coefficients of the full Fourier series. Note that these formulas are *not* exactly the same as (4) and (7).

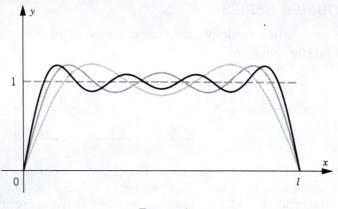

**Figure 1**

## Example 1.

Let $\phi(x) \equiv 1$ in the interval $[0, l]$. It has a Fourier sine series with coefficients

$$A_m = \frac{2}{l} \int_0^l \sin \frac{m\pi x}{l} \, dx = -\frac{2}{m\pi} \cos \frac{m\pi x}{l} \bigg|_0^l$$

$$= \frac{2}{m\pi}(1 - \cos m\pi) = \frac{2}{m\pi}[1 - (-1)^m].$$

Thus $A_m = 4/m\pi$ if $m$ is odd, and $A_m = 0$ if $m$ is even. Thus

$$1 = \frac{4}{\pi} \left( \sin \frac{\pi x}{l} + \frac{1}{3} \sin \frac{3\pi x}{l} + \frac{1}{5} \sin \frac{5\pi x}{l} + \cdots \right) \tag{11}$$

in $(0, l)$. (The factor $4/\pi$ is pulled out just for notational convenience.) See Figure 1 for a sketch of the first few partial sums.   □

## Example 2.

The same function $\phi(x) \equiv 1$ has a Fourier cosine series with coefficients

$$A_m = \frac{2}{l} \int_0^l \cos \frac{m\pi x}{l} \, dx = \frac{2}{m\pi} \sin \frac{m\pi x}{l} \bigg|_0^l$$

$$= \frac{2}{m\pi}(\sin m\pi - \sin 0) = 0 \qquad \text{for } m \neq 0.$$

So there is only one nonzero coefficient, namely, the one for $m = 0$. The Fourier cosine series is therefore trivial:

$$1 = 1 + 0 \cos \frac{\pi x}{l} + 0 \cos \frac{2\pi x}{l} + \cdots .$$

This is perfectly natural since the sum $1 = 1 + 0 + 0 + 0 + \cdots$ is obvious and the Fourier cosine expansion is unique. □

## Example 3.

Let $\phi(x) \equiv x$ in the interval $(0, l)$. Its Fourier sine series has the coefficients

$$A_m = \frac{2}{l} \int_0^l x \sin \frac{m\pi x}{l} \, dx$$

$$= -\frac{2x}{m\pi} \cos \frac{m\pi x}{l} + \frac{2l}{m^2\pi^2} \sin \frac{m\pi x}{l} \bigg|_0^l$$

$$= -\frac{2l}{m\pi} \cos m\pi + \frac{2l}{m^2\pi^2} \sin m\pi = (-1)^{m+1} \frac{2l}{m\pi}.$$

Thus in $(0, l)$ we have

$$x = \frac{2l}{\pi} \left( \sin \frac{\pi x}{l} - \frac{1}{2} \sin \frac{2\pi x}{l} + \frac{1}{3} \sin \frac{3\pi x}{l} - \cdots \right). \tag{12}$$

□

## Example 4.

Let $\phi(x) \equiv x$ in the interval $[0, l]$. Its Fourier cosine series has the coefficients

$$A_0 = \frac{2}{l} \int_0^l x \, dx = l$$

$$A_m = \frac{2}{l} \int_0^l x \cos \frac{m\pi x}{l} \, dx$$

$$= \frac{2x}{m\pi} \sin \frac{m\pi x}{l} + \frac{2l}{m^2\pi^2} \cos \frac{m\pi x}{l} \bigg|_0^l$$

$$= \frac{2l}{m\pi} \sin m\pi + \frac{2l}{m^2\pi^2}(\cos m\pi - 1) = \frac{2l}{m^2\pi^2}[(-1)^m - 1]$$

$$= \frac{-4l}{m^2\pi^2} \quad \text{for } m \text{ odd}, \quad \text{and} \quad 0 \quad \text{for } m \text{ even}.$$

Thus in $(0, l)$ we have

$$x = \frac{l}{2} - \frac{4l}{\pi^2} \left( \cos \frac{\pi x}{l} + \frac{1}{9} \cos \frac{3\pi x}{l} + \frac{1}{25} \cos \frac{5\pi x}{l} + \cdots \right). \tag{13}$$

□

## Example 5.

Let $\phi(x) \equiv x$ in the interval $[-l, l]$. Its full Fourier series has the coefficients

$$A_0 = \frac{1}{l} \int_{-l}^{l} x \, dx = 0$$

$$A_m = \frac{1}{l} \int_{-l}^{l} x \cos \frac{m\pi x}{l} \, dx$$

$$= \frac{x}{m\pi} \sin \frac{m\pi x}{l} + \frac{l}{m^2 \pi^2} \cos \frac{m\pi x}{l} \Big|_{-l}^{l}$$

$$= \frac{l}{m^2 \pi^2} (\cos m\pi - \cos(-m\pi)) = 0$$

$$B_m = \frac{1}{l} \int_{-l}^{l} x \sin \frac{m\pi x}{l} \, dx$$

$$= \frac{-x}{m\pi} \cos \frac{m\pi x}{l} + \frac{l}{m^2 \pi^2} \sin \frac{m\pi x}{l} \Big|_{-l}^{l}$$

$$= \frac{-l}{m\pi} \cos m\pi + \frac{-l}{m\pi} \cos(-m\pi) = (-1)^{m+1} \frac{2l}{m\pi}.$$

This gives us exactly the same series as (12), except that it is supposed to be valid in $(-l, l)$, which is not a surprising result because both sides of (12) are odd.    □

## Example 6.

Solve the problem

$$u_{tt} = c^2 u_{xx}$$
$$u(0, t) = u(l, t) = 0$$
$$u(x, 0) = x, \quad u_t(x, 0) = 0.$$

From Section 4.1 we know that $u(x, t)$ has an expansion

$$u(x, t) = \sum_{n=1}^{\infty} \left( A_n \cos \frac{n\pi ct}{l} + B_n \sin \frac{n\pi ct}{l} \right) \sin \frac{n\pi x}{l}.$$

Differentiating with respect to time yields

$$u_t(x, t) = \sum_{n=1}^{\infty} \frac{n\pi c}{l} \left( -A_n \sin \frac{n\pi ct}{l} + B_n \cos \frac{n\pi ct}{l} \right) \sin \frac{n\pi x}{l}.$$

Setting $t = 0$, we have

$$0 = \sum_{n=1}^{\infty} \frac{n\pi c}{l} B_n \sin \frac{n\pi x}{l}$$

so that all the $B_n = 0$. Setting $t = 0$ in the expansion of $u(x, t)$, we have

$$x = \sum_{n=1}^{\infty} A_n \sin \frac{n\pi x}{l}.$$

This is exactly the series of Example 3. Therefore, the complete solution is

$$u(x, t) = \frac{2l}{\pi} \sum_{n=1}^{\infty} \frac{(-1)^{n+1}}{n} \sin \frac{n\pi x}{l} \cos \frac{n\pi ct}{l}. \qquad \Box$$

## EXERCISES

1.  In the expansion $1 = \sum_{n \text{ odd}} (4/n\pi) \sin n\pi$, valid for $0 < x < \pi$, put $x = \pi/4$ to calculate the sum

    $$\left(1 - \tfrac{1}{5} + \tfrac{1}{9} - \tfrac{1}{13} + \cdots\right) + \left(\tfrac{1}{3} - \tfrac{1}{7} + \tfrac{1}{11} - \tfrac{1}{15} + \cdots\right)$$
    $$= 1 + \tfrac{1}{3} - \tfrac{1}{5} - \tfrac{1}{7} + \tfrac{1}{9} + \cdots$$

    (*Hint:* Since each of the series converges, they can be combined as indicated. However, they cannot be arbitrarily rearranged because they are only conditionally, not absolutely, convergent.)

2.  Let $\phi(x) \equiv x^2$ for $0 \leq x \leq 1 = l$.
    (a) Calculate its Fourier sine series.
    (b) Calculate its Fourier cosine series.

3.  Consider the function $\phi(x) \equiv x$ on $(0, l)$. On the same graph, *sketch* the following functions.
    (a) The sum of the first three (nonzero) terms of its Fourier sine series.
    (b) The sum of the first three (nonzero) terms of its Fourier cosine series.

4.  Find the Fourier cosine series of the function $|\sin x|$ in the interval $(-\pi, \pi)$. Use it to find the sums

    $$\sum_{n=1}^{\infty} \frac{1}{4n^2 - 1} \quad \text{and} \quad \sum_{n=1}^{\infty} \frac{(-1)^n}{4n^2 - 1}.$$

5.  Given the Fourier sine series of $\phi(x) \equiv x$ on $(0, l)$. Assume that the series can be integrated term by term, a fact that will be shown later.
    (a) Find the Fourier cosine series of the function $x^2/2$. Find the constant of integration that will be the first term in the cosine series.

(b)  Then by setting $x = 0$ in your result, find the *sum* of the series

$$\sum_{n=1}^{\infty} \frac{(-1)^{n+1}}{n^2}.$$

6.  (a)  By the same method, find the sine series of $x^3$.
    (b)  Find the cosine series of $x^4$.

7.  Put $x = 0$ in Exercise 6(b) to deduce the sum of the series

$$\sum_{1}^{\infty} \frac{(-1)^n}{n^4}.$$

8.  A rod has length $l = 1$ and constant $k = 1$. Its temperature satisfies the heat equation. Its left end is held at temperature 0, its right end at temperature 1. Initially (at $t = 0$) the temperature is given by

$$\phi(x) = \begin{cases} \dfrac{5x}{2} & \text{for } 0 < x < \frac{2}{3} \\ 3 - 2x & \text{for } \frac{2}{3} < x < 1. \end{cases}$$

Find the solution, including the coefficients. (*Hint:* First find the equilibrium solution $U(x)$, and then solve the heat equation with initial condition $u(x, 0) = \phi(x) - U(x)$.)

9.  Solve $u_{tt} = c^2 u_{xx}$ for $0 < x < \pi$, with the boundary conditions $u_x(0, t) = u_x(\pi, t) = 0$ and the initial conditions $u(x, 0) = 0$, $u_t(x, 0) = \cos^2 x$. (*Hint:* See (4.2.7).)

10.  A string (of tension $T$ and density $\rho$) with fixed ends at $x = 0$ and $x = l$ is hit by a hammer so that $u(x, 0) = 0$, and $\partial u / \partial t(x, 0) = V$ in $[-\delta + \frac{1}{2}l, \delta + \frac{1}{2}l]$ and $\partial u / \partial t(x, 0) = 0$ elsewhere. Find the solution explicitly in series form. Find the energy

$$E_n(h) = \frac{1}{2} \int_0^l \left[ \rho \left( \frac{\partial h}{\partial t} \right)^2 + T \left( \frac{\partial h}{\partial x} \right)^2 \right] dx$$

of the $n$th harmonic $h = h_n$. Conclude that if $\delta$ is small (a concentrated blow), each of the first few overtones has almost as much energy as the fundamental. We could say that the tone is saturated with overtones.

11.  On a string with fixed ends, show that if the center of a hammer blow is exactly at a node of the $n$th harmonic (a place where the $n$th eigenfunction vanishes), the $n$th overtone is absent from the solution.

## 5.2   EVEN, ODD, PERIODIC, AND COMPLEX FUNCTIONS

Each of the three kinds of Fourier series (sine, cosine, and full) of any given function $\phi(x)$ is now determined by the formula for its coefficients given in Section 5.1. We shall see shortly that almost any function $\phi(x)$ defined on the interval $(0, l)$ is the sum of its Fourier sine series and is *also* the sum of its Fourier cosine series. Almost any function defined on the interval $(-l, l)$ is the sum of its full Fourier series. Each of these series converges inside the interval, but not necessarily at the endpoints.

The concepts of oddness, evenness, and periodicity are closely related to the three kinds of Fourier series.

A function $\phi(x)$ that is defined for $-\infty < x < \infty$ is called *periodic* if there is a number $p > 0$ such that

$$\phi(x + p) = \phi(x) \qquad \text{for all } x. \tag{1}$$

A number $p$ for which this is true is called a *period* of $\phi(x)$. The graph of the function repeats forever horizontally. For instance, $\cos x$ has period $2\pi$, $\cos \lambda x$ has period $2\pi/\lambda$, and $\tan x$ has period $\pi$. Note that if $\phi(x)$ has period $p$, then $\phi(x + np) = \phi(x)$ for all $x$ and for all integers $n$. (Why?) The sum of two functions of period $p$ has period $p$. Notice that if $\phi(x)$ has period $p$, then $\int_a^{a+p} \phi(x)\, dx$ does not depend on $a$. (Why?)

For instance, the function $\cos(mx) + \sin 2mx$ is the sum of functions of periods $2\pi/m$ and $\pi/m$ and therefore itself has period $2\pi/m$, the larger of the two.

If a function is defined only on an interval of length $p$, it can be extended in only one way to a function of period $p$. The situation we care about for Fourier series is that of a function defined on the interval $-l < x < l$. Its *periodic extension* is

$$\phi_{\text{per}}(x) = \phi(x - 2lm) \qquad \text{for} \quad -l + 2lm < x < +l + 2lm \tag{2}$$

for all integers $m$. This definition does not specify what the periodic extension is at the endpoints $x = l + 2lm$. In fact, the extension has jumps at these points unless the one-sided limits are equal: $\phi(l-) = \phi(-l+)$ (see Figure 1). (See Section A.1 for the definition of one-sided limits.)

Since each term in the full Fourier series (5.1.8) has period $2l$, its sum (if it converges) also has to have period $2l$. Therefore, the full Fourier series can be regarded *either* as an expansion of an arbitrary function on the interval

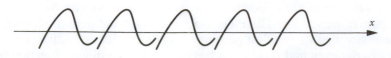

Figure 1

$(-l, l)$ *or* as an expansion of a periodic function of *period 2l* defined on the whole line $-\infty < x < +\infty$.                                                    □

An *even function* is a function that satisfies the equation

$$\phi(-x) = \phi(x). \tag{3}$$

That just means that its graph $y = \phi(x)$ is symmetric with respect to the $y$ axis. Thus the left and right halves of the graph are mirror images of each other. To make sense out of (3), we require that $\phi(x)$ be defined on some interval $(-l, +l)$ which is symmetric around $x = 0$.

An *odd function* is a function that satisfies the equation

$$\phi(-x) = -\phi(x). \tag{4}$$

That just means that its graph $y = \phi(x)$ is symmetric with respect to the origin. To make sense out of (4), we again require that $\phi(x)$ be defined on some interval $(-l, +l)$ which is symmetric around $x = 0$.

A monomial $x^n$ is an even function if $n$ is even and is an odd function if $n$ is odd. The functions $\cos x$, $\cosh x$, and any function of $x^2$ are even functions. The functions $\sin x$, $\tan x$, and $\sinh x$ are odd functions. In fact, the products of functions follow the usual rules: even × even = even, odd × odd = even, odd × even = odd. The sum of two odd functions is again odd, and the sum of two evens is even.

But the sum of an even and an odd function can be anything. Proof: Let $f(x)$ be any function at all defined on $(-l, l)$. Let $\phi(x) = \frac{1}{2}[f(x) + f(-x)]$ and $\psi(x) = \frac{1}{2}[f(x) - f(-x)]$. Then we easily check that $f(x) = \phi(x) + \psi(x)$, that $\phi(x)$ is even and that $\psi(x)$ is odd. The functions $\phi$ and $\psi$ are called the even and odd parts of $f$, respectively. For instance, cosh and sinh are the even and odd parts of exp since: $e^x = \cosh x + \sinh x$. If $p(x)$ is any polynomial, its even part is the sum of its even terms, and its odd part is the sum of its odd terms.

Integration and differentiation change the parity (evenness or oddness) of a function. That is, if $\phi(x)$ is even, then both $d\phi/dx$ and $\int_0^x \phi(s)\, ds$ are odd. If $\phi(x)$ is odd, then the derivative and integral are even. (Note that the lower limit of integration is at the origin.)

The graph of an odd function $\phi(x)$ must pass through the origin since $\phi(0) = 0$ follows directly from (4) by putting $x = 0$. The graph of an even function $\phi(x)$ must cross the $y$ axis horizontally, $\phi'(x) = 0$, since the derivative is odd (provided the derivative exists).

## Example 1.

$\tan x$ is the product of an odd function ($\sin x$) and an even function ($1/\cos x$), both of period $2\pi$. Therefore $\tan x$ is an odd and periodic function. But notice that its smallest period is $\pi$, not $2\pi$. Its derivative $\sec^2 x$ is necessarily even and periodic; it has period $\pi$. The dilated function $\tan ax$ is also odd and periodic and has period $\pi/a$ for any $a > 0$.                    □

Definite integrals around symmetric intervals have the useful properties:

$$\int_{-l}^{l} (\text{odd}) \, dx = 0 \qquad \text{and} \qquad \int_{-l}^{l} (\text{even}) \, dx = 2 \int_{0}^{l} (\text{even}) \, dx. \qquad (5)$$

Given any function defined on the interval $(0, l)$, it can be extended in only one way to be even or odd. The *even extension* of $\phi(x)$ is defined as

$$\phi_{\text{even}}(x) = \begin{cases} \phi(x) & \text{for} \quad 0 < x < l \\ \phi(-x) & \text{for} \quad -l < x < 0. \end{cases} \qquad (6)$$

This is just the mirror image. The even extension is not necessarily defined at the origin.

Its *odd extension* is

$$\phi_{\text{odd}}(x) = \begin{cases} \phi(x) & \text{for} \quad 0 < x < l \\ -\phi(-x) & \text{for} \quad -l < x < 0 \\ 0 & \text{for} \quad x = 0. \end{cases} \qquad (7)$$

This is its image through the origin.

## FOURIER SERIES AND BOUNDARY CONDITIONS

Now let's return to the Fourier sine series. Each of its terms, $\sin(n\pi x/l)$, is an odd function. Therefore, its sum (if it converges) also has to be odd. Furthermore, each of its terms has period $2l$, so that the same has to be true of its sum. Therefore, *the Fourier sine series can be regarded as an expansion of an arbitrary function that is odd and has period $2l$ defined on the whole line $-\infty < x < +\infty$.*

Similarly, since all the cosine functions are even, *the Fourier cosine series can be regarded as an expansion of an arbitrary function which is even and has period $2l$ defined on the whole line $-\infty < x < \infty$.*

From what we saw in Section 5.1, these concepts therefore have the following relationship to boundary conditions:

$u(0, t) = u(l, t) = 0$: Dirichlet BCs correspond to the odd extension. (8)

$u_x(0, t) = u_x(l, t) = 0$: Neumann BCs correspond to the even extension. (9)

$u(l, t) = u(-l, t), u_x(l, t) = u_x(-l, t)$: Periodic BCs correspond

to the periodic extension. (10)

## THE COMPLEX FORM OF THE FULL FOURIER SERIES

The eigenfunctions of $-d^2/dx^2$ on $(-l, l)$ with the periodic boundary conditions are $\sin(n\pi x/l)$ and $\cos(n\pi x/l)$. But recall the *DeMoivre formulas,* which express the sine and cosine in terms of the complex exponentials:

$$\sin \theta = \frac{e^{i\theta} - e^{-i\theta}}{2i} \quad \text{and} \quad \cos \theta = \frac{e^{i\theta} + e^{-i\theta}}{2}. \tag{11}$$

Therefore, instead of sine and cosine, we could use $e^{+in\pi x/l}$ and $e^{-in\pi x/l}$ as an alternative pair. But watch out: They're complex! If we do that, the collection of trigonometric functions $\{\sin n\theta, \cos n\theta\}$ is replaced by the collection of complex exponentials

$$\{1, e^{+i\pi x/l}, e^{+i2\pi x/l}, \ldots, e^{-i\pi x/l}, e^{-i2\pi x/l}, \ldots\}.$$

In other words, we get $\{e^{in\pi x/l}\}$, where $n$ is *any positive or negative* integer.

We should therefore be able to write the full Fourier series in the complex form

$$\phi(x) = \sum_{n=-\infty}^{\infty} c_n e^{in\pi x/l}. \tag{12}$$

This is the sum of two infinite series, one going from $n = 0$ to $+\infty$ and one going from $n = -1$ to $-\infty$. The magical fact in this case is

$$\int_{-l}^{l} e^{in\pi x/l} e^{-im\pi x/l} dx = \int_{-l}^{l} e^{i(n-m)\pi x/l} dx$$

$$= \frac{l}{i\pi(n-m)}[e^{i(n-m)\pi} - e^{i(m-n)\pi}]$$

$$= \frac{l}{i\pi(n-m)}[(-1)^{n-m} - (-1)^{m-n}] = 0$$

provided that $n \neq m$. Notice the extra minus sign in the second exponent of the first integral. When $n = m$, we have

$$\int_{-l}^{l} e^{i(n-n)\pi x/l} dx = \int_{-l}^{l} 1 \, dx = 2l.$$

It follows by the method of Section 5.1 that the coefficients are given by the formula

$$c_n = \frac{1}{2l} \int_{-l}^{l} \phi(x) e^{-in\pi x/l} dx. \tag{13}$$

The complex form is sometimes more convenient in calculations than the real form with sines and cosines. But it really is just the same series written in a different form.

## EXERCISES

1.  For each of the following functions, state whether it is even or odd or periodic. If periodic, what is its smallest period?
    (a)  $\sin ax$   $(a > 0)$
    (b)  $e^{ax}$   $(a > 0)$
    (c)  $x^m$   $(m = \text{integer})$
    (d)  $\tan x^2$
    (e)  $|\sin(x/b)|$   $(b > 0)$
    (f)  $x \cos ax$   $(a > 0)$

2.  Show that $\cos x + \cos \alpha x$ is periodic if $\alpha$ is a rational number. What is its period?

3.  Prove property (5) concerning the integrals of even and odd functions.

4.  (a)  Use (5) to prove that if $\phi(x)$ is an odd function, its full Fourier series on $(-l, l)$ has only sine terms.
    (b)  Also, if $\phi(x)$ is an even function, its full Fourier series on $(-l, l)$ has only cosine terms. (*Hint:* Don't use the series directly. Use the formulas for the coefficients to show that every second coefficient vanishes.)

5.  Show that the Fourier sine series on $(0, l)$ can be derived from the full Fourier series on $(-l, l)$ as follows. Let $\phi(x)$ be any (continuous) function on $(0, l)$. Let $\tilde{\phi}(x)$ be its odd extension. Write the full series for $\tilde{\phi}(x)$ on $(-l, l)$. [Assume that its sum is $\tilde{\phi}(x)$.] By Exercise 4, this series has only sine terms. Simply restrict your attention to $0 < x < l$ to get the sine series for $\phi(x)$.

6.  Show that the cosine series on $(0, l)$ can be derived from the full series on $(-l, l)$ by using the even extension of a function.

7.  Show how the full Fourier series on $(-l, l)$ can be derived from the full series on $(-\pi, \pi)$ by changing variables $w = (\pi/l)x$. (This is called a *change of scale;* it means that one unit along the $x$ axis becomes $\pi/l$ units along the $w$ axis.)

8.  (a)  Prove that differentiation switches even functions to odd ones, and odd functions to even ones.
    (b)  Prove the same for integration provided that we ignore the constant of integration.

9.  Let $\phi(x)$ be a function of period $\pi$. If $\phi(x) = \sum_{n=1}^{\infty} a_n \sin nx$ for all $x$, find the odd coefficients.

10. (a)  Let $\phi(x)$ be a continuous function on $(0, l)$. Under what conditions is its *odd* extension also a continuous function?
    (b)  Let $\phi(x)$ be a differentiable function on $(0, l)$. Under what conditions is its *odd* extension also a differentiable function?
    (c)  Same as part (a) for the *even* extension.
    (d)  Same as part (b) for the *even* extension.

11. Find the full Fourier series of $e^x$ on $(-l, l)$ in its real and complex forms. (*Hint:* It is convenient to find the complex form first.)

12. Repeat Exercise 11 for $\cosh x$. (*Hint:* Use the preceding result.)

13. Repeat Exercise 11 for $\sin x$. Assume that $l$ is not an integer multiple of $\pi$. (*Hint:* First find the series for $e^{ix}$.)

14. Repeat Exercise 11 for $|x|$.

15. Without any computation, predict which of the Fourier coefficients of $|\sin x|$ on the interval $(-\pi, \pi)$ must vanish.

16. Use the De Moivre formulas (11) to derive the standard formulas for $\cos(\theta + \phi)$ and $\sin(\theta + \phi)$.

17. Show that a complex-valued function $f(x)$ is real-valued if and only if its complex Fourier coefficients satisfy $c_n = \overline{c_{-n}}$, where $\overline{\phantom{xx}}$ denotes the complex conjugate.

## 5.3 ORTHOGONALITY AND GENERAL FOURIER SERIES

Let us try to understand what makes the beautiful method of Fourier series work. For the present let's stick with real functions. If $f(x)$ and $g(x)$ are two real-valued continuous functions defined on an interval $a \leq x \leq b$, we define their *inner product* to be the integral of their product:

$$(f, g) \equiv \int_a^b f(x)g(x) \, dx. \tag{1}$$

It is a real number. We'll call $f(x)$ and $g(x)$ *orthogonal* if $(f, g) = 0$. (This terminology is supposed to be analogous to the case of ordinary vectors and their inner or dot product.) Notice that no function is orthogonal to itself except $f(x) \equiv 0$. The key observation in each case discussed in Section 5.1 is that **every eigenfunction is orthogonal to every other eigenfunction**. Now we will explain why this fortuitous coincidence is in fact no accident.

We are studying the operator $A = -d^2/dx^2$ with some boundary conditions (either Dirichlet or Neumann or ... ). Let $X_1(x)$ and $X_2(x)$ be two different eigenfunctions. Thus

$$-X_1'' = \frac{-d^2 X_1}{dx^2} = \lambda_1 X_1$$

$$-X_2'' = \frac{-d^2 X_2}{dx^2} = \lambda_2 X_2, \tag{2}$$

where both functions satisfy the boundary conditions. Let's assume that $\lambda_1 \neq \lambda_2$. We now verify the identity

$$-X_1'' X_2 + X_1 X_2'' = (-X_1' X_2 + X_1 X_2')'.$$

(Work out the right side using the product rule and two of the terms will cancel.) We integrate to get

$$\int_a^b \left( -X_1'' X_2 + X_1 X_2'' \right) dx = \left. \left( -X_1' X_2 + X_1 X_2' \right) \right|_a^b. \tag{3}$$

This is sometimes called *Green's second identity*. If you wished, you could also think of it as the result of two integrations by parts.

On the left side of (3) we now use the differential equations (2). On the right side we use the boundary conditions to reach the following conclusions:

**Case 1: Dirichlet.** This means that both functions vanish at both ends: $X_1(a) = X_1(b) = X_2(a) = X_2(b) = 0$. So the right side of (3) is zero.

**Case 2: Neumann.** The first derivatives vanish at both ends. It is once again zero.

**Case 3: Periodic.** $X_j(a) = X_j(b)$, $X_j'(a) = X_j'(b)$ for both $j = 1, 2$. Again you get zero!

**Case 4: Robin.** Again you get zero! See Exercise 8.

Thus in all four cases, (3) reduces to

$$(\lambda_1 - \lambda_2) \int_a^b X_1 X_2 \, dx = 0. \tag{3a}$$

Therefore, $X_1$ and $X_2$ are orthogonal! This completely explains why Fourier's method works (at least if $\lambda_1 \neq \lambda_2$)!

The right side of (3) isn't always zero. For example, consider the different boundary conditions: $X(a) = X(b)$, $X'(a) = 2X'(b)$. Then the right side of (3) is $X_1'(b)X_2(b) - X_1(b)X_2'(b)$, which is *not* zero. So the method doesn't always work; the boundary conditions have to be right.

## SYMMETRIC BOUNDARY CONDITIONS

So now let us envision *any pair of boundary conditions*

$$\begin{aligned} \alpha_1 X(a) + \beta_1 X(b) + \gamma_1 X'(a) + \delta_1 X'(b) &= 0 \\ \alpha_2 X(a) + \beta_2 X(b) + \gamma_2 X'(a) + \delta_2 X'(b) &= 0 \end{aligned} \tag{4}$$

involving eight real constants. (Each of the examples above corresponds to a choice of these constants.) Such a set of boundary conditions is called *symmetric* if

$$\boxed{\left. f'(x)g(x) - f(x)g'(x) \right|_{x=a}^{x=b} = 0} \tag{5}$$

for any pair of functions $f(x)$ and $g(x)$ both of which satisfy the pair of boundary conditions (4). As we indicated above, each of the four standard boundary conditions (Dirichlet, etc.) is symmetric, but our fifth example is not. The most important thing to keep in mind is that all the standard boundary conditions are symmetric.

Green's second identity (3) then implies the following theorem. By an *eigenfunction* we now mean a solution of $-X'' = \lambda X$ that satisfies (4).

**Theorem 1.**    If you have *symmetric* boundary conditions, then any two eigenfunctions that correspond to distinct eigenvalues are orthogonal. Therefore, if any function is expanded in a series of these eigenfunctions, the coefficients are determined.

**Proof.**    Take two different eigenfunctions $X_1(x)$ and $X_2(x)$ with $\lambda_1 \neq \lambda_2$. We write Green's second identity (3). Because the boundary conditions are symmetric, the right side of (3) vanishes. Because of the different equations, the identity takes the form (3a), and the orthogonality is proven.

If $X_n(x)$ now denotes the eigenfunction with eigenvalue $\lambda_n$ and if

$$\phi(x) = \sum_n A_n X_n(x) \tag{6}$$

is a convergent series, where the $A_n$ are constants, then

$$(\phi, X_m) = \left( \sum_n A_n X_n, X_m \right) = \sum_n A_n (X_n, X_m) = A_m (X_m, X_m)$$

by the orthogonality. So if we denote $c_m = (X_m, X_m)$, we have

$$A_m = \frac{(\phi, X_m)}{c_m} \tag{7}$$

as the formula for the coefficients.    □

Two words of caution. First, we have so far avoided all questions of convergence. Second, if there are two eigenfunctions, say $X_1(x)$ and $X_2(x)$, but their eigenvalues are the same, $\lambda_1 = \lambda_2$, then they don't have to be orthogonal. But if they aren't orthogonal, they can be made so by the Gram–Schmidt orthogonalization procedure (see Exercise 10). For instance, in the case of periodic boundary conditions the two eigenfunctions $\sin(n\pi x/l)$ and $\cos(n\pi x/l)$ are orthogonal on $(-l, l)$, even though they have the same eigenvalue $(n\pi/l)^2$. But the two eigenfunctions $\sin(n\pi x/l)$ and $[\cos(n\pi x/l) + \sin(n\pi x/l)]$ are not orthogonal.

## COMPLEX EIGENVALUES

What about complex eigenvalues $\lambda$ and complex-valued eigenfunctions $X(x)$? If $f(x)$ and $g(x)$ are two complex-valued functions, we define the *inner product* on $(a, b)$ as

$$(f, g) = \int_a^b f(x)\overline{g(x)}\,dx. \tag{8}$$

The bar denotes the complex conjugate. The two functions are called *orthogonal* if $(f, g) = 0$. (This is exactly what is customary for ordinary complex vectors.)

Now suppose that you have the boundary conditions (4) with eight real constants. They are called *symmetric* (or *hermitian*) if

$$f'(x)\,\overline{g(x)} - f(x)\overline{g'(x)}\,\Big|_a^b = 0 \tag{9}$$

for all $f, g$ satisfying the BCs. Then Theorem 1 is true for complex functions without any change at all. But we also have the following important fact.

**Theorem 2.**   Under the same conditions as Theorem 1, *all the eigenvalues are real numbers.* Furthermore, all the eigenfunctions can be chosen to be real valued.

(This could be compared with the discussion at the end of Section 4.1, where complex eigenvalues were discussed explicitly.)

**Proof.**   Let $\lambda$ be an eigenvalue, possibly complex. Let $X(x)$ be its eigenfunction, also possibly complex. Then $-X'' = \lambda X$ plus BCs. Take the complex conjugate of this equation; thus $-\overline{X}'' = \overline{\lambda}\,\overline{X}$ plus BCs. So $\overline{\lambda}$ is also an eigenvalue. Now use Green's second identity with the functions $X$ and $\overline{X}$. Thus

$$\int_a^b (-X''\overline{X} + X\overline{X}'')\,dx = (-X'\overline{X} + X\overline{X}')\,\Big|_a^b = 0$$

since the BCs are symmetric. So

$$(\lambda - \overline{\lambda}) \int_a^b X\overline{X}\,dx = 0$$

But $X\overline{X} = |X|^2 \geq 0$ and $X(x)$ is not allowed to be the zero function. So the integral *cannot* vanish. Therefore, $\lambda - \overline{\lambda} = 0$, which means exactly that $\lambda$ *is real.*

Next, let's reconsider the same problem $-X'' = \lambda X$ together with (4), knowing that $\lambda$ is real. If $X(x)$ is complex, we write it as $X(x) = Y(x) + iZ(x)$, where $Y(x)$ and $Z(x)$ are real. Then $-Y'' - iZ'' = \lambda Y + i\lambda Z$. Equating the real and imaginary parts, we see that $-Y'' = \lambda Y$ and $-Z'' = \lambda Z$. The boundary conditions still hold for both $Y$ and $Z$ because the eight constants in (4) are real numbers. So the *real* eigenvalue $\lambda$ has the *real* eigenfunctions $Y$ and $Z$. We could therefore say that $X$ and $\overline{X}$ are replaceable by the $Y$ and $Z$. The linear combinations $aX + b\overline{X}$ are the same as the linear combinations $cY + dZ$, where $a$ and $b$ are somehow related to $c$ and $d$. This completes the proof of Theorem 2.     □

## NEGATIVE EIGENVALUES

We have seen that most of the eigenvalues turn out to be positive. An important question is whether *all* of them are positive. Here is a sufficient condition.

**Theorem 3.**   Assume the same conditions as in Theorem 1. If

$$f(x)f'(x)\Big|_{x=a}^{x=b} \leq 0 \tag{10}$$

for all (real-valued) functions $f(x)$ satisfying the BCs, then there is *no negative eigenvalue*.

This theorem is proved in Exercise 13. It is easy to verify that (10) is valid for Dirichlet, Neumann, and periodic boundary conditions, so that in these cases there are no negative eigenvalues (see Exercise 11). However, as we have already seen in Section 4.3, it could not be valid for certain Robin boundary conditions.

We have already noticed the close analogy of our analysis with linear algebra. Not only are functions acting as if they were vectors, but the operator $-d^2/dx^2$ is acting like a matrix; in fact, it *is* a linear transformation. Theorems 1 and 2 are like the corresponding theorems about real symmetric matrices. For instance, if $A$ is a real symmetric matrix and $f$ and $g$ are vectors, then $(Af, g) = (f, Ag)$. In our present case, $A$ is a differential operator with symmetric BCs and $f$ and $g$ are functions. The same identity $(Af, g) = (f, Ag)$ holds in our case [see (3)]. The two main differences from matrix theory are, first, that our vector space is infinite dimensional, and second, that the boundary conditions must comprise part of the definition of our linear transformation.

## EXERCISES

1.  (a)   Find the real vectors that are orthogonal to the given vectors $[1, 1, 1]$ and $[1, -1, 0]$.
    (b)   Choosing an answer to (a), expand the vector $[2, -3, 5]$ as a linear combination of these three mutually orthogonal vectors.
2.  (a)   On the interval $[-1, 1]$, show that the function $x$ is orthogonal to the constant functions.
    (b)   Find a quadratic polynomial that is orthogonal to both 1 and $x$.
    (c)   Find a cubic polynomial that is orthogonal to all quadratics. (These are the first few *Legendre polynomials*.)

3.  Consider $u_{tt} = c^2 u_{xx}$ for $0 < x < l$, with the boundary conditions $u(0, t) = 0$, $u_x(l, t) = 0$ and the initial conditions $u(x, 0) = x$, $u_t(x, 0) = 0$. Find the solution explicitly in series form.

4.  Consider the problem $u_t = k u_{xx}$ for $0 < x < l$, with the boundary conditions $u(0, t) = U$, $u_x(l, t) = 0$, and the initial condition $u(x, 0) = 0$, where $U$ is a constant.
    (a)  Find the solution in series form. (*Hint:* Consider $u(x, t) - U$.)
    (b)  Using a direct argument, show that the series converges for $t > 0$.
    (c)  If $\epsilon$ is a given margin of error, estimate how long a time is required for the value $u(l, t)$ at the endpoint to be approximated by the constant $U$ within the error $\epsilon$. (*Hint:* It is an alternating series with first term $U$, so that the error is less than the next term.)

5.  (a)  Show that the boundary conditions $u(0, t) = 0$, $u_x(l, t) = 0$ lead to the eigenfunctions $(\sin(\pi x / 2l), \ \sin(3\pi x / 2l), \ \sin(5\pi x / 2l), \dots)$.
    (b)  If $\phi(x)$ is any function on $(0, l)$, derive the expansion

$$\phi(x) = \sum_{n=0}^{\infty} C_n \sin \left\{ \left( n + \frac{1}{2} \right) \frac{\pi x}{l} \right\} \qquad (0 < x < l)$$

by the following method. Extend $\phi(x)$ to the function $\tilde{\phi}$ defined by $\tilde{\phi}(x) = \phi(x)$ for $0 \le x \le l$ and $\tilde{\phi}(x) = \phi(2l - x)$ for $l \le x \le 2l$. (This means that you are extending it *evenly across* $x = l$.) Write the Fourier sine series for $\tilde{\phi}(x)$ on the interval $(0, 2l)$ and write the formula for the coefficients.
    (c)  Show that every second coefficient vanishes.
    (d)  Rewrite the formula for $C_n$ as an integral of the original function $\phi(x)$ on the interval $(0, l)$.

6.  Find the complex eigenvalues of the first-derivative operator $d/dx$ subject to the single boundary condition $X(0) = X(1)$. Are the eigenfunctions orthogonal on the interval $(0, 1)$?

7.  Show *by direct integration* that the eigenfunctions associated with the Robin BCs, namely,

$$\phi_n(x) = \cos \beta_n x + \frac{a_0}{\beta_n} \sin \beta_n x \quad \text{where } \lambda_n = \beta_n^2,$$

are mutually orthogonal on $0 \le x \le l$, where $\beta_n$ are the positive roots of (4.3.8).

8.  Show directly that $(-X_1' X_2 + X_1 X_2')|_a^b = 0$ if both $X_1$ and $X_2$ satisfy the same Robin boundary condition at $x = a$ and the same Robin boundary condition at $x = b$.

9.  Show that the boundary conditions

$$X(b) = \alpha X(a) + \beta X'(a) \quad \text{and} \quad X'(b) = \gamma X(a) + \delta X'(a)$$

on an interval $a \le x \le b$ are symmetric if and only if $\alpha\delta - \beta\gamma = 1$.

10.  (*The Gram–Schmidt orthogonalization procedure*) If $X_1, X_2, \dots$ is any sequence (finite or infinite) of linearly independent vectors in any vector

space with an inner product, it can be replaced by a sequence of linear combinations that are mutually orthogonal. The idea is that at each step one subtracts off the components parallel to the previous vectors. The procedure is as follows. First, we let $Z_1 = X_1/\|X_1\|$. Second, we define

$$Y_2 = X_2 - (X_2, Z_1)Z_1 \quad \text{and} \quad Z_2 = \frac{Y_2}{\|Y_2\|}.$$

Third, we define

$$Y_3 = X_3 - (X_3, Z_2)Z_2 - (X_3, Z_1)Z_1 \quad \text{and} \quad Z_3 = \frac{Y_3}{\|Y_3\|},$$

and so on.

(a) Show that all the vectors $Z_1, Z_2, Z_3, \ldots$ are orthogonal to each other.

(b) Apply the procedure to the pair of functions $\cos x + \cos 2x$ and $3\cos x - 4\cos 2x$ in the interval $(0, \pi)$ to get an orthogonal pair.

11. (a) Show that the condition $f(x)f'(x)\big|_a^b \leq 0$ is valid for any function $f(x)$ that satisfies Dirichlet, Neumann, or periodic boundary conditions.

(b) Show that it is also valid for Robin BCs provided that the constants $a_0$ and $a_l$ are positive.

12. Prove *Green's first identity:* For every pair of functions $f(x)$, $g(x)$ on $(a, b)$,

$$\int_a^b f''(x)g(x)\,dx = -\int_a^b f'(x)g'(x)\,dx + f'g\bigg|_a^b.$$

13. Use Green's first identity to prove Theorem 3. (*Hint:* Substitute $f(x) = X(x) = g(x)$, a real eigenfunction.)

14. What do the terms in the series

$$\frac{\pi}{4} = \sin 1 + \frac{1}{3}\sin 3 + \frac{1}{5}\sin 5 + \cdots$$

look like? Make a graph of $\sin n$ for $n = 1, 2, 3, 4, \ldots, 20$ without drawing the intervening curve; that is, just plot the 20 points. Use a calculator; remember that we are using radians. In some sense the numbers $\sin n$ are *randomly* located in the interval $(-1, 1)$. There is a great deal of "random cancellation" in the series.

15. Use the same idea as in Exercises 12 and 13 to show that none of the eigenvalues of the fourth-order operator $+d^4/dx^4$ with the boundary conditions $X(0) = X(l) = X''(0) = X''(l) = 0$ are negative.    □


## 5.4 COMPLETENESS

In this section we state the basic theorems about the convergence of Fourier series. We discuss three senses of convergence of functions. The basic theorems

(Theorems 2, 3, and 4) state sufficient conditions on a function $f(x)$ that its Fourier series converge to it in these three senses. Most of the proofs are difficult, however, and we omit them for now. At the end of the section we discuss the mean-square convergence in greater detail and use it to define the notion of completeness.

Consider the eigenvalue problem

$$X'' + \lambda X = 0 \text{ in } (a, b) \text{ with any } \textit{symmetric BC.} \tag{1}$$

By Theorem 5.3.2, we know that all the eigenvalues $\lambda$ are real.

**Theorem 1.** There are an infinite number of eigenvalues. They form a sequence $\lambda_n \to +\infty$.

For a proof of Theorem 1, see Chapter 11 or [CL]. We may assume that the eigenfunctions $X_n(x)$ are pairwise orthogonal and real valued (see Section 5.3). For instance, if $k$ linearly independent eigenfunctions correspond to the *same* eigenvalue $\lambda_n$, then they can be rechosen to be orthogonal and real, and the sequence may be numbered so that $\lambda_n$ is repeated $k$ times. Thus we may list the eigenvalues as

$$\lambda_1 \leq \lambda_2 \leq \lambda_3 \leq \cdots \to +\infty \tag{2}$$

with the corresponding eigenfunctions

$$X_1, X_2, X_3, \ldots, \tag{3}$$

which are pairwise orthogonal. Some interesting examples were found in Section 4.3.

For any function $f(x)$ on $(a, b)$, its *Fourier coefficients* are defined as

$$\boxed{A_n = \frac{(f, X_n)}{(X_n, X_n)} = \frac{\int_a^b f(x)\overline{X_n(x)}\,dx}{\int_a^b |X_n(x)|^2\,dx}.} \tag{4}$$

Its *Fourier series* is the series $\sum_n A_n X_n(x)$.

In this section we present three convergence theorems. Just to convince the skeptic that convergence theorems are more than a pedantic exercise, we mention the curious fact that *there exists an integrable function $f(x)$ whose Fourier series diverges at every point $x$!* There even exists a *continuous* function whose Fourier series diverges at many points! See [Zy] for proofs.

To set the stage we need to introduce various notions of convergence. This is a good point for the reader to review the basic facts about infinite series (outlined in Section A.2).

## THREE NOTIONS OF CONVERGENCE

**Definition.** We say that an infinite series $\sum_{n=1}^{\infty} f_n(x)$ *converges* to $f(x)$ *pointwise* in $(a, b)$ if it converges to $f(x)$ for *each* $a < x < b$. That is, for each

$a < x < b$ we have

$$\left| f(x) - \sum_{n=1}^{N} f_n(x) \right| \to 0 \quad \text{as } N \to \infty. \tag{5}$$

**Definition.** We say that the series *converges uniformly* to $f(x)$ in $[a, b]$ if

$$\max_{a \le x \le b} \left| f(x) - \sum_{n=1}^{N} f_n(x) \right| \to 0 \quad \text{as } N \to \infty. \tag{6}$$

(Note that the endpoints are included in this definition.) That is, you take the biggest difference over all the $x$'s and *then* take the limit.

The two preceding concepts of convergence are also discussed in Section A.2. A third important concept is the following one.

**Definition.** We say the series *converges in the mean-square (or $L^2$) sense* to $f(x)$ in $(a, b)$ if

$$\int_a^b \left| f(x) - \sum_{n=1}^{N} f_n(x) \right|^2 dx \to 0 \quad \text{as } N \to \infty. \tag{7}$$

Thus we take the integral instead of the maximum. (The terminology $L^2$ refers to the square inside the integral.)

Notice that uniform convergence is stronger than both pointwise and $L^2$ convergence (see Exercise 2.) Figure 1 illustrates a typical uniformly convergent series by graphing both $f(x)$ and a partial sum for large $N$.

## Example 1.

Let $f_n(x) = (1 - x)x^{n-1}$ on the interval $0 < x < 1$. Then the series is "telescoping." The partial sums are

$$\sum_{n=1}^{N} f_n(x) = \sum_1^N (x^{n-1} - x^n) = 1 - x^N \to 1 \quad \text{as } N \to \infty$$

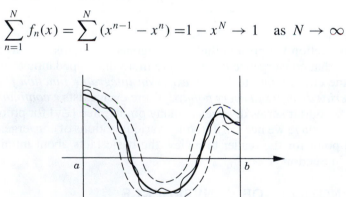

**Figure 1**

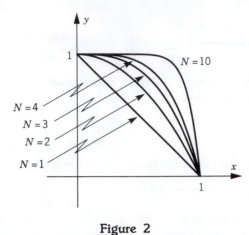

**Figure 2**

because $x < 1$. This convergence is valid for *each* $x$. Thus $\sum_{n=1}^{\infty} f_n(x) = 1$ pointwise. In words, the series *converges pointwise* to the function $f(x) \equiv 1$.

But the convergence is *not uniform* because max $[1 - (1 - x^N)] =$ max $x^N = 1$ for every $N$. However, it *does converge in mean-square* since

$$\int_0^1 |x^N|^2 \, dx = \frac{1}{2N + 1} \to 0.$$

Figure 2 is a sketch of a few partial sums of Example 1.    □

## Example 2.

Let

$$f_n(x) = \frac{n}{1 + n^2 x^2} - \frac{n - 1}{1 + (n - 1)^2 x^2}$$

in the interval $0 < x < l$. This series also telescopes so that

$$\sum_{n=1}^{N} f_n(x) = \frac{N}{1 + N^2 x^2} = \frac{1}{N[(1/N^2) + x^2]} \to 0 \text{ as } N \to \infty \text{ if } x > 0.$$

So the series *converges pointwise* to the sum $f(x) \equiv 0$.

On the other hand,

$$\int_0^l \left[ \sum_{n=1}^{N} f_n(x) \right]^2 dx = \int_0^l \frac{N^2}{(1 + N^2 x^2)^2} \, dx$$

$$= N \int_0^{Nl} \frac{1}{(1 + y^2)^2} \, dy \to +\infty \quad (\text{where } y = Nx)$$

because

$$\int_0^{Nl} \frac{1}{(1 + y^2)^2} \, dy \rightarrow \int_0^\infty \frac{1}{(1 + y^2)^2} \, dy.$$

So the series *does not converge in the mean-square sense*. Also, it *does not converge uniformly* because

$$\max_{(0, l)} \frac{1}{1 + N^2 x^2} = N,$$

which obviously does not tend to zero as $N \rightarrow \infty$.    □

## CONVERGENCE THEOREMS

Now let $f(x)$ be any function defined on $a \leq x \leq b$. Consider the Fourier series for the problem (1) with any given boundary conditions that are symmetric. We now state a convergence theorem for each of the three modes of convergence. They are partly proved in the next section.

**Theorem 2. Uniform Convergence**   The Fourier series $\Sigma A_n X_n(x)$ converges to $f(x)$ uniformly on $[a, b]$ provided that
   (i)   $f(x), f'(x)$, and $f''(x)$ exist and are continuous for $a \leq x \leq b$ and
   (ii)   $f(x)$ satisfies the given boundary conditions.
   Theorem 2 assures us of a very good kind of convergence provided that the conditions on $f(x)$ and its derivatives are met. For the classical Fourier series (full, sine, and cosine), it is not required that $f''(x)$ exist.

**Theorem 3. $L^2$ Convergence**   The Fourier series converges to $f(x)$ in the mean-square sense in $(a, b)$ provided only that $f(x)$ is any function for which

$$\int_a^b |f(x)|^2 \, dx \text{ is finite.} \tag{8}$$

Theorem 3 assures us of a certain kind of convergence under a very weak assumption on $f(x)$. [We could even use the very general Lebesgue integral here instead of the standard (Riemann) integral encountered in calculus courses. In fact, the Lebesgue integral was invented in order that Theorem 3 be true for the most general possible functions.]

Third, we present a theorem that is intermediate as regards the hypotheses on $f(x)$. It requires two more definitions. A function $f(x)$ has a *jump discontinuity* at a point $x = c$ if the one-sided limits $f(c+)$ and $f(c-)$ exist but are not equal. [It doesn't matter what $f(c)$ happens to be or even whether $f(c)$ is defined or not.] The *value* of the jump discontinuity is the number $f(c+) - f(c-)$. See Figure 3 for a function with two jumps.

A function $f(x)$ is called *piecewise continuous* on an interval $[a, b]$ if it is continuous at all but a finite number of points and has jump discontinuities

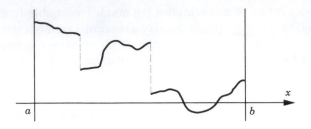

Figure 3

at those points. Another way to say this is that at every point in the interval (including the endpoints) the one-sided limits $f(c+)$ and $f(c-)$ exist; and except at a finite number of points they are equal. For these definitions, see also Section A.1. A typical piecewise continuous function is sketched in Figure 3. The function $Q(x, 0)$ in Section 2.4 is an example of a piecewise continuous function.

## Theorem 4. Pointwise Convergence of Classical Fourier Series

(i)  The classical Fourier series (full or sine or cosine) converges to $f(x)$ pointwise on $(a, b)$ provided that $f(x)$ is a continuous function on $a \le x \le b$ and $f'(x)$ is piecewise continuous on $a \le x \le b$.

(ii)  More generally, if $f(x)$ itself is only piecewise continuous on $a \le x \le b$ and $f'(x)$ is also piecewise continuous on $a \le x \le b$, then the classical Fourier series converges at every point $x(-\infty < x < \infty)$. The sum is

$$\sum_n A_n X_n(x) = \tfrac{1}{2}[f(x+) + f(x-)] \qquad \text{for all } a < x < b. \quad (9)$$

The sum is $\tfrac{1}{2}[f_{\text{ext}}(x+) + f_{\text{ext}}(x-)]$ for all $-\infty < x < \infty$, where $f_{\text{ext}}(x)$ is the extended function (periodic, odd periodic, or even periodic).

Thus at a jump discontinuity the series converges to the *average* of the limits from the right and from the left. In the case of the Fourier sine (or cosine) series on $(0, l)$, the extended function $f_{\text{ext}}(x)$ is the odd (or even) function of period $2l$. For the full series on $(-l, l)$, it is the periodic extension. The extension is piecewise continuous with a piecewise continuous derivative on $(-\infty, \infty)$.

It is convenient to restate Theorem 4 directly for functions that are already defined on the whole line. By considering the periodic, even, and odd extensions of functions, Theorem 4 is equivalent to the following statement.

## Theorem 4∞.  If $f(x)$ is a function of period $2l$ on the line for which $f(x)$ and $f'(x)$ are piecewise continuous, then the classical full Fourier series converges to $\tfrac{1}{2}[f(x+) + f(x-)]$ for $-\infty < x < \infty$.

The Fourier series of a continuous but nondifferentiable function $f(x)$ is not guaranteed to converge pointwise. By Theorem 3 it must converge to $f(x)$ in the $L^2$ sense. If we wanted to be sure of its pointwise convergence, we would have to know something about its derivative $f'(x)$.

## Example 3.

The Fourier sine series of the function $f(x) \equiv 1$ on the interval $(0, \pi)$ is

$$\sum_{n \text{ odd}} \frac{4}{n\pi} \sin nx. \tag{10}$$

Although it converges at each point, this series does not converge uniformly on $[0, \pi]$. One reason is that the series equals zero at both endpoints ($0$ and $\pi$) but the function is 1 there. Condition (ii) of Theorem 2 is not satisfied: the boundary conditions are Dirichlet and the function $f(x)$ does not vanish at the endpoints. However, Theorem 4(i) is applicable, so that the series does converge pointwise to $f(x)$. Thus (10) must sum to 1 for every $0 < x < \pi$. For instance, we get a true equation if we put $x = \pi/2$:

$$1 = f\left(\frac{\pi}{2}\right) = \sum_{n \text{ odd}} \frac{4}{n\pi}(-1)^{(n-1)/2} = \frac{4}{\pi}\sum_{m=0}^{\infty} \frac{(-1)^m}{2m+1}.$$

Therefore, we get the convergent series

$$\frac{\pi}{4} = 1 - \frac{1}{3} + \frac{1}{5} - \frac{1}{7} + \frac{1}{9} - \frac{1}{11} + \cdots.$$

Noting that $0 < 1 < \pi$, we may put $x = 1$ to get the convergent series

$$\frac{\pi}{4} = \sin 1 + \frac{1}{3}\sin 3 + \frac{1}{5}\sin 5 + \cdots.$$

Other amusing series are obtainable in this way.    □

Another important question, especially for our purposes, is whether a Fourier series can be *differentiated* term by term. Take the case of (10). On the left side the derivative is zero. On the right side we ought to get the series.

$$\frac{4}{\pi}\sum_{n \text{ odd}} \cos nx. \tag{11}$$

But this is clearly divergent because the terms don't even tend to zero as $n \to \infty$ (the *n*th *term test* for divergence)! So in this example you *cannot differentiate term by term*. For a more general viewpoint, however, see Example 8 in Section 12.1.

Differentiation of a Fourier series is a delicate matter. But integration term by term is not delicate and is usually valid (see Exercise 11).

The proofs of Theorems 1 to 4 are lengthy and will be postponed to the next section and to Chapter 11. For complete proofs of Theorems 2 and 3, see Section 7.4 of [CL]. For complete proofs of the classical cases of Theorems 2,

3, and 4, see [DM] or [CH]. Of the three convergence theorems, Theorem 3 is the easiest one to apply because $f'(x)$ does not have to exist and $f(x)$ itself does not even have to be continuous. We now pursue a set of ideas that is related to Theorem 3 and is important in quantum mechanics.

## THE $L^2$ THEORY

The main idea is to regard orthogonality as if it were a geometric property. We have already defined the inner product on $(a, b)$ as

$$(f, g) = \int_a^b f(x)\overline{g(x)}\, dx.$$

[In case the functions are real valued, we just ignore the complex conjugate $(^-)$.] We now define the $L^2$ *norm of* $f$ as

$$\|f\| = (f, f)^{1/2} = \left[\int_a^b |f(x)|^2\, dx\right]^{1/2}.$$

The quantity

$$\|f - g\| = \left[\int_a^b |f(x) - g(x)|^2\, dx\right]^{1/2} \tag{12}$$

is a measurement of the "distance" between two functions $f$ and $g$. It is sometimes called the $L^2$ *metric*. The concept of a metric was first mentioned in Section 1.5; the $L^2$ metric is the nicest one.

Theorem 3 can be restated as follows. If $\{X_n\}$ are the eigenfunctions associated with a set of symmetric BCs and if $\|f\| < \infty$, then

$$\left\| f - \sum_{n \leq N} A_n X_n \right\| \to 0 \qquad \text{as } N \to \infty. \tag{13}$$

That is, the partial sums get nearer and nearer to $f$.

**Theorem 5. Least-Square Approximation**   Let $\{X_n\}$ be any orthogonal set of functions. Let $\|f\| < \infty$. Let $N$ be a *fixed* positive integer. Among all possible choices of $N$ constants $c_1, c_2, \ldots, c_N$, the choice that minimizes

$$\left\| f - \sum_{n=1}^N c_n X_n \right\|$$

is $c_1 = A_1, \ldots, c_n = A_n$.

(These are the Fourier coefficients! It means that the linear combination of $X_1, \ldots, X_n$ which approximates $f$ most closely is the Fourier combination!)

**Proof.** For the sake of simplicity we assume in this proof that $f(x)$ and all the $X_n(x)$ are real valued. Denote the error (remainder) by

$$E_N = \left\| f - \sum_{n \leq N} c_n X_n \right\|^2 = \int_a^b \left| f(x) - \sum_{n \leq N} c_n X_n(x) \right|^2 dx. \tag{14}$$

Expanding the square, we have (assuming the functions are real valued)

$$E_N = \int_a^b |f(x)|^2 \, dx - 2 \sum_{n \leq N} c_n \int_a^b f(x) X_n(x) \, dx$$

$$+ \sum_n \sum_m c_n c_m \int_a^b X_n(x) X_m(x) \, dx.$$

Because of orthogonality, the last integral vanishes except for $n = m$. So the double sum reduces to $\Sigma c_n^2 \int |X_n|^2 \, dx$. Let us write this in the norm notation:

$$E_N = \|f\|^2 - 2 \sum_{n \leq N} c_n(f, X_n) + \sum_{n \leq N} c_n^2 \|X_n\|^2.$$

We may "complete the square":

$$E_N = \sum_{n \leq N} \|X_n\|^2 \left[ c_n - \frac{(f, X_n)}{\|X_n\|^2} \right]^2 + \|f\|^2 - \sum_{n \leq N} \frac{(f, X_n)^2}{\|X_n\|^2}. \tag{15}$$

Now the coefficients $c_n$ appear in only one place, inside the squared term. The expression is clearly smallest if the squared term vanishes. That is,

$$c_n = \frac{(f, X_n)}{\|X_n\|^2} \equiv A_n,$$

which proves Theorem 5.      □

The completion of the square has further consequences. Let's choose the $c_n$ to be the Fourier coefficients: $c_n = A_n$. The last expression (15) for the error $E_N$ becomes

$$0 \leq E_N = \|f\|^2 - \sum_{n \leq N} \frac{(f, X_n)^2}{\|X_n\|^2} = \|f\|^2 - \sum_{n \leq N} A_n^2 \|X_n\|^2. \tag{16}$$

Because this is positive, we have

$$\sum_{n \leq N} A_n^2 \int_a^b |X_n(x)|^2 \, dx \leq \int_a^b |f(x)|^2 \, dx. \tag{17}$$

On the left side we have the partial sums of a series of positive terms with bounded partial sums. Therefore, the corresponding infinite series converges and its sum satisfies

$$\sum_{n=1}^{\infty} A_n^2 \int_a^b |X_n(x)|^2 \, dx \leq \int_a^b |f(x)|^2 \, dx. \tag{18}$$

This is known as *Bessel's inequality*. It is valid as long as the integral of $|f|^2$ is finite.

**Theorem 6.** The Fourier series of $f(x)$ converges to $f(x)$ in the mean-square sense if and only if

$$\sum_{n=1}^{\infty} |A_n|^2 \int_a^b |X_n(x)|^2 \, dx = \int_a^b |f(x)|^2 \, dx \qquad (19)$$

(i.e., if and only if you have equality).

**Proof.** Mean-square convergence means that the remainder $E_N \to 0$. But from (16) this means that $\Sigma_{n \le N} |A_n|^2 \|X_n\|^2 \to \|f\|^2$, which in turn means (19), known as *Parseval's equality*.

**Definition.** The infinite orthogonal set of functions $\{X_1(x), X_2(x), \ldots\}$ is called *complete* if Parseval's equality (19) is true for all $f$ with $\int_a^b |f|^2 \, dx < \infty$.

Theorem 3 asserts that the set of eigenfunctions coming from (1) is always complete. Thus we have the following conclusion.

**Corollary 7.** If $\int_a^b |f(x)|^2 dx$ is finite, then the Parseval equality (19) is true.

## Example 4.

Consider once again the Fourier series (10). Parseval's equality asserts that

$$\sum_{n \text{ odd}} \left(\frac{4}{n\pi}\right)^2 \int_0^\pi \sin^2 nx \, dx = \int_0^\pi 1^2 \, dx.$$

This means that

$$\sum_{n \text{ odd}} \left(\frac{4}{n\pi}\right)^2 \frac{\pi}{2} = \pi.$$

In other words,

$$\sum_{n \text{ odd}} \frac{1}{n^2} = 1 + \frac{1}{9} + \frac{1}{25} + \frac{1}{49} + \cdots = \frac{\pi^2}{8},$$

another interesting numerical series. $\qquad \square$

For a full discussion of completeness using the concept of the Lebesgue integral, see [LL] for instance.

## EXERCISES

1. $\sum_{n=0}^{\infty}(-1)^n x^{2n}$ is a geometric series.
   - (a)   Does it converge pointwise in the interval $-1 < x < 1$?
   - (b)   Does it converge uniformly in the interval $-1 < x < 1$?
   - (c)   Does it converge in the $L^2$ sense in the interval $-1 < x < 1$?
     (*Hint:* You can compute its partial sums explicitly.)

2. Consider any series of functions on any finite interval. Show that if it converges uniformly, then it also converges in the $L^2$ sense and in the pointwise sense.

3. Let $\gamma_n$ be a sequence of constants tending to $\infty$. Let $f_n(x)$ be the sequence of functions defined as follows: $f_n(\frac{1}{2}) = 0$, $f_n(x) = \gamma_n$ in the interval $[\frac{1}{2} - \frac{1}{n}, \frac{1}{2})$, let $f_n(x) = -\gamma_n$ in the interval $(\frac{1}{2}, \frac{1}{2} + \frac{1}{n}]$ and let $f_n(x) = 0$ elsewhere. Show that:
   - (a)   $f_n(x) \to 0$ pointwise.
   - (b)   The convergence is not uniform.
   - (c)   $f_n(x) \to 0$ in the $L^2$ sense if $\gamma_n = n^{1/3}$.
   - (d)   $f_n(x)$ does not converge in the $L^2$ sense if $\gamma_n = n$.

4. Let

$$
g_n(x) = \begin{cases}
1 \text{ in the interval } \left[\dfrac{1}{4} - \dfrac{1}{n^2}, \dfrac{1}{4} + \dfrac{1}{n^2}\right) & \text{for odd } n \\[2ex]
1 \text{ in the interval } \left[\dfrac{3}{4} - \dfrac{1}{n^2}, \dfrac{3}{4} + \dfrac{1}{n^2}\right) & \text{for even } n \\[2ex]
0 & \text{for all other } x.
\end{cases}
$$

Show that $g_n(x) \to 0$ in the $L^2$ sense but that $g_n(x)$ does not tend to zero in the pointwise sense.

5. Let $\phi(x) = 0$ for $0 < x < 1$ and $\phi(x) = 1$ for $1 < x < 3$.
   - (a)   Find the first four nonzero terms of its Fourier cosine series explicitly.
   - (b)   For each $x$ ($0 \le x \le 3$), what is the sum of this series?
   - (c)   Does it converge to $\phi(x)$ in the $L^2$ sense? Why?
   - (d)   Put $x = 0$ to find the sum

   $$
   1 + \frac{1}{2} - \frac{1}{4} - \frac{1}{5} + \frac{1}{7} + \frac{1}{8} - \frac{1}{10} - \frac{1}{11} + \cdots .
   $$

6. Find the sine series of the function $\cos x$ on the interval $(0, \pi)$. For each $x$ satisfying $-\pi \le x \le \pi$, what is the sum of the series?

7. Let

$$
\phi(x) = \begin{cases}
-1 - x & \text{for } -1 < x < 0 \\
+1 - x & \text{for } 0 < x < 1.
\end{cases}
$$

(a) Find the full Fourier series of $\phi(x)$ in the interval $(-1, 1)$.
(b) Find the first three nonzero terms explicitly.
(c) Does it converge in the mean square sense?
(d) Does it converge pointwise?
(e) Does it converge uniformly to $\phi(x)$ in the interval $(-1, 1)$?

8. Consider the Fourier sine series of each of the following functions. In this exercise do not compute the coefficients but use the general convergence theorems (Theorems 2, 3, and 4) to discuss the convergence of each of the series in the pointwise, uniform, and $L^2$ senses.
   (a) $f(x) = x^3$ on $(0, l)$.
   (b) $f(x) = lx - x^2$ on $(0, l)$.
   (c) $f(x) = x^{-2}$ on $(0, l)$.

9. Let $f(x)$ be a function on $(-l, l)$ that has a continuous derivative and satisfies the periodic BCs. Let $a_n$ and $b_n$ be the Fourier coefficients of $f(x)$, and let $a'_n$ and $b'_n$ be the Fourier coefficients of its derivative $f'(x)$. Show that

$$a'_n = \frac{n\pi b_n}{l} \quad \text{and} \quad b'_n = \frac{-n\pi a_n}{l} \quad \text{for } n \neq 0.$$

(*Hint:* Write the formulas for $a'_n$ and $b'_n$ and integrate by parts.) This means that the Fourier series of $f'(x)$ is what you'd obtain as if you differentiated term by term. It does not mean that the differentiated series converges.

10. Deduce from Exercise 9 that there is a constant $k$ so that

$$|a_n| + |b_n| \leq \frac{k}{n} \quad \text{for all } n.$$

11. (*Term by term integration*)
    (a) If $f(x)$ is a piecewise continuous function in $[-l, l]$, show that its indefinite integral $F(x) = \int_{-l}^{x} f(s)\, ds$ has a full Fourier series that converges pointwise.
    (b) Write this convergent series for $F(x)$ explicitly in terms of the Fourier coefficients $a_n$, $b_n$ of $f(x)$ if $a_0 = 0$.
    (*Hint:* Apply a convergence theorem. Write the formulas for the coefficients and integrate by parts.)

12. Start with the Fourier sine series of $f(x) = x$ on the interval $(0, l)$. Apply Parseval's equality. Find the sum $\sum_{n=1}^{\infty} 1/n^2$.

13. Start with the Fourier cosine series of $f(x) = x^2$ on the interval $(0, l)$. Apply Parseval's equality. Find the sum $\sum_{n=1}^{\infty} 1/n^4$.

14. Find the sum $\sum_{n=1}^{\infty} 1/n^6$.

15. Let $\phi(x) \equiv 1$ for $0 < x < \pi$. Expand

$$1 = \sum_{n=0}^{\infty} B_n \cos\left[\left(n + \tfrac{1}{2}\right)x\right].$$

(a)  Find $B_n$.

(b)  Let $-2\pi < x < 2\pi$. For which such $x$ does this series converge? For each such $x$, what is the sum of the series? [*Hint:* Think of extending $\phi(x)$ beyond the interval $(0, \pi)$.]

(c)  Apply Parseval's equality to this series. Use it to calculate the sum

$$1 + \frac{1}{3^2} + \frac{1}{5^2} + \cdots.$$

16.  Let $\phi(x) = |x|$ in $(-\pi, \pi)$. If we approximate it by the function

$$f(x) = \tfrac{1}{2}a_0 + a_1 \cos x + b_1 \sin x + a_2 \cos 2x + b_2 \sin 2x,$$

what choice of coefficients will minimize the $L^2$ error?

17.  Modify the proofs of Theorems 5 and 6 for the case of complex-valued functions.

18.  Consider a solution of the wave equation with $c = 1$ on $[0, l]$ with homogeneous Dirichlet or Neumann boundary conditions.

(a)  Show that its energy $E = \frac{1}{2} \int_0^l (u_t^2 + u_x^2) \, dx$ is a constant.

(b)  Let $E_n(t)$ be the energy of its $n$th harmonic (the $n$th term in the expansion). Show that $E = \Sigma E_n$. (*Hint:* Use the orthogonality. Assume that you can integrate term by term.)

19.  Here is a general method to calculate the *normalizing constants*. Let $X(x, \lambda)$ be a family of real solutions of the ODE $-X'' = \lambda X$ which depends in a smooth manner on $\lambda$ as well as on $x$.

(a)  Find the ODE satisfied by $\partial X / \partial \lambda$.

(b)  Apply Green's second identity to the pair of functions $X$ and $\partial X / \partial \lambda$ in order to obtain a formula for $\int_a^b X^2 dx$ in terms of the boundary values.

(c)  As an example, use the result of part (b) and the Dirichlet boundary conditions to compute $\int_0^l \sin^2(m\pi x / l) \, dx$.

20.  Use the method of Exercise 19 to compute the normalizing constants $\int_0^l X^2 \, dx$ in the case of the Robin boundary conditions.

## 5.5   COMPLETENESS AND THE GIBBS PHENOMENON

Our purpose here is to *prove* the pointwise convergence of the classical full Fourier series. This will lead to the celebrated Gibbs phenomenon for jump discontinuities.

We may as well take the whole-line case, Theorem 4∞ of Section 5.4. To avoid technicalities, let us begin with a $C^1$ function $f(x)$ on the whole line of period $2l$. (A $C^1$ function is a function that has a continuous derivative in $(-\infty, \infty)$; see Section A.1.) We also assume that $l = \pi$, which can easily be arranged through a change of scale (see Exercise 5.2.7).

Thus the Fourier series is

$$f(x) = \tfrac{1}{2}A_0 + \sum_{n=1}^{\infty} (A_n \cos nx + B_n \sin nx) \tag{1}$$

with the coefficients

$$A_n = \int_{-\pi}^{\pi} f(y) \cos ny \frac{dy}{\pi} \qquad (n = 0, 1, 2, \ldots)$$

$$B_n = \int_{-\pi}^{\pi} f(y) \sin ny \frac{dy}{\pi} \qquad (n = 1, 2, \ldots).$$

The $N$th partial sum of the series is

$$S_N(x) = \tfrac{1}{2}A_0 + \sum_{n=1}^{N} (A_n \cos nx + B_n \sin nx). \tag{2}$$

We want to prove that $S_N(x)$ converges to $f(x)$ as $N \to \infty$. Pointwise convergence means that $x$ is kept fixed as we take the limit.

The first step of the proof is to stick the formulas for the coefficients into the partial sum and rearrange the terms. Doing this, we get

$$S_N(x) = \int_{-\pi}^{\pi} \left[ 1 + 2 \sum_{n=1}^{N} (\cos ny \cos nx + \sin ny \sin nx) \right] f(y) \frac{dy}{2\pi}.$$

Inside the parentheses is the cosine of a difference of angles, so we can summarize the formula as

$$S_N(x) = \int_{-\pi}^{\pi} K_N(x - y) f(y) \frac{dy}{2\pi}, \tag{3}$$

where

$$K_N(\theta) = 1 + 2 \sum_{n=1}^{N} \cos n\theta. \tag{4}$$

The second step is to study the properties of this function, called the *Dirichlet kernel*. Notice that $K_N(\theta)$ has period $2\pi$ and that

$$\int_{-\pi}^{\pi} K_N(\theta) \frac{d\theta}{2\pi} = 1 + 0 + 0 + \cdots + 0 = 1.$$

It is a remarkable fact that the series for $K_N$ can be summed! In fact,

$$K_N(\theta) = \frac{\sin \left( N + \tfrac{1}{2} \right) \theta}{\sin \tfrac{1}{2}\theta}. \tag{5}$$

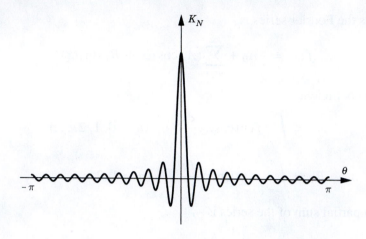

**Figure 1**

**Proof of (5).** The easiest proof is by complexification. By De Moivre's formula for complex exponentials,

$$K_N(\theta) = 1 + \sum_{n=1}^{N} (e^{in\theta} + e^{-in\theta}) = \sum_{n=-N}^{N} e^{in\theta}$$
$$= e^{-iN\theta} + \cdots + 1 + \cdots + e^{iN\theta}.$$

This is a finite geometric series with the first term $e^{-iN\theta}$, the ratio $e^{i\theta}$, and the last term $e^{iN\theta}$. So it adds up to

$$K_N(\theta) = \frac{e^{-iN\theta} - e^{i(N+1)\theta}}{1 - e^{i\theta}}$$

$$= \frac{e^{-i(N+\frac{1}{2})\theta} - e^{+i(N+\frac{1}{2})\theta}}{-e^{\frac{1}{2}i\theta} + e^{-\frac{1}{2}i\theta}}$$

$$= \frac{\sin\left[\left(N + \frac{1}{2}\right)\theta\right]}{\sin\frac{1}{2}\theta}. \qquad \square$$

Figure 1 is a sketch of $K_N(\theta)$. (It looks somewhat like the diffusion kernel, the source function of Section 2.4, except for its oscillatory tail.)

The third step is to combine (3) with (5). Letting $\theta = y - x$ and using the evenness of $K_N$, formula (3) takes the form

$$S_N(x) = \int_{-\pi}^{\pi} K_N(\theta) f(x + \theta) \frac{d\theta}{2\pi}.$$

The interval of integration really ought to be $[x - \pi, x + \pi]$, but since both $K_N$ and $f$ have period $2\pi$, any interval of length $2\pi$ will do. Next we subtract the constant $f(x) = f(x) \cdot 1$ and use formula (5) to get

$$S_N(x) - f(x) = \int_{-\pi}^{\pi} K_N(\theta) \left[ f(x + \theta) - f(x) \right] \frac{d\theta}{2\pi}$$

or

$$S_N(x) - f(x) = \int_{-\pi}^{\pi} g(\theta) \sin \left[ \left( N + \tfrac{1}{2} \right) \theta \right] \frac{d\theta}{2\pi}, \tag{6}$$

where

$$g(\theta) = \frac{f(x + \theta) - f(x)}{\sin \tfrac{1}{2}\theta} \tag{7}$$

Remember that $x$ remains fixed. All we have to show is that the integral (6) tends to zero as $N \to \infty$.

That is the fourth step. We notice that the functions

$$\phi_N(\theta) = \sin \left[ \left( N + \tfrac{1}{2} \right) \theta \right] \qquad (N = 1, 2, 3, \ldots) \tag{8}$$

form an *orthogonal* set on the interval $(0, \pi)$ because they correspond to mixed boundary conditions (see Exercise 5.3.5). Hence they are also orthogonal on the interval $(-\pi, \pi)$. Therefore, *Bessel's inequality* (5.4.18) is valid:

$$\sum_{N=1}^{\infty} \frac{|(g, \phi_N)|^2}{\|\phi_N\|^2} \le \|g\|^2. \tag{9}$$

By direct calculation, $\|\phi_N\|^2 = \pi$. If $\|g\| < \infty$, the series (9) is convergent and its terms tend to zero. So $(g, \phi_N) \to 0$, which says exactly that the integral in (6) tends to zero.

The final step is to check that $\|g\| < \infty$. We have

$$\|g\|^2 = \int_{-\pi}^{\pi} \frac{[f(x + \theta) - f(x)]^2}{\sin^2 \tfrac{1}{2}\theta} \, d\theta.$$

Since the numerator is continuous, the only possible difficulty could occur where the sine vanishes, namely at $\theta = 0$. At that point,

$$\lim_{\theta \to 0} g(\theta) = \lim_{\theta \to 0} \frac{f(x + \theta) - f(x)}{\theta} \cdot \frac{\theta}{\sin \tfrac{1}{2}\theta} = 2 f'(x) \tag{11}$$

by L'Hôpital's rule [since $f(x)$ is differentiable]. Therefore, $g(\theta)$ is everywhere continuous, so that the integral $\|g\|$ is finite. This completes the proof of pointwise convergence of the Fourier series of any $C^1$ function.   □

## PROOF FOR DISCONTINUOUS FUNCTIONS

If the periodic function $f(x)$ itself is only piecewise continuous and $f'(x)$ is also piecewise continuous on $-\infty < x < \infty$, we want to prove that the Fourier series converges and that its sum is $\tfrac{1}{2}[f(x+) + f(x-)]$ (see

Theorem 5.4.4∞). This means that we assume that $f(x)$ and $f'(x)$ are continuous except at a finite number of points, and at those points they have jump discontinuities.

The proof begins as before. However, we modify the third step, replacing (6) by

$$S_N(x) - \frac{1}{2}[f(x+) + f(x-)] = \int_0^\pi K_N(\theta)[f(x + \theta) - f(x+)]\frac{d\theta}{2\pi}$$

$$+ \int_{-\pi}^0 K_N(\theta)[f(x + \theta) - f(x-)]\frac{d\theta}{2\pi}$$

$$= \int_0^\pi g_+(\theta) \sin\left[\left(N + \tfrac{1}{2}\right)\theta\right] d\theta$$

$$+ \int_{-\pi}^0 g_-(\theta) \sin\left[\left(N + \tfrac{1}{2}\right)\theta\right] d\theta \qquad (12)$$

by (5), where

$$g_\pm(\theta) = \frac{f(x + \theta) - f(x\pm)}{\sin \frac{1}{2}\theta}. \qquad (13)$$

The    fourth    step    is    to    observe    that    the    functions $\sin[(N + \tfrac{1}{2})\theta]$ $(N = 1, 2, 3, \ldots)$ form an orthogonal set on the interval $(-\pi, 0)$, as well as on the interval $(-0, \pi)$. Using Bessel's inequality as before, we deduce (see Exercise 8) that both of the integrals in (12) tend to zero as $N \to \infty$ provided that $\int_0^\pi |g_+(\theta)|^2 d\theta$ and $\int_{-\pi}^0 |g_-(\theta)|^2 d\theta$ are finite.

That is the fifth step. The only possible reason for the divergence of these integrals would come from the vanishing of $\sin \frac{1}{2}\theta$ at $\theta = 0$. Now the *one-sided* limit of $g_+(\theta)$ is

$$\lim_{\theta \searrow 0} g_+(\theta) = \lim_{\theta \searrow 0} \frac{f(x + \theta) - f(x+)}{\theta} \cdot \frac{\theta}{\sin\left(\tfrac{1}{2}\theta\right)} = 2f'(x+) \qquad (14)$$

if $x$ is a point where the one-sided derivative $f'(x+)$ exists. If $f'(x+)$ does not exist (e.g., $f$ itself might have a jump at the point $x$), then $f$ still is differentiable at nearby points. By the mean value theorem, $[f(x + \theta) - f(x+)]/\theta = f'(\theta^*)$ for some point $\theta^*$ between $x$ and $x + \theta$. Since the derivative is bounded, it follows that $[f(x + \theta) - f(x)]/\theta$ is bounded as well for $\theta$ small and positive. So $g_+(\theta)$ is bounded and the integral $\int_0^\pi |g_+(\theta)|^2 d\theta$ is finite. It works the same way for $g_-(\theta)$.  □

## PROOF OF UNIFORM CONVERGENCE

This is Theorem 5.4.2, for the case of classical Fourier series. We assume again that $f(x)$ and $f'(x)$ are continuous functions of period $2\pi$. The idea of this proof is quite different from the preceding one. The main point is to show that the coefficients go to zero pretty fast. Let $A_n$ and $B_n$ be the Fourier coefficients of $f(x)$ and let $A'_n$ and $B'_n$ denote the Fourier coefficients of $f'(x)$. We integrate by parts to get

$$A_n = \int_{-\pi}^{\pi} f(x) \cos nx \, \frac{dx}{\pi}$$

$$= \frac{1}{n\pi} f(x) \sin nx \Big|_{-\pi}^{\pi} - \int_{-\pi}^{\pi} f'(x) \sin nx \, \frac{dx}{n\pi},$$

so that

$$A_n = -\frac{1}{n} B'_n \qquad \text{for } n \neq 0. \tag{15}$$

We have just used the periodicity of $f(x)$. Similarly,

$$B_n = \frac{1}{n} A'_n. \tag{16}$$

On the other hand, we know from Bessel's inequality [for the derivative $f'(x)$] that the infinite series

$$\sum_{n=1}^{\infty} \left( |A'_n|^2 + |B'_n|^2 \right) < \infty.$$

Therefore,

$$\sum_{n=1}^{\infty} \left( |A_n \cos nx| + |B_n \sin nx| \right) \leq \sum_{n=1}^{\infty} \left( |A_n| + |B_n| \right)$$

$$= \sum_{n=1}^{\infty} \frac{1}{n} \left( |B'_n| + |A'_n| \right)$$

$$\leq \left( \sum_{n=1}^{\infty} \frac{1}{n^2} \right)^{1/2} \left[ \sum_{n=1}^{\infty} 2 \left( |A'_n|^2 + |B'_n|^2 \right) \right]^{1/2} < \infty.$$

Here we have used Schwarz's inequality (see Exercise 5). The result means that the Fourier series converges *absolutely*.

We already know (from Theorem 5.4.4∞) that the sum of the Fourier series is indeed $f(x)$. So, again denoting by $S_N(x)$ the partial sum (2), we can write

$$\max|f(x) - S_N(x)| \leq \max \sum_{n=N+1}^{\infty} |A_n \cos nx + B_n \sin nx|$$

$$\leq \sum_{n=N+1}^{\infty} \left( |A_n| + |B_n| \right) < \infty. \tag{17}$$

The last sum is the tail of a convergent series of numbers so that it tends to zero as $N \to \infty$. Therefore, the Fourier series converges to $f(x)$ both absolutely and uniformly.                                                                        □

This proof is also valid if $f(x)$ is continuous but $f'(x)$ is merely piecewise continuous. An example is $f(x) = |x|$.

## THE GIBBS PHENOMENON

The Gibbs phenomenon is what happens to Fourier series at jump discontinuities. For a function with a jump, the partial sum $S_N(x)$ approximates the jump as in Figure 2 for a large value of $N$. Gibbs showed that $S_N(x)$ always differs from $f(x)$ near the jump by an "overshoot" of about 9 percent. The width of the overshoot goes to zero as $N \to \infty$ while the extra height remains at 9 percent (top and bottom). Thus

$$\lim_{N \to \infty} \max|S_N(x) - f(x)| \neq 0, \tag{18}$$

although $S_N(x) - f(x)$ does tend to zero for each $x$ where $f(x)$ does not jump.

We now verify the Gibbs phenomenon for an example. Let's take the simplest odd function with a jump of unity; that is,

$$f(x) = \begin{cases} \frac{1}{2} & \text{for } 0 < x < \pi \\ -\frac{1}{2} & \text{for } -\pi < x < 0, \end{cases}$$

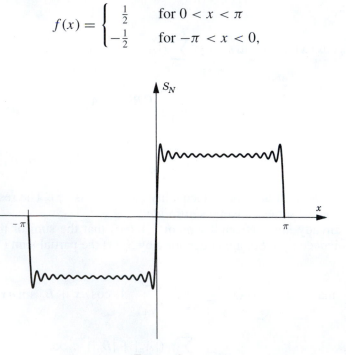

Figure 2

which has the Fourier series

$$\sum_{n \text{ odd}=1}^{\infty} \frac{2}{n\pi} \sin nx.$$

Figure 2 is a sketch of the partial sum $S_{16}(x)$. By (3) and (5), the partial sums are

$$S_N(x) = \left( \int_0^\pi - \int_{-\pi}^0 \right) K_N(x-y) \frac{dy}{4\pi}$$

$$= \left( \int_0^\pi - \int_{-\pi}^0 \right) \frac{\sin\left[ \left(N + \frac{1}{2}\right)(x-y) \right]}{\sin\left[\frac{1}{2}(x-y)\right]} \frac{dy}{4\pi}.$$

Let $M = N + \frac{1}{2}$. In the first integral let $\theta = M(x-y)$. In the second integral let $\theta = M(y-x)$. These changes of variables yield

$$S_N(x) = \left( \int_{M(x-\pi)}^{Mx} - \int_{-M(x+\pi)}^{-Mx} \right) \frac{\sin\theta}{2M \sin(\theta/2M)} \frac{d\theta}{2\pi}$$

$$= \left( \int_{-Mx}^{Mx} - \int_{-M\pi-Mx}^{-M\pi+Mx} \right) \frac{\sin\theta}{2M \sin(\theta/2M)} \frac{d\theta}{2\pi}$$

$$= \left( \int_{-Mx}^{Mx} - \int_{M\pi-Mx}^{M\pi+Mx} \right) \frac{\sin\theta}{2M \sin(\theta/2M)} \frac{d\theta}{2\pi}, \tag{19}$$

where we changed $\theta$ to $-\theta$ in the last step, the integrand being an even function.

We are interested in what happens near the jump, that is, where $x$ is small. Remember that $M$ is large. We will see that in (19) the first integral is the larger one because of the small denominator $\sin(\theta/2M)$. Where is the first integral in (19) maximized? Setting its derivative equal to zero, it is maximized where $\sin Mx = 0$. So we set $x = \pi/M$. Then (19) becomes

$$S_N\left(\frac{\pi}{M}\right) = \left( \int_{-\pi}^\pi - \int_{M\pi-\pi}^{M\pi+\pi} \right) \frac{\sin\theta}{2M \sin(\theta/2M)} \frac{d\theta}{2\pi}. \tag{20}$$

Inside the *second* integral in (20) the argument $\theta/2M$ is bounded both above and below, as follows:

$$\frac{\pi}{4} < \left[1 - \frac{1}{M}\right]\frac{\pi}{2} \le \frac{\theta}{2M} \le \left[1 + \frac{1}{M}\right]\frac{\pi}{2} < \frac{3\pi}{4}$$

for $M > 2$. Hence $\sin(\theta/2M) > 1/\sqrt{2}$, so that the second integral in (20) is less than

$$\int_{M\pi-\pi}^{M\pi+\pi} 1 \cdot \left[\frac{2M}{\sqrt{2}}\right]^{-1} \frac{d\theta}{2\pi} = \frac{1}{\sqrt{2}M},$$

which tends to zero as $M \to \infty$.

On the other hand, inside the *first* integral in (20) we have $|\theta| \le \pi$ and

$$2M \sin\frac{\theta}{2M} \to \theta \qquad \text{uniformly in } -\pi \le \theta \le \pi \quad \text{as } M \to \infty.$$

Hence, taking the limit in (20) as $M \to \infty$, we get

$$S_N\left(\frac{\pi}{M}\right) \to \int_{-\pi}^{\pi} \frac{\sin\theta}{\theta} \frac{d\theta}{2\pi} \simeq 0.59. \tag{21}$$

This is Gibbs's 9 percent overshoot (of the unit jump value).

## FOURIER SERIES SOLUTIONS

You could object, and you would be right, that we never showed that the Fourier series solutions actually solve the PDEs. Let's take a basic example to justify this final step. Consider the wave equation with Dirichlet boundary conditions and with initial conditions $u(x, 0) = \phi(x)$, $u_t(x, 0) = \psi(x)$ as in Section 4.1. The solution is supposed to be given by (4.1.9):

$$u(x, t) = \sum_n \left( A_n \cos\frac{n\pi ct}{l} + B_n \sin\frac{n\pi ct}{l} \right) \sin\frac{n\pi x}{l}. \tag{22}$$

However, we know that term-by-term differentiation of a Fourier series is *not* always valid (see Example 3, Section 5.4), so we cannot simply verify by direct differentiation that (22) is a solution.

Instead, let $\phi_{\text{ext}}$ and $\psi_{\text{ext}}$ denote the odd $2l$-periodic extensions of $\phi$ and $\psi$. Let us assume that $\phi$ and $\psi$ are continuous with piecewise continuous derivatives. We know that the function

$$u(x, t) = \frac{1}{2}[\phi_{\text{ext}}(x + ct) + \phi_{\text{ext}}(x - ct)] + \frac{1}{2c}\int_{x-ct}^{x+ct} \psi_{\text{ext}}(s)\,ds \tag{23}$$

solves the wave equation with $u(x, 0) = \phi_{\text{ext}}(x)$, $u_t(x, 0) = \psi_{\text{ext}}(x)$ for all $-\infty < x < \infty$. (Actually, it is a weak solution—see Section 12.1—but if we assume that $\phi_{\text{ext}}$ and $\psi_{\text{ext}}$ are twice differentiable, it is an ordinary twice-differentiable solution.) Since $\phi_{\text{ext}}$ and $\psi_{\text{ext}}$ agree with $\phi$ and $\psi$ on the interval $(0, l)$, $u$ satisfies the correct initial conditions on $(0, l)$. Since $\phi_{\text{ext}}$ and $\psi_{\text{ext}}$ are odd, it follows that $u(x, t)$ is also odd, so that $u(0, t) = u(l, t) = 0$, which is the correct boundary condition.

By Theorem 5.4.4(i), the Fourier sine series of $\phi_{\text{ext}}$ and $\psi_{\text{ext}}$, given by (4.1.10) and (4.1.11), converge pointwise. Substituting these series into (23), we get

$$u(x, t) = \frac{1}{2} \sum_{n=1}^{\infty} A_n \left( \sin \frac{n\pi(x + ct)}{l} + \sin \frac{n\pi(x - ct)}{l} \right)$$

$$+ \frac{1}{2c} \sum_{n=1}^{\infty} \int_{x-ct}^{x+ct} B_n \frac{n\pi c}{l} \sin \frac{n\pi s}{l} ds. \tag{24}$$

This series converges pointwise because term-by-term *integration* of a Fourier series is always valid, by Exercise 5.4.11. Now we use standard trigonometric identities and carry out the integrals explicitly. We get

$$u(x, t) = \sum_{n} \left( A_n \sin \frac{n\pi x}{l} \cos \frac{n\pi ct}{l} + B_n \sin \frac{n\pi x}{l} \sin \frac{n\pi ct}{l} \right). \tag{25}$$

This is precisely (22).

## EXERCISES

1.  Sketch the graph of the Dirichlet kernel

    $$K_N(\theta) = \frac{\sin \left( N + \frac{1}{2} \right) \theta}{\sin \frac{1}{2}\theta}$$

    in case $N = 10$. Use a computer graphics program if you wish.

2.  Prove the *Schwarz inequality* (for any pair of functions):

    $$|(f, g)| \leq \|f\| \cdot \|g\|.$$

    (*Hint:* Consider the expression $\|f + tg\|^2$, where $t$ is a scalar. This expression is a quadratic polynomial of $t$. Find the value of $t$ where it is a minimum. Play around and the Schwarz inequality will pop out.)

3.  Prove the inequality $l \int_0^l (f'(x))^2 dx \geq [f(l) - f(0)]^2$ for any real function $f(x)$ whose derivative $f'(x)$ is continuous. [*Hint:* Use Schwarz's inequality with the pair $f'(x)$ and 1.]

4.  (a)  Solve the problem $u_t = ku_{xx}$ for $0 < x < l$, $u(x, 0) = \phi(x)$, with the unusual boundary conditions

    $$u_x(0, t) = u_x(l, t) = \frac{u(l, t) - u(0, t)}{l}.$$

    Assume that there are no negative eigenvalues. (*Hint:* See Exercise 4.3.12.)

    (b)  Show that as $t \to \infty$,

    $$\lim u(x, t) = A + Bx,$$

    assuming that you can take limits term by term.

    (c)  Use Green's first identity and Exercise 3 to show that there are no negative eigenvalues.

(d) Find $A$ and $B$. (*Hint:* $A + Bx$ is the beginning of the series. Take the inner product of the series for $\phi(x)$ with each of the functions 1 and $x$. Make use of the orthogonality.)

5. Prove the *Schwarz inequality for infinite series:*

$$\sum a_n b_n \leq \left(\sum a_n^2\right)^{1/2} \left(\sum b_n^2\right)^{1/2}.$$

(*Hint:* See the hint in Exercise 2. Prove it first for finite series (ordinary sums) and then pass to the limit.)

6. Consider the diffusion equation on $[0, l]$ with Dirichlet boundary conditions and any continuous function as initial condition. Show from the series expansion that the solution is infinitely differentiable for $t > 0$. (*Hint:* Use the general theorem at the end of Section A.2 on the differentiability of series, together with the fact that the exponentials are very small for large $n$. See Section 3.5 for an analogous situation.)

7. Let $\int_{-\pi}^{\pi} [|f(x)|^2 + |g(x)|^2]\, dx$ be finite, where $g(x) = f(x)/(e^{ix} - 1)$. Let $c_n$ be the coefficients of the full complex Fourier series of $f(x)$. Show that $\sum_{n=-N}^{N} c_n \to 0$ as $N \to \infty$.

8. Prove that both integrals in (12) tend to zero.

9. Fill in the missing steps in the proof of uniform convergence.

10. Prove the theorem on uniform convergence for the case of the Fourier sine series and for the Fourier cosine series.

11. Prove that the classical full Fourier series of $f(x)$ converges *uniformly* to $f(x)$ if merely $f(x)$ is continuous of period $2\pi$ and its derivative $f'(x)$ is *piecewise* continuous. (*Hint:* Modify the discussion of uniform convergence in this section.)

12. Show that if $f(x)$ is a $C^1$ function in $[-\pi, \pi]$ that satisfies the periodic BC and if $\int_{-\pi}^{\pi} f(x)dx = 0$, then $\int_{-\pi}^{\pi} |f|^2 dx \leq \int_{-\pi}^{\pi} |f'|^2 dx$. (*Hint:* Use Parseval's equality.)

13. A very slick proof of the pointwise convergence of Fourier series, due to P. Chernoff (*American Mathematical Monthly*, May 1980), goes as follows.
    (a) Let $f(x)$ be a $C^1$ function of period $2\pi$. First show that we may as well assume that $f(0) = 0$ and we need only show that the Fourier series converges to zero at $x = 0$.
    (b) Let $g(x) = f(x)/(e^{ix} - 1)$. Show that $g(x)$ is a continuous function.
    (c) Let $C_n$ be the (complex) Fourier coefficients of $f(x)$ and $D_n$ the coefficients of $g(x)$. Show that $D_n \to 0$.
    (d) Show that $C_n = D_{n-1} - D_n$ so that the series $\Sigma C_n$ is telescoping.
    (e) Deduce that the Fourier series of $f(x)$ at $x = 0$ converges to zero.

14. Prove the validity of the Fourier series solution of the diffusion equation on $(0, l)$ with $u_x(x, 0) = u_x(x, l) = 0$, $u(x, 0) = \phi(x)$, where $\phi(x)$ is continuous with a piecewise continuous derivative. That is, prove that the series truly converges to the solution.

15. Carry out the step going from (24) to (25).

## 5.6  INHOMOGENEOUS BOUNDARY CONDITIONS

In this section we consider problems with sources given at the boundary. We shall see that naive use of the separation of variables technique will not work.

Let's begin with the *diffusion equation with sources at both endpoints.*

$$u_t = ku_{xx} \qquad 0 < x < l, \quad t > 0$$
$$\boldsymbol{u(0,t) = h(t)} \qquad \boldsymbol{u(l,t) = j(t)} \tag{1}$$
$$u(x, 0) \equiv 0.$$

A separated solution $u = X(x)T(t)$ just will not fit the boundary conditions. So we try a slightly different approach.

### EXPANSION METHOD

We already know that for the corresponding homogeneous problem the correct expansion is the Fourier sine series. For each $t$, we certainly can expand

$$\boxed{u(x, t) = \sum_{n=1}^{\infty} u_n(t) \sin \frac{n\pi x}{l}} \tag{2}$$

for *some* coefficients $u_n(t)$, because the completeness theorems guarantee that any function in $(0, l)$ can be so expanded. The coefficients are necessarily given by

$$u_n(t) = \frac{2}{l} \int_0^l u(x, t) \sin \frac{n\pi x}{l}\, dx. \tag{3}$$

You may object that each term in the series vanishes at both endpoints and thereby violates the boundary conditions. The answer is that we simply do not insist that the series converge at the endpoints but only inside the interval. In fact, we are exactly in the situation of Theorems 3 and 4 but not of Theorem 2 of Section 5.4.

Now differentiating the series (2) term by term, we get

$$0 = u_t - ku_{xx} = \sum \left[ \frac{du_n}{dt} + ku_n(t) \left( \frac{n\pi}{l} \right)^2 \right] \sin \frac{n\pi x}{l}.$$

So the PDE seems to require that $du_n/dt + k\lambda_n u_n = 0$, so that $u_n(t) = A_n e^{k\lambda_n t}$. There is no way for this to fit the boundary conditions. Our method fails! What's the moral? It is that you can't differentiate term by term. See Example 3 in Section 5.4 for the dangers of differentiation.

Let's start over again but avoid direct differentiation of the Fourier series. The expansion (2) with the coefficients (3) must be valid, by the completeness theorem 5.4.3, say, provided that $u(x, t)$ is a continuous function. Clearly, the initial condition requires that $u_n(0) = 0$. If the derivatives of $u(x, t)$ are also continuous, let's expand them, too. Thus

$$\frac{\partial u}{\partial t} = \sum_{n=1}^{\infty} v_n(t) \sin \frac{n\pi x}{l} \tag{4}$$

with

$$v_n(t) = \frac{2}{l} \int_0^l \frac{\partial u}{\partial t} \sin \frac{n\pi x}{l} \, dx = \frac{du_n}{dt}. \tag{5}$$

The last equality is valid since we can differentiate under an integral sign if the new integrand is continuous (see Section A.3). We also expand

$$\frac{\partial^2 u}{\partial x^2} = \sum_{n=1}^{\infty} w_n(t) \sin \frac{n\pi x}{l} \tag{6}$$

with the coefficients

$$w_n(t) = \frac{2}{l} \int_0^l \frac{\partial^2 u}{\partial x^2} \sin \frac{n\pi x}{l} \, dx. \tag{7}$$

By Green's second identity (5.3.3) the last expression equals

$$\frac{-2}{l} \int_0^l \left(\frac{n\pi}{l}\right)^2 u(x,t) \sin \frac{n\pi x}{l} \, dx + \frac{2}{l} \left(u_x \sin \frac{n\pi x}{l} - \frac{n\pi}{l} u \cos \frac{n\pi x}{l}\right)\Big|_0^l.$$

Here come the boundary conditions. The sine factor vanishes at both ends. The last term will involve the boundary conditions. Thus

$$w_n(t) = -\lambda_n u_n(t) - 2n\pi l^{-2}(-1)^n j(t) + 2n\pi l^{-2} h(t), \tag{8}$$

where $\lambda_n = (n\pi/l)^2$. Now by (5) and (7) the PDE requires

$$v_n(t) - k w_n(t) = \frac{2}{l} \int_0^l (u_t - k u_{xx}) \sin \frac{n\pi x}{l} \, dx = \int_0^l 0 = 0.$$

So from (5) and (8) we deduce that $u_n(t)$ satisfies

$$\frac{du_n}{dt} = k\{-\lambda_n u_n(t) - 2n\pi l^{-2}[(-1)^n j(t) - h(t)]\}. \tag{9}$$

This is just an ordinary differential equation, to be solved together with the initial condition $u_n(0) = 0$ from (1). The solution of (9) is

$$u_n(t) = C e^{-\lambda_n kt} - 2n\pi l^{-2} k \int_0^t e^{-\lambda_n k(t-s)} [(-1)^n j(s) - h(s)] \, ds. \tag{10}$$

As a second case, let's solve the *inhomogeneous wave problem*

$$u_{tt} - c^2 u_{xx} = f(x, t)$$
$$u(0, t) = h(t) \qquad u(l, t) = k(t) \qquad (11)$$
$$u(x, 0) = \phi(x) \qquad u_t(x, 0) = \psi(x).$$

Again we expand everything in the eigenfunctions of the corresponding homogeneous problem:

$$u(x, t) = \sum_{n=1}^{\infty} u_n(t) \sin \frac{n \pi x}{l},$$

$u_{tt}(x, t)$ with coefficients $v_n(t)$, $u_{xx}(x, t)$ with coefficients $w_n(t)$, $f(x, t)$ with coefficients $f_n(t)$, $\phi(x)$ with coefficients $\phi_n$, and $\psi(x)$ with coefficients $\psi_n$. Then

$$v_n(t) = \frac{2}{l} \int_0^l \frac{\partial^2 u}{\partial t^2} \sin \frac{n \pi x}{l} \, dx = \frac{d^2 u_n}{dt^2}$$

and, just as before,

$$w_n(t) = \frac{2}{l} \int_0^l \frac{\partial^2 u}{\partial x^2} \sin \frac{n \pi x}{l} \, dx$$
$$= -\lambda_n u_n(t) + 2n\pi l^{-2} [h(t) - (-1)^n k(t)].$$

From the PDE we also have

$$v_n(t) - c^2 w_n(t) = \frac{2}{l} \int_0^l (u_{tt} - c^2 u_{xx}) \sin \frac{n \pi x}{l} \, dx = f_n(t).$$

Therefore,

$$\frac{d^2 u_n}{dt^2} + c^2 \lambda_n u_n(t) = -2n\pi l^{-2} \left[ (-1)^n k(t) - h(t) \right] + f_n(t) \qquad (12)$$

with the initial conditions

$$u_n(0) = \phi_n \qquad u_n'(0) = \psi_n.$$

The solution can be written explicitly (see Exercise 11).

## METHOD OF SHIFTING THE DATA

By subtraction, the data can be shifted from the boundary to another spot in the problem. *The boundary conditions can be made homogeneous by subtracting any known function that satisfies them.* Thus for the problem (11) treated above, the function

$$\mathcal{U}(x, t) = \left(1 - \frac{x}{l}\right) h(t) + \frac{x}{l} k(t)$$

obviously satisfies the BCs. If we let

$$v(x, t) = u(x, t) - \mathcal{U}(x, t),$$

then $v(x, t)$ satisfies the same problem but with zero boundary data, with initial data $\phi(x) - \mathcal{U}(x, 0)$ and $\psi(x) - \mathcal{U}_t(x, 0)$, and with right-hand side $f$ replaced by $f - \mathcal{U}_{tt}$.

*The boundary condition and the differential equation can simultaneously be made homogeneous by subtracting any known function that satisfies them.* One case when this can surely be accomplished is the case of "stationary data" when $h$, $k$, and $f(x)$ all are independent of time. Then it is easy to find a solution of

$$-c^2 \mathcal{U}_{xx} = f(x) \qquad \mathcal{U}(0) = h \qquad \mathcal{U}(l) = k.$$

Then $v(x, t) = u(x, t) - \mathcal{U}(x)$ solves the problem with zero boundary data, zero right-hand side, and initial data $\phi(x) - \mathcal{U}(x)$ and $\psi(x)$.

For another example, take problem (11) for a simple periodic case:

$$f(x, t) = F(x)\cos\omega t \qquad h(t) = H\cos\omega t \qquad k(t) = K\cos\omega t,$$

that is, with the same time behavior in all the data. We wish to subtract a solution of

$$\mathcal{U}_{tt} - c^2\mathcal{U}_{xx} = F(x)\cos\omega t$$
$$\mathcal{U}(0, t) = H\cos\omega t \qquad \mathcal{U}(l, t) = K\cos\omega t.$$

A good guess is that $\mathcal{U}$ should have the form $\mathcal{U}(x, t) = \mathcal{U}_0(x)\cos\omega t$. This will happen if $\mathcal{U}_0(x)$ satisfies

$$-\omega^2\mathcal{U}_0 - c^2\mathcal{U}_0'' = F(x) \qquad \mathcal{U}_0(0) = H \qquad \mathcal{U}_0(l) = K. \qquad \square$$

There is also the method of Laplace transforms, which can be found in Section 12.5.

## EXERCISES

1.   (a)   Solve as a series the equation $u_t = u_{xx}$ in $(0, 1)$ with $u_x(0, t) = 0$, $u(1, t) = 1$, and $u(x, 0) = x^2$. Compute the first two coefficients explicitly.

    (b)   What is the equilibrium state (the term that does not tend to zero)?

2.   For problem (1), complete the calculation of the series in case $j(t) = 0$ and $h(t) = e^t$.

3.   Repeat problem (1) for the case of Neumann BCs.

4.   Solve $u_{tt} = c^2 u_{xx} + k$ for $0 < x < l$, with the boundary conditions $u(0, t) = 0$, $u_x(l, t) = 0$ and the initial conditions $u(x, 0) = 0$, $u_t(x, 0) = V$. Here $k$ and $V$ are constants.

5.   Solve $u_{tt} = c^2 u_{xx} + e^t \sin 5x$ for $0 < x < \pi$, with $u(0, t) = u(\pi, t) = 0$ and the initial conditions $u(x, 0) = 0$, $u_t(x, 0) = \sin 3x$.

6.  Solve $u_{tt} = c^2 u_{xx} + g(x)\sin\omega t$ for $0 < x < l$, with $u = 0$ at both ends and $u = u_t = 0$ when $t = 0$. For which values of $\omega$ can resonance occur? (Resonance means growth in time.)

7.  Repeat Exercise 6 for the damped wave equation $u_{tt} = c^2 u_{xx} - r u_t + g(x)\sin\omega t$, where $r$ is a positive constant.

8.  Solve $u_t = k u_{xx}$ in $(0, l)$, with $u(0, t) = 0$, $u(l, t) = At$, $u(x, 0) = 0$, where $A$ is a constant.

9.  Use the method of subtraction to solve $u_{tt} = 9 u_{xx}$ for $0 \leq x \leq 1 = l$, with $u(0, t) = h$, $u(1, t) = k$, where $h$ and $k$ are given constants, and $u(x, 0) = 0$, $u_t(x, 0) = 0$.

10. Find the temperature of a metal rod that is in the shape of a solid circular cone with cross-sectional area $A(x) = b(1 - x/l)^2$ for $0 \leq x \leq l$, where $b$ is a constant. Assume that the rod is made of a uniform material, is insulated on its sides, is maintained at zero temperature on its flat end ($x = 0$), and has an unspecified initial temperature distribution $\phi(x)$. Assume that the temperature is independent of $y$ and $z$. [*Hint:* Derive the PDE $(1 - x/l)^2 u_t = k\{(1 - x/l)^2 u_x\}_x$. Separate variables $u = T(t)X(x)$ and then substitute $v(x) = (1 - x/l)X(x)$.]

11. Write out the solution of problem (11) explicitly, starting from the discussion in Section 5.6.

12. Carry out the solution of (11) in the case that

$$f(x, t) = F(x)\cos\omega t \quad h(t) = H\cos\omega t \quad k(t) = K\cos\omega t.$$

13. If friction is present, the wave equation takes the form

$$u_{tt} - c^2 u_{xx} = -r u_t,$$

where the resistance $r > 0$ is a constant. Consider a periodic source at one end: $u(0, t) = 0$, $u(l, t) = Ae^{i\omega t}$.

(a) Show that the PDE and the BC are satisfied by

$$\mathcal{U}(x, t) = Ae^{i\omega t}\frac{\sin\beta x}{\sin\beta l}, \quad \text{where } \beta^2 c^2 = \omega^2 - ir\omega.$$

(b) No matter what the IC, $u(x, 0)$ and $u_t(x, 0)$, are, show that $\mathcal{U}(x, t)$ is the asymptotic form of the solution $u(x, t)$ as $t \to \infty$.

(c) Show that you can get resonance as $r \to 0$ if $\omega = m\pi c/l$ for some integer $m$.

(d) Show that friction can prevent resonance from occurring.

# 6

# HARMONIC FUNCTIONS

This chapter is devoted to the Laplace equation. We introduce two of its important properties, the maximum principle and the rotational invariance. Then we solve the equation in series form in rectangles, circles, and related shapes. The case of a circle leads to the beautiful Poisson formula.

## 6.1   LAPLACE'S EQUATION

If a diffusion or wave process is stationary (independent of time), then $u_t \equiv 0$ and $u_{tt} \equiv 0$. Therefore, both the diffusion and the wave equations reduce to the *Laplace equation*:

$$u_{xx} = 0 \qquad \text{in one dimension}$$
$$\nabla \cdot \nabla u = \Delta u = u_{xx} + u_{yy} = 0 \qquad \text{in two dimensions}$$
$$\nabla \cdot \nabla u = \Delta u = u_{xx} + u_{yy} + u_{zz} = 0 \qquad \text{in three dimensions}$$

A solution of the Laplace equation is called a *harmonic function.*

In *one* dimension, we have simply $u_{xx} = 0$, so the only harmonic functions in one dimension are $u(x) = A + Bx$. But this is so simple that it hardly gives us a clue to what happens in higher dimensions.

The inhomogeneous version of Laplace's equation

$$\boxed{\Delta u = f} \tag{1}$$

with $f$ a given function, is called *Poisson's equation.*

Besides stationary diffusions and waves, some other instances of Laplace's and Poisson's equations include the following.

1. **Electrostatics.** From Maxwell's equations, one has curl $\mathbf{E} = 0$ and div $\mathbf{E} = 4\pi\rho$, where $\rho$ is the charge density. The first equation implies $\mathbf{E} = -\text{grad } \phi$ for a scalar function $\phi$ (called the *electric potential*). Therefore,

$$\Delta\phi = \text{div(grad } \phi) = -\text{div } \mathbf{E} = -4\pi\rho,$$

which is Poisson's equation (with $f = -4\pi\rho$).

2. **Steady fluid flow**. Assume that the flow is irrotational (no eddies) so that curl $\mathbf{v} = 0$, where $\mathbf{v} = \mathbf{v}(x, y, z)$ is the velocity at the position $(x, y, z)$, assumed independent of time. Assume that the fluid is incompressible (e.g., water) and that there are no sources or sinks. Then div $\mathbf{v} = 0$. Hence $\mathbf{v} = -\text{grad } \phi$ for some $\phi$ (called the *velocity potential*) and $\Delta\phi = -\text{div } \mathbf{v} = 0$, which is Laplace's equation.

3. **Analytic functions of a complex variable.** Write $z = x + iy$ and

$$f(z) = u(z) + iv(z) = u(x + iy) + iv(x + iy),$$

where $u$ and $v$ are real-valued functions. An analytic function is one that is expressible as a power series in $z$. This means that the powers are not $x^m y^n$ but $z^n = (x + iy)^n$. Thus

$$f(z) = \sum_{n=0}^{\infty} a_n z^n$$

($a_n$ complex constants). That is,

$$u(x + iy) + iv(x + iy) = \sum_{n=0}^{\infty} a_n(x + iy)^n.$$

Formal differentiation of this series shows that

$$\frac{\partial u}{\partial x} = \frac{\partial v}{\partial y} \quad \text{and} \quad \frac{\partial u}{\partial y} = -\frac{\partial v}{\partial x}$$

(see Exercise 1). These are the *Cauchy–Riemann equations*. If we differentiate them, we find that

$$u_{xx} = v_{yx} = v_{xy} = -u_{yy},$$

so that $\Delta u = 0$. Similarly $\Delta v = 0$, where $\Delta$ is the two-dimensional laplacian. Thus the real and imaginary parts of an analytic function are harmonic.

4. **Brownian motion.** Imagine brownian motion in a container $D$. This means that particles inside $D$ move completely randomly until they hit the boundary, when they stop. Divide the boundary arbitrarily into two pieces, $C_1$ and $C_2$ (see Figure 1). Let $u(x, y, z)$ be the probability that a particle that begins at the point $(x, y, z)$ stops at some point of $C_1$. Then it can be deduced that

$$\Delta u = 0 \text{ in } D$$
$$u = 1 \text{ on } C_1 \qquad u = 0 \text{ on } C_2.$$

Thus $u$ is the solution of a Dirichlet problem.

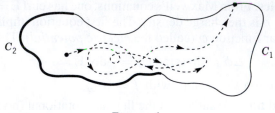

**Figure 1**

As we discussed in Section 1.4, the basic mathematical problem is to solve Laplace's or Poisson's equation in a given domain $D$ with a condition on bdy $D$:

$$\Delta u = f \text{ in } D$$
$$u = h \quad \text{or} \quad \frac{\partial u}{\partial n} = h \quad \text{or} \quad \frac{\partial u}{\partial n} + au = h \quad \text{on bdy } D.$$

In one dimension the only connected domain is an interval $\{a \leq x \leq b\}$. We will see that what is interesting about the two- and three-dimensional cases is the geometry.

## MAXIMUM PRINCIPLE

We begin our analysis with the maximum principle, which is easier for Laplace's equation than for the diffusion equation. By an *open set* we mean a set that includes none of its boundary points (see Section A.1).

**Maximum Principle.**   Let $D$ be a connected bounded open set (in *either* two- *or* three-dimensional space). Let *either* $u(x, y)$ or $u(x, y, z)$ be a harmonic function in $D$ that is continuous on $\overline{D} = D \cup$ (bdy $D$). *Then the maximum and the minimum values of $u$ are attained on* bdy $D$ *and nowhere inside (unless $u \equiv$ constant).*

In other words, a harmonic function is its biggest somewhere on the boundary and its smallest somewhere else on the boundary.

To understand the maximum principle, let us use the vector shorthand $\mathbf{x} = (x, y)$ in two dimensions or $\mathbf{x} = (x, y, z)$ in three dimensions. Also, the radial coordinate is written as $|\mathbf{x}| = (x^2 + y^2)^{1/2}$ or $|\mathbf{x}| = (x^2 + y^2 + z^2)^{1/2}$. The maximum principle asserts that there exist points $\mathbf{x}_M$ and $\mathbf{x}_m$ *on bdy $D$* such that

$$u(\mathbf{x}_m) \leq u(\mathbf{x}) \leq u(\mathbf{x}_M) \tag{2}$$

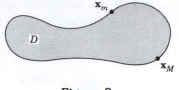

**Figure 2**

for all $\mathbf{x} \in D$ (see Figure 2). Also, there are no points inside $D$ with this property (unless $u \equiv$ constant). There could be several such points *on* the boundary.

The *idea* of the maximum principle is as follows, in two dimensions, say. At a maximum point inside $D$, if there were one, we'd have $u_{xx} \leq 0$ and $u_{yy} \leq 0$. (This is the second derivative test of calculus.) So $u_{xx} + u_{yy} \leq 0$. At most maximum points, $u_{xx} < 0$ and $u_{yy} < 0$. So we'd get a contradiction to Laplace's equation. However, since it *is* possible that $u_{xx} = 0 = u_{yy}$ at a maximum point, we have to work a little harder to get a proof.

Here we go. Let $\epsilon > 0$. Let $v(\mathbf{x}) = u(\mathbf{x}) + \epsilon |\mathbf{x}|^2$. Then, still in two dimensions, say,

$$\Delta v = \Delta u + \epsilon \Delta(x^2 + y^2) = 0 + 4\epsilon > 0 \quad \text{in } D.$$

But $\Delta v = v_{xx} + v_{yy} \leq 0$ at an interior maximum point, by the second derivative test in calculus! Therefore, $v(\mathbf{x})$ has no interior maximum in $D$.

Now $v(\mathbf{x})$, being a continuous function, has to have a maximum *somewhere* in the closure $\overline{D} = D \cup$ bdy $D$. Say that the maximum of $v(\mathbf{x})$ is attained at $\mathbf{x}_0 \in$ bdy $D$. Then, for all $\mathbf{x} \in D$,

$$u(\mathbf{x}) \leq v(\mathbf{x}) \leq v(\mathbf{x}_0) = u(\mathbf{x}_0) + \epsilon |\mathbf{x}_0|^2 \leq \max_{\text{bdy } D} u + \epsilon l^2,$$

where $l$ is the greatest distance from bdy $D$ to the origin. Since this is true for any $\epsilon > 0$, we have

$$u(\mathbf{x}) \leq \max_{\text{bdy } D} u \qquad \text{for all } \mathbf{x} \in D. \tag{3}$$

Now this maximum is attained at some point $\mathbf{x}_M \in$ bdy $D$. So $u(\mathbf{x}) \leq u(\mathbf{x}_M)$ for all $\mathbf{x} \in \overline{D}$, which is the desired conclusion.

The existence of a minimum point $x_m$ is similarly demonstrated. (The absence of such points inside $D$ will be proved by a different method in Section 6.3.) □

## UNIQUENESS OF THE DIRICHLET PROBLEM

To prove the uniqueness, suppose that

$$\Delta u = f \quad \text{in } D \qquad \Delta v = f \quad \text{in } D$$
$$u = h \quad \text{on bdy } D \qquad v = h \quad \text{on bdy } D.$$

We want to show that $u \equiv v$ in $D$. So we simply subtract equations and let $w = u - v$. Then $\Delta w = 0$ in $D$ and $w = 0$ on bdy $D$. By the maximum principle

$$0 = w(\mathbf{x}_m) \leq w(\mathbf{x}) \leq w(\mathbf{x}_M) = 0 \quad \text{for all } \mathbf{x} \in D.$$

Therefore, both the maximum and minimum of $w(\mathbf{x})$ are zero. This means that $w \equiv 0$ and $u \equiv v$.

## INVARIANCE IN TWO DIMENSIONS

The Laplace equation is invariant under all rigid motions. A rigid motion in the plane consists of translations and rotations. A *translation* in the plane is a transformation

$$x' = x + a \qquad y' = y + b.$$

Invariance under translations means simply that $u_{xx} + u_{yy} = u_{x'x'} + u_{y'y'}$.
A *rotation* in the plane through the angle $\alpha$ is given by

$$x' = x \cos \alpha + y \sin \alpha$$
$$y' = -x \sin \alpha + y \cos \alpha. \tag{4}$$

By the chain rule we calculate

$$u_x = u_{x'} \cos \alpha - u_{y'} \sin \alpha$$
$$u_y = u_{x'} \sin \alpha + u_{y'} \cos \alpha$$
$$u_{xx} = (u_{x'} \cos \alpha - u_{y'} \sin \alpha)_{x'} \cos \alpha - (u_{x'} \cos \alpha - u_{y'} \sin \alpha)_{y'} \sin \alpha$$
$$u_{yy} = (u_{x'} \sin \alpha + u_{y'} \cos \alpha)_{x'} \sin \alpha + (u_{x'} \sin \alpha + u_{y'} \cos \alpha)_{y'} \cos \alpha.$$

Adding, we have

$$u_{xx} + u_{yy} = (u_{x'x'} + u_{y'y'})(\cos^2 \alpha + \sin^2 \alpha) + u_{x'y'} \cdot (0)$$
$$= u_{x'x'} + u_{y'y'}.$$

This proves the invariance of the Laplace operator. In engineering the laplacian $\Delta$ is a model for *isotropic* physical situations, in which there is no preferred direction.

The rotational invariance suggests that the two-dimensional laplacian

$$\Delta_2 = \frac{\partial^2}{\partial x^2} + \frac{\partial^2}{\partial y^2}$$

should take a particularly simple form in *polar coordinates*. The transformation

$$x = r \cos \theta \qquad y = r \sin \theta$$

has the jacobian matrix

$$\mathcal{J} = \begin{pmatrix} \dfrac{\partial x}{\partial r} & \dfrac{\partial y}{\partial r} \\ \dfrac{\partial x}{\partial \theta} & \dfrac{\partial y}{\partial \theta} \end{pmatrix} = \begin{pmatrix} \cos\theta & \sin\theta \\ -r\sin\theta & r\cos\theta \end{pmatrix}$$

with the inverse matrix

$$\mathcal{J}^{-1} = \begin{pmatrix} \dfrac{\partial r}{\partial x} & \dfrac{\partial \theta}{\partial x} \\ \dfrac{\partial r}{\partial y} & \dfrac{\partial \theta}{\partial y} \end{pmatrix} = \begin{pmatrix} \cos\theta & \dfrac{-\sin\theta}{r} \\ \sin\theta & \dfrac{\cos\theta}{r} \end{pmatrix}.$$

(Beware, however, that $\partial r/\partial x \neq (\partial x/\partial r)^{-1}$.) So by the chain rule we have

$$\frac{\partial}{\partial x} = \cos\theta\,\frac{\partial}{\partial r} - \frac{\sin\theta}{r}\frac{\partial}{\partial \theta},$$

$$\frac{\partial}{\partial y} = \sin\theta\,\frac{\partial}{\partial r} + \frac{\cos\theta}{r}\frac{\partial}{\partial \theta}.$$

These *operators* are squared to give

$$\frac{\partial^2}{\partial x^2} = \left[\cos\theta\,\frac{\partial}{\partial r} - \frac{\sin\theta}{r}\frac{\partial}{\partial \theta}\right]^2$$

$$= \cos^2\theta\,\frac{\partial^2}{\partial r^2} - 2\left(\frac{\sin\theta\cos\theta}{r}\right)\frac{\partial^2}{\partial r\,\partial\theta}$$

$$+ \frac{\sin^2\theta}{r^2}\frac{\partial^2}{\partial\theta^2} + \frac{2\sin\theta\cos\theta}{r^2}\frac{\partial}{\partial\theta} + \frac{\sin^2\theta}{r}\frac{\partial}{\partial r}$$

$$\frac{\partial^2}{\partial y^2} = \left(\sin\theta\,\frac{\partial}{\partial r} + \frac{\cos\theta}{r}\frac{\partial}{\partial \theta}\right)^2$$

$$= \sin^2\theta\,\frac{\partial^2}{\partial r^2} + 2\left(\frac{\sin\theta\cos\theta}{r}\right)\frac{\partial^2}{\partial r\,\partial\theta}$$

$$+ \frac{\cos^2\theta}{r^2}\frac{\partial^2}{\partial\theta^2} - \frac{2\sin\theta\cos\theta}{r^2}\frac{\partial}{\partial\theta} + \frac{\cos^2\theta}{r}\frac{\partial}{\partial r}.$$

(The last two terms come from differentiation of the coefficients.) Adding these operators, we get (lo and behold!)

$$\Delta_2 = \frac{\partial^2}{\partial x^2} + \frac{\partial^2}{\partial y^2} = \frac{\partial^2}{\partial r^2} + \frac{1}{r}\frac{\partial}{\partial r} + \frac{1}{r^2}\frac{\partial^2}{\partial\theta^2}. \tag{5}$$

It is also natural to look for special *harmonic functions that themselves are rotationally invariant*. In two dimensions this means that we use polar coordinates $(r, \theta)$ and look for solutions depending only on $r$. Thus by (5)

$$0 = u_{xx} + u_{yy} = u_{rr} + \frac{1}{r} u_r$$

if $u$ does not depend on $\theta$. This ordinary differential equation is easy to solve:

$$(ru_r)_r = 0, \qquad ru_r = c_1, \qquad u = c_1 \log r + c_2.$$

The function **log $r$** will play a central role later.

## INVARIANCE IN THREE DIMENSIONS

The three-dimensional laplacian is invariant under all rigid motions in space. To demonstrate its rotational invariance we repeat the preceding proof using vector-matrix notation. Any rotation in three dimensions is given by

$$\mathbf{x}' = B\mathbf{x},$$

where $B$ is an *orthogonal* matrix ($^tBB = B^tB = I$). The laplacian is $\Delta u = \sum_{i=1}^{3} u_{ii} = \sum_{i,j=1}^{3} \delta_{ij} u_{ij}$ where the subscripts on $u$ denote partial derivatives. Therefore,

$$\Delta u = \sum_{k,l} \left( \sum_{i,j} b_{ki} \delta_{ij} b_{lj} \right) u_{k'l'} = \sum_{k,l} \delta_{kl} \, u_{k'l'}$$

$$= \sum_{k} u_{k'k'}$$

because the new coefficient matrix is

$$\sum_{i,j} b_{ki} \delta_{ij} b_{lj} = \sum_{i} b_{ki} b_{li} = (B^tB)_{kl} = \delta_{kl}.$$

So in the primed coordinates $\Delta u$ takes the usual form

$$\Delta u = u_{x'x'} + u_{y'y'} + u_{z'z'}.$$

For the three-dimensional laplacian

$$\Delta_3 = \frac{\partial^2}{\partial x^2} + \frac{\partial^2}{\partial y^2} + \frac{\partial^2}{\partial z^2}$$

it is natural to use *spherical coordinates* $(r, \theta, \phi)$ (see Figure 3). We'll use the notation

$$r = \sqrt{x^2 + y^2 + z^2} = \sqrt{s^2 + z^2}$$
$$s = \sqrt{x^2 + y^2}$$
$$x = s \cos \phi \qquad z = r \cos \theta$$
$$y = s \sin \phi \qquad s = r \sin \theta.$$

(*Watch out:* In some calculus books the letters $\phi$ and $\theta$ are switched.) The calculation, which is a little tricky, is organized as follows. The chain of

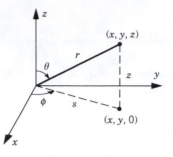

**Figure 3**

variables is $(x, y, z) \to (s, \phi, z) \to (r, \theta, \phi)$. By the two-dimensional Laplace calculation, we have both

$$u_{zz} + u_{ss} = u_{rr} + \frac{1}{r}u_r + \frac{1}{r^2}u_{\theta\theta}$$

and

$$u_{xx} + u_{yy} = u_{ss} + \frac{1}{s}u_s + \frac{1}{s^2}u_{\phi\phi}.$$

We add these two equations, and cancel $u_{ss}$, to get

$$\Delta_3 = u_{xx} + u_{yy} + u_{zz}$$

$$= u_{rr} + \frac{1}{r}u_r + \frac{1}{r^2}u_{\theta\theta} + \frac{1}{s}u_s + \frac{1}{s^2}u_{\phi\phi}.$$

In the last term we substitute $s^2 = r^2\sin^2\theta$ and in the next-to-last term

$$u_s = \frac{\partial u}{\partial s} = u_r\frac{\partial r}{\partial s} + u_\theta\frac{\partial\theta}{\partial s} + u_\phi\frac{\partial\phi}{\partial s}$$

$$= u_r \cdot \frac{s}{r} + u_\theta \cdot \frac{\cos\theta}{r} + u_\phi \cdot 0.$$

This leaves us with

$$\Delta_3 u = u_{rr} + \frac{2}{r}u_r + \frac{1}{r^2}\left[u_{\theta\theta} + (\cot\theta)u_\theta + \frac{1}{\sin^2\theta}u_{\phi\phi}\right], \qquad (6)$$

which may also be written as

$$\Delta_3 = \frac{\partial^2}{\partial r^2} + \frac{2}{r}\frac{\partial}{\partial r} + \frac{1}{r^2\sin\theta}\frac{\partial}{\partial\theta}\sin\theta\frac{\partial}{\partial\theta} + \frac{1}{r^2\sin^2\theta}\frac{\partial^2}{\partial\phi^2}. \qquad (7)$$

Finally, let's look for the special harmonic functions in three dimensions *which don't change under rotations*, that is, which depend only on $r$. By (7) they satisfy the ODE

$$0 = \Delta_3 u = u_{rr} + \frac{2}{r} u_r.$$

So $(r^2 u_r)_r = 0$. It has the solutions $r^2 u_r = c_1$. That is, $u = -c_1 r^{-1} + c_2$. This important harmonic function

$$\frac{1}{r} = (x^2 + y^2 + z^2)^{-1/2}$$

is the analog of the special two-dimensional function $\log(x^2 + y^2)^{1/2}$ found before. Strictly speaking, neither function is finite at the origin. In electrostatics the function $u(\mathbf{x}) = r^{-1}$ turns out to be the electrostatic potential when a unit charge is placed at the origin. For further discussion, see Section 12.2.

## EXERCISES

1. Show that a function which is a power series in the complex variable $x + iy$ must satisfy the Cauchy–Riemann equations and therefore Laplace's equation.

2. Find the solutions that depend only on $r$ of the equation $u_{xx} + u_{yy} + u_{zz} = k^2 u$, where $k$ is a positive constant. (*Hint:* Substitute $u = v/r$.)

3. Find the solutions that depend only on $r$ of the equation $u_{xx} + u_{yy} = k^2 u$, where $k$ is a positive constant. (*Hint:* Look up Bessel's differential equation in [MF] or in Section 10.5.)

4. Solve $u_{xx} + u_{yy} + u_{zz} = 0$ in the spherical shell $0 < a < r < b$ with the boundary conditions $u = A$ on $r = a$ and $u = B$ on $r = b$, where $A$ and $B$ are constants. (*Hint:* Look for a solution depending only on $r$.)

5. Solve $u_{xx} + u_{yy} = 1$ in $r < a$ with $u(x, y)$ vanishing on $r = a$.

6. Solve $u_{xx} + u_{yy} = 1$ in the annulus $a < r < b$ with $u(x, y)$ vanishing on both parts of the boundary $r = a$ and $r = b$.

7. Solve $u_{xx} + u_{yy} + u_{zz} = 1$ in the spherical shell $a < r < b$ with $u(x, y, z)$ vanishing on both the inner and outer boundaries.

8. Solve $u_{xx} + u_{yy} + u_{zz} = 1$ in the spherical shell $a < r < b$ with $u = 0$ on $r = a$ and $\partial u / \partial r = 0$ on $r = b$. Then let $a \to 0$ in your answer and interpret the result.

9. A spherical shell with inner radius 1 and outer radius 2 has a steady-state temperature distribution. Its inner boundary is held at $100°C$. Its outer boundary satisfies $\partial u / \partial r = -\gamma < 0$, where $\gamma$ is a constant.
   (a) Find the temperature. (*Hint:* The temperature depends only on the radius.)
   (b) What are the hottest and coldest temperatures?
   (c) Can you choose $\gamma$ so that the temperature on its outer boundary is $20°C$?

10. Prove the uniqueness of the Dirichlet problem $\Delta u = f$ in $D$, $u = g$ on bdy $D$ by the energy method. That is, after subtracting two solutions $w = u - v$, multiply the Laplace equation for $w$ by $w$ itself and use the divergence theorem.

11. Show that there is no solution of

$$\Delta u = f \quad \text{in } D, \qquad \frac{\partial u}{\partial n} = g \quad \text{on bdy } D$$

in three dimensions, unless

$$\iiint_D f \, dx \, dy \, dz = \iint_{\text{bdy}(D)} g \, dS.$$

(*Hint:* Integrate the equation.) Also show the analogue in one and two dimensions.

12. Check the validity of the maximum principle for the harmonic function $(1 - x^2 - y^2)/(1 - 2x + x^2 + y^2)$ in the disk $\overline{D} = \{x^2 + y^2 \le 1\}$. Explain.

13. A function $u(\mathbf{x})$ is *subharmonic* in $D$ if $\Delta u \ge 0$ in $D$. Prove that its maximum value is attained on bdy $D$. [Note that this is not true for the minimum value.]

## 6.2 RECTANGLES AND CUBES

Special geometries can be solved by separating the variables. The general procedure is the same as in Chapter 4.

(i) Look for separated solutions of the PDE.

(ii) Put in the homogeneous boundary conditions to get the eigenvalues. This is the step that requires the special geometry.

(iii) Sum the series.

(iv) Put in the inhomogeneous initial or boundary conditions.

It is important to do it in this order: homogeneous BC first, inhomogeneous BC last.

We begin with

$$\Delta_2 u = u_{xx} + u_{yy} = 0 \quad \text{in } D, \tag{1}$$

where $D$ is the rectangle $\{0 < x < a, \ 0 < y < b\}$ on each of whose sides one of the standard boundary conditions is prescribed (inhomogeneous Dirichlet, Neumann, or Robin).

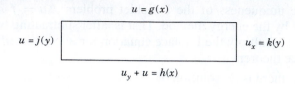

$u = g(x)$

$u = j(y)$ [ ] $u_x = k(y)$

$u_y + u = h(x)$

**Figure 1**

## Example 1.

Solve (1) with the boundary conditions indicated in Figure 1. If we call the solution $u$ with data $(g, h, j, k)$, then $u = u_1 + u_2 + u_3 + u_4$ where $u_1$ has data $(g, 0, 0, 0)$, $u_2$ has data $(0, h, 0, 0)$, and so on. For simplicity, let's assume that $h = 0$, $j = 0$, and $k = 0$, so that we have Figure 2. Now we *separate variables* $u(x, y) = X(x) \cdot Y(y)$. We get

$$\frac{X''}{X} + \frac{Y''}{Y} = 0.$$

Hence there is a constant $\lambda$ such that $X'' + \lambda X = 0$ for $0 \le x \le a$ and $Y'' - \lambda Y = 0$ for $0 \le y \le b$. Thus $X(x)$ satisfies a homogeneous one-dimensional problem which we well know how to solve: $X(0) = X'(a) = 0$. The solutions are

$$\beta_n^2 = \lambda_n = \left(n + \frac{1}{2}\right)^2 \frac{\pi^2}{a^2} \quad (n = 0, 1, 2, 3, \ldots) \tag{2}$$

$$X_n(x) = \sin\frac{(n + \frac{1}{2})\pi x}{a}. \tag{3}$$

Next we look at the $y$ variable. We have

$$Y'' - \lambda Y = 0 \quad \text{with } Y'(0) + Y(0) = 0.$$

(We shall save the *inhomogeneous* BCs for the last step.) From the previous part, we know that $\lambda = \lambda_n > 0$ for some $n$. The $Y$ equation has exponential solutions. As usual it is convenient to write them as

$$Y(y) = A \cosh \beta_n y + B \sinh \beta_n y.$$

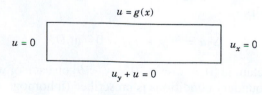

$u = g(x)$

$u = 0$ [ ] $u_x = 0$

$u_y + u = 0$

**Figure 2**

So $0 = Y'(0) + Y(0) = B\beta_n + A$. Without losing any information we may pick $B = -1$, so that $A = \beta_n$. Then

$$Y(y) = \beta_n \cosh \beta_n y - \sinh \beta_n y. \tag{4}$$

Because we're in the rectangle, this function is bounded. Therefore, the sum

$$u(x, y) = \sum_{n=0}^{\infty} A_n \sin \beta_n x \, (\beta_n \cosh \beta_n y - \sinh\beta_n y) \tag{5}$$

is a harmonic function in $D$ that satisfies all three homogeneous BCs. The remaining BC is $u(x, b) = g(x)$. It requires that

$$g(x) = \sum_{n=0}^{\infty} A_n(\beta_n \cosh \beta_n b - \sinh \beta_n b) \cdot \sin \beta_n x$$

for $0 < x < a$. This is simply a Fourier series in the eigenfunctions $\sin \beta_n x$.

By Chapter 5, the coefficients are given by the formula

$$A_n = \frac{2}{a}(\beta_n \cosh \beta_n b - \sinh \beta_n b)^{-1} \int_0^a g(x) \sin \beta_n x \, dx. \tag{6}$$

$\square$

## Example 2.

The same method works for a three-dimensional *box* $\{0 < x < a, 0 < y < b, 0 < z < c\}$ with boundary conditions on the six sides. Take Dirichlet conditions on a cube:

$$\Delta_3 u = u_{xx} + u_{yy} + u_{zz} = 0 \quad \text{in } D$$

$$D = \{0 < x < \pi, 0 < y < \pi, 0 < z < \pi\}$$

$$u(\pi, y, z) = g(y, z)$$

$$u(0, y, z) = u(x, 0, z) = u(x, \pi, z) = u(x, y, 0) = u(x, y, \pi) = 0.$$

To solve this problem we separate variables and use the five *homogeneous* boundary conditions:

$$u = X(x)Y(y)Z(z), \quad \frac{X''}{X} + \frac{Y''}{Y} + \frac{Z''}{Z} = 0$$

$$X(0) = Y(0) = Z(0) = Y(\pi) = Z(\pi) = 0.$$

Each quotient $X''/X$, $Y''/Y$, and $Z''/Z$ must be a constant. In the familiar way, we find

$$Y(y) = \sin my \quad (m = 1, 2, \ldots)$$

and

$$Z(z) = \sin nz \quad (n = 1, 2, \ldots),$$

so that

$$X'' = (m^2 + n^2)X, \quad X(0) = 0.$$

Therefore,

$$X(x) = A \sinh(\sqrt{m^2 + n^2}\, x).$$

Summing up, our complete solution is

$$u(x, y, z) = \sum_{n=1}^{\infty} \sum_{m=1}^{\infty} A_{mn} \sinh(\sqrt{m^2 + n^2}\, x) \sin my \sin nz. \qquad (7)$$

Finally, we plug in our inhomogeneous condition at $x = \pi$:

$$g(y, z) = \sum \sum A_{mn} \sinh(\sqrt{m^2 + n^2}\, \pi) \sin my \sin nz.$$

This is a *double* Fourier sine series in the variables $y$ and $z$! Its theory is similar to that of the single series. In fact, the eigenfunctions $\{\sin my \cdot \sin nz\}$ are mutually orthogonal on the square $\{0 < y < \pi, 0 < z < \pi\}$ (see Exercise 2). Their normalizing constants are

$$\int_0^{\pi} \int_0^{\pi} (\sin my \sin nz)^2 \, dy \, dz = \frac{\pi^2}{4}.$$

Therefore,

$$A_{mn} = \frac{4}{\pi^2 \sinh(\sqrt{m^2 + n^2}\, \pi)} \int_0^{\pi} \int_0^{\pi} g(y, z) \sin my \sin nz \, dy \, dz. \qquad (8)$$

Hence the solutions can be expressed as the doubly infinite series (7) with the coefficients $A_{mn}$. The complete solution to Example 2 is (7) and (8). With such a series, as with a double integral, one has to be careful about the order of summation, although in most cases any order will give the correct answer. □

## EXERCISES

1.  Solve $u_{xx} + u_{yy} = 0$ in the rectangle $0 < x < a$, $0 < y < b$ with the following boundary conditions:

$$u_x = -a \quad \text{on } x = 0 \qquad u_x = 0 \quad \text{on } x = a$$
$$u_y = b \quad \text{on } y = 0 \qquad u_y = 0 \quad \text{on } y = b.$$

(*Hint:* Note that the necessary condition of Exercise 6.1.11 is satisfied. A shortcut is to guess that the solution might be a quadratic polynomial in $x$ and $y$.)

2. Prove that the eigenfunctions $\{\sin my \sin nz\}$ are orthogonal on the square $\{0 < y < \pi, 0 < z < \pi\}$.

3. Find the harmonic function $u(x, y)$ in the square $D = \{0 < x < \pi, 0 < y < \pi\}$ with the boundary conditions:

$$u_y = 0 \quad \text{for } y = 0 \text{ and for } y = \pi, \quad u = 0 \quad \text{for } x = 0 \quad \text{and}$$
$$u = \cos^2 y = \tfrac{1}{2}(1 + \cos 2y) \quad \text{for } x = \pi.$$

4. Find the harmonic function in the square $\{0 < x < 1, 0 < y < 1\}$ with the boundary conditions $u(x, 0) = x$, $u(x, 1) = 0$, $u_x(0, y) = 0$, $u_x(1, y) = y^2$.

5. Solve Example 1 in the case $b = 1$, $g(x) = h(x) = k(x) = 0$ but $j(x)$ an arbitrary function.

6. Solve the following Neumann problem in the cube $\{0 < x < 1, 0 < y < 1, 0 < z < 1\}$: $\Delta u = 0$ with $u_z(x, y, 1) = g(x, y)$ and homogeneous Neumann conditions on the other five faces, where $g(x, y)$ is an arbitrary function with zero average.

7. (a) Find the harmonic function in the semi-infinite strip $\{0 \leq x \leq \pi, 0 \leq y < \infty\}$ that satisfies the "boundary conditions":

$$u(0, y) = u(\pi, y) = 0, \quad u(x, 0) = h(x), \quad \lim_{y \to \infty} u(x, y) = 0.$$

   (b) What would go awry if we omitted the condition at infinity?

## 6.3 POISSON'S FORMULA

A much more interesting case is the *Dirichlet problem for a circle*. The rotational invariance of $\Delta$ provides a hint that the circle is a natural shape for harmonic functions.

Let's consider the problem

$$\boxed{\begin{aligned} u_{xx} + u_{yy} &= 0 && \text{for } x^2 + y^2 < a^2 & (1) \\ u &= h(\theta) && \text{for } x^2 + y^2 = a^2 & (2) \end{aligned}}$$

with radius $a$ and any boundary data $h(\theta)$.

Our method, naturally, is to separate variables in *polar* coordinates: $u = R(r)\,\Theta(\theta)$ (see Figure 1). From (6.1.5) we can write

$$0 = u_{xx} + u_{yy} = u_{rr} + \frac{1}{r}u_r + \frac{1}{r^2}u_{\theta\theta}$$

$$= R''\Theta + \frac{1}{r}R'\Theta + \frac{1}{r^2}R\Theta''.$$

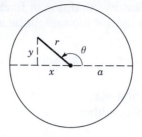

**Figure 1**

Dividing by $R\Theta$ and multiplying by $r^2$, we find that

$$\Theta'' + \lambda\Theta = 0 \tag{3}$$

$$r^2 R'' + rR' - \lambda R = 0. \tag{4}$$

These are ordinary differential equations, easily solved. What boundary conditions do we associate with them?

For $\Theta(\theta)$ we naturally require periodic BCs:

$$\Theta(\theta + 2\pi) = \Theta(\theta) \quad \text{for } -\infty < \theta < \infty. \tag{5}$$

Thus

$$\lambda = n^2 \quad \text{and} \quad \Theta(\theta) = A\cos n\theta + B\sin n\theta \quad (n = 1, 2, \ldots). \tag{6}$$

There is also the solution $\lambda = 0$ with $\Theta(\theta) = A$.

The equation for $R$ is also easy to solve because it is of the Euler type with solutions of the form $R(r) = r^\alpha$. Since $\lambda = n^2$ it reduces to

$$\alpha(\alpha - 1)r^\alpha + \alpha r^\alpha - n^2 r^\alpha = 0 \tag{7}$$

whence $\alpha = \pm n$. Thus $R(r) = Cr^n + Dr^{-n}$ and we have the separated solutions

$$u = \left( Cr^n + \frac{D}{r^n} \right)(A\cos n\theta + B\sin n\theta) \tag{8}$$

for $n = 1, 2, 3, \ldots$. In case $n = 0$, we need a second linearly independent solution of (4) (besides $R = \text{constant}$). It is $R = \log r$, as one learns in ODE courses. So we also have the solutions

$$u = C + D\log r. \tag{9}$$

(They are the same ones we observed back at the beginning of the chapter.)

All of the solutions (8) and (9) we have found are harmonic functions in the disk $D$, except that half of them are infinite at the origin ($r = 0$). But we haven't yet used any boundary condition at all in the $r$ variable. The interval is $0 < r < a$. At $r = 0$ some of the solutions ($r^{-n}$ and $\log r$) are infinite: We reject them. *The requirement that they are finite is the "boundary condition" at* $r = 0$. Summing the remaining solutions, we have

$$u = \tfrac{1}{2}A_0 + \sum_{n=1}^{\infty} r^n (A_n \cos n\theta + B_n \sin n\theta). \tag{10}$$

Finally, we use the inhomogeneous BCs at $r = a$. Setting $r = a$ in the series above, we require that

$$h(\theta) = \tfrac{1}{2}A_0 + \sum_{n=1}^{\infty} a^n (A_n \cos n\theta + B_n \sin n\theta).$$

This is precisely the full Fourier series for $h(\theta)$, so we know that

$$A_n = \frac{1}{\pi a^n} \int_0^{2\pi} h(\phi) \cos n\phi \, d\phi \tag{11}$$

$$B_n = \frac{1}{\pi a^n} \int_0^{2\pi} h(\phi) \sin n\phi \, d\phi. \tag{12}$$

Equations (10) to (12) constitute the full solution of our problem.     □

Now comes an amazing fact. *The series* (10) *can be summed explicitly!* In fact, let's plug (11) and (12) directly into (10) to get

$$u(r, \theta) = \int_0^{2\pi} h(\phi) \frac{d\phi}{2\pi}$$

$$+ \sum_{n=1}^{\infty} \frac{r^n}{\pi a^n} \int_0^{2\pi} h(\phi) \{\cos n\phi \cos n\theta + \sin n\phi \sin n\theta\} \, d\phi$$

$$= \int_0^{2\pi} h(\phi) \left\{ 1 + 2 \sum_{n=1}^{\infty} \left(\frac{r}{a}\right)^n \cos n(\theta - \phi) \right\} \frac{d\phi}{2\pi}.$$

The term in braces is *exactly* the series we summed before in Section 5.5 by writing it as a geometric series of complex numbers; namely,

$$1 + \sum_{n=1}^{\infty} \left(\frac{r}{a}\right)^n e^{in(\theta - \phi)} + \sum_{n=1}^{\infty} \left(\frac{r}{a}\right)^n e^{-in(\theta - \phi)}$$

$$= 1 + \frac{re^{i(\theta - \phi)}}{a - re^{i(\theta - \phi)}} + \frac{re^{-i(\theta - \phi)}}{a - re^{-i(\theta - \phi)}}$$

$$= \frac{a^2 - r^2}{a^2 - 2ar \cos(\theta - \phi) + r^2}.$$

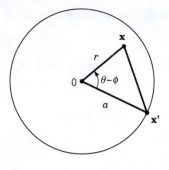

**Figure 2**

Therefore,

$$u(r, \theta) = (a^2 - r^2) \int_0^{2\pi} \frac{h(\phi)}{a^2 - 2ar \cos(\theta - \phi) + r^2} \frac{d\phi}{2\pi}. \qquad (13)$$

This single formula (13), known as *Poisson's formula*, replaces the triple of formulas (10)–(12). It expresses any harmonic function inside a circle in terms of its boundary values.

The Poisson formula can be written in a more geometric way as follows. Write $\mathbf{x} = (x, y)$ as a point with polar coordinates $(r, \theta)$ (see Figure 2). We could also think of $\mathbf{x}$ as the vector from the origin $\mathbf{0}$ to the point $(x, y)$. Let $\mathbf{x}'$ be a point on the boundary.

$$\mathbf{x}: \quad \text{polar coordinates } (r, \theta)$$
$$\mathbf{x}': \quad \text{polar coordinates } (a, \phi).$$

The origin and the points $\mathbf{x}$ and $\mathbf{x}'$ form a triangle with sides $r = |\mathbf{x}|, a = |\mathbf{x}'|$, and $|\mathbf{x} - \mathbf{x}'|$. By the law of cosines

$$|\mathbf{x} - \mathbf{x}'|^2 = a^2 + r^2 - 2ar \cos(\theta - \phi).$$

The arc length element on the circumference is $ds' = a \, d\phi$. Therefore, Poisson's formula takes the alternative form

$$u(\mathbf{x}) = \frac{a^2 - |\mathbf{x}|^2}{2\pi a} \int_{|\mathbf{x}'|=a} \frac{u(\mathbf{x}')}{|\mathbf{x} - \mathbf{x}'|^2} \, ds' \qquad (14)$$

for $\mathbf{x} \in D$, where we write $u(\mathbf{x}') = h(\phi)$. This is a line integral with respect to arc length $ds' = a \, d\phi$, since $s' = a\phi$ for a circle. For instance, in electrostatics this formula (14) expresses the value of the electric potential due to a given distribution of charges on a cylinder that are uniform along the length of the cylinder.

A careful mathematical statement of Poisson's formula is as follows. Its proof is given below, just prior to the exercises.

**Theorem 1.** Let $h(\phi) = u(\mathbf{x}')$ be any continuous function on the circle $C = \text{bdy } D$. Then the Poisson formula (13), or (14), provides the only harmonic function in $D$ for which

$$\lim_{\mathbf{x} \to \mathbf{x}_0} u(\mathbf{x}) = h(\mathbf{x}_0) \qquad \text{for all } \mathbf{x}_0 \in C. \tag{15}$$

This means that $u(\mathbf{x})$ is a continuous function on $\overline{D} = D \cup C$. It is also differentiable to all orders inside $D$.

The Poisson formula has several important consequences. The key one is the following.

## MEAN VALUE PROPERTY

Let $u$ be a harmonic function in a disk $D$, continuous in its closure $\overline{D}$. Then *the value of $u$ at the center of $D$ equals the average of $u$ on its circumference.*

**Proof.** Choose coordinates with the origin $\mathbf{0}$ at the center of the circle. Put $\mathbf{x} = \mathbf{0}$ in Poisson's formula (14), or else put $r = 0$ in (13). Then

$$u(\mathbf{0}) = \frac{a^2}{2\pi a} \int_{|\mathbf{x}'|=a} \frac{u(\mathbf{x}')}{a^2} \, ds'.$$

This is the average of $u$ on the circumference $|\mathbf{x}'| = a$.

## MAXIMUM PRINCIPLE

This was stated and partly proved in Section 6.1. *Here is a complete proof of its strong form.* Let $u(\mathbf{x})$ be harmonic in $D$. The maximum is attained somewhere (by the continuity of $u$ on $\overline{D}$), say at $\mathbf{x}_M \in \overline{D}$. We have to show that $\mathbf{x}_M \notin D$ unless $u \equiv$ constant. By definition of $M$, we know that

$$u(\mathbf{x}) \leq u(\mathbf{x}_M) = M \qquad \text{for all } \mathbf{x} \in D.$$

We draw a circle around $\mathbf{x}_M$ entirely contained in $D$ (see Figure 3). By the mean value property, $u(\mathbf{x}_M)$ is equal to its average around the circumference. Since the average is no greater than the maximum, we have the string of inequalities

$$M = u(\mathbf{x}_M) = \text{ average on circle } \leq M.$$

Therefore, $u(\mathbf{x}) = M$ for all $\mathbf{x}$ on the circumference. This is true for any such circle. So $u(\mathbf{x}) = M$ for all $\mathbf{x}$ in the diagonally shaded region (see Figure 3). Now we repeat the argument with a different center. We can fill the whole domain up with circles. In this way, using the assumption that $D$ is connected, we deduce that $u(\mathbf{x}) \equiv M$ throughout $D$. So $u \equiv$ constant.

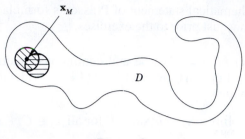

**Figure 3**

## DIFFERENTIABILITY

*Let u be a harmonic function in any open set D of the plane. Then $u(\mathbf{x}) = u(x, y)$ possesses all partial derivatives of all orders in D.*

This means that $\partial u/\partial x$, $\partial u/\partial y$, $\partial^2 u/\partial x^2$, $\partial^2 u/\partial x \partial y$, $\partial^{100} u/\partial x^{100}$, and so on, exist automatically. Let's show this first for the case where $D$ is a disk with its center at the origin. Look at Poisson's formula in its second form (14). The integrand is differentiable to all orders for $\mathbf{x} \in D$. Note that $\mathbf{x}' \in \text{bdy } D$ so that $\mathbf{x} \neq \mathbf{x}'$. By the theorem about differentiating integrals (Section A.3), we can differentiate under the integral sign. So $u(\mathbf{x})$ is differentiable to any order in $D$.

Second, let $D$ be any domain at all, and let $\mathbf{x}_0 \in D$. Let $B$ be a disk contained in $D$ with center at $\mathbf{x}_0$. We just showed that $u(\mathbf{x})$ is differentiable inside $B$, and hence at $\mathbf{x}_0$. But $\mathbf{x}_0$ is an arbitrary point in $D$. So $u$ is differentiable (to all orders) at all points of $D$.

This differentiability property is similar to the one we saw in Section 3.5 for the one-dimensional diffusion equation, but of course it is not at all true for the wave equation.

## PROOF OF THE LIMIT (15)

We begin the proof by writing (13) in the form

$$u(r, \theta) = \int_0^{2\pi} P(r, \theta - \phi) h(\phi) \frac{d\phi}{2\pi} \tag{16}$$

for $r < a$, where

$$P(r, \theta) = \frac{a^2 - r^2}{a^2 - 2ar \cos\theta + r^2} = 1 + 2 \sum_{n=1}^{\infty} \left(\frac{r}{a}\right)^n \cos n\theta \tag{17}$$

is the Poisson kernel. Note that $P$ has the following three properties.

(i)  $P(r, \theta) > 0$ for $r < a$. This property follows from the observation that $a^2 - 2ar \cos\theta + r^2 \geq a^2 - 2ar + r^2 = (a - r)^2 > 0$.

(ii)

$$\int_0^{2\pi} P(r, \theta)\frac{d\theta}{2\pi} = 1.$$

This property follows from the second part of (17) because $\int_0^{2\pi} \cos n\theta \, d\theta = 0$ for $n = 1, 2, \dots$.

(iii)   $P(r, \theta)$ is a harmonic function inside the circle. This property follows from the fact that each term $(r/a)^n \cos n\theta$ in the series is harmonic and therefore so is the sum.

Now we can differentiate under the integral sign (as in Appendix A.3) to get

$$u_{rr} + \frac{1}{r}u_r + \frac{1}{r^2}u_{\theta\theta} = \int_0^{2\pi} \left(P_{rr} + \frac{1}{r}P_r + \frac{1}{r^2}P_{\theta\theta}\right)(r, \theta - \phi)h(\phi)\frac{d\phi}{2\pi}$$

$$= \int_0^{2\pi} 0 \cdot h(\phi)\, d\phi = 0$$

for $r < a$. So $u$ is harmonic in $D$.

So it remains to prove (15). To do that, fix an angle $\theta_0$ and consider a radius $r$ near $a$. Then we will estimate the difference

$$u(r, \theta_0) - h(\theta_0) = \int_0^{2\pi} P(r, \theta_0 - \phi)[h(\phi) - h(\theta_0)]\frac{d\phi}{2\pi} \tag{18}$$

by Property (ii) of $P$. But $P(r, \theta)$ is concentrated near $\theta = 0$. This is true in the precise sense that, for $\delta \le \theta \le 2\pi - \delta$,

$$|P(r, \theta)| = \frac{a^2 - r^2}{a^2 - 2ar\cos\theta + r^2} = \frac{a^2 - r^2}{(a - r)^2 + 4ar\sin^2(\theta/2)} < \epsilon \tag{19}$$

for $r$ sufficiently close to $a$. Precisely, for each (small) $\delta > 0$ and each (small) $\epsilon > 0$, (19) is true for $r$ sufficiently close to $a$. Now from Property (i), (18), and (19), we have

$$|u(r, \theta_0) - h(\theta_0)| \le \int_{\theta_0 - \delta}^{\theta_0 + \delta} P(r, \theta_0 - \phi)\,\epsilon\frac{d\phi}{2\pi} + \epsilon\int_{|\phi - \theta_0| > \delta} |h(\phi) - h(\theta_0)|\frac{d\phi}{2\pi} \tag{20}$$

for $r$ sufficiently close to $a$. The $\epsilon$ in the first integral came from the continuity of $h$. In fact, there is some $\delta > 0$ such that $|h(\phi) - h(\theta_0)| < \epsilon$ for $|\phi - \theta_0| < \delta$. Since the function $|h| \le H$ for some constant $H$, and in view of Property (ii), we deduce from (20) that

$$|u(r, \theta_0) - h(\theta_0)| \le (1 + 2H)\epsilon$$

provided $r$ is sufficiently close to $a$. This is relation (15).

## EXERCISES

1.  Suppose that $u$ is a harmonic function in the disk $D = \{r < 2\}$ and that $u = 3 \sin 2\theta + 1$ for $r = 2$. Without finding the solution, answer the following questions.
    (a)  Find the maximum value of $u$ in $\overline{D}$.
    (b)  Calculate the value of $u$ at the origin.

2.  Solve $u_{xx} + u_{yy} = 0$ in the disk $\{r < a\}$ with the boundary condition
    $$u = 1 + 3 \sin \theta \quad \text{on } r = a.$$

3.  Same for the boundary condition $u = \sin^3 \theta$. (*Hint:* Use the identity $\sin 3\theta = 3 \sin \theta - 4 \sin^3 \theta$.)

4.  Show that $P(r, \theta)$ is a harmonic function in $D$ by using polar coordinates. That is, use (6.1.5) on the first expression in (17).

## 6.4  CIRCLES, WEDGES, AND ANNULI

The technique of separating variables in polar coordinates works for domains whose boundaries are made up of concentric circles and rays. The purpose of this section is to present several examples of this type. In each case we get the expansion as an infinite series. (But summing the series to get a Poisson-type formula is more difficult and works only in special cases.) The geometries we treat here are

> A wedge: $\{0 < \theta < \theta_0, 0 < r < a\}$
> An annulus: $\{0 < a < r < b\}$
> The exterior of a circle: $\{a < r < \infty\}$

We could do Dirichlet, Neumann, or Robin boundary conditions. This leaves us with a lot of possible examples!

### Example 1. The Wedge

Let us take the wedge with three sides $\theta = 0$, $\theta = \beta$, and $r = a$ and solve the Laplace equation with the homogeneous Dirichlet condition on the straight sides and the inhomogeneous Neumann condition on the curved side (see Figure 1). That is, using the notation $u = u(r, \theta)$, the BCs are

$$u(r, 0) = 0 = u(r, \beta), \qquad \frac{\partial u}{\partial r}(a, \theta) = h(\theta). \tag{1}$$

The separation-of-variables technique works just as for the circle, namely,

$$\Theta'' + \lambda \Theta = 0, \qquad r^2 R'' + r R' - \lambda R = 0.$$

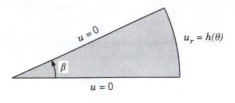

**Figure 1**

So the homogeneous conditions lead to

$$\Theta'' + \lambda\Theta = 0, \qquad \Theta(0) = \Theta(\beta) = 0. \tag{2}$$

This is our standard eigenvalue problem, which has the solutions

$$\lambda = \left(\frac{n\pi}{\beta}\right)^2, \qquad \Theta(\theta) = \sin\frac{n\pi\theta}{\beta} \tag{3}$$

As in Section 6.3, the radial equation

$$r^2 R'' + rR' - \lambda R = 0 \tag{4}$$

is an ODE with the solutions $R(r) = r^\alpha$, where $\alpha^2 - \lambda = 0$ or $\alpha = \pm\sqrt{\lambda} = \pm n\pi/\beta$. The negative exponent is rejected again because we are looking for a solution $u(r, \theta)$ that is continuous in the wedge as well as its boundary: the function $r^{-n\pi/\beta}$ is infinite at the origin (which is a boundary point of the wedge). Thus we end up with the series

$$u(r, \theta) = \sum_{n=1}^{\infty} A_n \, r^{n\pi/\beta} \sin\frac{n\pi\theta}{\beta}. \tag{5}$$

Finally, the inhomogeneous boundary condition requires that

$$h(\theta) = \sum_{n=1}^{\infty} A_n \frac{n\pi}{\beta} a^{-1+n\pi/\beta} \sin\frac{n\pi\theta}{\beta}.$$

This is just a Fourier sine series in the interval $[0, \beta]$, so its coefficients are given by the formula

$$A_n = a^{1-n\pi/\beta} \frac{2}{n\pi} \int_0^\beta h(\theta) \sin\frac{n\pi\theta}{\beta} \, d\theta. \tag{6}$$

The complete solution is given by (5) and (6).  □

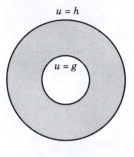

$u = h$

$u = g$

Figure 2

## Example 2. The Annulus

The Dirichlet problem for an annulus (see Figure 2) is

$$u_{xx} + u_{yy} = 0 \qquad \text{in } 0 < a^2 < x^2 + y^2 < b^2$$
$$u = g(\theta) \quad \text{for } x^2 + y^2 = a^2$$
$$u = h(\theta) \quad \text{for } x^2 + y^2 = b^2$$

The separated solutions are just the same as for a circle except that we don't throw out the functions $r^{-n}$ and $\log r$, as these functions are perfectly finite within the annulus. So the solution is

$$u(r, \theta) = \tfrac{1}{2}(C_0 + D_0 \log r) + \sum_{n=1}^{\infty}(C_n r^n + D_n r^{-n})\cos n\theta$$
$$+ (A_n r^n + B_n r^{-n})\sin n\theta. \tag{7}$$

The coefficients are determined by setting $r = a$ and $r = b$ (see Exercise 3). $\qquad \square$

## Example 3. The Exterior of a Circle

The Dirichlet problem for the exterior of a circle (see Figure 3) is

$$u_{xx} + u_{yy} = 0 \qquad \text{for } x^2 + y^2 > a^2$$
$$u = h(\theta) \quad \text{for } x^2 + y^2 = a^2$$

$$u \text{ bounded as } x^2 + y^2 \to \infty.$$

We follow the same reasoning as in the interior case. But now, instead of finiteness at the *origin*, we have imposed boundedness at *infinity*. Therefore, $r^{+n}$ is excluded and $r^{-n}$ is retained. So we have

$$u(r, \theta) = \tfrac{1}{2}A_0 + \sum_{n=1}^{\infty} r^{-n}(A_n \cos n\theta + B_n \sin n\theta). \tag{8}$$

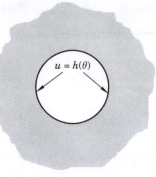

**Figure 3**

The boundary condition means

$$h(\theta) = \tfrac{1}{2}A_0 + \sum a^{-n}(A_n \cos n\theta + B_n \sin n\theta),$$

so that

$$A_n = \frac{a^n}{\pi} \int_{-\pi}^{\pi} h(\theta) \cos n\theta \, d\theta$$

and

$$B_n = \frac{a^n}{\pi} \int_{-\pi}^{\pi} h(\theta) \sin n\theta \, d\theta.$$

This is the complete solution but it is one of the rare cases when the series can actually be summed. Comparing it with the interior case, we see that the only difference between the two sets of formulas is that $r$ and $a$ are replaced by $r^{-1}$ and $a^{-1}$. Therefore, we get Poisson's formula with only this alteration. The result can be written as

$$u(r, \theta) = (r^2 - a^2) \int_0^{2\pi} \frac{h(\phi)}{a^2 - 2ar \cos(\theta - \phi) + r^2} \frac{d\phi}{2\pi} \qquad (9)$$

for $r > a$. $\qquad\qquad\qquad\qquad\qquad\qquad\qquad\qquad\qquad\qquad\qquad$ □

These three examples illustrate the technique of separating variables in polar coordinates. A number of other examples are given in the exercises. What is the most general domain that can be treated by this method?

## EXERCISES

1.  Solve $u_{xx} + u_{yy} = 0$ in the *exterior* $\{r > a\}$ of a disk, with the boundary condition $u = 1 + 3 \sin \theta$ on $r = a$, and the condition at infinity that $u$ be bounded as $r \to \infty$.

2. Solve $u_{xx} + u_{yy} = 0$ in the disk $r < a$ with the boundary condition

$$\frac{\partial u}{\partial r} - hu = f(\theta),$$

where $f(\theta)$ is an arbitrary function. Write the answer in terms of the Fourier coefficients of $f(\theta)$.

3. Determine the coefficients in the annulus problem of the text.

4. Derive Poisson's formula (9) for the exterior of a circle.

5. (a) Find the steady-state temperature distribution inside an annular plate $\{1 < r < 2\}$, whose outer edge ($r = 2$) is insulated, and on whose inner edge ($r = 1$) the temperature is maintained as $\sin^2 \theta$. (Find explicitly all the coefficients, etc.)
   (b) Same, except $u = 0$ on the outer edge.

6. Find the harmonic function $u$ in the semidisk $\{r < 1, 0 < \theta < \pi\}$ with $u$ vanishing on the diameter ($\theta = 0, \pi$) and

$$u = \pi \sin \theta - \sin 2\theta \quad \text{on } r = 1.$$

7. Solve the problem $u_{xx} + u_{yy} = 0$ in $D$, with $u = 0$ on the two straight sides, and $u = h(\theta)$ on the arc, where $D$ is the wedge of Figure 1, that is, a sector of angle $\beta$ cut out of a disk of radius $a$. Write the solution as a series, but don't attempt to sum it.

8. An annular plate with inner radius $a$ and outer radius $b$ is held at temperature $B$ at its outer boundary and satisfies the boundary condition $\partial u / \partial r = A$ at its inner boundary, where $A$ and $B$ are constants. Find the temperature if it is at a steady state. (*Hint:* It satisfies the two-dimensional Laplace equation and depends only on $r$.)

9. Solve $u_{xx} + u_{yy} = 0$ in the wedge $r < a, 0 < \theta < \beta$ with the BCs

$$u = \theta \quad \text{on } r = a, \quad u = 0 \quad \text{on } \theta = 0, \quad \text{and} \quad u = \beta \quad \text{on } \theta = \beta.$$

   (*Hint:* Look for a function independent of $r$.)

10. Solve $u_{xx} + u_{yy} = 0$ in the quarter-disk $\{x^2 + y^2 < a^2, x > 0, y > 0\}$ with the following BCs:

$$u = 0 \quad \text{on } x = 0 \text{ and on } y = 0 \quad \text{and} \quad \frac{\partial u}{\partial r} = 1 \quad \text{on } r = a.$$

   Write the answer as an infinite series and write the first two nonzero terms explicitly.

11. Prove the uniqueness of the Robin problem

$$\Delta u = f \quad \text{in } D, \quad \frac{\partial u}{\partial n} + au = h \quad \text{on bdy } D,$$

   where $D$ is any domain in three dimensions and where $a$ is a positive constant.

12. (a) Prove the following still stronger form of the maximum principle, called the Hopf form of the maximum principle. If $u(\mathbf{x})$ is a nonconstant harmonic function in a connected plane domain $D$ with a smooth boundary that has a maximum at $\mathbf{x}_0$ (necessarily on the boundary by the strong maximum principle), then $\partial u/\partial n > 0$ at $\mathbf{x}_0$ where $\mathbf{n}$ is the unit *outward* normal vector. (This is difficult: see [PW] or [Ev].)

   (b) Use part (a) to deduce the uniqueness of the Neumann problem in a connected domain, up to constants.

13. Solve $u_{xx} + u_{yy} = 0$ in the region $\{\alpha < \theta < \beta, a < r < b\}$ with the boundary conditions $u = 0$ on the two sides $\theta = \alpha$ and $\theta = \beta$, $u = g(\theta)$ on the arc $r = a$, and $u = h(\theta)$ on the arc $r = b$.

14. Answer the last question in the text.

# 7

# GREEN'S IDENTITIES AND GREEN'S FUNCTIONS

The Green's identities for the laplacian lead directly to the maximum principle and to Dirichlet's principle about minimizing the energy. The Green's function is a kind of universal solution for harmonic functions in a domain. All other harmonic functions can be expressed in terms of it. Combined with the method of reflection, the Green's function leads in a very direct way to the solution of boundary problems in special geometries. George Green was interested in the new phenomena of electricity and magnetism in the early 19th century.

## 7.1 GREEN'S FIRST IDENTITY

### NOTATION

In this chapter the divergence theorem and vector notation will be used extensively. Recall the notation (in three dimensions)

$$\text{grad } f = \nabla f = \text{the vector } (f_x, f_y, f_z)$$

$$\text{div } \mathbf{F} = \nabla \cdot \mathbf{F} = \frac{\partial F_1}{\partial x} + \frac{\partial F_2}{\partial y} + \frac{\partial F_3}{\partial z},$$

where $\mathbf{F} = (F_1, F_2, F_3)$ is a vector field. Also,

$$\Delta u = \text{div grad } u = \nabla \cdot \nabla u = u_{xx} + u_{yy} + u_{zz}$$

$$|\nabla u|^2 = |\text{grad } u|^2 = u_x^2 + u_y^2 + u_z^2.$$

Watch out which way you draw the triangle: in physics texts one often finds the laplacian $\nabla \cdot \nabla$ written as $\nabla^2$, but we write it as $\Delta$.

We will write almost everything in this chapter for the three-dimensional case. (However, using two dimensions is okay, too, even $n$ dimensions.) Thus we write

$$\iiint_D \cdots d\mathbf{x} = \iiint_D \cdots dx\,dy\,dz$$

if $D$ is a three-dimensional region (a solid), and

$$\iint_{\text{bdy } D} \cdots dS = \iint_S \cdots dS,$$

where $S = \text{bdy } D$ is the bounding surface for the solid region $D$. Here $dS$ indicates the usual surface integral, as in the calculus.

Our basic tool in this chapter will be the divergence theorem:

$$\boxed{\iiint_D \text{div }\mathbf{F}\,d\mathbf{x} = \iint_{\text{bdy } D} \mathbf{F} \cdot \mathbf{n}\,dS,} \tag{1}$$

where $\mathbf{F}$ is any vector function, $D$ is a bounded solid region, and $\mathbf{n}$ is the unit outer normal on bdy $D$ (see Figure 1) (see Section A.3).

## GREEN'S FIRST IDENTITY

We start from the product rule

$$(vu_x)_x = v_x u_x + vu_{xx}$$

and the same with $y$ and $z$ derivatives. Summing, this leads to the identity

$$\nabla \cdot (v\nabla u) = \nabla v \cdot \nabla u + v\Delta u.$$

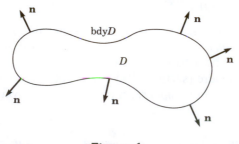

Figure 1

Then we integrate and use the divergence theorem on the left side to get

$$\iint\limits_{\text{bdy } D} v \frac{\partial u}{\partial n} \, dS = \iiint\limits_{D} \nabla v \cdot \nabla u \, d\mathbf{x} + \iiint\limits_{D} v \Delta u \, d\mathbf{x}, \tag{G1}$$

where $\partial u / \partial n = \mathbf{n} \cdot \nabla u$ is the directional derivative in the outward normal direction. This is *Green's first identity*. It is valid for any solid region $D$ and any pair of functions $u$ and $v$. For example, we could take $v \equiv 1$ to get

$$\iint\limits_{\text{bdy } D} \frac{\partial u}{\partial n} \, dS = \iiint\limits_{D} \Delta u \, d\mathbf{x}. \tag{2}$$

As an immediate application of (2), consider the *Neumann problem* in any domain $D$. That is,

$$\begin{cases} \Delta u = f(x) & \text{in } D \\ \dfrac{\partial u}{\partial n} = h(x) & \text{on bdy } D. \end{cases} \tag{3}$$

By (2) we have

$$\iint\limits_{\text{bdy } D} h \, dS = \iiint\limits_{D} f \, d\mathbf{x}. \tag{4}$$

It follows that the data ($f$ and $h$) are *not* arbitrary but are required to satisfy condition (4). Otherwise, there is no solution. In that sense the Neumann problem (3) is not completely well-posed. On the other hand, one can show that if (4) is satisfied, then (3) does have a solution—so the situation is not too bad.

What about uniqueness in problem (3)? Well, you could add any constant to any solution of (3) and still get a solution. So problem (3) lacks uniqueness as well as existence.

## MEAN VALUE PROPERTY

In three dimensions the mean value property states that *the average value of any harmonic function over any sphere equals its value at the center.* To prove this statement, let $D$ be a ball, $\{|\mathbf{x}| < a\}$, say; that is, $\{x^2 + y^2 + z^2 < a^2\}$. Then bdy $D$ is the sphere (surface) $\{|\mathbf{x}| = a\}$. Let $\Delta u = 0$ in any region that contains $D$ and bdy $D$. For a sphere, $\mathbf{n}$ points directly away from the origin, so that

$$\frac{\partial u}{\partial n} = \mathbf{n} \cdot \nabla u = \frac{\mathbf{x}}{r} \cdot \nabla u = \frac{x}{r} u_x + \frac{y}{r} u_y + \frac{z}{r} u_z = \frac{\partial u}{\partial r},$$

where $r = (x^2 + y^2 + z^2)^{1/2} = |\mathbf{x}|$ is the spherical coordinate, the distance of the point $(x, y, z)$ from the center $\mathbf{0}$ of the sphere. Therefore, (2) becomes

$$\iint_{\text{bdy } D} \frac{\partial u}{\partial r} \, dS = 0. \tag{5}$$

Let's write this integral in spherical coordinates, $(r, \theta, \phi)$. Explicitly, (5) takes the form

$$\int_0^{2\pi} \int_0^\pi u_r(a, \theta, \phi) \, a^2 \sin \theta \, d\theta \, d\phi = 0$$

since $r = a$ on bdy $D$. We divide this by the constant $4\pi a^2$ (the area of bdy $D$). This result is valid for all $a > 0$, so that we can think of $a$ as a variable and call it $r$. Then we pull $\partial/\partial r$ outside the integral (see Section A.3), obtaining

$$\frac{\partial}{\partial r} \left[ \frac{1}{4\pi} \int_0^{2\pi} \int_0^\pi u(r, \theta, \phi) \sin \theta \, d\theta \, d\phi \right] = 0.$$

Thus

$$\frac{1}{4\pi} \int_0^{2\pi} \int_0^\pi u(r, \theta, \phi) \sin \theta \, d\theta \, d\phi$$

is *independent of $r$*. This expression is precisely the average value of $u$ on the sphere $\{|\mathbf{x}| = r\}$. In particular, if we let $r \to 0$, we get

$$\frac{1}{4\pi} \int_0^{2\pi} \int_0^\pi u(\mathbf{0}) \sin \theta \, d\theta \, d\phi = u(\mathbf{0}).$$

That is,

$$\boxed{\frac{1}{\text{area of } S} \iint_S u \, dS = u(\mathbf{0}).} \tag{6}$$

This proves the mean value property in three dimensions.

Actually, the same idea works in $n$ dimensions. For $n = 2$ recall that we found another proof in Section 6.3 by a completely different method.

## MAXIMUM PRINCIPLE

Exactly as in two dimensions in Section 6.3, we deduce from the mean value property the maximum principle.

*If $D$ is any solid region, a nonconstant harmonic function in $D$ cannot take its maximum value inside $D$, but only on* bdy $D$.

It can also be shown that the outward normal derivative $\partial u/\partial n$ is strictly positive at a maximum point: $\partial u/\partial n > 0$ there. The last assertion is called the Hopf maximum principle. For a proof, see [PW].

## UNIQUENESS OF DIRICHLET'S PROBLEM

We gave one proof in Section 6.1 using the maximum principle. Now we give another proof by the *energy method*. If we have two harmonic functions $u_1$ and $u_2$ with the same boundary data, then their difference $u = u_1 - u_2$ is harmonic and has zero boundary data. We go back to (G1) and substitute $v = u$. Since $u$ is harmonic, we have $\Delta u = 0$ and

$$\iint_{\text{bdy } D} u \frac{\partial u}{\partial n} \, dS = \iiint_D |\nabla u|^2 \, d\mathbf{x}. \tag{7}$$

Since $u = 0$ on bdy $D$, the left side of (7) vanishes. Therefore, $\iiint_D |\nabla u|^2 d\mathbf{x} = 0$. By the first vanishing theorem in Section A.1, it follows that $|\nabla u|^2 \equiv 0$ in $D$. Now a function with vanishing gradient must be a constant (provided that $D$ is connected). So $u(\mathbf{x}) \equiv C$ throughout $D$. But $u$ vanishes somewhere (on bdy $D$), so $C$ must be 0. Thus $u(\mathbf{x}) \equiv 0$ in $D$. This proves the uniqueness of the Dirichlet problem.

*Uniqueness of Neumann's problem*: If $\Delta u = 0$ in $D$ and $\partial u / \partial n = 0$ on bdy $D$, then $u$ is a constant in $D$ (see Exercise 2).

## DIRICHLET'S PRINCIPLE

This is an important mathematical theorem based on the physical idea of energy. It states that among *all* the functions $w(\mathbf{x})$ in $D$ that satisfy the Dirichlet boundary condition

$$w = h(\mathbf{x}) \quad \text{on bdy } D, \tag{8}$$

the *lowest energy* occurs for the *harmonic* function satisfying (8).

In the present context the energy is defined as

$$\boxed{E[w] = \tfrac{1}{2} \iiint_D |\nabla w|^2 \, d\mathbf{x}.} \tag{9}$$

This is the pure potential energy, there being no kinetic energy because there is no motion. Now it is a general principle in physics that any system prefers to go to the state of lowest energy, called the *ground state*. Thus the harmonic function is the preferred physical stationary state. Mathematically, Dirichlet's principle can be stated precisely as follows:

*Let $u(\mathbf{x})$ be the unique harmonic function in $D$ that satisfies (8). Let $w(\mathbf{x})$ be any function in $D$ that satisfies (8). Then*

$$E[w] \geq E[u]. \tag{10}$$

To prove Dirichlet's principle, we let $v = u - w$ and expand the square in the integral

$$E[w] = \tfrac{1}{2} \iiint_D |\nabla(u - v)|^2 \, d\mathbf{x}$$

$$= E[u] - \iiint_D \nabla u \cdot \nabla v \, d\mathbf{x} + E[v]. \tag{11}$$

Next we apply Green's first identity (G1) to the pair of functions $u$ and $v$. In (G1) two of the three terms are zero because $v = 0$ on bdy $D$ and $\Delta u = 0$ in $D$. Therefore, the middle term in (11) is also zero. Thus

$$E[w] = E[u] + E[v].$$

Since it is obvious that $E[v] \geq 0$, we deduce that $E[w] \geq E[u]$. This means that the energy is smallest when $w = u$. This proves Dirichlet's principle.

An alternative proof goes as follows. Let $u(\mathbf{x})$ be a function that satisfies (8) and minimizes the energy (9). Let $v(\mathbf{x})$ be any function that vanishes on bdy $D$. Then $u + \epsilon v$ satisfies the boundary condition (8). So if the energy is smallest for the function $u$, we have

$$E[u] \leq E[u + \epsilon v] = E[u] - \epsilon \iiint_D \Delta u \, v \, d\mathbf{x} + \epsilon^2 E[v] \tag{12}$$

for any constant $\epsilon$. The minimum occurs for $\epsilon = 0$. By calculus,

$$\iiint_D \Delta u \, v \, d\mathbf{x} = 0. \tag{13}$$

This is valid for practically *all* functions $v$ in $D$. Let $D'$ be any strict subdomain of $D$; that is, $\overline{D'} \subset D$. Let $v(\mathbf{x}) \equiv 1$ for $\mathbf{x} \in D'$ and $v(\mathbf{x}) \equiv 0$ for $\mathbf{x} \in D - D'$. In (13) we choose this function $v$. (Because this $v$ is not smooth, an approximation argument is required that is omitted here.) Then (13) takes the form

$$\iiint_{D'} \Delta u \, d\mathbf{x} = 0 \qquad \text{for all } D'.$$

By the second vanishing theorem in Section A.1, it follows that $\Delta u = 0$ in $D$. Thus $u(\mathbf{x})$ is a harmonic function. By uniqueness, it is the only function satisfying (8) that can minimize the energy.

## EXERCISES

1. Derive the three-dimensional maximum principle from the mean value property.

2. Prove the uniqueness up to constants of the Neumann problem using the energy method.

3. Prove the uniqueness of the Robin problem $\partial u / \partial n + a(\mathbf{x}) u(\mathbf{x}) = h(\mathbf{x})$ provided that $a(\mathbf{x}) > 0$ on the boundary.

4. Generalize the energy method to prove uniqueness for the diffusion equation with Dirichlet boundary conditions in three dimensions.

5. Prove Dirichlet's principle for the Neumann boundary condition. It asserts that among *all* real-valued functions $w(\mathbf{x})$ on $D$ the quantity

$$E[w] = \tfrac{1}{2} \iiint\limits_{D} |\nabla w|^2 \, d\mathbf{x} - \iint\limits_{\text{bdy } D} hw \, dS$$

is the smallest for $w = u$, where $u$ is the solution of the Neumann problem

$$-\Delta u = 0 \quad \text{in } D, \qquad \frac{\partial u}{\partial n} = h(\mathbf{x}) \quad \text{on bdy } D.$$

It is required to assume that the average of the given function $h(\mathbf{x})$ is zero (by Exercise 6.1.11).

   Notice three features of this principle:
   (i)   There is *no constraint at all* on the trial functions $w(\mathbf{x})$.
   (ii)  The function $h(\mathbf{x})$ appears in the energy.
   (iii) The functional $E[w]$ does not change if a constant is added to $w(\mathbf{x})$.
   (*Hint:* Follow the method in Section 7.1.)

6. Let $A$ and $B$ be two disjoint bounded spatial domains, and let $D$ be their exterior. So bdy $D =$ bdy $A \cup$ bdy $B$. Consider a harmonic function $u(\mathbf{x})$ in $D$ that tends to zero at infinity, which is *constant* on bdy $A$ and *constant* on bdy $B$, and which satisfies

$$\iint\limits_{\text{bdy } A} \frac{\partial u}{\partial n} \, dS = Q > 0 \qquad \text{and} \qquad \iint\limits_{\text{bdy } B} \frac{\partial u}{\partial n} \, dS = 0.$$

[*Interpretation:* The harmonic function $u(\mathbf{x})$ is the electrostatic potential of two conductors, $A$ and $B$; $Q$ is the charge on $A$, while $B$ is uncharged.]
   (a) Show that the solution is unique. (*Hint:* Use the Hopf maximum principle.)
   (b) Show that $u \geq 0$ in $D$. [*Hint:* If not, then $u(\mathbf{x})$ has a negative minimum. Use the Hopf principle again.]
   (c) Show that $u > 0$ in $D$.

7. (*Rayleigh-Ritz approximation* to the harmonic function $u$ in $D$ with $u = h$ on bdy $D$.) Let $w_0, w_1, \ldots, w_n$ be arbitrary functions such that $w_0 = h$ on bdy $D$ and $w_1 = \cdots = w_n = 0$ on bdy $D$. The problem is to find

constants $c_1, \ldots, c_n$ so that

$$w_0 + c_1 w_1 + \cdots + c_n w_n \quad \text{has the least possible energy.}$$

Show that the constants must solve the linear system

$$\sum_{k=1}^{n} (\nabla w_j, \nabla w_k) c_k = -(\nabla w_0, \nabla w_j) \quad \text{for } j = 1, 2, \ldots, n.$$

8. Consider the problem $u_{xx} + u_{yy} = 0$ in the triangle $\{x > 0,\, y > 0,\, 3x + y < 3\}$ with the boundary conditions

$$u(x, 0) = 0 \qquad u(0, y) = y(3 - y) \qquad u(x, 3 - 3x) = 0$$

Choose $w_0 = y(3 - 3x - y)$ and $w_1 = xy(3 - 3x - y)$. Find the Rayleigh–Ritz approximation $w_0 + c_1 w_1$ to $u$. That is, use Exercise 7 to find the constant $c_1$.

9. Repeat Exercise 8 with the same choice of $w_0$ and $w_1$ and with $w_2 = x^2 y(3 - 3x - y)$. That is, find the Rayleigh–Ritz approximation $w_0 + c_1 w_1 + c_2 w_2$ to $u$.

10. Let $u(x, y)$ be the harmonic function in the unit disk with the boundary values $u(x, y) = x^2$ on $\{x^2 + y^2 = 1\}$. Find its Rayleigh–Ritz approximation of the form $x^2 + c_1(1 - x^2 - y^2)$.

## 7.2 GREEN'S SECOND IDENTITY

Green's second identity is the higher-dimensional version of the identity (5.3.3). It leads to a basic representation formula for harmonic functions that we require in the next section.

The middle term in (G1) does not change if $u$ and $v$ are switched. So if we write (G1) for the pair $u$ and $v$, and again for the pair $v$ and $u$, and then subtract, we get

$$\iiint_D (u \, \Delta v - v \, \Delta u) \, d\mathbf{x} = \iint_{\text{bdy } D} \left( u \frac{\partial v}{\partial n} - v \frac{\partial u}{\partial n} \right) dS. \tag{G2}$$

This is *Green's second identity*. Just like (G1), it is valid for any pair of functions $u$ and $v$.

It leads to the following natural definition. A boundary condition is called *symmetric* for the operator $\Delta$ if the right side of (G2) vanishes for all pairs of functions $u$, $v$ that satisfy the boundary condition. Each of the three classical boundary conditions (Dirichlet, Neumann, and Robin) is symmetric.

## REPRESENTATION FORMULA

This formula represents any *harmonic* function as an integral over the boundary. It states the following: *If $\Delta u = 0$ in $D$, then*

$$u(\mathbf{x}_0) = \iint\limits_{\text{bdy } D} \left[ -u(\mathbf{x}) \frac{\partial}{\partial n}\left(\frac{1}{|\mathbf{x} - \mathbf{x}_0|}\right) + \frac{1}{|\mathbf{x} - \mathbf{x}_0|} \frac{\partial u}{\partial n} \right] \frac{dS}{4\pi} \tag{1}$$

What is involved here is the same fundamental radial solution $r^{-1}$ that we found in Section 6.1, but translated by the vector $\mathbf{x}_0$.

**Proof of (1).** The representation formula (1) is the special case of (G2) with the choice $v(\mathbf{x}) = (-4\pi |\mathbf{x} - \mathbf{x}_0|)^{-1}$. Clearly, the right side of (G2) agrees with (1). Also, $\Delta u = 0$ and $\Delta v = 0$, which kills the left side of (G2). So where does the left side of (1) come from? From the fact that the function $v(\mathbf{x})$ is *infinite* at the point $\mathbf{x}_0$. Therefore, it is forbidden to apply (G2) in the whole domain $D$. So let's take a pair of scissors and cut out a small ball around $\mathbf{x}_0$. Let $D_\epsilon$ be the region $D$ with this ball (of radius $\epsilon$ and center $\mathbf{x}_0$) excised (see Figure 1).

For simplicity let $\mathbf{x}_0$ be the origin. Then $v(\mathbf{x}) = -1/(4\pi r)$, where $r = (x^2 + y^2 + z^2)^{1/2} = |\mathbf{x}|$. Writing down (G2) with this choice of $v$, we have, since $\Delta u = 0 = \Delta v$ in $D_\epsilon$,

$$-\iint\limits_{\text{bdy } D_\epsilon} \left[ u \cdot \frac{\partial}{\partial n}\left(\frac{1}{r}\right) - \frac{\partial u}{\partial n} \cdot \frac{1}{r} \right] dS = 0.$$

But bdy $D_\epsilon$ consists of two parts: the original boundary bdy $D$ and the sphere $\{r = \epsilon\}$. On the sphere, $\partial/\partial n = -\partial/\partial r$. Thus the surface integral breaks into two pieces,

$$-\iint\limits_{\text{bdy } D} \left[ u \cdot \frac{\partial}{\partial n}\left(\frac{1}{r}\right) - \frac{\partial u}{\partial n} \cdot \frac{1}{r} \right] dS = -\iint\limits_{r=\epsilon} \left[ u \cdot \frac{\partial}{\partial r}\left(\frac{1}{r}\right) - \frac{\partial u}{\partial r} \cdot \frac{1}{r} \right] dS. \tag{2}$$

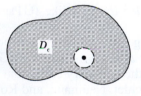

Figure 1

This identity (2) is valid for any small $\epsilon > 0$. Our representation formula (1) would follow provided that we could show that the right side of (2) tended to $4\pi u(\mathbf{0})$ as $\epsilon \to 0$.

Now, on the little spherical surface $\{r = \epsilon\}$, we have

$$\frac{\partial}{\partial r}\left(\frac{1}{r}\right) = -\frac{1}{r^2} = -\frac{1}{\epsilon^2},$$

so that the right side of (2) equals

$$\frac{1}{\epsilon^2}\iint_{r=\epsilon} u \, dS + \frac{1}{\epsilon}\iint_{r=\epsilon} \frac{\partial u}{\partial r} \, dS = 4\pi\bar{u} + 4\pi\epsilon\frac{\overline{\partial u}}{\partial r}, \tag{3}$$

where $\bar{u}$ denotes the average value of $u(\mathbf{x})$ on the sphere $|\mathbf{x}| = r = \epsilon$, and $\overline{\partial u/\partial r}$ denotes the average value of $\partial u/\partial n$ on this sphere. As $\epsilon \to 0$, the expression (3) approaches

$$4\pi u(\mathbf{0}) + 4\pi \cdot 0 \cdot \frac{\partial u}{\partial r}(\mathbf{0}) = 4\pi u(\mathbf{0}) \tag{4}$$

because $u$ is continuous and $\partial u/\partial r$ is bounded. Thus (2) turns into (1), and this completes the proof.

The corresponding formula in *two dimensions* is

$$\boxed{u(\mathbf{x_0}) = \frac{1}{2\pi}\int_{\text{bdy } D}\left[u(\mathbf{x})\frac{\partial}{\partial n}(\log|\mathbf{x} - \mathbf{x_0}|) - \frac{\partial u}{\partial n}\log|\mathbf{x} - \mathbf{x_0}|\right]ds} \tag{5}$$

whenever $\Delta u = 0$ in a plane domain $D$ and $\mathbf{x_0}$ is a point within $D$. The right side is a line integral over the boundary curve with respect to arc length. Log denotes the natural logarithm and $ds$ the arc length on the bounding curve.

## EXERCISES

1.  Derive the representation formula for harmonic functions (7.2.5) in two dimensions.

2.  Let $\phi(\mathbf{x})$ be any $C^2$ function defined on all of three-dimensional space that vanishes outside some sphere. Show that

    $$\phi(\mathbf{0}) = -\iiint \frac{1}{|\mathbf{x}|}\Delta\phi(\mathbf{x})\frac{d\mathbf{x}}{4\pi}.$$

    The integration is taken over the region where $\phi(\mathbf{x})$ is not zero.

3.  Give yet another derivation of the mean value property in three dimensions by choosing $D$ to be a ball and $\mathbf{x_0}$ its center in the representation formula (1).

## 7.3    GREEN'S FUNCTIONS

We now use Green's identities to study the Dirichlet problem. The representation formula (7.2.1) used exactly two properties of the function $v(\mathbf{x}) = (-4\pi |\mathbf{x} - \mathbf{x}_0|)^{-1}$: that it is harmonic except at $\mathbf{x}_0$ and that it has a certain singularity there. Our goal is to modify this function so that one of the terms in (7.2.1) disappears. The modified function is called the Green's function for $D$.

**Definition.**    The *Green's function* $G(\mathbf{x})$ for the operator $-\Delta$ and the domain $D$ at the point $\mathbf{x}_0 \in D$ is a function defined for $\mathbf{x} \in D$ such that:

    (i)   $G(\mathbf{x})$ possesses continuous second derivatives and $\Delta G = 0$ in $D$, except at the point $\mathbf{x} = \mathbf{x}_0$.

    (ii)   $G(\mathbf{x}) = 0$ for $x \in$ bdy $D$.

    (iii)   The function $G(\mathbf{x}) + 1/(4\pi |\mathbf{x} - \mathbf{x}_0|)$ is finite at $\mathbf{x}_0$ and has continuous second derivatives everywhere and is harmonic at $\mathbf{x}_0$.

It can be shown that a Green's function exists. Also, it is unique by Exercise 1. The usual notation for the Green's function is $G(\mathbf{x}, \mathbf{x}_0)$.

**Theorem 1.**    If $G(\mathbf{x}, \mathbf{x}_0)$ is the Green's function, then the solution of the Dirichlet problem is given by the formula

$$u(\mathbf{x}_0) = \iint\limits_{\text{bdy } D} u(\mathbf{x}) \frac{\partial G(\mathbf{x}, \mathbf{x}_0)}{\partial n} \, dS. \tag{1}$$

**Proof.**    Let us go back to the representation formula (7.2.1):

$$u(\mathbf{x}_0) = \iint\limits_{\text{bdy } D} \left( u \frac{\partial v}{\partial n} - \frac{\partial u}{\partial n} v \right) dS, \tag{2}$$

where $v(\mathbf{x}) = -(4\pi |\mathbf{x} - \mathbf{x}_0|)^{-1}$, as before. Now let's write $G(\mathbf{x}, \mathbf{x}_0) = v(\mathbf{x}) + H(\mathbf{x})$. [This is the definition of $H(\mathbf{x})$.] Then $H(\mathbf{x})$ is a harmonic function throughout the domain $D$ [by (iii) and (i)]. We apply Green's second identity (G2) to the pair of harmonic functions $u(\mathbf{x})$ and $H(\mathbf{x})$:

$$0 = \iint\limits_{\text{bdy } D} \left( u \frac{\partial H}{\partial n} - \frac{\partial u}{\partial n} H \right) dS. \tag{3}$$

Adding (2) and (3), we get

$$u(\mathbf{x}_0) = \iint_{\text{bdy } D} \left( u \frac{\partial G}{\partial n} - \frac{\partial u}{\partial n} G \right) dS.$$

But by (ii), $G$ vanishes on bdy $D$, so the last term vanishes and we end up with formula (1).  □

The only thing wrong with this beautiful formula is that it is not usually easy to find $G$ explicitly. Nevertheless, in the next section we'll see how to use the reflection method to find $G$ in some situations and thereby solve the Dirichlet problem for some special geometries.

## SYMMETRY OF THE GREEN'S FUNCTION

For any region $D$ we have a Green's function $G(\mathbf{x}, \mathbf{x}_0)$. It is always symmetric:

$$G(\mathbf{x}, \mathbf{x}_0) = G(\mathbf{x}_0, \mathbf{x}) \qquad \text{for } \mathbf{x} \neq \mathbf{x}_0. \tag{4}$$

In order to prove (4), we apply Green's second identity (G2) to the pair of functions $u(\mathbf{x}) = G(\mathbf{x}, \mathbf{a})$ and $v(\mathbf{x}) = G(\mathbf{x}, \mathbf{b})$ and to the domain $D_\epsilon$. By $D_\epsilon$ we denote the domain $D$ with two little spheres of radii $\epsilon$ cut out around the points $\mathbf{a}$ and $\mathbf{b}$ (see Figure 1). So the boundary of $D_\epsilon$ consists of three parts: the original boundary bdy $D$ and the two spheres $|\mathbf{x} - \mathbf{a}| = \epsilon$ and $|\mathbf{x} - \mathbf{b}| = \epsilon$. Thus

$$\iiint_{D_\epsilon} (u \Delta v - v \Delta u) \, d\mathbf{x} = \iint_{\text{bdy } D} \left( u \frac{\partial v}{\partial n} - v \frac{\partial u}{\partial n} \right) dS + A_\epsilon + B_\epsilon, \tag{5}$$

where

$$A_\epsilon = \iint_{|\mathbf{x}-\mathbf{a}|=\epsilon} \left( u \frac{\partial v}{\partial n} - v \frac{\partial u}{\partial n} \right) dS$$

and $B_\epsilon$ is given by the same formula at $\mathbf{b}$. Because both $u$ and $v$ are harmonic in $D_\epsilon$, the left side of (5) vanishes. Since both $u$ and $v$ vanish on bdy $D$, the integral over bdy $D$ also vanishes. Therefore,

$$A_\epsilon + B_\epsilon = 0 \quad \text{for each } \epsilon.$$

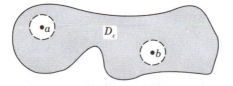

Figure 1

Let's calculate the limits as $\epsilon \to 0$. We shall then have $\lim A_\epsilon + \lim B_\epsilon = 0$. For $A_\epsilon$, denote $r = |\mathbf{x} - \mathbf{a}|$. Then

$$\lim_{\epsilon \to 0} A_\epsilon = \lim_{\epsilon \to 0} \iint_{r=\epsilon} \left\{ \left( -\frac{1}{4\pi r} + H \right) \frac{\partial v}{\partial n} - v \frac{\partial}{\partial n} \left( -\frac{1}{4\pi r} + H \right) \right\} r^2 \sin\theta \, d\theta \, d\phi$$

where $\theta$ and $\phi$ are the spherical angles for $\mathbf{x} - \mathbf{a}$, and $H$ is a continuous function. Now $\partial/\partial n = -\partial/\partial r$ for the sphere. Among the four terms in the last integrand, only the third one contributes a nonzero expression to the limit [for the same reason as in the derivation of (7.2.1)]. Thus

$$\lim_{\epsilon \to 0} A_\epsilon = \lim_{\epsilon \to 0} \int_0^{2\pi} \int_0^{\pi} v \frac{1}{4\pi\epsilon^2} \epsilon^2 \sin\theta \, d\theta \, d\phi = v(\mathbf{a})$$

by cancellation of the $\epsilon^2$. A quite similar calculation shows that $\lim B_\epsilon = -u(\mathbf{b})$. Therefore,

$$0 = \lim(A_\epsilon + B_\epsilon) = v(\mathbf{a}) - u(\mathbf{b}) = G(\mathbf{a}, \mathbf{b}) - G(\mathbf{b}, \mathbf{a}).$$

This proves the symmetry (4).    □

In electrostatics, $G(\mathbf{x}, \mathbf{x}_0)$ is interpreted as the electric potential inside a conducting surface $S = $ bdy $D$ due to a charge at a single point $\mathbf{x}_0$. The symmetry (4) is known as the *principle of reciprocity*. It asserts that a source located at the point $\mathbf{a}$ produces at the point $\mathbf{b}$ the same effect as a source at $\mathbf{b}$ would produce at $\mathbf{a}$.

The Green's function also allows us to solve *Poisson's equation*.

**Theorem 2.**    The solution of the problem

$$\Delta u = f \quad \text{in } D \qquad u = h \quad \text{on bdy } D \tag{6}$$

is given by

$$u(\mathbf{x}_0) = \iint_{\text{bdy } D} h(\mathbf{x}) \frac{\partial G(\mathbf{x}, \mathbf{x}_0)}{\partial n} \, dS + \iiint_D f(\mathbf{x}) G(\mathbf{x}, \mathbf{x}_0) \, d\mathbf{x}. \tag{7}$$

The proof is left as an exercise.

## EXERCISES

1.  Show that the Green's function is unique. (*Hint:* Take the difference of two of them.)

2.  Prove Theorem 2, which gives the solution of Poisson's equation in terms of the Green's function.

3.  Verify the limit of $A_\epsilon$ as claimed in the proof of the symmetry of the Green's function.

## 7.4  HALF-SPACE AND SPHERE

We solve for the harmonic functions in a half-space and a sphere by combining the Green's function with the method of reflection.

### THE HALF-SPACE

We first determine the Green's function for a half-space. A half-space is the region lying on one side of a plane. Although it is an infinite domain, all the ideas involving Green's functions are still valid if we impose the "boundary condition at infinity" that the functions and their derivatives tend to 0 as $|\mathbf{x}| \to \infty$.

We write the coordinates as $\mathbf{x} = (x, y, z)$. Say that the half-space is $D = \{z > 0\}$, the domain that lies above the $xy$ plane (see Figure 1). Each point $\mathbf{x} = (x, y, z)$ in $D$ has a *reflected point* $\mathbf{x}^* = (x, y, -z)$ that is not in $D$.

Now we already know that the function $1/(4\pi|\mathbf{x} - \mathbf{x}_0|)$ satisfies two of the three conditions—(i) and (iii)—required of the Green's function: We want to modify it to get (ii) as well.

We assert that the Green's function for $D$ is

$$G(\mathbf{x}, \mathbf{x}_0) = -\frac{1}{4\pi|\mathbf{x} - \mathbf{x}_0|} + \frac{1}{4\pi|\mathbf{x} - \mathbf{x}_0^*|}. \tag{1}$$

In coordinates,

$$G(\mathbf{x}, \mathbf{x}_0) = -\frac{1}{4\pi}[(x - x_0)^2 + (y - y_0)^2 + (z - z_0)^2]^{-1/2}$$

$$+ \frac{1}{4\pi}[(x - x_0)^2 + (y - y_0)^2 + (z + z_0)^2]^{-1/2}.$$

Notice that the two terms differ only in the $(z \pm z_0)$ factors. Let's verify the assertion (1) by checking each of the three properties of $G$.

(i)  Clearly, $G$ is finite and differentiable except at $\mathbf{x}_0$. Also, $\Delta G = 0$.

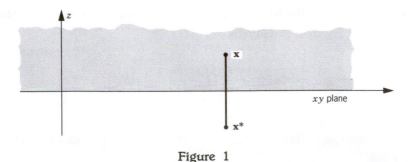

Figure 1

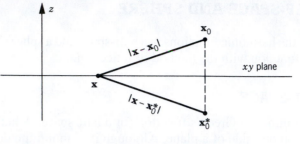

**Figure 2**

(ii)   This is the main property to check. Let $\mathbf{x} \in$ bdy $D$, so that $z = 0$.
From Figure 2 we see that $|\mathbf{x} - \mathbf{x}_0| = |\mathbf{x} - \mathbf{x}_0^*|$. Thus $G(\mathbf{x}, \mathbf{x}_0) = 0$.

(iii)  Because $\mathbf{x}_0^*$ is outside our domain $D$, the function $-1/(4\pi |\mathbf{x} - \mathbf{x}_0^*|)$
has no singularity inside the domain, so that $G$ has the proper singularity at $\mathbf{x}_0$.

These three properties prove that $G(\mathbf{x}, \mathbf{x}_0)$ is the Green's function for this
domain. Let's now use it to solve the Dirichlet problem

$$\Delta u = 0 \quad \text{for } z > 0, \qquad u(x, y, 0) = h(x, y). \tag{2}$$

We use formula (7.3.1). Notice that $\partial G/\partial n = -\partial G/\partial z|_{z=0}$ because $\mathbf{n}$ points
downward (outward from the domain). Furthermore,

$$-\frac{\partial G}{\partial z} = \frac{1}{4\pi} \left( \frac{z + z_0}{|\mathbf{x} - \mathbf{x}_0^*|^3} - \frac{z - z_0}{|\mathbf{x} - \mathbf{x}_0|^3} \right)$$

$$= \frac{1}{2\pi} \frac{z_0}{|\mathbf{x} - \mathbf{x}_0|^3}$$

on $z = 0$. Therefore, the solution of (2) is

$$u(x_0, y_0, z_0) = \frac{z_0}{2\pi} \iint [(x - x_0)^2 + (y - y_0)^2 + (z_0)^2]^{-3/2} h(x, y)\, dx\, dy, \tag{3}$$

where both integrals run over $(-\infty, \infty)$, noting that $z = 0$ in the integrand.
In vector notation, (3) takes the form

$$u(\mathbf{x}_0) = \frac{z_0}{2\pi} \iint_{\text{bdy } D} \frac{h(\mathbf{x})}{|\mathbf{x} - \mathbf{x}_0|^3}\, dS. \tag{4}$$

This is the complete formula that solves the Dirichlet problem for the half-space.

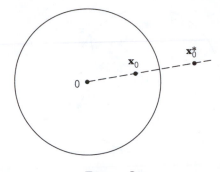

**Figure 3**

## THE SPHERE

The Green's function for the ball $D = \{|\mathbf{x}| < a\}$ of radius $a$ can also be found by the reflection method. In this case, however, the reflection is across the sphere $\{|\mathbf{x}| = a\}$, which is the boundary of $D$ (see Figure 3).

Fix any nonzero point $\mathbf{x}_0$ in the ball (that is, $0 < |\mathbf{x}_0| < a$). The *reflected point* $\mathbf{x}_0^*$ is defined by two properties. It is collinear with the origin $\mathbf{0}$ and the point $\mathbf{x}_0$. Its distance from the origin is determined by the formula $|\mathbf{x}_0|\,|\mathbf{x}_0^*| = a^2$. Thus

$$\mathbf{x}_0^* = \frac{a^2 \mathbf{x}_0}{|\mathbf{x}_0|^2}. \tag{5}$$

If $\mathbf{x}$ is any point at all, let's denote $|\mathbf{x} - \mathbf{x}_0| = \rho$ and $|\mathbf{x} - \mathbf{x}_0^*| = \rho^*$. Then the Green's function of the ball is

$$\boxed{G(\mathbf{x}, \mathbf{x}_0) = -\frac{1}{4\pi\rho} + \frac{a}{|\mathbf{x}_0|}\frac{1}{4\pi\rho^*}} \tag{6}$$

if $\mathbf{x}_0 \neq \mathbf{0}$. To verify this formula, we need only check the three conditions (i), (ii), and (iii). We'll consider the case $\mathbf{x}_0 = \mathbf{0}$ separately.

First of all, $G$ has no singularity except at $\mathbf{x} = \mathbf{x}_0$ because $\mathbf{x}_0^*$ lies outside the ball. The functions $1/\rho$ and $1/\rho^*$ are harmonic in $D$ except at $\mathbf{x}_0$ because they are just translates of $1/r$. Therefore, (i) and (iii) are true.

To prove (ii), we show that $\rho^*$ is proportional to $\rho$ for all points $\mathbf{x}$ on the spherical surface $|\mathbf{x}| = a$. To do this, we notice from the congruent triangles in Figure 4 that

$$\left|\frac{r_0}{a}\mathbf{x} - \frac{a}{r_0}\mathbf{x}_0\right| = |\mathbf{x} - \mathbf{x}_0|, \tag{7}$$

where $r_0 = |\mathbf{x}_0|$. The left side of (7) equals

$$\frac{r_0}{a}\left|\mathbf{x} - \frac{a^2}{r_0^2}\mathbf{x}_0\right| = \frac{r_0}{a}\rho^*$$

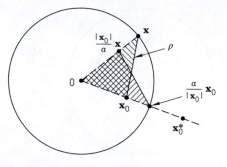

**Figure 4**

Thus

$$\frac{r_0}{a}\rho^* = \rho \qquad \text{for all } |\mathbf{x}| = a. \tag{8}$$

Therefore, the function

$$-\frac{1}{4\pi\rho} + \frac{a}{|\mathbf{x}_0|}\frac{1}{4\pi\rho^*}, \tag{9}$$

defined above, is zero on the sphere $|\mathbf{x}| = a$. This is condition (ii). This proves formula (6).

We can also write (6) in the form

$$\boxed{G(\mathbf{x}, \mathbf{x}_0) = -\frac{1}{4\pi|\mathbf{x} - \mathbf{x}_0|} + \frac{1}{4\pi|r_0\mathbf{x}/a - a\mathbf{x}_0/r_0|}.} \tag{10}$$

In case $\mathbf{x}_0 = \mathbf{0}$, the formula for the Green's function is

$$\boxed{G(\mathbf{x}, \mathbf{0}) = -\frac{1}{4\pi|\mathbf{x}|} + \frac{1}{4\pi a}} \tag{11}$$

(see Exercise 10).

Let's now use (6) to write the formula for the solution of the Dirichlet problem in a ball:

$$\boxed{\Delta u = 0 \quad \text{in } |\mathbf{x}| < a, \qquad u = h \quad \text{on } |\mathbf{x}| = a.} \tag{12}$$

We already know from Chapter 6 that $u(\mathbf{0})$ is the average of $h(\mathbf{x})$ on the sphere, so let's consider $\mathbf{x}_0 \neq \mathbf{0}$. To apply (7.3.1), we need to calculate $\partial G/\partial n$ on $|\mathbf{x}| = a$. (Let's not forget that $\mathbf{x}_0$ is considered to be fixed, and the derivatives are with respect to $\mathbf{x}$.)

We note that $\rho^2 = |x - x_0|^2$. Differentiating, we have $2\rho \nabla \rho = 2(x - x_0)$. So $\nabla \rho = (x - x_0)/\rho$ and $\nabla(\rho^*) = (x - x_0^*)/\rho^*$. Hence differentiating (6), we have

$$\nabla G = \frac{x - x_0}{4\pi \rho^3} - \frac{a}{r_0} \frac{x - x_0^*}{4\pi \rho^{*3}}. \tag{13}$$

Remember that $x_0^* = (a/r_0)^2 x_0$. If $|x| = a$, we showed above that $\rho^* = (a/r_0)\rho$. Substituting these expressions into the last term of $\nabla G$, we get

$$\nabla G = \frac{1}{4\pi \rho^3} \left[ x - x_0 - \left(\frac{r_0}{a}\right)^2 x + x_0 \right] \tag{14}$$

on the surface, so that

$$\frac{\partial G}{\partial n} = \frac{x}{a} \cdot \nabla G = \frac{a^2 - r_0^2}{4\pi a \rho^3}. \tag{15}$$

Thus (7.3.1) takes the form

$$u(x_0) = \frac{a^2 - |x_0|^2}{4\pi a} \iint\limits_{|x|=a} \frac{h(x)}{|x - x_0|^3} \, dS. \tag{16}$$

This is the solution to (12). It is the three-dimensional version of the Poisson formula. In more classical notation, it would be written in the usual spherical coordinates as

$$u(r_0, \theta_0, \phi_0) = \frac{a(a^2 - r_0^2)}{4\pi} \int_0^{2\pi} \int_0^{\pi} \frac{h(\theta, \phi)}{(a^2 + r_0^2 - 2ar_0\cos\psi)^{3/2}} \sin\theta \, d\theta \, d\phi, \tag{17}$$

where $\psi$ denotes the angle between $x_0$ and $x$.

In almost the same way, we can use the method of reflection in *two dimensions* to recover the Poisson formula for

$$u_{xx} + u_{yy} = 0 \quad \text{in } x^2 + y^2 < a^2, \qquad u = h \quad \text{on } x^2 + y^2 = a^2.$$

Beginning with the function $(1/2\pi) \log r$, we find (see Exercise 11) that

$$G(x, x_0) = \frac{1}{2\pi} \log \rho - \frac{1}{2\pi} \log\left(\frac{r_0}{a} \rho^*\right) \tag{18}$$

and hence that

$$u(x_0) = \frac{a^2 - |x_0|^2}{2\pi a} \int_{|x|=a} \frac{h(x)}{|x - x_0|^2} \, ds,$$

which is exactly the same as the Poisson formula (6.3.14), which we found earlier in a completely different way!

## EXERCISES

1. Find the one-dimensional Green's function for the interval $(0, l)$. The three properties defining it can be restated as follows.
   (i)   It solves $G''(x) = 0$ for $x \neq x_0$ ("harmonic").
   (ii)  $G(0) = G(l) = 0$.
   (iii) $G(x)$ is continuous at $x_0$ and $G(x) + \frac{1}{2}|x - x_0|$ is harmonic at $x_0$.

2. Verify directly from (3) or (4) that the solution of the half-space problem satisfies the condition at infinity:

$$u(\mathbf{x}) \to 0 \qquad \text{as } |\mathbf{x}| \to \infty.$$

   Assume that $h(x, y)$ is a continuous function that vanishes outside some circle.

3. Show directly from (3) that the boundary condition is satisfied: $u(x_0, y_0, z_0) \to h(x_0, y_0)$ as $z_0 \to 0$. Assume $h(x, y)$ is continuous and bounded. [*Hint:* Change variables $s^2 = [(x - x_0)^2 + (y - y_0)^2]/z_0^2$ and use the fact that $\int_0^\infty s(s^2 + 1)^{-3/2} ds = 1$.]

4. Verify directly from (3) that the solution has derivatives of all orders in $\{z > 0\}$. Assume that $h(x, y)$ is a continuous function that vanishes outside some circle. (*Hint:* See Section A.3 for differentiation under an integral sign.)

5. Notice that the function $xy$ is harmonic in the half-plane $\{y > 0\}$ and vanishes on the boundary line $\{y = 0\}$. The function 0 has the same properties. Does this mean that the solution is not unique? Explain.

6. (a) Find the Green's function for the half-plane $\{(x, y): y > 0\}$.
   (b) Use it to solve the Dirichlet problem in the half-plane with boundary values $h(x)$.
   (c) Calculate the solution with $u(x, 0) = 1$.

7. (a) If $u(x, y) = f(x/y)$ is a harmonic function, solve the ODE satisfied by $f$.
   (b) Show that $\partial u/\partial r \equiv 0$, where $r = \sqrt{x^2 + y^2}$ as usual.
   (c) Suppose that $v(x, y)$ is any function in $\{y > 0\}$ such that $\partial v/\partial r \equiv 0$. Show that $v(x, y)$ is a function of the quotient $x/y$.
   (d) Find the boundary values $\lim_{y \to 0} u(x, y) = h(x)$.
   (e) Show that your answer to parts (c) and (d) agrees with the general formula from Exercise 6.

8. (a) Use Exercise 7 to find the harmonic function in the half-plane $\{y > 0\}$ with the boundary data $h(x) = 1$ for $x > 0$, $h(x) = 0$ for $x < 0$.
   (b) Do the same as part (a) for the boundary data $h(x) = 1$ for $x > a$, $h(x) = 0$ for $x < a$. (*Hint:* Translate the preceding answer.)

(c)  Use part (b) to solve the same problem with the boundary data $h(x)$, where $h(x)$ is any step function. That is,

$$h(x) = c_j \quad \text{for } a_{j-1} < x < a_j \quad \text{for } 1 \le j \le n,$$

where $-\infty = a_0 < a_1 < \cdots < a_{n-1} < a_n = \infty$ and the $c_j$ are constants.

9.  Find the Green's function for the tilted half-space $\{(x, y, z): ax + by + cz > 0\}$. (*Hint:* Either do it from scratch by reflecting across the tilted plane, or change variables in the double integral (3) using a linear transformation.)

10.  Verify the formula (11) for $G(\mathbf{x}, \mathbf{0})$, the Green's function with its second argument at the center of the sphere.

11.  (a)  Verify that (18) is the Green's function for the disk.
    (b)  Use it to recover the Poisson formula.

12.  Find the potential of the electrostatic field due to a point charge located outside a grounded sphere. (*Hint:* This is just the Green's function for the exterior of the sphere. Find it by the method of reflection.)

13.  Find the Green's function for the half-ball $D = \{x^2 + y^2 + z^2 < a^2, z > 0\}$. (*Hint:* The easiest method is to use the solution for the whole ball and reflect it across the plane.)

14.  Do the same for the eighth of a ball

$$D = \{x^2 + y^2 + z^2 < a^2, \ x > 0, \ y > 0, \ z > 0\}.$$

15.  (a)  Show that if $v(x, y)$ is harmonic, so is $u(x, y) = v(x^2 - y^2, 2xy)$.
    (b)  Show that the transformation $(x, y) \longmapsto (x^2 - y^2, 2xy)$ maps the first quadrant onto the half-plane $\{y > 0\}$. (*Hint:* Use either polar coordinates or complex variables.)

16.  Use Exercises 15 and 7 to find the harmonic function $u(x, y)$ in the first quadrant that has the boundary values $u(x, 0) = A$, $u(0, y) = B$, where $A$ and $B$ are constants. (*Hint:* $u(x, 0) = v(x^2, 0)$, etc.)

17.  (a)  Find the Green's function for the quadrant

$$Q = \{(x, y): x > 0, y > 0\}.$$

(*Hint:* Either use the method of reflection or reduce to the half-plane problem by the transformation in Exercise 15.)

    (b)  Use your answer in part (a) to solve the Dirichlet problem

$$u_{xx} + u_{yy} = 0 \text{ in } Q, \quad u(0, y) = g(y) \text{ for } y > 0,$$
$$u(x, 0) = h(x) \text{ for } x > 0.$$

18.  (a)  Find the Green's function for the octant $\mathcal{O} = \{(x, y, z): x > 0, y > 0, z > 0\}$. (*Hint:* Use the method of reflection.)

(b)   Use your answer in part (a) to solve the Dirichlet problem

$$\begin{cases} u_{xx} + u_{yy} + u_{zz} = 0 & \text{in } \mathbb{O} \\ u(0, y, z) = 0, \, u(x, 0, z) = 0, & u(x, y, 0) = h(x, y) \\ \qquad\qquad\qquad\qquad\qquad\qquad \text{for } x > 0, \, y > 0, \, z > 0. \end{cases}$$

19.   Consider the four-dimensional laplacian $\Delta u = u_{xx} + u_{yy} + u_{zz} + u_{ww}$. Show that its fundamental solution is $r^{-2}$, where $r^2 = x^2 + y^2 + z^2 + w^2$.

20.   Use Exercise 19 to find the Green's function for the half-hyperspace $\{(x, \, y, \, z, \, w) : w > 0\}$.

21.   The *Neumann function* $N(x, y)$ for a domain $D$ is defined exactly like the Green's function in Section 7.3 except that (ii) is replaced by the Neumann boundary condition

(ii)\*
$$\frac{\partial N}{\partial n} = c \quad \text{for } x \in \text{bdy } D.$$

for a suitable constant $c$.
   (a)   Show that $c = 1/A$, where $A$ is the area of bdy $D$. ($c = 0$ if $A = \infty$)
   (b)   State and prove the analog of Theorem 7.3.1, expressing the solution of the Neumann problem in terms of the Neumann function.

22.   Solve the Neumann problem in the half-plane: $\Delta u = 0$ in $\{y > 0\}$, $\partial u / \partial y = h(x)$ on $\{y = 0\}$ with $u(x, y)$ bounded at infinity. (*Hint:* Consider the problem satisfied by $v = \partial u / \partial y$.)

23.   Solve the Neumann problem in the quarter-plane $\{x > 0, \, y > 0\}$.

24.   Solve the Neumann problem in the half-space $\{z > 0\}$.

25.   Let the nonconstant function $u(\mathbf{x})$ satisfy the inequality $\Delta u \geq 0$ in a domain $D$ in three dimensions. Prove that it cannot assume its maximum inside $D$. This is the maximum principle for *subharmonic functions.* (*Hint:* Let $f = \Delta u$, and let $h$ denote $u$ restricted to the boundary bdy $D$. Let $B \subset D$ be any ball and let $\mathbf{x}_0$ be its center. Use (11) and (16) together with (7.3.7) in the ball $B$. Show that $u(\mathbf{x}_0)$ is at most the average of $h$ on bdy $B$. Continue the proof as in Section 6.3.)

# 8

# COMPUTATION OF SOLUTIONS

We have found formulas for many solutions to PDEs, but other problems encountered in practice are not as simple and cannot be solved by formula. Even when there is a formula, it might be so complicated that we would prefer to visualize a typical solution by looking at its graph. The opportunity presented in this chapter is to reduce the process of solving a PDE with its auxiliary conditions to a finite number of arithmetical calculations that can be carried out by computer. All the problems we have studied can be so reduced. However, there are dangers in doing so. If the method is not carefully chosen, the numerically computed solution may not be anywhere close to the true solution. The other danger is that the computation (for a difficult problem) could easily take so long that it would take more computer time than is practical to carry out (years, millenia, ...). The purpose of this chapter is to illustrate the most important techniques of computation using quite simple equations as examples.

## 8.1 OPPORTUNITIES AND DANGERS

The best known method, *finite differences*, consists of replacing each derivative by a difference quotient. Consider, for instance, a function $u(x)$ of one variable. Choose a *mesh size* $\Delta x$. Let's approximate the value $u(j\Delta x)$ for $x = j\Delta x$ by a number $u_j$ indexed by an integer $j$:

$$u_j \sim u(j\Delta x).$$

Then the three standard approximations for the *first derivative* $\dfrac{\partial u}{\partial x}(j\Delta x)$ are:

> The backward difference: $\dfrac{u_j - u_{j-1}}{\Delta x}$        (1)
>
> The forward difference: $\dfrac{u_{j+1} - u_j}{\Delta x}$        (2)
>
> The centered difference: $\dfrac{u_{j+1} - u_{j-1}}{2\Delta x}.$        (3)

Each of them is a correct approximation because of the Taylor expansion:

$$u(x + \Delta x) = u(x) + u'(x)\Delta x + \tfrac{1}{2}u''(x)(\Delta x)^2 + \tfrac{1}{6}u'''(x)(\Delta x)^3 + O(\Delta x)^4.$$

[It is valid if $u(x)$ is a $C^4$ function.] Replacing $\Delta x$ by $-\Delta x$, we get

$$u(x - \Delta x) = u(x) - u'(x)\Delta x + \tfrac{1}{2}u''(x)(\Delta x)^2 - \tfrac{1}{6}u'''(x)(\Delta x)^3 + O(\Delta x)^4.$$

From these two expansions we deduce that

$$u'(x) = \frac{u(x) - u(x - \Delta x)}{\Delta x} + O(\Delta x)$$

$$= \frac{u(x + \Delta x) - u(x)}{\Delta x} + O(\Delta x)$$

$$= \frac{u(x + \Delta x) - u(x - \Delta x)}{2\Delta x} + O(\Delta x)^2.$$

We have written $O(\Delta x)$ to mean any expression that is bounded by a constant times $\Delta x$, and so on. Replacing $x$ by $j\,\Delta x$, we see that (1) and (2) are correct approximations to the order $O(\Delta x)$ and (3) is correct to the order $O(\Delta x)^2$.

For the *second derivative*, the simplest approximation is the

> centered second difference:   $u''(j\Delta x) \sim \dfrac{u_{j+1} - 2u_j + u_{j-1}}{(\Delta x)^2}.$     (4)

This is justified by the same two Taylor expansions given above which, when added, give

$$u''(x) = \frac{u(x + \Delta x) - 2u(x) + u(x - \Delta x)}{(\Delta x)^2} + O(\Delta x)^2.$$

That is, (4) is valid with an error of $O(\Delta x)^2$.

For functions of two variables $u(x, t)$, we choose a mesh size for both variables. We write

$$u(j\Delta x, n\,\Delta t) \sim u_j^n,$$

where the $n$ is a superscript, not a power. Then we can approximate, for instance,

$$\frac{\partial u}{\partial t}(j\,\Delta x,\, n\,\Delta t) \sim \frac{u_j^{n+1} - u_j^n}{\Delta t}, \tag{5}$$

the forward difference for $\partial u / \partial t$. Similarly, the forward difference for $\partial u / \partial x$ is

$$\frac{\partial u}{\partial x}(j\,\Delta x,\, n\,\Delta t) \sim \frac{u_{j+1}^n - u_j^n}{\Delta x}, \tag{6}$$

and we can write similar expressions for the differences (1)–(4) in either the $t$ or $x$ variables.   □

Two kinds of errors can be introduced in a computation using such approximations. *Truncation error* refers to the error introduced in the solutions by the approximations themselves, that is, the $O(\Delta x)$ terms. Although the error in the equation may be $O(\Delta x)$, the error in the solutions (the truncation error) may or may not be small. This error is a complicated combination of many small errors. We want the truncation error to tend to zero as the mesh size tends to zero. Thinking of $\Delta x$ as a very small number, it is clear that $O(\Delta x)^2$ is a much smaller error than $O(\Delta x)$. The errors written in (1)–(4) are, strictly speaking, called *local* truncation errors. They occur in the approximation of the individual terms in a differential equation. *Global* truncation error is the error introduced in the actual solutions of the equation by the cumulative effects of the local truncation errors. The passage from local to global errors is usually too complicated to follow in any detail.

*Roundoff error* occurs in a real computation because only a certain number of digits, typically 8 or 16, are retained by the computer at each step of the computation. For instance, if all numbers are rounded to eight digits, the dropping of the ninth digit could introduce big cumulative errors in a large computation. We have to prevent these little errors from accumulating.

## Example 1.

Let's solve the very simple problem

$$u_t = u_{xx}, \qquad u(x, 0) = \phi(x)$$

using finite differences. We use a forward difference for $u_t$ and a centered difference for $u_{xx}$. Then the *difference equation* is

$$\boxed{\frac{u_j^{n+1} - u_j^n}{\Delta t} = \frac{u_{j+1}^n - 2u_j^n + u_{j-1}^n}{(\Delta x)^2}.} \tag{7}$$

It has a local truncation error of $O(\Delta t)$ (from the left side) and $O(\Delta x)^2$ (from the right side).

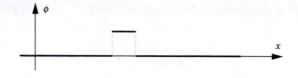

**Figure 1**

Suppose that we choose a very small value for $\Delta x$ and choose $\Delta t = (\Delta x)^2$. Then (7) simplifies to

$$u_j^{n+1} = u_{j+1}^n - u_j^n + u_{j-1}^n. \tag{8}$$

Let's take $\phi(x)$ to be the very simple step function (see Figure 1), which is to be approximated by the values $\phi_j$:

$$0 \quad 0 \quad 0 \quad 0 \quad 1 \quad 0 \quad 0 \quad 0 \quad 0 \quad 0 \quad \rightarrow x.$$

A sample calculation with these simple initial data can be done by hand by simply "marching in time." That is, $\phi(x)$ provides $u_j^0$, then the "scheme" (8) gives $u_j^1$, then (8) gives $u_j^2$, and so on. We can summarize (8) schematically using the diagram

$$\overset{*}{\phantom{x}}$$
$$\bullet +1 \quad \bullet -1 \quad \bullet +1$$

(called a *template*), which means that in order to get $u_j^{n+1}$ you just add or subtract its three lower neighbors as indicated. Thus simple arithmetic gives us the result shown in Figure 2. (Verify it!) The values of $u_j^n$ are written in the $(j, n)$ location. This is supposed to be an approximate solution.

The result is horrendous! It is nowhere near the true solution of the PDE. We know that by the maximum principle, the true solution of the diffusion equation will always be between zero and one, but the difference equation has given us an "approximation" with the value 19 and growing! □

In the next section we analyze what went wrong.

| $n = 4$ | 1 | $-4$ | 10 | $-16$ | 19 | $-16$ | 10 | $-4$ | 1 | |
|---|---|---|---|---|---|---|---|---|---|---|
| $n = 3$ | 0 | 1 | $-3$ | 6 | $-7$ | 6 | $-3$ | 1 | 0 | |
| $n = 2$ | 0 | 0 | 1 | $-2$ | 3 | $-2$ | 1 | 0 | 0 | |
| $n = 1$ | 0 | 0 | 0 | 1 | $-1$ | 1 | 0 | 0 | 0 | |
| $n = 0$ | 0 | 0 | 0 | 0 | 1 | 0 | 0 | 0 | 0 | $\rightarrow x$ |

**Figure 2**

## EXERCISES

1.  The Taylor expansion written in Section 8.1 is valid if $u$ is a $C^4$ function. If $u(x)$ is merely a $C^3$ function, the best we can say is that the Taylor expansion is valid only with a $o(\Delta x)^3$ error. [This notation means that the error is $(\Delta x)^3$ times a factor that tends to zero as $\Delta x \to 0$.] If merely a $C^2$ function, it is only valid with a $o(\Delta x)^2$ error, and so on.
    (a)  If $u(x)$ is merely a $C^3$ function, what is the error in the first derivative due to its approximation by the centered difference?
    (b)  What if $u(x)$ is merely a $C^2$ function?
2.  (a)  If $u(x)$ is merely a $C^3$ function, what is the error in the *second* derivative due to its approximation by a centered second difference?
    (b)  What if $u(x)$ is merely a $C^2$ function?
3.  Suppose that we wish to approximate the first derivative $u'(x)$ of a very smooth function with an error of only $O(\Delta x)^4$. Which difference approximation could we use?

## 8.2  APPROXIMATIONS OF DIFFUSIONS

We take up our discussion of the diffusion equation $u_t = u_{xx}$ again. There is nothing obviously wrong with the scheme we used, as each derivative is appropriately approximated with a small local truncation error. Somehow the little errors have accumulated! What turns out to be wrong, but this is *not* obvious at this point, is the choice of the mesh $\Delta t$ relative to the mesh $\Delta x$. Let's make no assumption now about these meshes; in fact, let

$$s = \frac{\Delta t}{(\Delta x)^2}. \tag{1}$$

As before, we can solve the scheme (8.1.7) for $u_j^{n+1}$:

$$u_j^{n+1} = s\left(u_{j+1}^n + u_{j-1}^n\right) + (1 - 2s)u_j^n. \tag{2}$$

The scheme is said to be *explicit* because the values at the $(n + 1)$st time step are given explicitly in terms of the values at the earlier times.

### Example 1.

To be specific, let's consider the standard problem:

$$u_t = u_{xx} \qquad \text{for } 0 < x < \pi, t > 0$$
$$u = 0 \qquad \text{at } x = 0, \pi$$

$$u(x, 0) = \phi(x) = \begin{cases} x & \text{in } \left(0, \frac{\pi}{2}\right) \\ \pi - x & \text{in } \left(\frac{\pi}{2}, \pi\right). \end{cases}$$

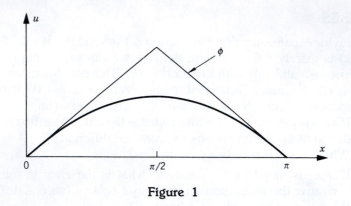

**Figure 1**

Its exact solution from Section 5.1 is

$$u(x, t) = \sum_{k=1}^{\infty} b_k \sin kx \, e^{-k^2 t}, \tag{3}$$

where $b_k = 4(-1)^{(k+1)/2}/\pi k^2$ for odd $k$, and $b_k = 0$ for even $k$. It looks like Figure 1 for some $t > 0$ ($t = 3\pi^2/80$).

We approximate this problem by the scheme (2) for $j = 0, 1, \ldots,$ $J - 1$ and $n = 0, 1, 2, \ldots$ together with the discrete boundary and initial conditions

$$u_0^n = u_J^n = 0 \qquad \text{and} \qquad u_j^0 = \phi(j\Delta x).$$

For $J = 20$, $\Delta x = \pi/20$, and $s = \frac{5}{11}$, the result of the calculation (from page 6 of [RM]) is shown in Figure 2 (exactly on target!). However, if we repeat the calculation for $J = 20$, $\Delta x = \pi/20$, and $s = \frac{5}{9}$, the result is as shown in Figure 3 (wild oscillations as in Section 8.1!). Thus the choice $s = \frac{5}{11}$ is stable, whereas $s = \frac{5}{9}$ is clearly unstable. $\square$

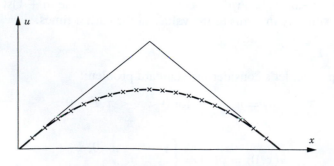

**Figure 2**

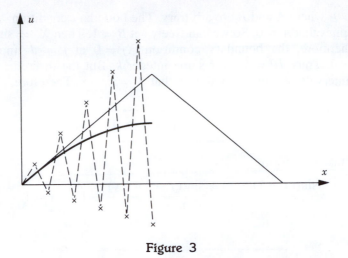

**Figure 3**

## STABILITY CRITERION

The primary distinction between these two calculations turns out to be whether $s$ is bigger or smaller than $\frac{1}{2}$. We might have gotten a suspicion of this from the scheme (2) itself, because when $s < \frac{1}{2}$, the coefficients in (2) are positive. But to actually demonstrate that this is the stability condition, we *separate the variables in the difference equation*. Thus we look for solutions of equation (2) of the form

$$u_j^n = X_j T_n. \tag{4}$$

Thus

$$\frac{T_{n+1}}{T_n} = 1 - 2s + s\frac{X_{j+1} + X_{j-1}}{X_j}. \tag{5}$$

Both sides of (4) must be a constant $\xi$ independent of $j$ and $n$. Therefore,

$$\boxed{T_n = \xi^n T_0} \tag{6}$$

and

$$s\frac{X_{j+1} + X_{j-1}}{X_j} + 1 - 2s = \xi. \tag{7}$$

To solve the spatial equation (7), we argue that it is a discrete version of a second-order ODE which has sine and cosine solutions. Therefore, we guess solutions of (7) of the form

$$X_j = A\cos j\theta + B\sin j\theta$$

for some $\theta$, where $A$ and $B$ are arbitrary. The boundary condition $X_0 = 0$ at $j = 0$ implies that $A = 0$. So we can freely set $B = 1$. Then $X_j = \sin j\theta$.

Furthermore, the boundary condition $X_J = 0$ at $j = J$ implies that $\sin J\theta = 0$. Thus $J\theta = k\pi$ for some integer $k$. But the discretization into $J$ equal intervals of length $\Delta x$ means that $J = \pi/\Delta x$. Therefore, $\theta = k\,\Delta x$ and

$$\boxed{X_j = \sin(jk\Delta x).} \tag{8}$$

Now (7) takes the form

$$s\frac{\sin((j+1)k\Delta x) + \sin((j-1)k\Delta x)}{\sin(jk\Delta x)} + 1 - 2s = \xi$$

or

$$\boxed{\xi = \xi(k) = 1 - 2s[1 - \cos(k\Delta x)].} \tag{9}$$

According to (6), the growth in time $t = n\,\Delta t$ at the *wave number* $k$ is governed by the powers $\xi(k)^n$. So

*unless $|\xi(k)| \le 1$ for all $k$, the scheme is unstable*

and could not possibly approximate the true (exact) solution. (Recall that the true solution tends to zero as $t \to \infty$.) Now we analyze (9) to determine whether $|\xi(k)| \le 1$ or not. Since the factor $1 - \cos(k\Delta x)$ ranges between 0 and 2, we have $1 - 4s \le \xi(k) \le 1$. So stability requires that $1 - 4s \ge -1$, which means that

$$\boxed{\frac{\Delta t}{(\Delta x)^2} = s \le \frac{1}{2}.} \tag{10}$$

Thus (10) is the condition required for stability.

This condition explains the instability that we observed in Section 8.1. It means that in practice the time steps must be taken quite short. For instance, if $\Delta x = 0.01$, an apparently reasonable choice, then $\Delta t$ can be at most 0.00005. Then solving up to time $t = 1$ would require 20,000 time steps!

The analysis above shows that it is precisely the wave number $k$ for which $\xi(k) = -1$, which is the most dangerous for stability. That critical situation happens when $\cos(k\Delta x) = -1$, that is, when $k = \pi/\Delta x$. In practice, this is a fairly high wave number.

By the way, the complete solution of the difference scheme (2), together with the discrete boundary conditions, is the "Fourier series"

$$u_j^n = \sum_{k=-\infty}^{\infty} b_k \sin(jk\Delta x)\,[\xi(k)]^n. \tag{11}$$

Let's see how it could be that this "discrete" series converges to the "true" series (3). In fact, the Taylor series of (9) is

$$\xi(k) = 1 - 2sk^2(\Delta x)^2/2! + \cdots \simeq 1 - k^2\Delta t$$

if $k\Delta x$ is small. Taking the $n$th power and letting $j\Delta x = x$ and $n\Delta t = t$, we have

$$\xi(k)^n \simeq (1 - k^2\Delta t)^{t/\Delta t} \simeq e^{-k^2 t}$$

in the limit as $\Delta t \to 0$, using the well-known limit for the exponential. So the series (11) looks like it tends to the series (3), as it should. Of course, this could not possibly be a proof of convergence (since we know it does not even converge at all if $s > \frac{1}{2}$). An actual proof for $s \leq \frac{1}{2}$, which we omit, would require a careful analysis of the approximations.

The example discussed above indicates that the general procedure to determine stability in a diffusion or wave problem is to separate the variables in the difference equation. For the time factor we obtain a simple equation like (6) which has an *amplification factor* $\xi(k)$. In the analysis above we used the stability condition $|\xi(k)| \leq 1$. More precisely, it can be shown that the correct condition necessary for stability is

$$\boxed{|\xi(k)| \leq 1 + O(\Delta t) \qquad \text{for all } k} \tag{12}$$

and for small $\Delta t$. (We omit the proof.) This is the *von Neumann stability condition* [RM]. The extra term in (12) is irrelevant for the example above but important for problems where the exact solution may grow in time (as in Exercise 11).

In practice we could go more quickly from (7) to (9) simply by assuming that

$$\boxed{X_j = (e^{ik\Delta x})^j} \tag{13}$$

is an exponential. (*This is the procedure to be followed in doing the exercises.*) Plugging (13) into (7), we immediately have

$$\xi = 1 - 2s + s(e^{ik\Delta x} + e^{-ik\Delta x}).$$

Thus we again recover equation (9) for the amplification factor $\xi$.

## NEUMANN BOUNDARY CONDITIONS

Suppose that the interval is $0 \leq x \leq l$ and the boundary conditions are

$$u_x(0, t) = g(t) \qquad \text{and} \qquad u_x(l, t) = h(t).$$

Although the simplest approximations would be

$$\frac{u_1^n - u_0^n}{\Delta x} = g^n \quad \text{and} \quad \frac{u_J^n - u_{J-1}^n}{\Delta x} = h^n,$$

they would introduce local truncation errors of order $O(\Delta x)$, bigger than the $O(\Delta x)^2$ errors in the equation. To introduce $O(\Delta x)^2$ errors only, we prefer to use centered differences for the derivatives on the boundary.

To accomplish this, we introduce "ghost points" $u_{-1}^n$ and $u_{J+1}^n$ in addition to $u_0^n, \ldots, u_J^n$. The discrete boundary conditions then are

$$\frac{u_1^n - u_{-1}^n}{2\,\Delta x} = g^n \quad \text{and} \quad \frac{u_{J+1}^n - u_{J-1}^n}{2\,\Delta x} = h^n. \tag{14}$$

At the $n$th time step, we can calculate $u_0^n, \ldots, u_J^n$ using the scheme for the PDE, and then calculate the values at the ghost points using (14).

## CRANK-NICOLSON SCHEME

We could try to avoid the restrictive stability condition (10) by using another scheme. There is a class of schemes that is stable no matter what the value of $s$. In fact, let's denote the centered second difference by

$$\frac{u_{j+1}^n - 2u_j^n + u_{j-1}^n}{(\Delta x)^2} = (\delta^2 u)_j^n.$$

*Pick a number $\theta$ between 0 and 1.* Consider the scheme

$$\boxed{\frac{u_j^{n+1} - u_j^n}{\Delta t} = (1 - \theta)(\delta^2 u)_j^n + \theta(\delta^2 u)_j^{n+1}.} \tag{15}$$

We'll call it *the $\theta$ scheme*. If $\theta = 0$, it reduces to the previous scheme. If $\theta > 0$, the scheme is *implicit*, since $u^{n+1}$ appears on both sides of the equation. Therefore, (15) means that we solve a set ($n = 1$) of algebraic linear equations to get $u_j^1$, another set ($n = 2$) to get $u_j^2$, and so on.

Let us analyze the stability of this scheme by plugging in a separated solution

$$u_j^n = (e^{ik\Delta x})^j (\xi(k))^n$$

as before. Then

$$\xi(k) = \frac{1 - 2(1 - \theta)s(1 - \cos k\Delta x)}{1 + 2\theta s(1 - \cos k\Delta x)},$$

where $s = \Delta t/(\Delta x)^2$ (see Exercise 9).

Again we look for the stability condition: $|\xi(k)| \leq 1$ for all $k$. It is always true that $\xi(k) \leq 1$, but the condition $\xi(k) \geq -1$ requires that

$$s(1 - 2\theta)(1 - \cos k\Delta x) \leq 1.$$

(Why?) If $1 - 2\theta \leq 0$, it is always true! This means that

$$\boxed{\text{if } \tfrac{1}{2} \leq \theta \leq 1, \text{ there is no restriction on the size of } s} \tag{16}$$

for stability to hold. Such a scheme is called *unconditionally stable*.

The special case $\theta = \tfrac{1}{2}$ is called the *Crank–Nicolson scheme*. It has the template

$$
\begin{array}{ccc}
\dfrac{1}{2}\dfrac{s}{1+s}\,\bullet & * & \bullet\,\dfrac{1}{2}\dfrac{s}{1+s} \\[2ex]
\dfrac{1}{2}\dfrac{s}{1+s}\,\bullet & \dfrac{1-s}{1+s}\,\bullet & \bullet\,\dfrac{1}{2}\dfrac{s}{1+s}
\end{array}
$$

On the other hand, in case $\theta < \tfrac{1}{2}$, a necessary condition for stability is $s \leq (2 - 4\theta)^{-1}$. Thus (15) is expected to be a stable scheme if

$$\boxed{\dfrac{\Delta t}{(\Delta x)^2} = s < \dfrac{1}{2 - 4\theta}.} \tag{17}$$

## EXERCISES

1. (a) Solve the problem $u_t = u_{xx}$ in the interval $[0, 4]$ with $u = 0$ at both ends and $u(x, 0) = x(4 - x)$, using the forward difference scheme with $\Delta x = 1$ and $\Delta t = 0.25$. Calculate four time steps (up to $t = 1$).
   (b) Do the same with $\Delta x = 0.50$ and $\Delta t = 0.0625 = \tfrac{1}{16}$. Calculate four time steps (up to $t = 0.25$).
   (c) Compare your answers with each other. How close are they at $x = 2.0$, $t = 0.25$?

2. Do the same with $\Delta x = 1$ and $\Delta t = 1$. Calculate by hand or by computer up to $t = 7$.

3. Solve $u_t = u_{xx}$ in the interval $[0, 5]$ with $u(0, t) = 0$ and $u(5, t) = 1$ for $t \geq 0$, and with $u(x, 0) = 0$ for $0 < x < 5$.
   (a) Compute $u(3, 3)$ using the mesh sizes $\Delta x = 1$ and $\Delta t = 0.5$.
   (b) Write the exact solution as an infinite series. Calculate $u(3, 3)$ to three decimal places exactly. Compare with your answer in (a).

4. Solve by hand the problem $u_t = u_{xx}$ in the interval $[0, 1]$ with $u_x = 0$ at both ends. Use the forward scheme (2) for the PDE, and the scheme (14) for the boundary conditions. Assume $\Delta x = \tfrac{1}{5}$, $\Delta t = \tfrac{1}{100}$, and start with the initial data: 0  0  64  0  0  0. Compute for four time steps.

5. Using the forward scheme (2), solve $u_t = u_{xx}$ in $[0, 5]$ with the mixed boundary conditions $u(0, t) = 0$ and $u_x(5, t) = 0$ for $t \geq 0$, and the initial condition $u(x, 0) = 25 - x^2$ for $0 < x < 5$. Use $\Delta x = 1$ and $\Delta t = \tfrac{1}{2}$. Compute $u(3, 3)$ approximately.

6. Do the same with the conditions $u_x(0, t) = u(5, t) = 0$ and $u(x, 0) = x$.

7. Show that the local truncation error in the Crank-Nicolson scheme is $O((\Delta x)^2 + (\Delta t)^2)$.

8. (a) Write down the Crank–Nicolson scheme ($\theta = \frac{1}{2}$) for $u_t = u_{xx}$.
   (b) Consider the solution in the interval $0 \le x \le 1$ with $u = 0$ at both ends. Assume $u(x, 0) = \phi(x)$ and $\phi(1 - x) = \phi(x)$. Show, using uniqueness, that the exact solution must be even around the midpoint $x = \frac{1}{2}$. [That is, $u(x, t) = u(1 - x, t)$.]
   (c) Let $\Delta x = \Delta t = \frac{1}{6}$. Let the initial data be 0  0  0  1  0  0  0. Compute the solution by the Crank–Nicolson scheme for one time step ($t = \frac{1}{6}$). (*Hint:* Use part (b) to halve the computation.)

9. For the $\theta$ scheme (15) for the diffusion equation, provide the details of the derivation of the stability conditions (16) and (17).

10. For the diffusion equation $u_t = u_{xx}$, use centered differences for both $u_t$ and $u_{xx}$.
    (a) Write down the scheme. Is it explicit or implicit?
    (b) Show that it is unstable, no matter what $\Delta x$ and $\Delta t$ are.

11. Consider the equation $u_t = au_{xx} + bu$, where $a$ and $b$ are constants and $a > 0$. Use forward differences for $u_t$, use centered differences for $u_{xx}$, and use $bu_j^n$ for the last term.
    (a) Write the scheme. Let $s = \Delta t / (\Delta x)^2$.
    (b) Find the condition on $s$ for numerical stability. (*Hint:* check condition (12).)

12. (a) Solve by hand the nonlinear PDE $u_t = u_{xx} + (u)^3$ for all $x$ using the standard forward difference scheme with $(u)^3$ treated as $(u_j^n)^3$. Use $s = \frac{1}{4}$, $\Delta t = 1$, and initial data $u_j^0 = 1$ for $j = 0$ and $u_j^0 = 0$ for $j \ne 0$. Solve for $u_0^3$.
    (b) Compare your answer to the same problem without the nonlinear term.
    (c) Exactly solve the ODE $dv/dt = (v)^3$ with the condition $v(0) = 1$. Use it to explain why $u_0^3$ is so large in part (a).
    (d) Repeat part (a) with the same initial data but for the PDE $u_t = u_{xx} - (u)^3$. Compare with the answer in part (a) and explain.

13. Consider the following scheme for the diffusion equation:

$$\frac{u_j^{n+1} - u_j^{n-1}}{2\,\Delta t} = \frac{u_{j+1}^n + u_{j-1}^n - u_j^{n+1} - u_j^{n-1}}{(\Delta x)^2}.$$

It uses a centered difference for $u_t$ and a modified form of the centered difference for $u_{xx}$.
    (a) Solve it for $u_j^{n+1}$ in terms of $s$ and the previous time steps.
    (b) Show that it is stable for all $s$.

14. (a) Formulate an explicit scheme for $u_t = u_{xx} + u_{yy}$.

(b) What is the stability condition for your scheme in terms of $s_1 = \Delta t/(\Delta x)^2$ and $s_2 = \Delta t/(\Delta y)^2$?

15. Formulate the Crank-Nicolson scheme for $u_t = u_{xx} + u_{yy}$.

## 8.3 APPROXIMATIONS OF WAVES

In this section we continue our discussion of finite difference approximations for some very simple PDEs. Although the PDEs are simple, the methods we develop can be used for more difficult, even nonlinear, equations. For the one-dimensional wave equation $u_{tt} = c^2 u_{xx}$ the simplest scheme is the one using centered differences for both terms:

$$\frac{u_j^{n+1} - 2u_j^n + u_j^{n-1}}{(\Delta t)^2} = c^2 \frac{u_{j+1}^n - 2u_j^n + u_{j-1}^n}{(\Delta x)^2}. \tag{1}$$

It is explicit since the $(n+1)$st time step appears only on the left side. Thus

$$u_j^{n+1} = s\left(u_{j+1}^n + u_{j-1}^n\right) + 2(1-s)u_j^n - u_j^{n-1}, \tag{2}$$

where we now denote $s = c^2(\Delta t)^2/(\Delta x)^2$. Its template diagram is

$$
\begin{array}{ccc}
n+1 & & * \\
\\
n & \bullet \quad \bullet \quad \bullet \\
& s \quad 2-2s \quad s \\
n-1 & \bullet \\
& -1
\end{array}
$$

Notice that the value at the $(n+1)$st time step depends on the *two* previous steps, because the wave equation has time derivatives of second order. Therefore, the first *two* rows $u_j^0$ and $u_j^1$ must be given as initial conditions.

### Example 1.

If we pick $s = 2$, the scheme simplifies to

$$u_j^{n+1} = 2\left(u_{j+1}^n + u_{j-1}^n - u_j^n\right) - u_j^{n-1} \tag{3}$$

and it is easy to compute by hand the solution shown in Figure 1, given its first two rows. This horrendous solution bears no relationship to the true solution of the wave equation, which is a pair of waves traveling to the left and right, $u(x, t) = \frac{1}{2}[\phi(x + ct) + \phi(x - ct)]$. The scheme for $s = 2$ is highly unstable. $\square$

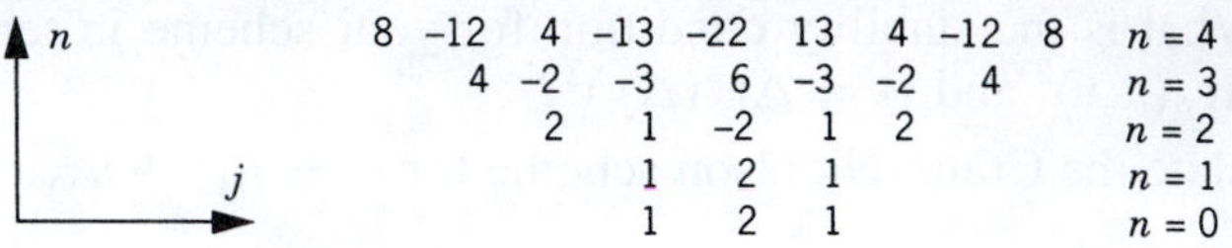

| $n$ | | | | | | | | | | |
|---|---|---|---|---|---|---|---|---|---|---|
| 8 | –12 | 4 | –13 | –22 | 13 | 4 | –12 | 8 | $n = 4$ |
| | 4 | –2 | –3 | 6 | –3 | –2 | 4 | | $n = 3$ |
| | | 2 | 1 | –2 | 1 | 2 | | | $n = 2$ |
| | | | 1 | 2 | 1 | | | | $n = 1$ |
| | | | 1 | 2 | 1 | | | | $n = 0$ |

Figure 1

## Example 2.

For $s = 1$ we have $\Delta x = c\,\Delta t$ and the scheme

$$u_j^{n+1} = u_{j+1}^n + u_{j-1}^n - u_j^{n-1}. \qquad (4)$$

The same initial data as above lead to the solution shown in Figure 2. This is an excellent approximation to the true solution!    □

## INITIAL CONDITIONS

How do we handle the initial conditions? We approximate the conditions $u(x, 0) = \phi(x)$ and $\partial u/\partial t(x, 0) = \psi(x)$ by

$$u_j^0 = \phi(j\Delta x), \qquad \frac{u_j^1 - u_j^{-1}}{2\,\Delta t} = \psi(j\Delta x). \qquad (5)$$

This approximation is chosen to have a $O(\Delta x)^2$ local truncation error in order to match the $O(\Delta x)^2 + O(\Delta t)^2$ truncation error of the scheme (2). (If we only used a simpler approximation with a $O(\Delta x)$ error, the initial conditions would contaminate the solution with too big an error.) Let's abbreviate $\phi_j = \phi(j\Delta x)$ and $\psi_j = \psi(j\Delta x)$. Now (2) in the case $n = 0$ is

$$u_j^1 + u_j^{-1} = s\left(u_{j+1}^0 + u_{j-1}^0\right) + 2(1 - s)u_j^0.$$

Together with (5), this gives us the starting values

$$\begin{aligned} u_j^0 &= \phi_j, \\ u_j^1 &= \frac{s}{2}(\phi_{j+1} + \phi_{j-1}) + (1 - s)\phi_j + \psi_j\Delta t, \end{aligned} \qquad (6)$$

the first two rows of the computation. Then we march ahead in time to get $u_j^2$, $u_j^3$, and so on, using (2).

```
1 1 0 0 0 0 0 1 1
  1 1 0 0 0 1 1                      .
    1 1 0 1 1              +1  0  +1
      1 2 1                    –1
      1 2 1
```

Figure 2

$$\begin{array}{l}
\tfrac{1}{2}\ 1\ \tfrac{1}{2}\ 0\ 0\ 0\ 0\ 0\ 0\ 0\ \tfrac{1}{2}\ 1\ \tfrac{1}{2} \qquad n = 5\\[4pt]
\tfrac{1}{2}\ 1\ \tfrac{1}{2}\ 0\ 0\ 0\ 0\ 0\ \tfrac{1}{2}\ 1\ \tfrac{1}{2} \qquad n = 4\\[4pt]
\tfrac{1}{2}\ 1\ \tfrac{1}{2}\ 0\ 0\ 0\ \tfrac{1}{2}\ 1\ \tfrac{1}{2} \qquad n = 3\\[4pt]
\tfrac{1}{2}\ 1\ \tfrac{1}{2}\ 0\ \tfrac{1}{2}\ 1\ \tfrac{1}{2} \qquad n = 2\\[4pt]
\tfrac{1}{2}\ 1\ 1\ 1\ \tfrac{1}{2} \qquad n = 1\\[4pt]
1\ 2\ 1 \qquad n = 0
\end{array}$$

**Figure 3**

## Example 3.

For instance, let the initial data be

$$\phi(x) = 0\ \ 0\ \ 0\ \ 0\ \ 0\ \ 0\ \ 1\ \ 2\ \ 1\ \ 0\ \ 0\ \ 0\ \ 0\ \ 0\ \ 0$$

and $\psi(x) \equiv 0$. Let $s = 1$. Then from (6) we get the starting values (the first two rows)

$$0\ \ 0\ \ 0\ \ 0\ \ 0\ \ \tfrac{1}{2}\ \ 1\ \ 1\ \ 1\ \ \tfrac{1}{2}\ \ 0\ \ 0\ \ 0\ \ 0\ \ 0$$
$$0\ \ 0\ \ 0\ \ 0\ \ 0\ \ 0\ \ 1\ \ 2\ \ 1\ \ 0\ \ 0\ \ 0\ \ 0\ \ 0\ \ 0.$$

If we use (4), we get the solution shown in Figure 3. This is an even better approximation to the true solution than that shown in Figure 2. □

## STABILITY CRITERION

Now let's analyze the stability by the method of Section 8.2. Again, a clue may be found in the values of the coefficients. None are negative if $s \leq 1$. Once again this simple observation turns out to be the correct stability condition. However, proceeding more logically, we separate the variables

$$u_j^n = (\eta)^j (\xi)^n \qquad \text{where } \eta = e^{ik\Delta x}.$$

From (1) we get

$$\xi + \frac{1}{\xi} - 2 = s\left(\eta + \frac{1}{\eta} - 2\right) = 2s\,[\cos(k\,\Delta x) - 1]. \tag{7}$$

Letting $p = s[\cos(k\Delta x) - 1]$ for the sake of brevity, (7) can be written as

$$\xi^2 - 2(1 + p)\xi + 1 = 0, \text{ which has the roots } \xi = 1 + p \pm \sqrt{p^2 + 2p}. \tag{8}$$

Note that $p \leq 0$. If $p < -2$, then $p^2 + 2p > 0$ and there are two real roots, one of which is less than $-1$. Thus for one of the roots we have $|\xi| > 1$, so that the scheme is unstable. On the other hand, if $p > -2$, then $p^2 + 2p < 0$ and there are two complex conjugate roots $1 + p \pm i\sqrt{-p^2 - 2p}$. These complex roots satisfy

$$|\xi|^2 = (1 + p)^2 - p^2 - 2p = 1.$$

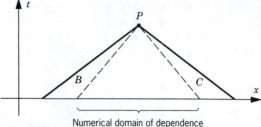

Numerical domain of dependence

**Figure 4**

So $\xi = \cos\theta + i\sin\theta$ for some real number $\theta$. In this case the solutions oscillate in time (just as they ought to for the wave equation). Finally, if $p = -2$, then $\xi = -1$.

Thus a *necessary condition for stability* is that $p \geq -2$ for all $k$. This means that

$$s \leq \frac{2}{1 - \cos(k\,\Delta x)}$$

for all $k$. Thus stability requires that

$$s = c^2\frac{(\Delta t)^2}{(\Delta x)^2} \leq 1. \tag{9}$$

There is a nice way to understand this condition (9). At each time step $\Delta t$ the values of the numerical solution spread out by one unit $\Delta x$. So the ratio $\Delta x/\Delta t$ is the propagation speed of the numerical scheme. The propagation speed for the exact wave equation is $c$. So the stability condition requires the numerical propagation speed to be at least as large as the continuous propagation speed. In Figure 4 we have sketched the domains of dependence of the true and the computed solutions for the case $c = 1$ and $\Delta t/\Delta x = 2$ (so that $s = 4$). The computed solution at the point $P$ does not make use of the initial data in the regions $B$ and $C$ as it ought to. Therefore, the scheme leads to entirely erroneous values of the solution.

On the other hand, even the stable schemes do not do a very good job at resolving singularities in the true solution. For instance, one solution of the nice scheme (4) with $s = 1$ is shown in Figure 5. This initial condition is

|  |  |  |  |  |  |  |  |  |
|---|---|---|---|---|---|---|---|---|
| 1 | -1 | 1 | -1 | 1 | -1 | 1 |  | $n = 4$ |
|  | 1 | -1 | 1 | -1 | 1 |  |  | $n = 3$ |
|  |  | 1 | -1 | 1 |  |  |  | $n = 2$ |
|  |  |  | 1 |  |  |  |  | $n = 1$ |
|  |  |  | 1 |  |  |  |  | $n = 0$ |

**Figure 5**

"singular" because it has a sudden up and down jump. The solution in Figure 5 isn't as unstable as the one in Figure 1, but it surely is a poor approximation to the true solution. (It's a good approximation only for someone who wears fuzzy glasses.) The difficulty here is that the initial function $\phi(x)$ has a significant "jump" at one point; the earlier cases illustrated in Figures 2 and 3 were at least slightly gradual. More sophisticated schemes must be used to solve problems with singularities, as in shock wave problems.

There are also implicit schemes for the wave equation (like the Crank–Nicolson scheme) but they are less urgently needed here since the stability condition (9) for the explicit scheme does not require $\Delta t$ to be so much smaller than $\Delta x$.

## Example 4.

For a more interesting PDE, let's consider the *nonlinear* wave equation

$$u_{tt} - \Delta u + u + [u]^7 = 0 \qquad (10)$$

in three dimensions $(x, y, z)$, where $[u]^7$ denotes the seventh power. Let $(r, \theta, \phi)$ be the usual spherical coordinates. We shall make the calculation manageable by computing only those solutions that are independent of $\theta$ and $\phi$. Then the equation takes the form

$$u_{tt} - u_{rr} - \frac{2}{r} u_r + u + [u]^7 = 0$$

by (6.1.7), which is a modification of the one-dimensional wave equation. To get rid of the middle term, it is convenient to change variables $v(r, t) = ru(r, t)$ to get

$$\begin{cases} v_{tt} - v_{rr} + v + r^{-6}[v]^7 = 0 & (0 < r < \infty) \\ v(0, t) = 0. \end{cases} \qquad (11)$$

The last condition comes from the definition of $v$.

Now we use the scheme (1) with $s = 1$ and with suitable additional terms to get

$$\frac{v_j^{n+1} - 2v_j^n + v_j^{n-1}}{(\Delta t)^2} = \frac{v_{j+1}^n - 2v_j^n + v_{j-1}^n}{(\Delta r)^2}$$

$$- \frac{1}{2}\left(v_j^{n+1} + v_j^{n-1}\right) - \frac{1}{8}(j\Delta r)^{-6} \frac{\left(v_j^{n+1}\right)^8 - \left(v_j^{n-1}\right)^8}{v_j^{n+1} - v_j^{n-1}} \qquad (12)$$

One reason for this treatment of the additional terms is that this scheme has a constant energy (independent of $n$), an analog of the continuous energy of Section 2.2 (see Exercise 9).

Using the mesh sizes $\Delta r = \Delta t = 0.002$ and certain initial data, the computed solution is graphically presented in Figure 6 (see [SV]). The

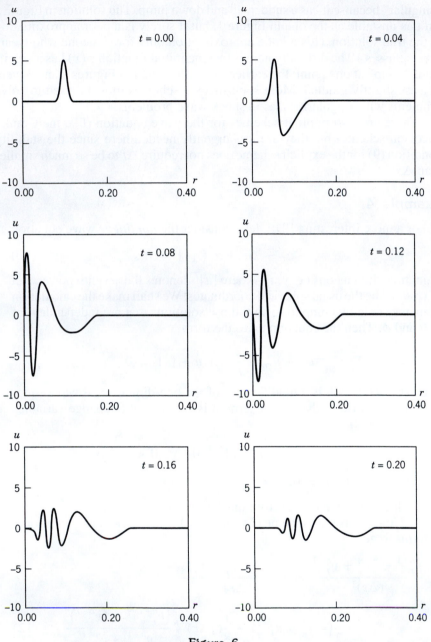

Figure 6

effect of the nonlinear term is visible in the oscillations of fairly large amplitude which reflect at the origin.    □

## EXERCISES

1. (a) Write the scheme (2) for the wave equation in the case $s = \frac{1}{4}$ and draw the template.
   (b) Compute the solution by hand for five time levels with the same starting values as in Figure 2.
   (c) Convince yourself that the computed solution is not too accurate but is "in the right ballpark." When interpreting the solution remember that $\Delta x / \Delta t = 2$.

2. Solve by hand for a few time steps the numerical scheme (2) for $u_{tt} = u_{xx}$, with $u(x, 0) \equiv 0$, with

$$\psi_j = \quad 0 \ 0 \ 0 \ 0 \ 1 \ 2 \ 1 \ 0 \ 0 \ 0 \ 0$$

   and with the starting scheme (6).
   (a) First use $\Delta t = 1$ and $\Delta x = 0.5$.
   (b) Then use $\Delta t = 1$ and $\Delta x = 1$.
   (c) Compare your answers to parts (a) and (b).

3. (a) Use the scheme (2) with $\Delta x = \Delta t = 0.2$ to approximately solve $u_{tt} = u_{xx}$ with $u(x, 0) = x^2$ and $u_t(x, 0) = 1$. Solve it in the region $\{0 \le t \le 1, |x| \le 2 - t\}$.
   (b) Solve the problem exactly and compare the exact and approximate solutions.

4. (a) Use the scheme (2) with $\Delta x = \Delta t = 0.25$ to solve $u_{tt} = u_{xx}$ approximately in the interval $0 \le x \le 1$ with $u = 0$ at both ends and $u(x, 0) = \sin \pi x$ and $u_t(x, 0) = 0$. Show that the solution is periodic.
   (b) Compare your answer to the exact solution. What is its period?

5. Solve by hand for a few time steps the equation $u_{tt} = u_{xx}$ in the finite interval $0 \le x \le 1$, with $u_x = 0$ at both ends, using $\Delta t = \Delta x = \frac{1}{6}$ and the initial conditions

$$u(x, 0) = \quad 0 \ 0 \ 0 \ 1 \ 2 \ 1 \ 0 \ 0 \ 0 \quad \text{and} \quad u_t(x, 0) \equiv 0.$$

   Use central differences for the boundary derivatives as in (8.2.14) and use second-order-accurate initial conditions as in (6). Do you see the reflections at the boundary?

6. Consider the wave equation on the half-line $0 < x < \infty$, with the boundary condition $u = 0$ at $x = 0$. With the starting values $u_4^0 = u_5^0 = u_4^1 = u_5^1 = 1$ and $u_j^0 = u_j^1 = 0$ for all other $j$ ($j = 1, 2, \ldots$), compute the solution by hand up to 10 time steps. Observe the reflection at the boundary and compare with Section 3.2.

7.  Solve by hand the nonlinear equation $u_{tt} = u_{xx} + u^3$ up to $t = 4$, using the same initial conditions as in Figure 3, replacing the cubic term by $(u_j^n)^3$, and using $\Delta x = \Delta t = 1$. What is the effect of the nonlinear term? Compare with the linear problem in Figure 3.

8.  Repeat Exercise 7 by computer for the equation $u_{tt} = u_{xx} - u^3$ using an implicit scheme like (12) with $\Delta t = \Delta x = 1$.

9.  Consider the scheme (12) for the nonlinear wave equation (10). Let the *discrete energy* be defined as

$$\frac{E_n}{\Delta r} = \frac{1}{2} \sum_j \left( \frac{v_j^{n+1} - v_j^n}{\Delta t} \right)^2 + \frac{1}{2} \sum_j \left( \frac{v_{j+1}^{n+1} - v_j^{n+1}}{\Delta r} \right) \left( \frac{v_{j+1}^n - v_j^n}{\Delta r} \right)$$

$$+ \frac{1}{4} \sum_j \left[ \left( v_j^{n+1} \right)^2 + \left( v_j^n \right)^2 \right] + \frac{1}{16} \sum_j \frac{\left( v_j^{n+1} \right)^8 + \left( v_j^n \right)^8}{(j \Delta r)^6}.$$

By multiplying (12) by $\frac{1}{2}(v_j^{n+1} - v_j^{n-1})$, show that $E_n = E_{n-1}$. Conclude that $E_n$ does not depend on $n$.

10. Consider the equation $u_t = u_x$. Use forward differences for both partial derivatives.
    (a) Write down the scheme.
    (b) Draw the template.
    (c) Find the separated solutions.
    (d) Show that the scheme is stable if $0 < \Delta t / \Delta x \leq 1$.

11. Consider the first-order equation $u_t + a u_x = 0$.
    (a) Solve it exactly with the initial condition $u(x, 0) = \phi(x)$.
    (b) Write down the finite difference scheme which uses the forward difference for $u_t$ and the centered difference for $u_x$.
    (c) For which values of $\Delta x$ and $\Delta t$ is the scheme stable?

## 8.4   APPROXIMATIONS OF LAPLACE'S EQUATION

For Dirichlet's problem in a domain of irregular shape, it may be more convenient to compute numerically than to try to find the Green's function. As with the other equations, the same ideas of numerical computation can easily be carried over to more complicated equations. For Laplace's equation

$$u_{xx} + u_{yy} = 0$$

the natural approximation is that of centered differences,

$$\frac{u_{j+1,k} - 2u_{j,k} + u_{j-1,k}}{(\Delta x)^2} + \frac{u_{j,k+1} - 2u_{j,k} + u_{j,k-1}}{(\Delta y)^2} = 0. \tag{1}$$

Here $u_{j,k}$ is an approximation to $u(j\Delta x, k\Delta y)$. The relative choice of mesh sizes turns out not to be critical so we just choose $\Delta x = \Delta y$. Then (1) can be written as

$$u_{j,k} = \tfrac{1}{4}(u_{j+1,k} + u_{j-1,k} + u_{j,k+1} + u_{j,k-1}). \tag{2}$$

Thus $u_{j,k}$ is the *average* of the values at the four neighboring sites. The template is

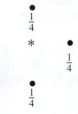

The scheme (2) has some nice properties. The most obvious one is the *mean value property*, the exact analog of the same property for the Laplace equation. In its discrete version (2), the difference equation and the mean value property become identical! It follows that a solution $u_{jgk}$ cannot take its maximum or minimum value at an interior point, unless it is a constant; for otherwise it couldn't be the average of its neighbors. Thus, if (2) is valid in a region, the maximum and minimum values can be taken only at the boundary.

To solve the Dirichlet problem for $u_{xx} + u_{yy} = 0$ in $D$ with given boundary values, we draw a grid covering $D$ and approximate $D$ by a union of squares (see Figure 1). Then the discrete solution is specified on the boundary of the "discrete region." Our task is to fill in the interior values so as to satisfy (2). In contrast to time-dependent problems, no marching method is available. If $N$ is the number of interior grid points, *the equations (2) form a system of $N$ linear equations in $N$ unknowns*. For instance, if we divide $x$ and $y$ each into 100 parts, we get about 10,000 little squares. Thus $N$ can be very large.

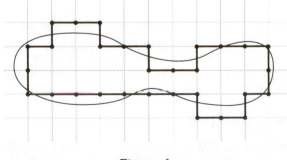

Figure 1

|   |   |   |   |   |   |   |   |   |   |
|---|---|---|---|---|---|---|---|---|---|
| 0 | 0 | 0 | 0 |   |   | 0 | 0 | 0 | 0 |
| 0 |   | 24 |   |   |   | 0 | 2 | 7 | 24 |
| 0 |   | 0 |   |   |   | 0 | 1 | 2 | 0 |
| 0 | 0 | 0 | 0 |   |   | 0 | 0 | 0 | 0 |
|   | (a) |   |   |   |   |   | (b) |   |   |

**Figure 2**

The system we get in this way has *exactly one* solution. To prove this, suppose that there were two solutions, $\{u_{j,k}\}$ and $\{v_{j,k}\}$, of (2) in $D$ with identical boundary values. Their difference $\{u_{j,k} - v_{j,k}\}$ also satisfies (2) in $D$ but with zero boundary values. By the maximum principle stated above, $u_{j,k} - v_{j,k} \le 0$, and by the minimum principle, $u_{j,k} - v_{j,k} \ge 0$. Hence $u_{j,k} = v_{j,k}$. So there is at most one solution. But this means that the determinant of the linear system of $N$ equations is not zero, which means that there exists exactly one solution.

## Example 1.

As a baby example, consider solving (2) in the square with the boundary values indicated in Figure 2(a). This is a set of four linear equations, one for each interior point. The solution is given in Figure 2(b). Notice that each interior entry is indeed the average of its four neighbors.  □

## JACOBI ITERATION

In the absence of a marching method to solve (2), several techniques are available. One is called Jacobi iteration. We start from any reasonable first approximation $u_{j,k}^{(1)}$. Then we successively solve

$$u_{j,k}^{(n+1)} = \frac{1}{4}\left[u_{j+1,k}^{(n)} + u_{j-1,k}^{(n)} + u_{j,k+1}^{(n)} + u_{j,k-1}^{(n)}\right]. \tag{3}$$

It can be shown that $u_{j,k} = \lim u_{j,k}^{(n)}$ converges as $n \to \infty$ to a limit which is a solution of (2). It converges, however, very slowly and so Jacobi iteration is never used in practice. Since $N$ is usually quite large, a more efficient method is needed.

It might be noticed that (3) is exactly the same calculation as if one were solving the two-dimensional heat equation $v_t = v_{xx} + v_{yy}$ using centered differences for $v_{xx}$ and $v_{yy}$ and using the forward difference for $v_t$, with $\Delta x = \Delta y$ and $\Delta t = (\Delta x)^2/4$ (see Exercise 11). In effect, we are solving the Dirichlet problem by taking the limit of the discretized $v(x, t)$ as $t \to \infty$.

## GAUSS–SEIDEL METHOD

This method improves the rate of convergence. Here it is important to specify the order of operations. Let's compute $u_{j,k}^{(n+1)}$ one row at a time starting at the bottom row and let's go from left to right. But every time a calculation is completed, we'll throw out the old value and update it by its newly calculated one. This procedure means that

$$u_{j,k}^{(n+1)} = \tfrac{1}{4}\left[u_{j+1,k}^{(n)} + u_{j-1,k}^{(n+1)} + u_{j,k+1}^{(n)} + u_{j,k-1}^{(n+1)}\right].$$ (4)

The new values (with superscript $n + 1$) are used to the *left* and *below* the $(j, k)$ location. It turns out that Gauss–Seidel works about twice as fast as Jacobi.

## SUCCESSIVE OVERRELAXATION

This method is still faster. It is the scheme

$$u_{j,k}^{(n+1)} = u_{j,k}^{(n)} + \omega\left[u_{j+1,k}^{(n)} + u_{j-1,k}^{(n+1)} + u_{j,k+1}^{(n)} + u_{j,k-1}^{(n+1)} - 4u_{j,k}^{(n)}\right].$$ (5)

If $\omega = \tfrac{1}{4}$, it is the same as Gauss–Seidel. It is quite surprising that a different choice of $\omega$ could make a significant improvement, but it does. But how to choose the relaxation factor $\omega$ in practice is an art whose discussion we leave to more specialized texts. Note again that once we know that $u_{j,k} = \lim u_{j,k}^{(n)}$ exists, the limit must satisfy

$$u_{j,k} = u_{j,k} + \omega(u_{j+1,k} + u_{j-1,k} + u_{j,k+1} + u_{j,k-1} - 4u_{j,k})$$

and hence it satisfies (2).

## EXERCISES

1. Set up the linear equations to find the four unknown values in Figure 2($a$), write them in vector-matrix form, and solve them. You should deduce the answer in Figure 2($b$).

2. Apply Jacobi iteration to the example of Figure 2($a$) with zero initial values in the interior. Compute six iterations.

3. Apply four Gauss–Seidel iterations to the example of Figure 2($a$).

4. Solve the example of Figure 2($a$) but with the boundary conditions (by rows, top to bottom) 0, 48, 0, 0;   0, *, *, 24;   0, *, *, 0;   0, 0, 0, 0.

5. Consider the PDE $u_{xx} + u_{yy} = 0$ in the unit square $\{0 \le x \le 1, 0 \le y \le 1\}$ with the boundary conditions:

$$u = 0 \qquad\qquad \text{on } x = 0, \text{ on } x = 1, \text{ and on } y = 1$$
$$u = 324\,x^2(1 - x) \qquad \text{on } y = 0.$$

Calculate the approximation to the solution using finite differences (2) with the very coarse grid $\Delta x = \Delta y = \tfrac{1}{3}$.
(*Hint:* You may use Figure 2 if you wish.)

6. (a) Write down the scheme using centered differences for the equation $u_{xx} + u_{yy} = f(x, y)$.

(b)  Use it with $\Delta x = \Delta y = 0.5$ to solve the problem $u_{xx} + u_{yy} = 1$ in the square $0 \le x \le 1, 0 \le y \le 1$ with $u = 0$ on the boundary.

(c)  Repeat with $\Delta x = \Delta y = \frac{1}{3}$.

(d)  Compute the exact value at the center of the square and compare with your answer to part (b).

7.  Solve $u_{xx} + u_{yy} = 0$ in the unit square $\{0 \le x \le 1, 0 \le y \le 1\}$ with the boundary conditions: $u(x, 0) = u(0, y) = 0$, $u(x, 1) = x$, $u(1, y) = y$. Use $\Delta x = \Delta y = \frac{1}{4}$, so that there are nine interior points for the scheme (2).

(a)  Use two steps of Jacobi iteration, with the initial guess that the value at each of the nine points equals 1.

(b)  Use two steps of Gauss–Seidel iteration, with the same initial guess.

(c)  Compare parts (a) and (b) and the exact solution.

8.  Formulate a finite difference scheme for $u_{xx} + u_{yy} = f(x, y)$ in the unit square $\{0 \le x \le 1, 0 \le y \le 1\}$ with Neumann conditions $\partial u / \partial n = g(x, y)$ on the boundary. Use $u_{j,k}$ for $-1 \le j \le N + 1$ and $-1 \le k \le N + 1$ and use centered differences for the normal derivative, such as $(u_{j,N+1} - u_{j,N-1})/2\,\Delta y$. [That is, use ghost points as in (8.2.14).]

9.  Apply Exercise 8 to approximately find the harmonic function in the unit square with the boundary conditions $u_x(0, y) = 0$, $u_x(1, y) = -1$, $u_y(x, 0) = 0, u_y(x, 1) = 1$. Formulate a Gauss–Seidel method of solving the difference scheme and compute two iterations with $\Delta x = \Delta y = \frac{1}{3}$. Compare with the exact solution $u = \frac{1}{2}y^2 - \frac{1}{2}x^2$. You may use a computer program.

10.  Try to do the same with the boundary conditions $u_x(0, y) = 0$, $u_x(1, y) = 1, u_y(x, 0) = 0, u_y(x, 1) = 1$. What's wrong?

11.  Show that performing Jacobi iteration (3) is the same as solving the two-dimensional diffusion equation $v_t = v_{xx} + v_{yy}$ using centered differences for $v_{xx}$ and $v_{yy}$ and using the forward difference for $v_t$, with $\Delta x = \Delta y$ and $\Delta t = (\Delta x)^2/4$.

12.  Do the same (solving the diffusion equation) with $\Delta t = \omega(\Delta x)^2$ and compare with successive overrelaxation.

## 8.5   FINITE ELEMENT METHOD

All computational methods reduce PDEs to discrete form. But there are other methods besides finite differences. Here we briefly discuss the finite element method. The idea is to divide the domain into simple pieces (polygons) and to approximate the solution by extremely simple functions on these pieces. In one of its incarnations, the one we shall discuss, the simple pieces are triangles and the simple functions are linear. The finite element method was developed by engineers to handle curved or irregularly shaped domains. If $D$ is a circle, say, they were having trouble using finite differences, which are not

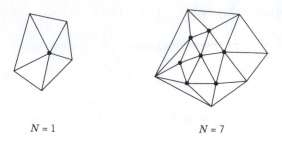

$N = 1$                              $N = 7$

**Figure 1**

particularly efficient simply because a circle is not very accurately partitioned into rectangles.

Let's consider the Dirichlet problem for Poisson's equation in the plane

$$-\Delta u = f \quad \text{in } D, \quad u = 0 \quad \text{on bdy} D. \tag{1}$$

First, $D$ is *triangulated*; that is, $D$ is approximated by a region $D_N$ which is the union of a finite number of triangles (see Figure 1). Let the interior vertices be denoted by $V_1, \ldots, V_N$.

Next, we pick $N$ *trial functions*, $v_1(x, y), \ldots, v_N(x, y)$, one for each interior vertex. Each trial function $v_i(x, y)$ is chosen to equal 1 at "its" vertex $V_i$ and to equal 0 at all the other vertices (see Figure 2). Inside each triangle, each trial function is a *linear* function: $v_i(x, y) = a + bx + cy$. (The coefficients $a$, $b$, $c$ are different for each trial function and for each triangle.) This prescription determines $v_i(x, y)$ uniquely. In fact, its graph is simply a pyramid of unit height with its summit at $V_i$ and it is identically zero on all the triangles that do not touch $V_i$.

We shall approximate the solution $u(x, y)$ by a linear combination of the $v_i(x, y)$:

$$u_N(x, y) = U_1 v_1(x, y) + \cdots + U_N v_N(x, y). \tag{2}$$

How do we choose the coefficients $U_1, \ldots, U_N$?

To motivate our choice we need a digression. Let's rewrite the problem (1) using Green's first identity [formula (G1) from Section 7.1]. We multiply

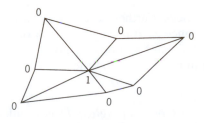

**Figure 2**

Poisson's equation by any function $v(x, y)$ that vanishes on the boundary. Then

$$\iint_D \nabla u \cdot \nabla v \, dx \, dy = \iint_D f v \, dx \, dy. \tag{3}$$

Rather than requiring (3) to be valid for $u_N(x, y)$ for *all* functions $v(x, y)$, we ask only that it be valid for the first $N$ special trial functions $v = v_j \, (j = 1, \ldots, N)$. Thus, with $u(x, y) = u_N(x, y)$ and $v(x, y) = v_j(x, y)$, (3) becomes

$$\sum_{i=1}^{N} U_i \left( \iint_D \nabla v_i \cdot \nabla v_j \, dx \, dy \right) = \iint_D f v_j \, dx \, dy.$$

This is a system of $N$ linear equations $(j = 1, \ldots, N)$ in the $N$ unknowns $U_1, \ldots, U_N$. If we denote

$$m_{ij} = \iint_D \nabla v_i \cdot \nabla v_j \, dx \, dy \quad \text{and} \quad f_j = \iint_D f v_j \, dx \, dy, \tag{4}$$

then the system takes the form

$$\boxed{\sum_{i=1}^{N} m_{ij} U_i = f_j \qquad (j = 1, \ldots, N).} \tag{5}$$

*The finite element method consists of calculating $m_{ij}$ and $f_j$ from (4) and solving (5). The approximate value of the solution $u(x, y)$ then is given by (2).*

The trial functions $v_j$ are completely explicit and depend only on the geometry. The approximate solution $u_N$ automatically vanishes on the boundary of $D_N$. Notice also that, at a vertex $V_k = (x_k, y_k)$,

$$u_N(x_k, y_k) = U_1 v_1(x_k, y_k) + \cdots + U_N v_N(x_k, y_k) = U_k,$$

since $v_i(x_k, y_k)$ equals 0 for $i \neq k$ and equals 1 for $i = k$. Thus *the coefficients are precisely the values of the approximate solution at the vertices.*

Another way to understand $u_N(x, y)$ is that it is a continuous and piecewise-linear function (linear on each triangle), simply because it is a sum of such functions. In fact, it is the unique piecewise-linear continuous function on the triangulation such that $u_N(x_k, y_k) = U_k \, (k = 1, \ldots, N)$.

Notice also that the matrix $m_{ij}$ is "sparse": $m_{ij} = 0$ whenever $V_i$ and $V_j$ are not neighboring vertices. Furthermore, for a pair of neighboring vertices, $m_{ij}$ is easy to compute since each $v_i(x, y)$ is linear on each triangle.

Triangulations with linear functions are not the only versions of the finite element method used in practice. Two other versions in two variables are as follows.

(i)  *Bilinear elements on rectangles:* $D$ is divided into rectangles on each of which the solution is approximated using trial functions

$v_i(x, y) = a + bx + cy + dxy$. Each trial function is associated with a corner of a rectangle.

(ii) *Quadratic elements on triangles: D* is triangulated and the trial functions have the form $v_i(x, y) = a + bx + cy + dx^2 + exy + fy^2$. Each trial function is associated with one of the six "nodes" of a triangle, namely, the three vertices and the midpoints of the three sides.

For a reference, see [TR].

As a further example, consider solving the diffusion equation

$$u_t = \kappa u_{xx} + f(x, t); \qquad u = 0 \text{ at } x = 0, l; \quad u = \phi(x) \text{ at } t = 0.$$

Suppose, for simplicity, that $l$ is an integer. Partition the interval $[0, l]$ into $l$ equal subintervals. We assign the trial function $v_j(x)$ to each of the $N = l - 1$ interior vertices, where $v_j(x)$ is the linear element of Exercise 3. Now we multiply the diffusion equation by any function $v(x)$ that vanishes on the boundary. Integrating by parts, we get

$$\frac{d}{dt}\int_0^l uv \, dx = -\kappa \int_0^l \frac{\partial u}{\partial x}\frac{dv}{dx} \, dx + \int_0^l f(x, t)\, v(x)\, dx. \tag{6}$$

In order to use the finite-element method, we look for a solution of the form

$$u(x, t) = \sum_{i=1}^{N} U_i(t)\, v_i(x)$$

and we merely require (6) to hold for $v = v_1, \ldots, v_N$. Then

$$\sum_{i=1}^{N}\left(\int_0^l v_i v_j \, dx\right)\frac{dU_i}{dt} = -\kappa \sum_{i=1}^{N}\left(\int_0^l \frac{dv_i}{dx}\frac{dv_j}{dx} dx\right)U_i(t) + \int_0^l f(x, t)v_j(x)\, dx.$$

This is a system of ODEs for $U_1, \ldots, U_N$ that can be written as a vector equation as follows.

Let $U$ be the column vector $[U_1, \ldots, U_N]$ and let $F$ be the column vector $[F_1(t), \ldots, F_N(t)]$ with $F_j(t) = \int_0^l f(x, t)v_j(x)dx$. Let $M$ be the matrix with entries $m_{ij}$ and $K$ be the matrix with entries $k_{ij}$ where

$$k_{ij} = \int_0^l v_i v_j \, dx, \qquad m_{ij} = \int_0^l \frac{dv_i}{dx}\frac{dv_j}{dx}\, dx.$$

Then the system of $N$ ODEs in $N$ unknowns takes the simple vector form

$$K\frac{dU}{dt} = -\kappa M U(t) + F(t). \tag{7}$$

$M$ is called the stiffness matrix, $K$ the mass matrix, and $F$ the forcing vector. In Exercise 3(a), the stiffness and mass matrices are computed to be

$$
M = \begin{pmatrix} 2 & -1 & 0 & \cdots & 0 \\ -1 & 2 & -1 & \cdots & 0 \\ & \cdots & & & \\ 0 & \cdots & 0 & -1 & 2 \end{pmatrix}, \quad K = \begin{pmatrix} \frac{2}{3} & \frac{1}{6} & 0 & \cdots & 0 \\ \frac{1}{6} & \frac{2}{3} & \frac{1}{6} & \cdots & 0 \\ & \cdots & & & \\ 0 & \cdots & 0 & \frac{1}{6} & \frac{2}{3} \end{pmatrix}
$$

The matrices $M$ and $K$ have two important features. They are sparse and they depend only on the trial functions. So they remain the same as we change the data. We also have the initial condition

$$
U_i(0) = \Phi_i \equiv \int_0^l \phi(x) v_i(x)\, dx. \tag{8}
$$

This ODE system (7)-(8) can be solved numerically by any of a number of methods. One simple method is Euler's. One chooses $t_p = p \Delta t$ for $p = 0, 1, 2, \ldots$ and then solves

$$
U^{(p+1)} = U^{(p)} + \Delta t\, W^{(p)},
$$
$$
K W^{(p)} = -\kappa M U^{(p)} + F(t_p).
$$

Another method is the backwards Euler method, in which we solve

$$
K \left[ \frac{U^{(p+1)} - U^{(p)}}{\Delta t} \right] = -\kappa M U^{(p+1)} + F(t_{p+1}).
$$

This is the same as

$$
[K + \kappa \Delta t\, M]\, U^{(p+1)} = K U^{(p)} + \Delta t\, F(t_{p+1}),
$$

which is solved recursively for $U^{(1)}, U^{(2)}, \ldots$.

## EXERCISES

1.  Consider the problem $u_{xx} + u_{yy} = -4$ in the unit square with $u(0, y) = 0$, $u(1, y) = 0$, $u(x, 0) = 0$, $u(x, 1) = 0$. Partition the square into four triangles by drawing its diagonals. Use the finite element method to find the approximate value $u(\frac{1}{2}, \frac{1}{2})$ at the center.

2.  (a) Find the area $A$ of the triangle with three given vertices $(x_1, y_1)$, $(x_2, y_2)$, and $(x_3, y_3)$.
    (b) Let $(x_1, y_1)$ be a vertex in the finite element method and let $v(x, y)$ be its trial function. Let $T$ be one of the triangles with that vertex and let $(x_2, y_2)$ and $(x_3, y_3)$ be its other two vertices. Find the formula for $v(x, y)$ on $T$.

3.  (*Linear elements on intervals*) In one dimension the geometric building blocks of the finite element method are the intervals. Let the trial function $v_j(x)$ be the "tent" function defined by $v_j(x) = 1 - j + x$ for $j - 1 \le x \le j$, $v_j(x) = 1 + j - x$ for $j \le x \le j + 1$, and $v_j(x) = 0$ elsewhere.

That is, $v_j(x)$ is continuous and piecewise-linear with $v_j(j) = 1$ and $v_j(k) = 0$ for all integers $k \neq j$.

(a) Show that $\int [v_j(x)]^2 dx = 2$ and $\int v_j(x) v_{j+1}(x)\, dx = -1$.

(b) Deduce that the one-dimensional analog of the matrix $m_{ij}$ is the tridiagonal matrix with 2 along the diagonal and $-1$ next to the diagonal.

4. (*Finite elements for the wave equation*) Consider the problem $u_{tt} = u_{xx}$ in $[0, l]$, with $u = 0$ at both ends, and some initial conditions. For simplicity, suppose that $l$ is an integer and partition the interval into $l$ equal sub-intervals. Each of the $l - 1 = N$ interior vertices has the trial function defined in Exercise 3. The approximate solution is defined by the formula $u_N(x) = U_1(t)v_1(x) + \cdots + U_N(t)v_N(x)$, where the coefficients are unknown functions of $t$.

(a) Show that a reasonable requirement is that

$$\sum_{i=1}^{N} U_i''(t) \int_0^l v_i(x) v_j(x)\, dx + \sum_{i=1}^{N} U_i(t) \int_0^l \frac{\partial v_i}{\partial x} \frac{\partial v_j}{\partial x} dx = 0$$

for $j = 1, \ldots, N$.

(b) Show that the finite element method reduces in this case to a system of ODEs: $K\, d^2U/dt^2 + MU = 0$ with an initial condition $U(0) = \Phi$. Here $K$ and $M$ are $N \times N$ matrices, $U(t)$ is an $N$-vector function, and $\Phi$ is an $N$-vector.

5. (*Bilinear elements on rectangles*) On the rectangle with vertices $(0, 0), (A, 0), (0, B),$ and $(A, B)$, find the bilinear function $v(x, y) = a + bx + cy + dxy$ with the values $U_1, U_2, U_3,$ and $U_4$, respectively.

# 9

# WAVES IN SPACE

In two and three dimensions we derive the energy and causality principles and then solve the wave equation in the absence of boundaries. In Section 9.3 we study the geometry of the characteristics. We also solve the wave equation with a source term. In Section 9.4 we solve the diffusion and Schrödinger equations and the harmonic oscillator. In the final section we derive the energy levels of the hydrogen atom.

## 9.1 ENERGY AND CAUSALITY

Our goal now is to study the wave equation

$$u_{tt} - c^2 \, \Delta u = 0 \tag{1}$$

in two and three dimensions in the absence of boundaries. As before, we concentrate on the three-dimensional case

$$u_{tt} = c^2(u_{xx} + u_{yy} + u_{zz}).$$

This equation is invariant under (i) translations in space and time, (ii) rotations in space, and (iii) Lorentz transformations (see Exercise 4).

## THE CHARACTERISTIC CONE

The notion of characteristics is as fundamental as it was in one dimension, but now *the characteristics are surfaces*. Take, for example, a characteristic line in one dimension $x - x_0 = c(t - t_0)$ and rotate it around the $t = t_0$ axis. You get the "hypercone"

$$|\mathbf{x} - \mathbf{x}_0| = [(x - x_0)^2 + (y - y_0)^2 + (z - z_0)^2]^{1/2} = c|t - t_0|, \tag{2}$$

which is a cone in four-dimensional space-time. The set in space-time defined by equation (2) is called the *characteristic cone* or *light cone* at $(\mathbf{x}_0, t_0)$. The reason for the latter term is that if $c$ is the speed of light in electromagnetics, the cone is the union of all the light rays that emanate from the point $(\mathbf{x}_0, t_0)$.

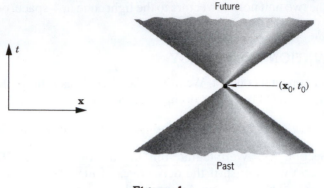

Figure 1

The *solid light cone* is the "inside" of the cone, namely, $\{|\mathbf{x} - \mathbf{x}_0| < c|t - t_0|\}$. It is the union of the future and the past half-cones (see Figure 1). At any fixed time $t$, the light cone is just an ordinary sphere, and the future is just the ball consisting of the points that can be reached in time $t$ by a particle traveling from $(\mathbf{x}_0, t_0)$ at less than the speed of light. As $t \to +\infty$ the sphere grows concentrically at the speed $c$. The light cone is the quintessential characteristic surface; general characteristic surfaces are discussed in Section 9.3.

As an exercise in geometry, let's calculate the *unit normal vector to the light cone* (2). It is the 3-surface in 4-space given by the equation

$$\phi(t, x, y, z) \equiv -c^2(t - t_0)^2 + (x - x_0)^2 + (y - y_0)^2 + (z - z_0)^2 = 0.$$

This is a level surface of $\phi$, so a normal vector is the gradient vector of $\phi(x, y, z, t)$. (We're talking here about vectors with *four* components.) Now

$$\text{grad } \phi = (\phi_x, \phi_y, \phi_z, \phi_t) = 2(x - x_0, y - y_0, z - z_0, -c^2(t - t_0)).$$

The unit normal vectors are

$$\mathbf{n} = \pm \frac{\text{grad } \phi}{|\text{grad } \phi|}$$

$$= \pm \frac{(x - x_0, \ y - y_0, \ z - z_0, -c^2(t - t_0))}{(c^4(t - t_0)^2 + (x - x_0)^2 + (y - y_0)^2 + (z - z_0)^2)^{1/2}}.$$

Let $r^2 = (x - x_0)^2 + (y - y_0)^2 + (z - z_0)^2$. With this notation the equation of the cone is $r = \pm c(t - t_0)$. We can use it to simplify the formula for $\mathbf{n}$ to

$$\mathbf{n} = \pm \left( \frac{x - x_0}{\sqrt{(c^2 + 1)r^2}}, \dots, \frac{-c^2(t - t_0)}{\sqrt{(c^4 + c^2)(t - t_0)^2}} \right)$$

or

$$\mathbf{n} = \pm \frac{c}{\sqrt{c^2 + 1}} \left( \frac{x - x_0}{cr}, \frac{y - y_0}{cr}, \frac{z - z_0}{cr}, -\frac{t - t_0}{|t - t_0|} \right). \tag{3}$$

These are the two unit normal vectors to the light cone in 4-space, one pointing in and one pointing out.

## CONSERVATION OF ENERGY

This is a fundamental concept. We mimic Section 2.2 as follows. Multiplying the wave equation (1) by $u_t$ and doing some algebra, we obtain

$$0 = (u_{tt} - c^2 \Delta u)u_t = \left(\tfrac{1}{2}u_t^2 + \tfrac{1}{2}c^2|\nabla u|^2\right)_t - c^2 \nabla \cdot (u_t \nabla u) \tag{4}$$

(also see Section 7.1). We integrate this identity over 3-space. The integral of the last term will vanish if the derivatives of $u(\mathbf{x}, t)$ tend to zero (in an appropriate sense) as $|\mathbf{x}| \to \infty$. Assuming this, we get

$$0 = \iiint \frac{\partial}{\partial t}\left(\frac{1}{2}u_t^2 + \frac{1}{2}c^2|\nabla u|^2\right) d\mathbf{x} \tag{5}$$

(integration over all 3-space $\mathbb{R}^3$). But the time derivative can be pulled out of the integral (by Section A.3). Therefore, the (*total*) *energy*

$$\boxed{E = \frac{1}{2}\iiint \left(u_t^2 + c^2|\nabla u|^2\right) d\mathbf{x}} \tag{6}$$

*is a constant* (independent of $t$). The first term is the *kinetic energy*, the second the *potential energy*.

## PRINCIPLE OF CAUSALITY

Consider a solution of the wave equation with any initial conditions

$$u(\mathbf{x}, 0) = \phi(\mathbf{x}) \qquad u_t(\mathbf{x}, 0) = \psi(\mathbf{x}).$$

Let $\mathbf{x}_0$ be any point and $t_0 > 0$ any time. The principle of causality asserts that the value of $u(\mathbf{x}_0, t_0)$ depends only on the values of $\phi(\mathbf{x})$ and $\psi(\mathbf{x})$ in the ball $\{|\mathbf{x} - \mathbf{x}_0| \le ct_0\}$. This ball is the intersection of the solid light cone with the initial hyperplane $\{t = 0\}$ (see Figure 2).

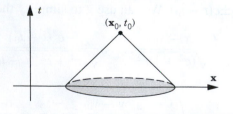

**Figure 2**

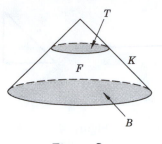

**Figure 3**

**Proof.** We start from the energy identity (4) written in the explicit form

$$\partial_t\left(\tfrac{1}{2}u_t^2 + \tfrac{1}{2}c^2|\nabla u|^2\right) + \partial_x(-c^2 u_t u_x) + \partial_y(-c^2 u_t u_y) + \partial_z(-c^2 u_t u_z) = 0,$$
(7)

abbreviating $\partial_t = \partial/\partial t$, and so on. This time, however, we integrate (7) over a solid cone frustum $F$ in *four*-dimensional space-time, with top $T$, bottom $B$, and side $K$. $F$ is just a piece of a solid light cone (see Figure 3).

We regard the identity (7) as stating that the divergence of a certain four-dimensional vector field vanishes. This is perfectly set up for the *four-dimensional divergence theorem* (see Section A.3)! The frustum $F$ is four-dimensional and its boundary bdy $F$ is three-dimensional. Let $(n_x, n_y, n_z, n_t)$ denote the unit outward normal 4-vector on bdy $F$ and let $dV$ denote the *three*-dimensional volume integral over bdy $F$. Then we get

$$\iiint\limits_{\text{bdy } F} \left[ n_t\left(\tfrac{1}{2}u_t^2 + \tfrac{1}{2}c^2|\nabla u|^2\right) - n_x(c^2 u_t u_x) - n_y(c^2 u_t u_y) - n_z(c^2 u_t u_z) \right] dV = 0.$$
(8)

Now bdy $F = T \cup B \cup K$, which means that the integral in (8) has three parts. So (8) takes the form

$$\iiint\limits_{T} + \iiint\limits_{B} + \iiint\limits_{K} = 0.$$

On the *top* $T$, the normal vector points straight up, so that $\mathbf{n} = (n_x, n_y, n_z, n_t) = (0, 0, 0, 1)$ and we get simply

$$\iiint\limits_{T} \left(\tfrac{1}{2}u_t^2 + \tfrac{1}{2}c^2|\nabla u|^2\right) d\mathbf{x}.$$

On the *bottom* $B$, it points straight down, so that $\mathbf{n} = (0, 0, 0, -1)$ and we get simply

$$\iiint\limits_{B} (-1)\left(\tfrac{1}{2}u_t^2 + \tfrac{1}{2}c^2|\nabla u|^2\right) d\mathbf{x} = -\iiint\limits_{B} \left(\tfrac{1}{2}\psi^2 + \tfrac{1}{2}c^2|\nabla\phi|^2\right) d\mathbf{x}.$$

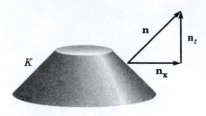

**Figure 4**

The integral over the *mantle K* is more complicated, but we claim that *it is positive* (or zero). To prove this, we plug the formula (3) for **n** into the *K* integral. We use the plus sign in (3) because the *outward* normal vector has a positive *t* component on *K* (see Figure 4). Note that $t < t_0$. As before, $r = |\mathbf{x} - \mathbf{x}_0|$. So the integral is

$$\frac{c}{\sqrt{c^2 + 1}} \iiint_K \left[ \frac{1}{2}u_t^2 + \frac{1}{2}c^2|\nabla u|^2 + \frac{x - x_0}{cr}(-c^2 u_t u_x) \right.$$
$$\left. + \frac{y - y_0}{cr}(-c^2 u_t u_y) + \frac{(z - z_0)}{cr}(-c^2 u_t u_z) \right] dV. \tag{9}$$

The last integrand can be written more concisely as

$$I = \tfrac{1}{2}u_t^2 + \tfrac{1}{2}c^2|\nabla u|^2 - cu_t u_r, \tag{10}$$

where $\nabla u = (u_x, u_y, u_z)$,

$$\hat{\boldsymbol{r}} = \frac{\mathbf{x} - \mathbf{x}_0}{|\mathbf{x} - \mathbf{x}_0|} = \left( \frac{x - x_0}{r}, \frac{y - y_0}{r}, \frac{z - z_0}{r} \right),$$

and the radial derivative is

$$u_r = u_x \frac{\partial x}{\partial r} + u_y \frac{\partial y}{\partial r} + u_z \frac{\partial z}{\partial r} = \hat{\boldsymbol{r}} \cdot \nabla u.$$

Completing the square in (10), we get

$$I = \tfrac{1}{2}(u_t - cu_r)^2 + \tfrac{1}{2}c^2 \left( |\nabla u|^2 - u_r^2 \right) = \tfrac{1}{2}(u_t - cu_r)^2 + \tfrac{1}{2}c^2|\nabla u - u_r \hat{\boldsymbol{r}}|^2, \tag{11}$$

which is clearly positive. Since the integrand is positive, the integral (9) is also positive, as we wished to show.

Hence from (8) we end up with the inequality

$$\boxed{\iiint_T \left( \tfrac{1}{2}u_t^2 + \tfrac{1}{2}c^2|\nabla u|^2 \right) d\mathbf{x} \leq \iiint_B \left( \tfrac{1}{2}\psi^2 + \tfrac{1}{2}c^2|\nabla \phi|^2 \right) d\mathbf{x}.} \tag{12}$$

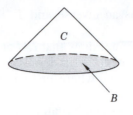

**Figure 5**

Suppose now that $\psi$ and $\phi$ vanish in $B$. By (12) and the first vanishing theorem in Section A.1, the integrand $\frac{1}{2}u_t^2 + \frac{1}{2}c^2|\nabla u|^2$ vanishes in $T$. Therefore, $u_t$ and $\nabla u$ also vanish in $T$. But we can vary the height of the frustum $F$ at will. Therefore, $u_t$ and $\nabla u$ vanish in the whole piece of solid cone $C$ that lies above $B$ (see Figure 5). So $u$ is a constant in the solid cone $C$. Since $u = 0$ on the bottom $B$, the constant is zero. So $u \equiv 0$ in all of $C$. In particular, $u(x_0, y_0, z_0, t_0) = 0$.

By taking the difference of two solutions as in Section 2.2, we easily deduce that if $u$ and $v$ are any two solutions of (1) and if $u = v$ in $B$, then $u(x_0, y_0, z_0, t_0) = v(x_0, y_0, z_0, t_0)$. This completes the proof of the principle of causality.  □

The solid cone $C$ is called the *domain of dependence* or the *past history* of the vertex $(\mathbf{x}_0, t_0)$. As in Section 2.2, we can restate the result as follows. We let $t_0 = 0$.

**Corollary.** The initial data $\phi$, $\psi$ at a spatial point $\mathbf{x}_0$ can influence the solution only in the solid light cone with vertex at $(\mathbf{x}_0, 0)$.

That is, *the* domain of influence *of a point is at most the solid light cone emanating from that point*. Thus we have proved, *from the PDE alone*, that no signal can travel faster than the speed of light!

The same causality principle is true in two dimensions.

## EXERCISES

1. Find all the three-dimensional plane waves; that is, all the solutions of the wave equation of the form $u(\mathbf{x}, t) = f(\mathbf{k} \cdot \mathbf{x} - ct)$, where $\mathbf{k}$ is a fixed vector and $f$ is a function of one variable.

2. Verify that $(c^2t^2 - x^2 - y^2 - z^2)^{-1}$ satisfies the wave equation except on the light cone.

3. Verify that $(c^2t^2 - x^2 - y^2)^{-1/2}$ satisfies the two-dimensional wave equation except on the cone $\{x^2 + y^2 = c^2t^2\}$.

4. (*Lorentz invariance of the wave equation*) Thinking of the coordinates of space-time as 4-vectors $(x, y, z, t)$, let $\Gamma$ be the diagonal matrix with the diagonal entries 1, 1, 1, $-1$. Another matrix $L$ is called a *Lorentz*

*transformation* if $L$ has an inverse and $L^{-1} = \Gamma\,{}^tL\Gamma$, where ${}^tL$ is the transpose.

(a)   If $L$ and $M$ are Lorentz, show that $LM$ and $L^{-1}$ also are.

(b)   Show that $L$ is Lorentz if and only if $m(L\mathbf{v}) = m(\mathbf{v})$ for all 4-vectors $\mathbf{v} = (x, y, z, t)$, where $m(\mathbf{v}) = x^2 + y^2 + z^2 - t^2$ is called the *Lorentz metric*.

(c)   If $u(x, y, z, t)$ is any function and $L$ is Lorentz, let $U(x, y, z, t) = u(L(x, y, z, t))$. Show that

$$U_{xx} + U_{yy} + U_{zz} - U_{tt} = u_{xx} + u_{yy} + u_{zz} - u_{tt}.$$

(d)   Explain the meaning of a Lorentz transformation in more geometrical terms. (*Hint:* Consider the level sets of $m(\mathbf{v})$.)

5.   Prove the principle of causality in two dimensions.

6.   (a)   Derive the conservation of energy for the wave equation in a domain $D$ with homogeneous Dirichlet or Neumann boundary conditions.

   (b)   What about the Robin condition?

7.   For the boundary condition $\partial u/\partial n + b\,\partial u/\partial t = 0$ with $b > 0$, show that the energy defined by (6) decreases.

8.   Consider the equation $u_{tt} - c^2 \Delta u + m^2 u = 0$, where $m > 0$, known as the *Klein–Gordon equation*.

   (a)   What is the energy? Show that it is a constant.

   (b)   Prove the causality principle for it.

## 9.2   THE WAVE EQUATION IN SPACE-TIME

We are looking for an explicit formula for the solution of

$$u_{tt} = c^2(u_{xx} + u_{yy} + u_{zz}) \tag{1}$$

$$u(\mathbf{x}, 0) = \phi(\mathbf{x}) \qquad u_t(\mathbf{x}, 0) = \psi(\mathbf{x}). \tag{2}$$

[like d'Alembert's formula (2.1.8)]. The answer is

$$u(\mathbf{x}, t_0) = \frac{1}{4\pi c^2 t_0} \iint_S \psi(\mathbf{x})\, dS + \frac{\partial}{\partial t_0}\left[\frac{1}{4\pi c^2 t_0}\iint_S \phi(\mathbf{x})\, dS\right], \tag{3}$$

where $S$ is the sphere of center $\mathbf{x}_0$ and radius $ct_0$. This famous formula is due to Poisson but is known as *Kirchhoff's formula*.

   We will derive (3) shortly, but first let's compare the result with the causality principle. The value of $u(\mathbf{x}_0, t_0)$ depends, according to (3), just on the values of $\psi(\mathbf{x})$ and $\phi(\mathbf{x})$ for $\mathbf{x}$ on the spherical *surface* $S = \{|\mathbf{x} - \mathbf{x}_0| = ct_0\}$ but not on the values of $\psi(\mathbf{x})$ and $\phi(\mathbf{x})$ *inside* this sphere. This statement can be

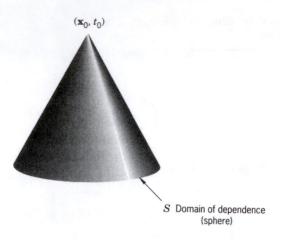

$(\mathbf{x}_0, t_0)$

$S$  Domain of dependence
(sphere)

**Figure 1**

inverted to say that the values of $\psi$ and $\phi$ at a spatial point $\mathbf{x}_1$ influence the solution only on the *surface* $\{|\mathbf{x} - \mathbf{x}_1| = ct\}$ of the light cone that emanates from $(\mathbf{x}_1, 0)$. This fact, illustrated in Figures 1 and 2, is called *Huygens's principle*. It means that any solution of the three-dimensional wave equation (e.g., any electromagnetic signal in a vacuum) propagates at *exactly* the speed $c$ of light, no faster and *no slower*.

This is the principle that allows us to see sharp images. It also means that any sound is carried through the air at exactly a fixed speed $c$ without "echoes," assuming the absence of walls or inhomogeneities in the air. Thus at any time $t$ a listener hears exactly what has been played at the time $t - d/c$, where $d$ is the distance to the musical instrument, rather than hearing a mixture of the notes played at various earlier times.

**Proof of Kirchhoff's Formula (3).**   We shall use the *method of spherical means*. Let the average (the mean) of $u(\mathbf{x}, t)$ on the sphere $\{|\mathbf{x}| = r\}$, of center

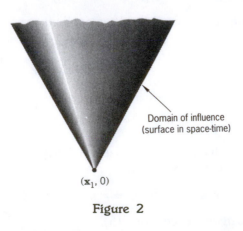

Domain of influence
(surface in space-time)

$(\mathbf{x}_1, 0)$

**Figure 2**

**0** and any radius $r$, be denoted by $\bar{u}(r, t)$. That is,

$$\bar{u}(r, t) = \frac{1}{4\pi r^2} \iint\limits_{|\mathbf{x}|=r} u(\mathbf{x}, t) \, dS$$

$$= \frac{1}{4\pi} \int_0^{2\pi} \int_0^\pi u(\mathbf{x}, t) \sin\theta \, d\theta \, d\phi, \tag{4}$$

where $x$, $y$, and $z$ are expressed in terms of the spherical coordinates $r$, $\theta$, and $\phi$. We'll now show that $\bar{u}$ itself satisfies the PDE

$$\boxed{(\bar{u})_{tt} = c^2 (\bar{u})_{rr} + 2c^2 \frac{1}{r} (\bar{u})_r \, .} \tag{5}$$

**Proof of (5).**   For simplicity, assume $c = 1$. Equation (5) follows from the rotational invariance of $\Delta$. Indeed, by Exercise 1, we have $\Delta(\bar{u}) = (\overline{\Delta u})$. That is, the laplacian of the mean is the mean of the laplacian. Therefore,

$$\Delta(\bar{u}) = (\overline{\Delta u}) = (\overline{u_{tt}}) = (\bar{u})_{tt}.$$

So $\bar{u}$ satisfies exactly the same PDE that $u$ does. Now in spherical coordinates we know that

$$\Delta \bar{u} = \bar{u}_{rr} + \frac{2}{r} \bar{u}_r + \text{angular derivative terms}$$

from (6.1.7). For $\bar{u}$, which depends only on $r$, the angular derivatives must vanish, so (5) is proved.

To give an alternative proof of (5), we apply the divergence theorem to the equation $u_{tt} = \Delta u$ over the domain $D = \{|\mathbf{x}| \le r\}$. Thus

$$\iiint\limits_D u_{tt} \, d\mathbf{x} = \iiint\limits_D \Delta u \, d\mathbf{x} = \iint\limits_{\partial D} \frac{\partial u}{\partial n} \, dS. \tag{6}$$

In spherical coordinates, (6) can be written explicitly as

$$\int_0^r \int_0^{2\pi} \int_0^\pi u_{tt} \, \rho^2 \sin\theta \, d\theta \, d\phi \, d\rho = \int_0^{2\pi} \int_0^\pi \frac{\partial u}{\partial r} r^2 \sin\theta \, d\theta \, d\phi$$

or

$$\int_0^r \rho^2 \overline{u_{tt}}(\rho, t) \, d\rho = r^2 \frac{\partial \bar{u}(r, t)}{\partial r}. \tag{7}$$

Differentiating (7) with respect to $r$, we get the integrand on the left side and two terms on the right side, as follows:

$$r^2 \overline{u_{tt}} = (r^2 \bar{u}_r)_r = r^2 \bar{u}_{rr} + 2r \bar{u}_r.$$

Dividing by $r^2$, we *again* get equation (5).

Continuing with the proof of (3), we now substitute

$$v(r, t) = r\overline{u}(r, t)$$

into the PDE (5). Then $v_r = r\overline{u}_r + \overline{u}$ and $v_{rr} = r\overline{u}_{rr} + 2\overline{u}_r$. So in terms of $v(r, t)$, (5) simplifies to

$$\boxed{v_{tt} = c^2 v_{rr}.}\tag{8}$$

Of course, equation (8) is valid only for $0 \leq r < \infty$. The function $v = r\overline{u}$ clearly vanishes at $r = 0$:

$$v(0, t) = 0 \qquad (\text{at } r = 0)\tag{9}$$

and satisfies the initial conditions

$$v(r, 0) = r\overline{\phi}(r) \qquad v_r(r, 0) = r\overline{\psi}(r) \qquad (\text{at } r = 0).\tag{10}$$

Thus we are reduced to a half-line problem in one dimension: the PDE (8), the BC (9), and the IC (10). This problem for $v$ was solved back in Section 3.2. Its solution is given by the formula [from (3.2.3)]

$$v(r, t) = \frac{1}{2c} \int_{ct-r}^{ct+r} s\overline{\psi}(s)\, ds + \frac{\partial}{\partial t}\left[ \frac{1}{2c} \int_{ct-r}^{ct+r} s\overline{\phi}(s)\, ds \right]\tag{11}$$

for $0 \leq r \leq ct$ and by a different formula for $r \geq ct$.

The next step is to recover $u$ at the origin $r = 0$:

$$u(\mathbf{0}, t) = \overline{u}(0, t) = \lim_{r \to 0} \frac{v(r, t)}{r}$$

$$= \lim_{r \to 0} \frac{v(r, t) - v(0, t)}{r} = \frac{\partial v}{\partial r}(0, t).\tag{12}$$

Differentiating (11), we have

$$\frac{\partial v}{\partial r} = \frac{1}{2c}[(ct + r)\overline{\psi}(ct + r) + (ct - r)\overline{\psi}(ct - r)] + \cdots,$$

where $\cdots$ denotes a similar term depending on $\overline{\phi}$. When we put $r = 0$, it simplifies to $[\partial v/\partial r](0, t) = (1/2c)[2ct\overline{\psi}(ct)] = t\overline{\psi}(ct)$. This is the right side of (12). Therefore,

$$u(\mathbf{0}, t) = t\overline{\psi}(ct) = \frac{1}{4\pi c^2 t} \iint_{|\mathbf{x}|=ct} \psi(\mathbf{x})\, dS + \cdots.\tag{13}$$

This is precisely the first term in formula (3) (in the case that the spatial point is the origin and the time is denoted by $t$). It is just the time $t$ multiplied by the average of $\psi$ on the sphere of center $\mathbf{0}$ and radius $ct$.

Next we translate the result (13). If $\mathbf{x}_0$ is *any* spatial point at all, let

$$w(\mathbf{x}, t) = u(\mathbf{x} + \mathbf{x}_0, t).$$

This is the solution of the wave equation whose initial data are $\phi(\mathbf{x} + \mathbf{x}_0)$ and $\psi(\mathbf{x} + \mathbf{x}_0)$. So we can apply the result (13) to $w(\mathbf{x}, t)$, in order to obtain the formula

$$u(\mathbf{x}_0, t) = w(\mathbf{0}, t) = \frac{1}{4\pi c^2 t} \iint\limits_{|\mathbf{x}|=ct} \psi(\mathbf{x} + \mathbf{x}_0) \, dS + \cdots$$

$$= \frac{1}{4\pi c^2 t} \iint\limits_{|\mathbf{x} - \mathbf{x}_0| = ct} \psi(\mathbf{x}) \, dS + \cdots. \tag{14}$$

This is precisely the first term of (3).

A little thought shows that the second term in (3) works in the same way. In fact, if we replace $\psi$ by $\phi$ in the first term of (11) and take the time derivative, we get the second term. The two terms in (3) must have the same relationship. $\qquad\square$

## SOLUTION IN TWO DIMENSIONS

We shall see that Huygens's principle is *not* valid in two dimensions! What we want to solve is

$$u_{tt} = c^2(u_{xx} + u_{yy}) \tag{15}$$

$$u(x, y, 0) = \phi(x, y), \quad u_t(x, y, 0) = \psi(x, y). \tag{16}$$

The key idea is to regard $u(x, y, t)$ as a solution of the three-dimensional problem *which just happens not to depend on z*. So it must be given by the Kirchhoff formula. Let's again assume for the sake of simplicity that $\phi \equiv 0$ and that $(x_0, y_0) = (0, 0)$. By the three-dimensional formula (13) we have

$$u(0, 0, t) = \frac{1}{4\pi c^2 t} \iint\limits_{x^2 + y^2 + z^2 = c^2 t^2} \psi(x, y) \, dS.$$

This is a correct formula for the solution of (15), (16), but we can simplify it as follows. It is twice the integral over the top hemisphere $z = \sqrt{c^2 t^2 - x^2 - y^2}$. On the hemisphere (see Figure 3) we can use the usual formula for the surface element $dS$ in terms of the coordinates $(x, y)$, getting the double integral

$$u(0, 0, t) = \frac{1}{2\pi c^2 t} \iint\limits_{x^2 + y^2 \leq c^2 t^2} \psi(x, y) \left[ 1 + \left(\frac{\partial z}{\partial x}\right)^2 + \left(\frac{\partial z}{\partial y}\right)^2 \right]^{1/2} dx \, dy.$$

$$\tag{17}$$

The term in brackets inside (17) equals

$$[\,\cdot\,] = 1 + \left(-\frac{x}{z}\right)^2 + \left(-\frac{y}{z}\right)^2 = \frac{c^2 t^2}{z^2} = \frac{c^2 t^2}{c^2 t^2 - x^2 - y^2}.$$

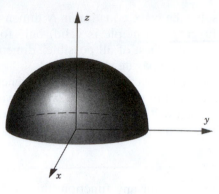

**Figure 3**

Hence (17) becomes

$$u(0,\ 0,\ t) = \frac{1}{2\pi c} \iint\limits_{x^2+y^2\le c^2t^2} \frac{\psi(x,\ y)}{(c^2t^2 - x^2 - y^2)^{1/2}}\ dx\,dy. \tag{18}$$

This is the solution formula at the point $(0, 0, t)$. At a general point, the formula is

$$u(x_0, y_0, t_0) = \iint\limits_{D} \frac{\psi(x,\ y)}{\left[c^2t_0^2 - (x - x_0)^2 - (y - y_0)^2\right]^{1/2}} \frac{dx\,dy}{2\pi c}$$

$$+ \frac{\partial}{\partial t_0}(\text{same expression with } \phi). \tag{19}$$

where $D$ is the disk $\{(x - x_0)^2 + (y - y_0)^2 \le c^2t_0^2\}$. (Why?)

Our formula (19) shows that the value of $u(x_0, y_0, t_0)$ does depend on the values of $\phi(x, y)$ and $\psi(x, y)$ *inside* the cone:

$$(x - x_0)^2 + (y - y_0)^2 \le c^2t_0^2.$$

This means that *Huygens's principle is false in two dimensions*. For instance, when you drop a pebble onto a calm pond, surface waves are created that (approximately) satisfy the two-dimensional wave equation with a certain speed $c$, where $x$ and $y$ are horizontal coordinates. A water bug whose distance from the point of impact is $\delta$ experiences a wave first at time $t = \delta/c$ but thereafter continues to feel ripples. These ripples die down, like $t^{-1}$ according to Exercise 18, but theoretically continue forever. (Physically, when the ripples become small enough, the wave equation is not really valid anymore as other physical effects begin to dominate.)

One can speculate what it would be like to live in Flatland, a two-dimensional world. Communication would be difficult because light and sound waves would not propagate sharply. It would be a noisy world! It turns out that if you solve the wave equation in $N$ dimensions, signals propagate sharply (i.e., Huygens's principle is valid) only for dimensions $N = 3, 5, 7, \ldots$. Thus three is the "best of all possible" dimensions, the smallest dimension in which signals propagate sharply!

In fact, the method of spherical means can be generalized to any odd dimension $\geq 5$. For each odd dimension $n = 2m + 1$ we can "descend" to the even dimension $2m$ below it to get a formula that shows that Huygens's principle is false in $2m$ dimensions [CH].

## EXERCISES

1. Prove that $\Delta(\bar{u}) = \overline{(\Delta u)}$ for any function; that is, the laplacian of the average is the average of the laplacian. (*Hint:* Write $\Delta u$ in spherical coordinates and show that the angular terms have zero average on spheres centered at the origin.)

2. Verify that (3) is correct in the case of the example $u(x, y, z, t) \equiv t$.

3. Solve the wave equation in three dimensions with the initial data $\phi \equiv 0$, $\psi(x, y, z) = y$, by use of (3).

4. Solve the wave equation in three dimensions with the initial data $\phi \equiv 0$, $\psi(x, y, z) = x^2 + y^2 + z^2$. (*Hint:* Use (5).)

5. Where does a three-dimensional wave have to vanish if its initial data $\phi$ and $\psi$ vanish outside a sphere?

6. (a) Let $S$ be the sphere of center $\mathbf{x}$ and radius $R$. What is the surface area of $S \cap \{|\mathbf{x}| < \rho\}$, the portion of $S$ that lies within the sphere of center $\mathbf{0}$ and radius $\rho$?

   (b) Solve the wave equation in three dimensions for $t > 0$ with the initial conditions $\phi(\mathbf{x}) \equiv 0$, $\psi(\mathbf{x}) = A$ for $|\mathbf{x}| < \rho$, and $\psi(\mathbf{x}) = 0$ for $|\mathbf{x}| > \rho$, where $A$ is a constant. Sketch the regions in spacetime that illustrate your answer. (This is like the hammer blow of Section 2.1.)

   (c) Sketch the graph of the solution ($u$ versus $|\mathbf{x}|$) for $t = \frac{1}{2}, 1$, and $2$, assuming that $\rho = c = A = 1$. (This is a "movie" of the solution.)

   (d) Sketch the graph of $u$ versus $t$ for $|\mathbf{x}| = \frac{1}{2}$ and $2$, assuming that $\rho = c = A = 1$. (This is what a stationary observer sees.)

   (e) Let $|\mathbf{x}_0| < \rho$. Ride the wave along a light ray emanating from $(\mathbf{x}_0, 0)$. That is, look at $u(\mathbf{x}_0 + t\mathbf{v}, t)$ where $|\mathbf{v}| = c$. Prove that
   $$t \cdot u(\mathbf{x}_0 + t\mathbf{v}, t) \text{ converges as } t \to \infty.$$

   (*Hint:* (a) Divide into cases depending on whether one sphere contains the other or not. Use the law of cosines. (b) Use Kirchhoff's formula.)

7. (a) Solve the wave equation in three dimensions for $t > 0$ with the initial conditions $\phi(\mathbf{x}) = A$ for $|\mathbf{x}| < \rho$, $\phi(\mathbf{x}) = 0$ for $|\mathbf{x}| > \rho$, and $\psi|\mathbf{x}| \equiv 0$, where $A$ is a constant. (This is somewhat like the plucked string.) (*Hint:* Differentiate the solution in Exercise 6(b).)

(b)  Sketch the regions in space-time that illustrate your answer. Where does the solution have jump discontinuities?

(c)  Let $|\mathbf{x}_0| < \rho$. Ride the wave along a light ray emanating from $(\mathbf{x}_0, 0)$. That is, look at $u(\mathbf{x}_0 + t\mathbf{v}, t)$ where $|\mathbf{v}| = c$. Prove that

$$t \cdot u(\mathbf{x}_0 + t\mathbf{v}, t) \text{ converges as } t \to \infty.$$

8.  Carry out the derivation of the second term in (3).

9.  (a)  For any solution of the three-dimensional wave equation with initial data vanishing outside some sphere, show that $u(x, y, z, t) = 0$ for fixed $(x, y, z)$ and large enough $t$.

(b)  Prove that $u(x, y, z, t) = O(t^{-1})$ *uniformly* as $t \to \infty$; that is, prove that $t \cdot u(x, y, z, t)$ is a bounded function of $x, y, z$, and $t$. (*Hint:* Use Kirchhoff's formula.)

10.  Derive the mean value property of harmonic functions $u(x, y, z)$ by the following method. A harmonic function is a wave that happens not to depend on time, so that its mean value $\bar{u}(r, t) = \bar{u}(r)$ satisfies (5). Deduce that $\bar{u}(r) = u(\mathbf{0})$.

11.  Find all the spherical solutions of the three-dimensional wave equation; that is, find the solutions that depend only on $r$ and $t$. (*Hint:* See (5).)

12.  Solve the three-dimensional wave equation in $\{r \neq 0, t > 0\}$ with zero initial conditions and with the limiting condition

$$\lim_{r \to 0} 4\pi r^2 u_r(r, t) = g(t).$$

Assume that $g(0) = g'(0) = g''(0) = 0$.

13.  Solve the wave equation in the half-space $\{(x, y, z, t): z > 0\}$ with the Neumann condition $\partial u / \partial z = 0$ on $z = 0$, and with initial data $\phi(x, y, z) \equiv 0$ and general $\psi(x, y, z)$. (*Hint:* See (3) and use the method of reflection.)

14.  Why doesn't the method of spherical means work for two-dimensional waves?

15.  Obtain the general solution formula (19) in two dimensions from the special case (18).

16.  (a)  Solve the wave equation in two dimensions for $t > 0$ with the initial conditions $\phi(\mathbf{x}) \equiv 0$, $\psi(\mathbf{x}) = A$ for $|\mathbf{x}| < \rho$, and $\psi(\mathbf{x}) = 0$ for $|\mathbf{x}| > \rho$, where $A$ is a constant. Do not carry out the integral.

(b)  Under the same conditions find a simple formula for the solution $u(\mathbf{0}, t)$ at the origin by carrying out the integral.

17.  Use the result of Exercise 16 to compute the limit of $t \cdot u(\mathbf{0}, t)$ as $t \to \infty$.

18.  For any solution of the two-dimensional wave equation with initial data vanishing outside some circle, prove that $u(x, y, t) = O(t^{-1})$ for fixed $(x, y)$ as $t \to \infty$; that is, $t \cdot u(x, y, t)$ is a bounded function of $t$ for fixed $x$ and $y$. Note the contrast to three dimensions. (*Hint:* Use formula (19).)

19. (*difficult*) Show, however, that if we are interested in uniform convergence, that $u(x, y, t) = O(t^{-1/2})$ *uniformly* as $t \to \infty$.

20. "Descend" from two dimensions to one as follows. Let $u_{tt} = c^2 u_{xx}$ with initial data $\phi(x) \equiv 0$ and general $\psi(x)$. Imagine that we don't know d'Alembert's solution formula. Think of $u(x, t)$ as a solution of the two-dimensional equation that happens not to depend on $y$. Plug it into (19) and carry out the integration.

## 9.3    RAYS, SINGULARITIES, AND SOURCES

In this section we discuss the geometry of the characteristics, the geometric concepts occurring in relativity theory, and the fact that wave singularities travel along the characteristics. We also solve the inhomogeneous wave equation.

### CHARACTERISTICS

A *light ray* is the path of a point in three dimensions moving in a straight line at speed $c$. That is, $|d\mathbf{x}/dt| = c$, or

$$\mathbf{x} = \mathbf{x}_0 + \mathbf{v}_0 t \qquad \text{where } |\mathbf{v}_0| = c. \tag{1}$$

Such a ray is orthogonal to the sphere $|\mathbf{x} - \mathbf{x}_0| = c|t|$.

We saw earlier in this chapter that the basic geometry of the wave equation is the light cone $|\mathbf{x}| = c|t|$. It is made up of all the light rays (1) with $\mathbf{x}_0 = \mathbf{0}$.

Now consider any surface $S$ in space-time. Its *time slices* are denoted by $S_t = S \cap \{t = \text{constant}\}$. Thus $S$ is a three-dimensional surface sitting in four-dimensional space-time and each $S_t$ is an ordinary two-dimensional surface. $S$ is called a *characteristic surface* if it is a union of light rays each of which is orthogonal in three-dimensional space to the time slices $S_t$.

For a more analytical description of a characteristic surface, let's suppose that $S$ is the level surface of a function of the form $t - \gamma(\mathbf{x})$. That is, $S = \{(\mathbf{x}, t): t - \gamma(\mathbf{x}) = k\}$ for some constant $k$. Then the time slices are $S_t = \{\mathbf{x}: t - \gamma(\mathbf{x}) = k\}$. Here is the analytical description.

**Theorem 1.**   All the level surfaces of $t - \gamma(\mathbf{x})$ are characteristic if and only if $|\nabla \gamma(\mathbf{x})| = 1/c$.

**Proof.**   First suppose that all the level surfaces of $t - \gamma(\mathbf{x})$ are characteristic. Let $\mathbf{x}_0$ be any spatial point. Let $S$ be the level surface of $t - \gamma(\mathbf{x})$ that contains the point $(\mathbf{x}_0, 0)$. Thus $S = \{(x, t): t - \gamma(\mathbf{x}) = -\gamma(\mathbf{x}_0)\}$. Since $S$ is characteristic and $(\mathbf{x}_0, 0) \in S$, there is a ray of the form (1) that is contained in $S$ for which $\mathbf{v}_0$ is orthogonal to $S_t$ for all $t$. Since the ray lies on $S$, it satisfies the equation

$$t - \gamma(\mathbf{x}_0 + \mathbf{v}_0 t) = -\gamma(\mathbf{x}_0) \tag{2}$$

for all $t$. Differentiating this equation with respect to $t$, we find that $1 - \mathbf{v}_0 \cdot \nabla\gamma(\mathbf{x}_0 + \mathbf{v}_0 t) = 0$. Setting $t = 0$, we get $\mathbf{v}_0 \cdot \nabla\gamma(\mathbf{x}_0) = 1$.

On the other hand, the time slice $S_0 = \{\mathbf{x}: \gamma(\mathbf{x}) = \gamma(\mathbf{x}_0)\}$ has $\nabla\gamma(\mathbf{x}_0)$ as a normal vector. Another normal vector is $\mathbf{v}_0$, so that $\nabla\gamma(\mathbf{x}_0)$ and $\mathbf{v}_0$ are parallel. Therefore, $1 = |\mathbf{v}_0 \cdot \nabla\gamma(\mathbf{x}_0)| = |\mathbf{v}_0||\nabla\gamma(\mathbf{x}_0)| = c|\nabla\gamma(\mathbf{x}_0)|$. Hence $|\nabla\gamma(\mathbf{x}_0)| = 1/c$. This is what we wanted to prove. For the converse, see Exercise 2.  $\square$

### Example 1.

Starting from any surface $S_0$ at all in 3-space at $t = 0$, we could draw straight lines (1) of slope $c$ with $\mathbf{x}_0 \in S_0$ to construct a characteristic surface $S$. For instance, the *plane* $S_0 = \{\mathbf{x}: a_1 x + a_2 y + a_3 z = b\}$ with $a_1^2 + a_2^2 + a_3^2 = 1$ gives rise in this manner to the plane characteristic surface $S = \{(\mathbf{x}, t): a_1 x + a_2 y + a_3 z - ct = b\}$. It also gives rise to the plane characteristic surface $S' = \{(\mathbf{x}, t): a_1 x + a_2 y + a_3 z + ct = b\}$. Similarly, the *sphere* $S_0 = \{\mathbf{x}: |\mathbf{x} - \mathbf{x}_0| = R\}$ gives rise to the pair of characteristic surfaces $S = \{(\mathbf{x}, t): |\mathbf{x} - \mathbf{x}_0| = R \pm ct\}$.  $\square$

### RELATIVISTIC GEOMETRY

In relativity theory the following terminology is commonly used. The *past* (or *past history*) of the point $(\mathbf{0}, 0)$ is the set $\{ct < -|\mathbf{x}|\}$, its *future* is $\{ct > -|\mathbf{x}|\}$, and its *present* is $\{-|\mathbf{x}| < ct < |\mathbf{x}|\}$. A four-dimensional vector $(\mathbf{v}, v^0)$ is called (see Figure 1)

*Timelike* if $|v^0| > c|\mathbf{v}|$

*Spacelike* if $|v^0| < c|\mathbf{v}|$

*Null* (or *characteristic*) if $|v^0| = c|\mathbf{v}|$.

Thus a timelike vector points into either the future or the past. A straight line in space-time is called a *ray* (or *bicharacteristic*) if its tangent vector is null; it projects onto a light ray [as defined by (1)].

Still another description of a surface in space-time being characteristic is that its (four-dimensional) normal vector is a null vector. Indeed, if the surface is represented as $S = \{t = \gamma(\mathbf{x})\}$, then a normal 4-vector is $(\nabla\gamma(\mathbf{x}), -1)$. $S$ is characteristic if this vector is null. That is, $1 = |v^0| = c|\mathbf{v}| = c|\nabla\gamma(\mathbf{x})|$, in agreement with Theorem 1.

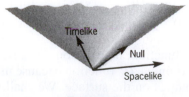

Figure  1

A surface is *spacelike* if all its normal vectors are timelike, that is, if $|\nabla\gamma(\mathbf{x})| < 1/c$. For instance, the initial surface $\{t=0\}$, considered as a surface in space-time, is spacelike since $\gamma \equiv 0$. *The spacelike surfaces are just the ones that naturally carry initial conditions*, as stated in the following theorem.

**Theorem 2.**   If $S$ is any spacelike surface, then one can uniquely solve the initial-value problem

$$u_{tt} = c^2 \Delta u \qquad \text{in all of space-time}$$

$$u = \phi \qquad \text{and} \qquad \frac{\partial u}{\partial n} = \psi \text{ on } S, \tag{3}$$

where $\partial/\partial n$ indicates the derivative in the direction normal to $S$.

If $S$ is represented as $\{t = \gamma(\mathbf{x})\}$, the second initial condition in (3) means explicitly that

$$u_t - \nabla\gamma \cdot \nabla u = [1 + |\nabla\gamma|^2]^{1/2}\psi \qquad \text{for } t = \gamma(\mathbf{x}). \tag{4}$$

(Why?) We omit the proof of Theorem 2.

## SINGULARITIES

Here is another basic property of characteristic surfaces that is also proved in advanced texts [CH].

**Theorem 3.**   Characteristic surfaces are the only surfaces that can carry the singularities of solutions of the wave equation.

The idea is that information gets transported along light rays (cf. Section 2.5) and a singularity is a very specific bit of information. A *singularity* of a solution is any point where the solution, or a derivative of it of some order, is not continuous. For instance, in the plucked string of Section 2.1, the singularity is the jump discontinuity in the first derivative; it clearly occurs along a characteristic.

## Example 2.

A more elaborate example of a singularity is the following. Let

$$\begin{aligned} u(\mathbf{x}, t) &= \tfrac{1}{2}v(\mathbf{x}, t)[t - \gamma(\mathbf{x})]^2 & \text{for } \gamma(\mathbf{x}) \le t \\ u(\mathbf{x}, t) &= 0 & \text{for } \gamma(\mathbf{x}) \ge t, \end{aligned} \tag{5}$$

where $v(\mathbf{x}, t)$ is a $C^2$ function, nonzero on the surface $S = \{t = \gamma(\mathbf{x})\}$. This function $u(\mathbf{x}, t)$ is only a $C^1$ function because its second derivatives have jump discontinuities on the surface. We shall show that *if $u(\mathbf{x}, t)$ solves the wave equation, then the surface must be characteristic.*

Indeed, on one side $\{\gamma(\mathbf{x}) < t\}$ of the surface, we calculate

$$u_t = v(t - \gamma) + \tfrac{1}{2}v_t(t - \gamma)^2,$$

$$u_{tt} = v + 2v_t(t - \gamma) + \tfrac{1}{2}v_{tt}(t - \gamma)^2,$$

$$\nabla u = -v\nabla\gamma(t - \gamma) + \tfrac{1}{2}\nabla v(t - \gamma)^2,$$

$$\Delta u = \nabla\cdot\nabla u = v|\nabla\gamma|^2 - v\Delta\gamma(t-\gamma) - 2\nabla v\cdot\nabla\gamma(t-\gamma) + \tfrac{1}{2}\Delta v(t-\gamma)^2.$$

Hence, on the side $\{\gamma(\mathbf{x}) < t\}$, we have

$$0 = u_u - c^2\Delta u = (v)(1 - c^2|\nabla\gamma|^2)$$
$$+ (2v_t + c^2 v\Delta\gamma + 2c^2\nabla v\cdot\nabla\gamma)(t - \gamma) + \tfrac{1}{2}(v_{tt} - c^2\Delta v)(t - \gamma)^2. \quad (6)$$

Of course, everything is zero on the other side $\{\gamma(\mathbf{x}) > t\}$. So for $u(\mathbf{x}, t)$ to be a solution *across* the surface, the expression (6) must be zero *on* the surface $\{t - \gamma(\mathbf{x}) = 0\}$. Set $t = \gamma(\mathbf{x})$ in (6). Then on $S$ we have $(v)(1 - c^2|\nabla\gamma|^2) = 0$, or $|\nabla\gamma| = 1/c$, which means that the surface $S$ is characteristic. This proves the assertion made above. In diffraction theory the equation $|\nabla\gamma| = 1/c$ is called the *eikonal equation*. It is a nonlinear first-order PDE satisfied by $\gamma$.

Because the first term on the right side of (6) is zero, (6) may be divided by $(t - \gamma)$. So it also implies that

$$0 = (2v_t + c^2 v\Delta\gamma + 2c^2\nabla v\cdot\nabla\gamma) + \tfrac{1}{2}(v_{tt} - c^2\Delta v)(t - \gamma) \quad (7)$$

on one side of $S$. Matching across $S$ again, it follows that (7) must be valid when $t = \gamma(\mathbf{x})$, which means that

$$\boxed{v_t + c^2\nabla\gamma\cdot\nabla v = -\tfrac{1}{2}c^2(\Delta\gamma)v.} \quad (8)$$

This is called the *transport equation*; it is a linear first-order PDE satisfied by $v$ on $S$.

To understand it, notice that $\mathcal{D} = \partial_t + c^2\nabla\gamma\cdot\nabla$ is a derivative in a direction tangent to $S$. In fact, $\mathcal{D}$ is the derivative in the direction of the ray $d\mathbf{x}/dt = c^2\nabla\gamma$ with $|d\mathbf{x}/dt| = c^2|\nabla\gamma| = c$. Thus $v(\mathbf{x}, t)$ is "transported" along the ray by the differential equation (8). Equation (8) can be solved by the methods of Section 1.2. Equation (8) also implies that $v \neq 0$ everywhere along the ray because $v \neq 0$ where the ray meets $S$, by assumption.  $\square$

## WAVE EQUATION WITH A SOURCE

Now we shall solve the three-dimensional problem

$$u_{tt} - c^2\Delta u = f(\mathbf{x}, t)$$
$$u(\mathbf{x}, 0) \equiv 0, \quad u_t(\mathbf{x}, 0) \equiv 0 \quad (9)$$

using the operator method of Section 3.4.

The solution we found in Section 9.2 for the *homogeneous* problem with initial data $\phi$ and $\psi$ was

$$(\partial_t \mathscr{S}(t_0)\phi)(\mathbf{x}_0) + (\mathscr{S}(t_0)\psi)(\mathbf{x}_0),$$

where

$$(\mathscr{S}(t_0)\psi)(\mathbf{x}_0) = \frac{1}{4\pi c^2 t_0} \iint\limits_S \psi(\boldsymbol{\xi}) \, dS_{\boldsymbol{\xi}} \tag{10}$$

and $S = \{\boldsymbol{\xi}: |\boldsymbol{\xi} - \mathbf{x}_0| = ct_0\}$ is a sphere. Now let's drop the subscripts "0." The operator $\mathscr{S}(t)$ is the *source operator*.

Just as in Section 3.4, the unique solution of (9) is expressible in terms of the source operator as

$$\boxed{\; u(\mathbf{x}, t) = \int_0^t \mathscr{S}(t - s) f(\mathbf{x}, s) \, ds. \;} \tag{11}$$

This is sometimes called the *Duhamel formula*. Inside the integral (11), the operator $\mathscr{S}(t - s)$ acts on $f(\mathbf{x}, s)$ as a function of $\mathbf{x}$, with $s$ merely playing the role of a parameter. Formula (11) means that in (10) we must replace $t_0$ by $(t - s)$, $\mathbf{x}_0$ by $\mathbf{x}$, and $\psi(\boldsymbol{\xi})$ by $f(\boldsymbol{\xi}, s)$. Thus

$$u(\mathbf{x}, t) = \int_0^t \frac{1}{4\pi c^2(t - s)} \iint\limits_{\{|\boldsymbol{\xi}-\mathbf{x}|=c(t-s)\}} f(\boldsymbol{\xi}, s) \, dS_{\boldsymbol{\xi}} \, ds$$

$$= \frac{1}{4\pi c} \int_0^t \iint\limits_{\{|\boldsymbol{\xi}-\mathbf{x}|=c(t-s)\}} \frac{f(\boldsymbol{\xi}, t - |\boldsymbol{\xi} - \mathbf{x}|/c)}{|\boldsymbol{\xi} - \mathbf{x}|} \, dS_{\boldsymbol{\xi}} \, ds, \tag{12}$$

where we have substituted the value of $s = t - |\boldsymbol{\xi} - \mathbf{x}|/c$ on the sphere $S$.

Now the last expression is exactly an iterated integral in spherical coordinates. The region of integration in space-time is the backward cone surface sketched in Figure 2. The coordinates $\boldsymbol{\xi}$ run over the base of the conical surface, which is the ball $\{|\boldsymbol{\xi} - \mathbf{x}| = c(t - s)\}$. The volume element $d\boldsymbol{\xi}$ is the ordinary one $d\boldsymbol{\xi} = c \, dS_{\boldsymbol{\xi}} \, ds$. Thus the iterated integral combines into a triple

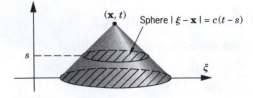

**Figure 2**

integral to produce the solution formula

$$u(\mathbf{x}, t) = \frac{1}{4\pi c^2} \iiint_{\{|\boldsymbol{\xi}-\mathbf{x}|\le ct\}} \frac{f(\boldsymbol{\xi}, t - |\boldsymbol{\xi} - \mathbf{x}|/c)}{|\boldsymbol{\xi} - \mathbf{x}|}\, d\boldsymbol{\xi}. \tag{13}$$

This result says that, in order to solve (9), you just multiply $f(\boldsymbol{\xi}, s)$ by the "potential" $1/(4\pi c|\boldsymbol{\xi} - \mathbf{x}|)$ and integrate it over the backward cone. The backward cone consists exactly of the domain of dependence of the given point $(\mathbf{x}, t)$; that is, those points that can reach $(\mathbf{x}, t)$ via a light ray from some time $s$ in the past $(0 \le s \le t)$.

It is interesting to compare this formula with the solution of Poisson's equation in the whole of three-dimensional space. See (7.3.7) without the boundary term and with $G = -1/(4\pi r)$. Changing $\mathbf{x}_0$ to $\mathbf{x}$, and $\mathbf{x}$ to $\boldsymbol{\xi}$, formula (7.3.7) says that the bounded solution of Poisson's equation $-\Delta w = f$ in all of 3-space is

$$w(\mathbf{x}) = \frac{1}{4\pi c} \iiint \frac{f(\boldsymbol{\xi})}{|\boldsymbol{\xi} - \mathbf{x}|}\, d\boldsymbol{\xi}. \tag{14}$$

The only difference between (13) and (14) is that time is "retarded" by the amount $|\boldsymbol{\xi} - \mathbf{x}|/c$. So in the formula (13) the potential is called *retarded*.

## EXERCISES

1. Let $S$ be a characteristic surface for which $S \cap \{(x, y, z): t = 0\}$ is the sphere $\{x^2 + y^2 + z^2 = a^2\}$. Describe $S$ geometrically.

2. Prove the converse of Theorem 1. That is, prove that a level surface of $t - \gamma(\mathbf{x})$ is characteristic if $\gamma(\mathbf{x})$ satisfies the nonlinear PDE

$$|\nabla\gamma(\mathbf{x})| \equiv \frac{1}{c}. \tag{$*$}$$

   (*Hint:* Differentiate the equation ($*$) to get $\Sigma \gamma_{ij}(\mathbf{x})\gamma_j(\mathbf{x}) = 0$, where subscripts denote partial derivatives. Show that a curve, which satisfies the ODE $d\mathbf{x}/dt = c^2 \nabla\gamma(\mathbf{x})$, also satisfies $d^2\mathbf{x}/dt^2 = 0$ and hence is a ray. Show that $t - \gamma(\mathbf{x})$ is constant along a ray. Deduce that any level surface of $t - \gamma(\mathbf{x})$ is characteristic.)

3. Prove Theorem 2 in the one-dimensional case. That is, if $\mathcal{C}$ is a spacelike curve in the $xt$ plane, there is a unique solution of $u_{tt} = c^2 u_{xx}$ with $u = \phi$ and $\partial u/\partial n = \psi$ on $\mathcal{C}$.

4. Verify that the solution given by (5) has second derivatives which have jump discontinuities on the surface $S = \{(\mathbf{x}, t): t = \gamma(\mathbf{x})\}$.

5. Verify the correctness of (13) for the example $u(x, y, z, t) = t^2$ and $f(x, y, z, t) \equiv 2$.

6.  Show that the unique solution of (9) is expressible in terms of the source operator by the simple formula (11).

7.  (*difficult*) Solve $u_{tt} - c^2 \Delta u = f(\mathbf{x})$, where $f(\mathbf{x}) = A$ for $|\mathbf{x}| < \rho$, $f(\mathbf{x}) = 0$ for $|\mathbf{x}| > \rho$, $A$ is a constant, and the initial data are identically zero. Sketch the regions in space-time that illustrate your answer. (*Hint:* Use (13) and find the volume of intersection of two balls, or use (11) and Exercise 9.2.6(b).)

8.  Carry out the passage from (11) to (13) more explicitly using spherical coordinates.

9.  Simplify formula (13) for the solution of $u_{tt} - c^2 \Delta u = f(\mathbf{x}, t)$ in the special case that $f$ is spherically symmetric $[f = f(r, t)]$.


## 9.4   THE DIFFUSION AND SCHRÖDINGER EQUATIONS

### THREE-DIMENSIONAL DIFFUSION EQUATION

Consider the diffusion equation in all of 3-space,

$$\frac{\partial u}{\partial t} = k \Delta u = k \left( \frac{\partial^2 u}{\partial x^2} + \frac{\partial^2 u}{\partial y^2} + \frac{\partial^2 u}{\partial z^2} \right) \tag{1}$$

$$u(\mathbf{x}, 0) = \phi(\mathbf{x}). \tag{2}$$

It is very easy to solve, using our knowledge from Chapter 2.

**Theorem 1.**   For any bounded continuous function $\phi(\mathbf{x})$, the solution of (1), (2) is

$$u(\mathbf{x}, t) = \frac{1}{(4\pi kt)^{3/2}} \iiint \exp\left( -\frac{|\mathbf{x} - \mathbf{x}'|^2}{4kt} \right) \phi(\mathbf{x}') \, d\mathbf{x}' \tag{3}$$

for all $t > 0$. The dummy variables of integration, $\mathbf{x}' = (x', y', z')$, run over all of 3-space.

**Proof.**   To derive (3), let's denote by

$$S(z, t) = \frac{1}{(4\pi kt)^{1/2}} e^{-z^2/4kt}$$

the *one*-dimensional source function. Let

$$S_3(x, y, z, t) = S(x, t) S(y, t) S(z, t) \tag{4}$$

be the *product of three such functions in the different variables*. Then

$$\frac{\partial S_3}{\partial t} = \frac{\partial S}{\partial t}(x, t) \cdot S(y, t) \cdot S(z, t) + \text{(two similar terms)}$$

$$= k\frac{\partial^2 S}{\partial x^2}(x, t) \cdot S(y, t) \cdot S(z, t) + S(x, t) \cdot k\frac{\partial^2 S}{\partial y^2}(y, t) \cdot S(z, t)$$

$$+ S(x, t) \cdot S(y, t) \cdot k\frac{\partial^2 S}{\partial z^2}(z, t)$$

$$= k\left(\frac{\partial^2}{\partial x^2} + \frac{\partial^2}{\partial y^2} + \frac{\partial^2}{\partial z^2}\right)(S(x, t)S(y, t)S(z, t))$$

$$= k\Delta S_3.$$

So $S_3(\mathbf{x}, t)$ satisfies the three-dimensional diffusion equation.

We claim that $S_3$ is the source function. To prove it, note that

$$\iiint S_3(\mathbf{x}, t)\, d\mathbf{x} = \left(\int S(x, t)\, dx\right)\left(\int S(y, t)\, dy\right)\left(\int S(z, t)\, dz\right)$$

$$= 1^3 = 1. \tag{5}$$

Now in the special case that $\phi(x, y, z)$ depends only on $z$, we have

$$\lim_{t \to 0} \iiint S_3(\mathbf{x} - \mathbf{x}', t)\, \phi(z')\, d\mathbf{x}'$$

$$= \left[\int S(x - x', t)\, dx'\right] \cdot \left[\int S(y - y', t)\, dy'\right] \cdot \left[\lim_{t \to 0}\int S(z - z', t)\phi(z')\, dz'\right]$$

$$= 1 \cdot 1 \cdot \lim_{t \to 0}\int S(z - z', t)\phi(z')\, dz' = \phi(z)$$

by Theorem 3.5.1. In a similar way, we can show that

$$\lim_{t \to 0} \iiint S_3(\mathbf{x} - \mathbf{x}', t)\phi(\mathbf{x}')\, d\mathbf{x}' = \phi(\mathbf{x}) \tag{6}$$

if $\phi(\mathbf{x})$ is a product $\phi(x)\psi(y)\zeta(z)$. Therefore, (6) is also true of any linear combination of such products. It can be deduced (Exercise 2) that the same is true for any bounded continuous function $\phi(\mathbf{x})$. Equations (5) and (6) imply that $S_3(\mathbf{x}, t)$ is the source function.

Consequently, the unique bounded solution of (1), (2) is

$$u(\mathbf{x}, t) = \iiint S_3(\mathbf{x} - \mathbf{x}', t)\phi(\mathbf{x}')\, d\mathbf{x}'.$$

But the explicit formula for $S_3$ is

$$S_3(\mathbf{x}, t) = \left( \frac{1}{\sqrt{4\pi kt}} e^{-x^2/4kt} \right) \cdot \left( \frac{1}{\sqrt{4\pi kt}} e^{-y^2/4kt} \right) \cdot \left( \frac{1}{\sqrt{4\pi kt}} e^{-z^2/4kt} \right)$$

$$= \frac{1}{(4\pi kt)^{3/2}} e^{-(x^2+y^2+z^2)/4kt}. \tag{7}$$

and therefore we have derived (3). The complete proof, including the convergence of the triple integral and so on, can also be carried out directly just as in Section 3.5. $\qquad\square$

## SCHRÖDINGER'S EQUATION

We saw in Chapter 1 how the simplest atom is described by the PDE

$$-ihu_t = \frac{h^2}{2m} \Delta u + \frac{e^2}{r} u. \tag{8}$$

The potential $e^2/r$ is a variable coefficient.

So, as a simple warm-up problem, let's take the *free* Schrödinger equation

$$\boxed{ -i\frac{\partial u}{\partial t} = \frac{1}{2} \Delta u } \tag{9}$$

in three dimensions, where we've set $h = m = 1$ and dropped the potential term. It looks suspiciously like the diffusion equation. In fact, the only difference is that $k = i/2$ is imaginary instead of real. The presence of the $i = \sqrt{-1}$ implies that the solutions are "waves" because the temporal factor (see below) has the form

$$T(t) = e^{i\lambda t} = \cos \lambda t + i \sin \lambda t,$$

which is oscillatory.

We are looking for solutions of (9) that tend to zero as $|\mathbf{x}| \to \infty$. It is not difficult to show that the solution of (9) with the initial condition $u(\mathbf{x}, 0) = \phi(\mathbf{x})$ is

$$\boxed{ u(\mathbf{x}, t) = \frac{1}{(2\pi it)^{3/2}} \iiint \exp\left( -\frac{|\mathbf{x} - \mathbf{x}'|^2}{2it} \right) \phi(\mathbf{x}')\, d\mathbf{x}', } \tag{10}$$

the same as for the diffusion equation except for the $i$.

Because complex numbers have two square roots, which one do we take in the first factor here? To answer this question, as well as to justify (10), we use the following reasoning. With either one of the choices of the square root, (10) appears to be correct. Let's assume that $\phi(\mathbf{x}')$ vanishes for large $|\mathbf{x}'|$. But let's first solve the nearby equation

$$\frac{\partial u_\epsilon}{\partial t} = \frac{+\epsilon + i}{2}\Delta u_\epsilon, \qquad u_\epsilon(\mathbf{x}, 0) = \phi(\mathbf{x}), \tag{11}$$

whose solution depends on the real number $\epsilon > 0$. Equation (11) can be solved *exactly* like the diffusion equation with $k = (\epsilon + i)/2$. The formula is

$$u_\epsilon(\mathbf{x}, t) = \frac{1}{(2\pi t)^{3/2}(\epsilon + i)^{3/2}} \iiint \exp\left[-\frac{|\mathbf{x} - \mathbf{x}'|^2}{2(\epsilon + i)t}\right]\phi(\mathbf{x}')\, d\mathbf{x}'. \tag{12}$$

Here $(\epsilon + i)^{1/2}$ denotes the unique square root with the *positive* real part. Because $\epsilon > 0$, it contributes to a negative exponent and the integral converges. For that reason there is no difficulty in justifying the formula (12). (We need to take the *positive* real part because otherwise the exponent would be too large as $|\mathbf{x} - \mathbf{x}'| \to \infty$ and we wouldn't get a bounded solution.)

As $\epsilon \searrow 0$, we get the solution of (9), given by formula (10), where the $i^{1/2}$ factor is the unique square root with the positive real part. That is,

$$\lim_{\epsilon \searrow 0}(\epsilon + i)^{1/2} = \frac{1+i}{\sqrt{2}}.$$

This is the correct factor in front of the formula (10). [It is not exactly a rigorous proof of (10) but it does provide the correct answer.]

A different method we could envision to solve (9) would be to separate variables: $u(\mathbf{x}, t) = T(t)X(\mathbf{x})$. Then, in the one-dimensional case, say, we'd have

$$-2i\frac{T'}{T} = \frac{X''}{X} = -\lambda. \tag{13}$$

There are *no* solutions of $X'' + \lambda X = 0$ that satisfy the required condition at infinity, that $X(x) \to 0$ as $x \to \pm\infty$. (It can be satisfied at $+\infty$, but then not $-\infty$, and vice versa.) So the condition at infinity prevents the existence of any eigenvalues, and the method of Chapter 5 fails. (The method of separation of variables is salvaged, however, by the use of the Fourier transform; see Section 12.3.)

## HARMONIC OSCILLATOR

The addition of a potential term to Schrödinger's equation, as in equation (8), sometimes leads to the occurrence of eigenvalues. As an example, we now study the quantum-mechanical harmonic oscillator equation in one dimension, which in appropriate units is

$$-iu_t = u_{xx} - x^2 u \qquad (-\infty < x < \infty). \tag{14}$$

For our eigenfunctions we will require the condition that

$$u \to 0 \qquad \text{as } x \to \pm\infty.$$

We separate variables $u = T(t)v(x)$ to get

$$-i\frac{T'}{T} = \frac{v'' - x^2 v}{v} = -\lambda.$$

The constant $\lambda$ is interpreted as the "energy" of the harmonic oscillator. Thus $v(x)$ satisfies the ODE

$$v'' + (\lambda - x^2)v = 0. \tag{15}$$

Because of its variable coefficient, (15) is not easily solvable. The simplest case turns out to be $\lambda = 1$, in which case the solutions are $e^{-x^2/2}$. (Check it!) So for *any* $\lambda$ it is natural to attempt the substitution

$$v(x) = w(x)e^{-x^2/2}.$$

This leads to an equation for $w$,

$$(x^2 - \lambda)e^{-x^2/2}w = (x^2 - \lambda)v = v'' = [w'' - 2xw' + (x^2 - 1)w]e^{-x^2/2},$$

or

$$w'' - 2xw' + (\lambda - 1)w = 0, \tag{16}$$

which is known as *Hermite's differential equation*.

We shall solve (16) by the method of power series. Substituting

$$w(x) = a_0 + a_1 x + a_2 x^2 + \cdots = \sum_{k=0}^{\infty} a_k x^k \tag{17}$$

into (16), we get

$$\sum_{k=0}^{\infty} k(k-1)a_k x^{k-2} - \sum_{k=0}^{\infty}(2k - \lambda + 1)a_k x^k = 0.$$

Matching the like powers of $x$, we get

$$2a_2 = (1 - \lambda)a_0, \qquad 6a_3 = (3 - \lambda)a_1, \qquad \text{etc.}$$

In general,

$$(k + 2)(k + 1)a_{k+2} = (2k + 1 - \lambda)a_k \qquad (k = 0, 1, 2, 3, \ldots). \tag{18}$$

This "recursion formula" yields all the coefficients provided that we know the first two, $a_0$ and $a_1$. The first two are arbitrary. If $a_0 = 0$, the solution is an odd function; if $a_1 = 0$, the solution is even.

There is one particularly simple case. *In case $\lambda = 2k + 1$ for some integer k*, then (18) shows that $a_{k+2} = 0$, $a_{k+4} = 0$, and so on. Then we get an even or an odd polynomial (depending on whether $k$ is even or odd) of degree $k$. It is called the *Hermite polynomial $H_k(x)$* (with an appropriate normalization). The first five Hermite polynomials are:

$$
\begin{aligned}
H_0(x) &= 1 & (\lambda = 1, a_1 = a_2 = 0) \\
H_1(x) &= 2x & (\lambda = 3, a_0 = a_3 = 0) \\
H_2(x) &= 4x^2 - 2 & (\lambda = 5, a_1 = a_4 = 0) \\
H_3(x) &= 8x^3 - 12x & (\lambda = 7, a_0 = a_5 = 0) \\
H_4(x) &= 16x^4 - 48x^2 + 12 & (\lambda = 9, a_1 = a_6 = 0).
\end{aligned}
$$

Thus we have found some separated solutions of equation (15) of the form

$$
v_k(x) = H_k(x)\, e^{-x^2/2} \qquad \text{if } \lambda = 2k + 1.
$$

The corresponding solutions of (14) are

$$
u_k(x, t) = e^{-i(2k+1)t}\, H_k(x)\, e^{-x^2/2}
$$

for $k = 0, 1, 2, \ldots$. Note that $u_k(x, t) \to 0$ as $x \to \pm\infty$, as required.

If we go back to the full power series (17), it can be shown that if $\lambda \neq 2k + 1$, no power series solution can satisfy the condition at infinity, and therefore the only eigenvalues (energy levels) are the positive odd integers (see Exercise 7).

## EXERCISES

1. Find a simple formula for the solution of the three-dimensional diffusion equation with $\phi(x, y, z) = xy^2z$. (*Hint:* See Exercise 2.4.9 or 2.4.10.)

2. (a) Prove that (6) is valid for products of the form $\phi(x)\psi(y)\zeta(z)$ and hence for any finite sum of such products.
   (b) Deduce (6) for any bounded continuous function $\phi(\mathbf{x})$. You may use the fact that there is a sequence of finite sums of products as in part (a) which converges uniformly to $\phi(\mathbf{x})$.

3. Find the solution of the diffusion equation in the half-space $\{(x, y, z, t): z > 0\}$ with the Neumann condition $\partial u/\partial z = 0$ on $z = 0$. (*Hint:* Use the method of reflection.)

4. Derive the first four Hermite polynomials from scratch.

5. Show that all the Hermite polynomials are given by the formula

$$
H_k(x) = (-1)^k e^{x^2} \frac{d^k}{dx^k} e^{-x^2}
$$

up to a constant factor.

6.  Show directly from the ODE (15) that the functions $H_k(x)e^{-x^2/2}$ are mutually orthogonal on the interval $(-\infty, \infty)$. That is

$$\int_{-\infty}^{\infty} H_k(x)H_l(x)e^{-x^2}\,dx = 0 \qquad \text{for } k \neq l.$$

(*Hint:* See Section 5.3.)

7.  (a) Show that if $\lambda \neq 2k + 1$, any solution of Hermite's ODE is a power series but not a polynomial.
    (b) Deduce that in this case no solution of Hermite's ODE can satisfy the condition at infinity. (*Hint:* Use the recursion relation (18) to find the behavior of $a_k$ as $k \to \infty$. Compare with the power series expansion of $e^{x^2}$. Deduce that $u(x, t)$ behaves like $e^{x^2}$ as $|x| \to \infty$.)

## 9.5    THE HYDROGEN ATOM

Now let's return to the hydrogen atom, which we are modeling by the PDE

$$iu_t = -\frac{1}{2}\Delta u - \frac{1}{r}u \tag{1}$$

with the units chosen so that $e = m = h = 1$. Equation (1) is supposed to be satisfied in all of space $\mathbf{x} = (x, y, z)$. We have written $r = |\mathbf{x}| = (x^2 + y^2 + z^2)^{1/2}$. We also require that

$$\iiint |u(\mathbf{x}, t)|^2 d\mathbf{x} < \infty \tag{2}$$

which may be interpreted as a condition of *vanishing at infinity* (see Section 1.3, Example 7).

Although this is a whole-space problem, let's separate variables anyway. It turns out, as with the harmonic oscillator, that the potential term does lead to some eigenvalues. Writing $u(\mathbf{x}, t) = T(t)v(\mathbf{x})$ as usual, we have

$$2i\frac{T'}{T} = \frac{-\Delta v - \frac{2}{r}v}{v} = \lambda,$$

a constant. Thus $u = v(\mathbf{x})e^{-i\lambda t/2}$, where

$$-\Delta v - \frac{2}{r}v = \lambda v. \tag{3}$$

In quantum mechanics, $\lambda$ *is the energy of the bound state* $u(\mathbf{x}, t)$. Bohr observed in 1913 that the energy levels of the electron in a hydrogen atom occur only at special values (related to squares of integers). We shall verify Bohr's observation mathematically!

We look for solutions of (3) that are spherically symmetric: $v(\mathbf{x}) = R(r)$. Later (in Section 10.7) we shall look for the others. By (6.1.7), equation (3) reduces to the ODE

$$\boxed{-R_{rr} - \frac{2}{r}R_r - \frac{2}{r}R = \lambda R} \tag{4}$$

in $0 < r < \infty$ with the *condition at infinity* that

$$\int_0^\infty |R(r)|^2 r^2 dr < \infty. \tag{5}$$

It is also understood that

$$R(0) \text{ is finite.} \tag{6}$$

As with the harmonic oscillator, this ODE is not easily solved. After some changes of variable, (4) is known as *Laguerre's differential equation*. It turns out that all of the eigenvalues $\lambda$ are negative. For the time being, let's just *assume that $\lambda < 0$*.

It is quite convenient to first make a couple of changes of variables. Notice that if the second and third terms in (4) were absent (which is true "at infinity"), the equation would simply be $-R'' = \lambda R$, which has the solutions $e^{\pm\beta r}$ with $\beta = \sqrt{-\lambda}$. We are interested only in solutions that vanish at infinity, so we choose the negative exponent. We could consider $e^{-\beta r}$ as an approximation to a solution of equation (4). At any rate, we are thus motivated to try the change of variables

$$w(r) = e^{+\beta r} R(r) \qquad \text{where } \beta = \sqrt{-\lambda}. \tag{7}$$

Then $R = we^{-\beta r}$, $R_r = (w_r - \beta w)e^{-\beta r}$, and $R_{rr} = (w_{rr} - 2\beta w_r + \beta^2 w)e^{-\beta r}$ so that (4) is converted to the equation

$$-w_{rr} + 2\left(\beta - \frac{1}{r}\right)w_r + \left(2(\beta - 1)\frac{1}{r}\right)w = 0$$

or the equation

$$\tfrac{1}{2}rw_{rr} - \beta rw_r + w_r + (1 - \beta)w = 0. \tag{8}$$

To understand the ODE (8), we observe that $r = 0$ is a regular singular point. For this terminology, see Section A.4. We shall solve it by the power series method. (This will provide some, but not all, of the solutions.) We look for a solution of (8) of the form

$$w(r) = \sum_{k=0}^\infty a_k r^k = a_0 + a_1 r + a_2 r^2 + \cdots$$

whose coefficients are to be determined. Substituting into (8), we get

$$\frac{1}{2}\sum_{k=0}^{\infty}k(k-1)a_k r^{k-1} - \beta\sum_{k=0}^{\infty}ka_k r^k + \sum_{k=0}^{\infty}ka_k r^{k-1} + (1-\beta)\sum_{k=0}^{\infty}a_k r^k = 0.$$

In the second and fourth terms, we change the dummy variable $k$ to $(k-1)$, so that

$$\sum_{k=0}^{\infty}[\tfrac{1}{2}k(k-1)+k]a_k r^{k-1} + \sum_{k=1}^{\infty}[-\beta(k-1)+(1-\beta)]a_{k-1}r^{k-1} = 0.$$

Each coefficient must vanish:

$$\frac{k(k+1)}{2}a_k = (\beta k - 1)a_{k-1} \qquad (k=1,2,\ldots). \qquad (9)$$

This means

$$a_1 = (\beta - 1)a_0 \qquad\qquad 3a_2 = (2\beta - 1)a_1$$
$$6a_3 = (3\beta - 1)a_2 \qquad\quad 10a_4 = (4\beta - 1)a_3$$
$$15a_5 = (5\beta - 1)a_4 \qquad\quad 21a_6 = (6\beta - 1)a_5 \qquad \text{etc.}$$

*If $\beta$ happens to be the reciprocal of a positive integer, the sequence of coefficients terminates and we have a polynomial solution of (8)!*

Since $v(\mathbf{x}) = R(r) = w(r)e^{-\beta r}$, we have a polynomial times a decaying exponential. This tends to zero as $r \to \infty$, so the condition at infinity (2) is also satisfied.

The first few solutions of (8) and (3) are

| $n$ | $\beta$ | $\lambda$ | $w(r)$ | $v(\mathbf{x})$ |
|---|---|---|---|---|
| 1 | 1 | $-1$ | $1$ | $e^{-r}$ |
| 2 | $\frac{1}{2}$ | $-\frac{1}{4}$ | $1 - \frac{1}{2}r$ | $e^{-r/2}(1 - \frac{1}{2}r)$ |
| 3 | $\frac{1}{3}$ | $-\frac{1}{9}$ | $1 - \frac{2}{3}r + \frac{2}{27}r^2$ | $e^{-r/3}[1 - \frac{2}{3}r + \frac{2}{27}r^2]$ |
| 4 | $\frac{1}{4}$ | $-\frac{1}{16}$ | | |

The lowest energy state (the ground state) $v(\mathbf{x}) = e^{-r}$ drops off exponentially with the distance from the proton and vanishes nowhere. The second state corresponds to $n = 2$ and vanishes for a single value of $r$ (has one *node*). *The nth state has $(n-1)$ nodes.* Its energy is $\lambda = -\beta^2 = -1/n^2$. Thus the lowest possible energy levels are

$$-1, -\tfrac{1}{4}, -\tfrac{1}{9}, \ldots,$$

which agrees with the experiments of Bohr. These energy levels lead to spectral lines whose frequencies are proportional to the differences between the energy levels.

**Other Solutions.**   For $\beta = 1/n$, there is, of course, another, linearly independent solution of the second-order ODE (8). However, that solution is singular at $r = 0$ and it does not interest us.

What happens when $\beta \neq 1/n$? Then the factor $(\beta k - 1)$ never vanishes, so that the recursion relation (9) looks approximately like $(k^2/2)a_k \sim (\beta k)a_{k-1}$ for large $k$, or like $a_k \sim (2\beta/k)a_{k-1}$. These are the coefficients in the Taylor expansion of $e^{2\beta r}$. So $R(r)$ looks approximately like

$$e^{-\beta r}e^{+2\beta r} = e^{+\beta r}.$$

Such a solution would not satisfy the condition at infinity (2). So we see that the only eigenvalues are $\lambda = 1/n$ for $n = 1, 2, 3, \ldots$ . (This argument is not rigorous but could be made so.)

Are the eigenfunctions complete? By no means, for two reasons. First, there are plenty of eigenfunctions that possess angular dependence (spin) (see Section 10.7). Second, there is plenty of *continuous spectrum* as a consequence of our domain $D$ being all of space, rather than a bounded part of it (see Section 13.4). Physically, the continuous spectrum corresponds to the "unbound states" which are scattered by the potential. See a good book on quantum mechanics, such as [St], [MF], or [AJS].

## EXERCISES

1.  Verify the formulas for the first three solutions of the hydrogen atom.
2.  For the hydrogen atom if $\lambda > 0$, why would you expect equation (4) not to have a solution that satisfies the condition at infinity?

# 10

# BOUNDARIES IN THE PLANE AND IN SPACE

In Chapters 4 and 5 we used separation of variables and Fourier series to solve one-dimensional wave and diffusion problems. This chapter is devoted to extending the same methods to higher dimensions. We begin with a general review of these methods. Then Section 2 is devoted to the circular disk and Section 3 to the spherical ball. The problems with circular symmetry lead inexorably to Bessel functions and (in three dimensions) to Legendre functions, which are the topics of Sections 5 and 6. In Section 4 we discuss the nodal sets of the eigenfunctions. Finally, in Section 7 we complete our analysis of the hydrogen atom by discussing the states that have angular momentum.

## 10.1 FOURIER'S METHOD, REVISITED

We would like to solve the wave and diffusion equations

$$u_{tt} = c^2 \Delta u \quad \text{and} \quad u_t = k \, \Delta u$$

in any bounded domain $D$ with one of the classical homogeneous conditions on bdy $D$ and with the standard initial condition. We denote

$$\Delta = \frac{\partial^2}{\partial x^2} + \frac{\partial^2}{\partial y^2} \quad \text{or} \quad \frac{\partial^2}{\partial x^2} + \frac{\partial^2}{\partial y^2} + \frac{\partial^2}{\partial z^2}$$

in two or three dimensions, respectively. For brevity, we continue to use the vector notation $\mathbf{x} = (x, y)$ or $(x, y, z)$. The general discussion that follows works in either dimension, but for definiteness, let's say that we're in three dimensions. Then $D$ is a solid domain and bdy $D$ is a surface.

The first step is to separate the time variable only,

$$u(x, y, z, t) = T(t)v(x, y, z). \tag{1}$$

Then

$$-\lambda = \frac{T''}{c^2 T} = \frac{\Delta v}{v} \quad \text{or} \quad -\lambda = \frac{T'}{kT} = \frac{\Delta v}{v}, \tag{2}$$

depending on whether we are doing waves or diffusions. In *either* case we get the eigenvalue problem

$$\boxed{\begin{aligned} -\Delta v = \lambda v \quad \text{in } D \\ v \text{ satisfies (D), (N), (R)} \quad \text{on bdy } D. \end{aligned}} \tag{3}$$

Therefore, if this problem has eigenvalues $\lambda_n$ (all positive, say) and eigenfunctions $v_n(x, y, z) = v_n(\mathbf{x})$, then the solutions of the wave equation are

$$\boxed{u(\mathbf{x}, t) = \sum_n [A_n \cos (\sqrt{\lambda_n}\, ct) + B_n \sin (\sqrt{\lambda_n}\, ct)]\, v_n(\mathbf{x})} \tag{4}$$

and the solutions of the diffusion equation are

$$\boxed{u(\mathbf{x}, t) = \sum_n A_n e^{-\lambda_n kt} v_n(\mathbf{x}).} \tag{5}$$

As usual, the coefficients will be determined by the initial conditions. However, to carry this out, we'll need to know that the eigenfunctions are orthogonal. This is our next goal. One point of notation in (4) and (5): In three dimensions the index $n$ in the sums (4) and (5) will be a *triple index* $[(l, m, n)$, say] and the various series will be *triple series*, one sum for each coordinate.

## ORTHOGONALITY

Our discussion of orthogonality and completeness is practically a repetition of Section 5.3. We define the *inner product*

$$(f, g) = \iiint_D f(\mathbf{x}) \, \overline{g(\mathbf{x})} \, d\mathbf{x} \quad (\text{where } d\mathbf{x} = dx \, dy \, dz)$$

as a triple integral. (In two dimensions it would be a double integral.) If $\nabla \cdot$ denotes the divergence, the identity

$$u(\Delta v) - (\Delta u)v = \nabla \cdot [u(\nabla v) - (\nabla u)v] \tag{6}$$

(check it!) is integrated over $D$. Using the divergence theorem (Section A.3), we obtain *Green's second identity:*

$$\iiint_D [u(\Delta v) - (\Delta u)v] \, d\mathbf{x} = \iint_{\text{bdy } D} \left( u \frac{\partial v}{\partial n} - \frac{\partial u}{\partial n} v \right) dS. \tag{G2}$$

The right side of (G2) is a surface integral and $\partial u / \partial n = n \cdot \nabla u$ is the directional derivative in the normal direction.

If $u$ and $v$ both satisfy homogeneous Dirichlet conditions ($u = v = 0$ on bdy $D$), the surface integral must vanish. The same is true for Neumann or Robin boundary conditions. For instance, if

$$\frac{\partial u}{\partial n} + au = 0 = \frac{\partial v}{\partial n} + av \qquad \text{on bdy } D,$$

then $u(\partial v / \partial n) - (\partial u / \partial n)v = -uav + auv = 0$. We therefore say that *each of the three classical BCs is symmetric* since in each case

$$(u, \Delta v) = (\Delta u, v) \quad \text{for all functions that satisfy the BCs.}$$

Now suppose that both $u$ and $v$ are real eigenfunctions:

$$-\Delta u = \lambda_1 u \quad \text{and} \quad -\Delta v = \lambda_2 v \quad \text{in } D, \tag{7}$$

where $u$ and $v$ both satisfy (D) [or (N) or (R)] on bdy $D$. By (G2),

$$(\lambda_1 - \lambda_2)(u, v) = (u, \Delta v) - (\Delta u, v) = 0. \tag{8}$$

Therefore, $u$ and $v$ are orthogonal provided that $\lambda_1 \neq \lambda_2$. As in Section 5.3, a similar argument shows that all the eigenvalues are necessarily real. We summarize these observations in the following theorem.

**Theorem 1.** Consider any one of the problems (3). Then all the eigenvalues are real. The eigenfunctions can be chosen to be real valued. The eigenfunctions that correspond to distinct eigenvalues are necessarily orthogonal. All the eigenfunctions may be chosen to be orthogonal.

## MULTIPLICITY

An eigenvalue $\lambda$ is *double* (*triple*, . . . ) if there are two (three, . . . ) linearly independent eigenfunctions for it. It has *multiplicity m* if it has $m$ linearly independent eigenfunctions. In other words, the "eigenspace" for $\lambda$ has dimension $m$.

If a given eigenvalue $\lambda$ has multiplicity $m$, let $w_1, \ldots, w_m$ be linearly independent eigenfunctions. They are not necessarily orthogonal. But we can always choose a new set of eigenfunctions that *is* orthogonal. The step-by-step procedure to accomplish this is the *Gram–Schmidt orthogonalization method*, which works as follows.

Let $w_1, \ldots, w_m$ be any (finite or infinite) set of linearly independent vectors in any vector space $V$ that has an inner product. First we normalize:

$$u_1 = \frac{w_1}{\|w_1\|}.$$

Second, we subtract from $w_2$ the component parallel to $u_1$ and then normalize. That is, we define

$$v_2 = w_2 - (w_2, u_1)u_1 \quad \text{and} \quad u_2 = \frac{v_2}{\|v_2\|}. \tag{9}$$

That $u_2$ and $u_1$ are orthogonal is easy to see either by a calculation or from a sketch. Third, we subtract off from $w_3$ the component in the $u_1u_2$ plane and then normalize. That is, we define

$$v_3 = w_3 - (w_3, u_2)u_2 - (w_3, u_1)u_1 \quad \text{and} \quad u_3 = \frac{v_3}{\|v_3\|}, \tag{10}$$

and so on. At each stage we subtract off the components in all the previous directions. Then $\{u_1, u_2, u_3, \ldots\}$ is an orthogonal set of vectors. In fact,

$$(v_2, u_1) = (w_2 - (w_2, u_1)u_1, u_1) = (w_2, u_1) - (w_2, u_1)(u_1, u_1) = 0$$
$$(v_3, u_1) = (w_3 - (w_3, u_2)u_2 - (w_3, u_1)u_1, u_1)$$
$$= (w_3, u_1) - (w_3, u_2)(u_2, u_1) - (w_3, u_1)(u_1, u_1)$$
$$= (w_3, u_1) - (w_3, u_2) \cdot 0 - (w_3, u_1) \cdot 1 = 0,$$

and so on. See the exercises for some examples with large multiplicity.

## GENERAL FOURIER SERIES

Because of Theorem 1, we can talk about general Fourier series which are made up of the eigenfunctions in $D$. If

$$\phi(\mathbf{x}) = \sum_n A_n v_n(\mathbf{x}), \tag{11}$$

where $v_n(\mathbf{x})$ denote orthogonal eigenfunctions of (3), then

$$A_n = \frac{(\phi, v_n)}{(v_n, v_n)} = \frac{\iiint_D \phi(\mathbf{x})\overline{v_n(\mathbf{x})} \, d\mathbf{x}}{\iiint_D |v_n(\mathbf{x})|^2 \, d\mathbf{x}}. \tag{12}$$

The question of the positivity of the eigenvalues is addressed in the next theorem.

**Theorem 2.** All the eigenvalues are positive in the Dirichlet case. All the eigenvalues are positive or zero in the Neumann case, as well as in the Robin case $\partial u/\partial n + au = 0$ provided that $a \geq 0$.

**Proof.** We use Green's first identity (G1) with $u \equiv v$, namely,

$$\iiint_D (-\Delta v)\overline{v} \, d\mathbf{x} = \iiint_D |\nabla v|^2 \, d\mathbf{x} - \iint_{\text{bdy } D} \overline{v}\frac{\partial v}{\partial n} \, dS.$$

In the Dirichlet case, with $v$ an eigenfunction of (3), we get

$$\lambda \iiint\limits_D |v|^2 \, d\mathbf{x} = \iiint\limits_D |\nabla v|^2 \, d\mathbf{x} \geq 0.$$

In fact, the last integral cannot be zero, because if it were, $\nabla v(\mathbf{x})$ would be identically zero and $v(\mathbf{x}) \equiv C$ would be a constant function and $C = 0$ by the boundary condition. Therefore, $\lambda > 0$ in the Dirichlet case. See Exercise 7 for the other cases.    □

Besides orthogonality, the other property the eigenfunctions had better have is completeness. The discussion of completeness is left for Chapter 11. Suffice it to say that completeness is always true as long as the boundary bdy $D$ of the domain is not too wild (i.e., for any domain one normally encounters in scientific problems). Completeness in the mean-square sense for (1) means that

$$\left\| \phi - \sum_{n \leq N} A_n v_n \right\|^2 = \iiint\limits_D \left| \phi(\mathbf{x}) - \sum_{n \leq N} A_n v_n(\mathbf{x}) \right|^2 \, d\mathbf{x} \to 0 \qquad (13)$$

as $N \to \infty$.

What we have just shown is how a wave or diffusion problem with boundary and initial conditions is reducible to the eigenvalue problem (3). But we are still left with finding the solutions of (3). If we expect to carry out a specific computation, we will need to assume that $D$ has a very special geometry in which we can separate the space variables (in cartesian, polar, or some other coordinate system). We already did this for harmonic functions. What we are dealing with at present is similar to the harmonic case except for the parameter $\lambda$.

## Example 1.

Take the *cube* $Q = \{0 < x < \pi, 0 < y < \pi, 0 < z < \pi\}$ and solve the problem

$$
\begin{aligned}
\text{DE: } & u_t = k \, \Delta u \quad \text{in } Q \\
\text{BC: } & u = 0 \qquad \text{on bdy } Q \\
\text{IC: } & u = \phi(\mathbf{x}) \quad \text{when } t = 0.
\end{aligned}
\qquad (14)
$$

Separating the time variable as in the general discussion above, we are led to the eigenvalue problem

$$-\Delta v = \lambda v \quad \text{in } Q, \qquad v = 0 \quad \text{on bdy } Q. \qquad (15)$$

Because the sides of $Q$ are parallel to the axes, we can successfully separate the $x$, $y$, and $z$ variables: $v = X(x)Y(y)Z(z)$,

$$\frac{X''}{X} + \frac{Y''}{Y} + \frac{Z''}{Z} = -\lambda.$$

The separated BCs are

$$X(0) = X(\pi) = Y(0) = Y(\pi) = Z(0) = Z(\pi) = 0.$$

Clearly, the solutions are

$$v(x, y, z) = \sin lx \sin my \sin nz = v_{lmn}(\mathbf{x}), \tag{16}$$

where

$$\lambda = l^2 + m^2 + n^2 = \lambda_{lmn} \quad (1 \le m,l,n < \infty). \tag{17}$$

Note the triple index! Therefore,

$$u(\mathbf{x}, t) = \sum_n \sum_m \sum_l A_{lmn}\, e^{-(l^2+m^2+n^2)kt}\, \sin lx \sin my \sin nz. \tag{18}$$

Orthogonality in Example 1 implies that

$$A_{lmn} = (2/\pi)^3 \int_0^\pi \int_0^\pi \int_0^\pi \phi(x, y, z) \sin lx \sin my \sin nz\, dx\, dy\, dz. \tag{19}$$

Notice that the orthogonality of the functions $v_{lmn}(x, y, z)$ is, in this case, a direct consequence of the *separate* orthogonalities of the separated eigenfunctions $\sin lx$, $\sin my$, and $\sin nz$. Namely,

$$\iiint_Q v_{lmn}(\mathbf{x})\, v_{l'm'n'}(\mathbf{x})\, d\mathbf{x} = \left( \int_0^\pi \sin lx \sin l'x\, dx \right)$$

$$\left( \int_0^\pi \sin my \sin m'y\, dy \right)\left( \int_0^\pi \sin nz \sin n'z\, dz \right) = 0$$

unless all three indices exactly match.    □

We shall observe the same phenomenon in the polar, cylindrical, and spherical coordinate systems. To study the cases of circular symmetry is the subject of our next investigation.

## EXERCISES

1. Solve the wave equation in the square $S = \{0 < x < \pi, 0 < y < \pi\}$, with homogeneous Neumann conditions on the boundary, and the initial conditions $u(x, y, 0) \equiv 0$, $u_t(x, y, 0) = \sin^2 x$.

2.  Solve the wave equation in the rectangle $R = \{0 < x < a, 0 < y < b\}$, with homogeneous Dirichlet conditions on the boundary, and the initial conditions $u(x, y, 0) = xy(b - y)(a - x)$, $u_t(x, y, 0) \equiv 0$.

3.  In the cube $(0, a)^3$, a substance is diffusing whose molecules multiply at a rate proportional to the concentration. It therefore satisfies the PDE $u_t = k \, \Delta u + \gamma u$, where $\gamma$ is a constant. Assume that $u = 0$ on all six sides. What is the condition on $\gamma$ so that the concentration does not grow without bound?

4.  Consider the eigenvalue problem $-\Delta v = \lambda v$ in the unit square $D = \{0 < x < 1, 0 < y < 1\}$ with the Dirichlet BC $v = 0$ on the bottom and both vertical sides, and the Robin BC $\partial v / \partial y = -v$ on the top $\{y = 1\}$.
    (a)  Show that all the eigenvalues are positive.
    (b)  Find an equation for the eigenvalues $\lambda$. Show that they can be expressed in terms of the roots of the equation $s + \tan s = 0$.
    (c)  Find the solutions of the last equation graphically. Find an approximate formula for the $(m, n)$th eigenvalue for large $(m, n)$.

5.  Find the dimension of each of the following vector spaces.
    (a)  The space of all the solutions of $u'' + x^2 u = 0$.
    (b)  The eigenspace with eigenvalue $(2\pi / l)^2$ of the operator $-d^2/dt^2$ on the interval $(-l, l)$ with the periodic boundary conditions.
    (c)  The space of harmonic functions in the unit disk with the homogeneous Neumann BCs.
    (d)  The eigenspace with eigenvalue $\lambda = 25\pi^2$ of $-\Delta$ in the unit square $(0,1)^2$ with the homogeneous Neumann BCs on all four sides.
    (e)  The space of all the solutions of $u_{tt} = c^2 u_{xx}$ in $-\infty < x < \infty$, $-\infty < t < \infty$.

6.  Illustrate the Gram–Schmidt orthogonality method by sketching two linearly independent vectors $w_1$ and $w_2$ in the plane that are not orthogonal. Then do it with three vectors in space.

7.  Prove Theorem 2 in the Neumann and Robin cases.

## 10.2    VIBRATIONS OF A DRUMHEAD

Consider a membrane stretched across the top of a circular drum $D = \{x^2 + y^2 < a^2\}$ of radius $a$. Its small transverse vibrations satisfy the two-dimensional wave equation in $D$ with Dirichlet boundary conditions. Therefore, we want to solve the problem

$$\begin{cases} u_{tt} = c^2(u_{xx} + u_{yy}) & \text{in } D \\ u = 0 & \text{on bdy } D \\ u, u_t \text{ are given functions when } t = 0. \end{cases} \quad (1)$$

To solve this problem, of course we'll use polar coordinates just as we did in Section 6.3. We write

$$c^{-2}u_{tt} = u_{rr} + \frac{1}{r}u_r + \frac{1}{r^2}u_{\theta\theta}. \tag{2}$$

Separating the variables $u(r, \theta, t) = T(t)R(r)\Theta(\theta)$ gives

$$\frac{T''}{c^2 T} = \frac{R''}{R} + \frac{R'}{rR} + \frac{\Theta''}{r^2 \Theta}. \tag{3}$$

It follows by the usual argument that $T''/c^2 T$ is a constant (call it $-\lambda$) and $\Theta''/\Theta$ is a constant (call it $-\gamma$). Thus we have the three ODEs

$$T'' + c^2 \lambda T = 0 \tag{4}$$

$$\Theta'' + \gamma \Theta = 0 \tag{5}$$

$$R'' + \frac{1}{r}R' + \left(\lambda - \frac{\gamma}{r^2}\right)R = 0. \tag{6}$$

We'll save (4) for last, because it involves the inhomogeneous (initial) conditions.

As for (5), we have the periodic boundary conditions, $\Theta(\theta + 2\pi) = \Theta(\theta)$, exactly as in Section 6.3. Therefore,

$$\gamma = n^2 \quad \text{and} \quad \Theta(\theta) = A_n \cos n\theta + B_n \sin n\theta \quad (n = 1, 2, \ldots) \tag{7}$$

or else $\gamma = 0$ and $\Theta(\theta) = \frac{1}{2}A_0$.

As for the radial part (6), we have the ODE

$$R_{rr} + \frac{1}{r}R_r + \left(\lambda - \frac{n^2}{r^2}\right)R = 0 \tag{8}$$

for $0 < r < a$ together with the boundary conditions

$$\begin{cases} R(0) \text{ finite} \\ R(a) = 0. \end{cases} \tag{9}$$

From (7) we know that $n$ must be an integer. If $\lambda = 0$, equation (8) is of the Euler type and we already solved it in Section 6.3, but since $R(a) = 0$ we would get only the trivial solution $R(r) \equiv 0$. [In fact, by Theorem 2 of Section 10.1, we already knew that all the eigenvalues of $-\Delta$ are positive.] So let $\lambda > 0$. We can transform (8) into a standard form by changing scale $\rho = \sqrt{\lambda}r$, so that

$$R_r = R_\rho \frac{d\rho}{dr} = \sqrt{\lambda}\, R_\rho, \quad R_{rr} = \lambda R_{\rho\rho}$$

and

$$R_{\rho\rho} + \frac{1}{\rho}R_{\rho} + \left(1 - \frac{n^2}{\rho^2}\right)R = 0. \tag{10}$$

This is *Bessel's differential equation* of order $n$. It is the third time we have encountered an ODE that is not directly solvable, the first times being in Sections 9.4 and 9.5.

## SOLUTION OF BESSEL'S EQUATION (10)

It is a second-order linear ODE and, as such, has a two-dimensional space of solutions. At $\rho = 0$ its coefficients become infinite; this is a *singular point*. However, it is the least troublesome kind of singular point, a so-called *regular singular point* (see Section A.4). Recall that an ODE of Euler type also has a regular singular point and generally has solutions of the form $R(\rho) = C\rho^{\alpha} + D\rho^{\beta}$. To solve Bessel's equation, we guess a solution of the form

$$R(\rho) = \rho^{\alpha} \sum_{k=0}^{\infty} a_k \rho^k, \quad a_0 \neq 0, \tag{11}$$

with coefficients $a_k$ to be determined. Plugging (11) into (10), we get

$$\rho^{\alpha} \sum_{k=0}^{\infty} [(\alpha + k)(\alpha + k - 1)a_k\rho^{k-2} + (\alpha + k)a_k\rho^{k-2} + a_k\rho^k - n^2 a_k\rho^{k-2}] = 0. \tag{12}$$

The third sum can be rewritten as

$$\sum_{k=0}^{\infty} a_k \rho^k = \sum_{k=2}^{\infty} a_{k-2}\rho^{k-2}$$

by changing the name of the dummy variable. Therefore, equating the like powers of $\rho$, we get

For $k = 0$, $[\alpha(\alpha - 1) + \alpha - n^2] a_0 = 0$
For $k = 1$, $[(\alpha + 1)\alpha + \alpha + 1 - n^2] a_1 = 0$
For $k \geq 2$, $[(\alpha + k)(\alpha + k - 1) + \alpha + k - n^2] a_k + a_{k-2} = 0.$

The first equation gives us $\alpha = \pm n$ (since $a_0 \neq 0$). We thus have two choices, $\alpha = +n$ and $\alpha = -n$. Let us begin with the case $\alpha = +n$. The second equation gives $[(\alpha + 1)^2 - n^2]a_1 = 0$, whence $a_1 = 0$. The infinite set of equations for $k = 2, 3, 4, \ldots$ (called the *recursion relations*) determine $a_k$ from $a_{k-2}$:

$$a_k = -\frac{a_{k-2}}{(\alpha + k)^2 - n^2} \quad (k = 2, 3, \ldots) \tag{13}$$

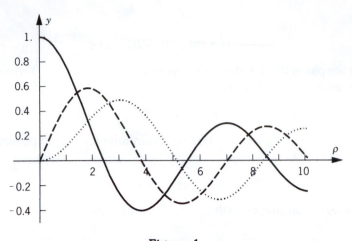

**Figure 1**

and therefore determine all the succeeding coefficients. Therefore, $a_k = 0$ for $k$ odd. Making the conventional choice $a_0 = 2^{-n}/n!$, we end up with the particular solution

$$J_n(\rho) = \frac{\rho^n}{2^n n!} \left[ 1 - \frac{\rho^2}{2^2(n+1)} + \frac{\rho^4}{2!2^4(n+1)(n+2)} - \cdots \right]$$

$$= \sum_{j=0}^{\infty} (-1)^j \frac{\left(\frac{1}{2}\rho\right)^{n+2j}}{j!(n+j)!}. \tag{14}$$

This particular solution is called the *Bessel function of order n* (see Figure 1). It crosses the axis an infinite number of times. In fact, one can prove that $J_n(\rho)$ has the asymptotic form

$$J_n(\rho) \sim \sqrt{\frac{2}{\pi\rho}} \cos\left(\rho - \frac{\pi}{4} - \frac{n\pi}{2}\right) + O\left(\frac{1}{\rho^{3/2}}\right) \tag{15}$$

as $\rho \to \infty$. The Bessel function $J_n(\rho)$ is the only solution (except for a constant factor) of Bessel's ODE that is finite at the singular point $\rho = 0$. All the *other* solutions of (6) look like a constant times $\rho^{-n}$ near $\rho = 0$ in case $n > 0$ (and also have a term $\rho^n \log \rho$ in their expansions). In the case $n = 0$, the other solutions look like $C \log \rho$ near $\rho = 0$. For a discussion of these properties and for more information about Bessel's ODE, see Section 10.5.

## THE EIGENFUNCTION EXPANSION

Now let's return to the drumhead that led to (8) and (9). Since $R(0)$ is to be finite, $R = cJ_n(\rho)$ for any constant $c$. Therefore, we have the separated solutions

$$J_n(\sqrt{\lambda}r)(A_n \cos n\theta + B_n \sin n\theta). \tag{16}$$

Next we put in the homogeneous *boundary condition* $u = 0$ where $r = a$. Thus $\lambda$ must be chosen as a root of

$$J_n(\sqrt{\lambda}a) = 0. \tag{17}$$

From Figure 1, each Bessel function has an infinite number of positive roots. Call the roots of (17)

$$0 < \lambda_{n1} < \lambda_{n2} < \lambda_{n3} < \cdots.$$

Finally, we can sum everything up. The full solution of (1) is

$$u(r, \theta, t) = \sum_{m=1}^{\infty} J_0(\sqrt{\lambda_{0m}}r)(A_{0m} \cos \sqrt{\lambda_{0m}}ct + C_{0m} \sin \sqrt{\lambda_{0m}}ct)$$

$$+ \sum_{m,n=1}^{\infty} J_n(\sqrt{\lambda_{nm}}r)\Big[(A_{nm} \cos n\theta + B_{nm} \sin n\theta) \cos \sqrt{\lambda_{nm}}ct$$

$$+ (C_{nm} \cos n\theta + D_{nm} \sin n\theta) \sin \sqrt{\lambda_{nm}}ct\Big]. \tag{18}$$

This is quite formidable! We shall give some manageable examples shortly.

Before we do that, let's put in the *initial conditions* $u(r, \theta, 0) = \phi(r, \theta)$ and $u_t(r, \theta, 0) = \psi(r, \theta)$. Abbreviating $\beta_{nm} = \sqrt{\lambda_{nm}}$, we must have

$$\phi(r, \theta) = \sum_{m=1}^{\infty} A_{0m} J_0(\beta_{0m}r) + \sum_{m,n=1}^{\infty} J_n(\beta_{nm}r)(A_{nm} \cos n\theta + B_{nm} \sin n\theta)$$

$$\tag{19}$$

and

$$\psi(r, \theta) = \sum_{m=1}^{\infty} c\beta_{0m} C_{0m} J_0(\beta_{0m}r)$$

$$+ \sum_{m,n=1}^{\infty} c\beta_{nm} J_n(\beta_{nm}r)(C_{nm} \cos n\theta + D_{nm} \sin n\theta).$$

(Why?) These expansions are an example of the general Fourier series discussed in Section 10.1. So the coefficients are given by the formulas

$$A_{0m} = \frac{1}{2\pi j_{0m}} \int_0^a \int_{-\pi}^{\pi} \phi(r, \theta) J_0(\beta_{0m}r) \, r \, d\theta \, dr$$

$$A_{nm} = \frac{1}{\pi j_{nm}} \int_0^a \int_{-\pi}^{\pi} \phi(r, \theta) J_n(\beta_{nm}r) \cos n\theta \, r \, d\theta \, dr \tag{20}$$

$$B_{nm} = \text{(same formula with } \sin n\theta\text{)},$$

and similarly for the other coefficients in terms of $\psi(r, \theta)$, where for relative brevity we have denoted

$$j_{nm} = \int_0^a [J_n(\beta_{nm}r)]^2 r \; dr = \tfrac{1}{2}a^2[J_n'(\beta_{nm}a)]^2. \tag{21}$$

The evaluation of the last integral comes from (10.5.9). The formulas (21) come from the orthogonality of the eigenfunctions

$$J_n(\beta_{nm}r)\binom{\cos}{\sin}(n\theta) \tag{22}$$

with respect to the inner product on the disk $D$,

$$(f, g) = \iint_D f\,\overline{g}\;dx\;dy = \int_{-\pi}^{\pi}\int_0^a f\,\overline{g}\,r\;dr\;d\theta. \tag{23}$$

The formulas (21) are just special cases of the formulas in Section 10.1.

Not only do we have the orthogonality in the disk $D$, but we also have the separate orthogonalities of $\sin n\theta$ and $\cos n\theta$ on the interval $-\pi < \theta < \pi$ and of the Bessel functions on the interval $0 < r < a$. The last orthogonality statement is

$$\int_0^a J_n(\beta_{nm}r)J_n(\beta_{np}r)r\;dr = 0 \quad \text{for } m \neq p. \tag{24}$$

Note that the $n$ is the same in both factors and that the extra factor $r$ is retained from (24).

### Example 1. The Radial Vibrations of a Drumhead

You beat the center of the drum with the baton at time $t = 0$ and listen for the ensuing vibrations. This means the initial conditions are

$$u(x, y, 0) = 0 \quad \text{and} \quad u_t(x, y, 0) = \psi(r) \tag{25}$$

[where $\psi(r)$ is concentrated near $r = 0$]. Because $\phi(r, \theta) = 0$, all the $A_{nm}$ and $B_{nm}$ equal zero (see Exercise 1). Furthermore, because $\psi(r)$ does not depend on $\theta$, we have $C_{nm} = D_{nm} = 0$ for $n \neq 0$. So all that remains from (18) is the series

$$u(r, t) = \sum_{m=1}^{\infty} C_{0m} J_0(\beta_{0m}r) \sin(\beta_{0m}ct), \tag{26}$$

where

$$c\beta_{0m}C_{0m} = \frac{(\psi, J_0)}{(J_0, J_0)}$$

$$= \frac{\int_0^a \psi(r) J_0(\beta_{0m}r)r\;dr}{\int_0^a [J_0(\beta_{0m}r)]^2 r\;dr}.$$

By (10.5.9) the coefficients are given by

$$C_{0m} = \frac{\int_0^a \psi(r) J_0(\beta r) r \, dr}{\frac{1}{2} a^2 c \beta [J_1(\beta a)]^2}, \tag{27}$$

where we have written $\beta = \beta_{0m}$.

The lowest note you'll hear is the *fundamental frequency* $\beta_{01}c = z_1 c/a$, where $z_1$ denotes the smallest positive root of $J_0(z_1) = 0$. Numerically, $z_1 = 2.405$. [This is also the fundamental frequency of the general (nonradial) vibrations of the drumhead.] It is interesting to compare this with the one-dimensional string, whose lowest frequency was $\pi c/l$, where of course $\pi = 3.142$.    □

## EXERCISES

1.  Show that with the initial conditions (26), all the $\cos \sqrt{\lambda} ct$ terms in the series (18) are missing. Also show that $D_{nm} = C_{nm} = 0$ for $n \neq 0$.

2.  Determine the vibrations of a circular drumhead (held fixed on the boundary) with the initial conditions $u = 1 - r^2/a^2$ and $u_t \equiv 0$ when $t = 0$.

3.  Suppose that you had a circular drum with wave speed $c_d$ and radius $a$ and a violin string with wave speed $c_v$ and length $l$. In order to make the fundamental frequencies of the drum and the violin the same, how would you choose the length $l$?

4.  Find all the solutions of the wave equation of the form $u = e^{-i\omega t} f(r)$ that are finite at the origin, where $r = \sqrt{x^2 + y^2}$.

5.  Solve the diffusion equation in the disk of radius $a$, with $u = B$ on the boundary and $u = 0$ when $t = 0$, where $B$ is a constant. (*Hint:* The answer is radial.)

6.  Do the same for the annulus $\{a^2 < x^2 + y^2 < b^2\}$ with $u = B$ on the whole boundary.

7.  Let $D$ be the semidisk $\{x^2 + y^2 < b^2, y > 0\}$. Consider the diffusion equation in $D$ with the conditions: $u = 0$ on bdy $D$ and $u = \phi(r, \theta)$ when $t = 0$. Write the complete expansion for the solution $u(r, \theta, t)$, including the formulas for the coefficients.

## 10.3    SOLID VIBRATIONS IN A BALL

We take $D$ to be the ball with its center at the origin and radius $a$. We consider the wave equation with Dirichlet BCs. Upon separating out the time by writing $u(\mathbf{x}, t) = T(t)v(\mathbf{x})$, we get the eigenvalue problem

$$\begin{cases} -\Delta v = \lambda v & \text{in } D \\ \quad v = 0 & \text{on } \partial D, \end{cases} \tag{1}$$

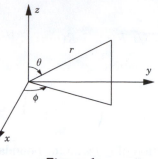

**Figure 1**

as described in Section 10.1. Naturally, we shall separate variables in spherical coordinates (see Figure 1)

$$0 \leq r < a \qquad x = r \sin \theta \cos \phi$$
$$0 \leq \phi < 2\pi \qquad y = r \sin \theta \sin \phi$$
$$0 \leq \theta \leq \pi \qquad z = r \cos \theta.$$

(*Watch out:* In some math books $\theta$ and $\phi$ are switched!) In these coordinates the equation looks like

$$0 = \Delta v + \lambda v$$
$$= v_{rr} + \frac{2}{r} v_r + \frac{1}{r^2} \left[ \frac{1}{\sin^2 \theta} v_{\phi\phi} + \frac{1}{\sin \theta} (\sin \theta \, v_\theta)_\theta \right] + \lambda v.$$

Now we separate the $r$ coordinate only:

$$v = R(r) \cdot Y(\theta, \phi),$$

so that

$$\lambda r^2 + \frac{r^2 R_{rr} + 2r R_r}{R} + \frac{(1/\sin^2 \theta) Y_{\phi\phi} + (1/\sin \theta)(\sin \theta \, Y_\theta)_\theta}{Y} = 0.$$

We get two equations, because the first two terms depend only on $r$ and the last expression only on the angles. So the $R$ equation is

$$R_{rr} + \frac{2}{r} R_r + \left( \lambda - \frac{\gamma}{r^2} \right) R = 0 \tag{2}$$

and the $Y$ equation is

$$\frac{1}{\sin^2 \theta} Y_{\phi\phi} + \frac{1}{\sin \theta} (\sin \theta \, Y_\theta)_\theta + \gamma Y = 0, \tag{3}$$

where $\gamma$ is the separation constant.

Equation (2) is similar to the Bessel equation, but it misses because of the coefficient $2/r$ instead of $1/r$. To handle that coefficient, we change dependent variables by

$$w(r) = \sqrt{r}\, R(r), \qquad R(r) = r^{-1/2} w(r), \tag{4}$$

which converts (2) to the equation

$$w_{rr} + \frac{1}{r} w_r + \left( \lambda - \frac{\gamma + \frac{1}{4}}{r^2} \right) w = 0. \tag{5}$$

We are looking for a solution of (5) with the boundary conditions

$$w(0) \text{ finite} \qquad \text{and} \qquad w(a) = 0. \tag{6}$$

As in Section 10.2, the solution is any constant multiple of the Bessel function

$$w(r) = J_{\sqrt{\gamma + \frac{1}{4}}}(\sqrt{\lambda}\, r). \tag{7}$$

Here the "order" of the Bessel function is $n = \sqrt{\gamma + \frac{1}{4}}$. (See the discussion in Section 10.5 for Bessel's equation with any real order.) Thus the whole radial factor is

$$R(r) = \frac{J_{\sqrt{\gamma + \frac{1}{4}}}(\sqrt{\lambda}\, r)}{\sqrt{r}} \tag{8}$$

Let's go on to the angular functions $Y(\theta, \phi)$. We wish to solve equation (3) with the "boundary conditions"

$$\begin{cases} Y(\theta, \phi) & \text{of period } 2\pi \text{ in } \phi \\ Y(\theta, \phi) & \text{finite at } \theta = 0, \pi. \end{cases} \tag{9}$$

Such a function is called an eigenfunction of the spherical surface, or a *spherical harmonic*. (The reason for the name is this: A harmonic function $v$ in $D$, which corresponds to the case $\lambda = 0$, will have an expansion in the $r$, $\theta$, $\phi$ variables which is an infinite series in the spherical harmonics.)

To solve (3) with the boundary conditions (9), we separate a final time:

$$Y(\theta, \phi) = p(\theta)\, q(\phi).$$

Thus

$$\frac{q''}{q} + \frac{\sin \theta (\sin \theta\, p_\theta)_\theta}{p} + \gamma \sin^2 \theta = 0. \tag{10}$$

The first term in (10) must be a constant, which we call $(-\alpha)$. This means that the $\phi$ equation is

$$q_{\phi\phi} + \alpha q = 0, \qquad q(\phi) \text{ of period } 2\pi. \tag{11}$$

This is a familiar problem. We know that the eigenvalues are $\alpha = m^2$ ($m = 0, 1, 2, \ldots$) and the eigenfunctions are

$$q(\phi) = A \cos m\phi + B \sin m\phi.$$

Finally, the $\theta$ equation is, from (10) with the first term $(-m^2)$,

$$\frac{(d/d\theta)[\sin\theta(dp/d\theta)]}{\sin\theta} + \left(\gamma - \frac{m^2}{\sin^2\theta}\right)p = 0 \qquad (12)$$

with the conditions

$$p \text{ finite at } \theta = 0, \pi. \qquad (13)$$

Let's introduce the variable $s = \cos\theta$ so that $\sin^2\theta = 1 - \cos^2\theta = 1 - s^2$. Then equation (12) is converted to the form

$$\frac{d}{ds}\left[(1 - s^2)\frac{dp}{ds}\right] + \left(\gamma - \frac{m^2}{1 - s^2}\right)p = 0 \qquad (14)$$

with

$$p(s) \text{ finite at } s = \pm 1. \qquad (15)$$

Note the singular behavior of the coefficients of (14) at $s = \pm 1$. This is a consequence of the degeneration of the coordinate system at the north and south poles and is the reason for the unusual boundary conditions (15).

The ODE (14) is called the *associated Legendre equation*. It too can be solved most readily by the method of power series. The details are given in Section 10.6. The main fact that we need to know about it is the following. The eigenvalues of problem (14)-(15) are

$$\gamma = l(l+1), \quad \text{where } l \text{ is an } integer \geq m \qquad (16)$$

and the eigenfunctions are (any constant times)

$$P_l^m(s) = \frac{(-1)^m}{2^l l!}(1 - s^2)^{m/2}\frac{d^{l+m}}{ds^{l+m}}[(s^2 - 1)^l]. \qquad (17)$$

The function (17) is called the *associated Legendre function*. Notice that it is merely a polynomial multiplied by a power of $\sqrt{1 - s^2}$. Also notice that it is finite at $s = \pm 1$.

Finally, let's put the whole problem together. The separated solutions of (1) are

$$v = R(r)p(\theta)q(\phi)$$

$$= \frac{J_{l+\frac{1}{2}}(\sqrt{\lambda}r)}{\sqrt{r}}P_l^m(\cos\theta)\,(A\cos m\phi + B\sin m\phi)$$

because $\sqrt{\gamma + \frac{1}{4}} = \sqrt{l(l+1) + \frac{1}{4}} = l + \frac{1}{2}$. As usual, we could replace the last factors of cosine and sine by $e^{im\phi}$ and $e^{-im\phi}$. When we finally insert the boundary condition $v = 0$ at $r = a$, we get the eigenvalue equation

$$J_{l+\frac{1}{2}}(\sqrt{\lambda}\,a) = 0. \tag{18}$$

Let's call its roots $\lambda = \lambda_{l1}, \lambda_{l2}, \lambda_{l3}, \ldots$. Then the eigenfunctions that correspond to the eigenvalue $\lambda_{lj}$ can be rewritten as

$$v_{lmj}(r, \theta, \phi) = \frac{J_{l+\frac{1}{2}}(\sqrt{\lambda_{lj}}\,r)}{\sqrt{r}} \cdot P_l^{|m|}(\cos\theta) \cdot e^{im\phi}, \tag{19}$$

where we allow $m = -l, \ldots, 0, \ldots, +l$ since we've replaced the sines and cosines by complex exponentials. Thus we see that the eigenvalue $\lambda_{lj}$ has multiplicity $(2l + 1)$, since there are that many different $m$'s. The whole set of eigenfunctions for

$$m = -l, \ldots, l; \quad l = 0, \ldots, \infty; \quad j = 1, \ldots, \infty \tag{20}$$

is orthogonal and complete! What does *orthogonality* mean for this case? It means that

$$\int_0^{2\pi} \int_0^{\pi} \int_0^a v_{lmj}(r, \theta, \phi) \cdot v_{l'm'j'}(r, \theta, \phi) \cdot r^2 \sin\theta \, dr \, d\theta \, d\phi = 0 \tag{21}$$

for all the different triplets $(l, m, j) \neq (l', m', j')$.

## Example 1.

*Solve the heat equation in the ball of radius $a$ with $u = 0$ on bdy $D$ and with a given initial condition $u(\mathbf{x}, 0) = g(\mathbf{x})$. The exact solution is*

$$u(x, t) = \sum_{l=0}^{\infty} \sum_{j=1}^{\infty} \sum_{m=-l}^{l} A_{lmj} e^{-k\lambda_{lj}t} \frac{J_{l+\frac{1}{2}}(\sqrt{\lambda_{lj}}\,r)}{\sqrt{r}} P_l^{|m|}(\cos\theta)\, e^{im\phi} \tag{22}$$

*where the coefficients are*

$$A_{lmj} = \frac{\iiint_D \overline{v_{lmj}(\mathbf{x})}\, g(\mathbf{x})\, d\mathbf{x}}{\iiint_D |v_{lmj}(\mathbf{x})|^2 \, d\mathbf{x}}.$$

What does the solution look like for very large $t$? It looks approximately like its first term, the one with the smallest $\lambda_{lj}$.    □

Returning to the general properties of the eigenfunctions, we can also verify the *separate orthogonality conditions* in each variable. In $\phi$,

$$\int_0^{2\pi} e^{im\phi} e^{-im'\phi}\, d\phi = 0 \quad \text{for } m \neq m'.$$

In $\theta$,

$$\int_0^\pi P_l^m(\cos\theta) P_{l'}^m(\cos\theta) \sin\theta\, d\theta = 0 \quad \text{for } l \neq l'$$

with the same index $m$; or in terms of $s = \cos\theta$,

$$\int_{-1}^1 P_l^m(s) P_{l'}^m(s)\, ds = 0 \quad \text{for } l \neq l'.$$

In $r$,

$$\int_0^a J_{l+\frac{1}{2}}(\sqrt{\lambda_{lj}}\, r) J_{l+\frac{1}{2}}(\sqrt{\lambda_{lj'}}\, r)\, r\, dr = 0 \quad \text{for } j \neq j'$$

with the same $l$. The normalizing constants

$$\int_0^\pi \left[ P_l^m(\cos\theta) \right]^2 \sin\theta\, d\theta = \frac{2}{2l+1} \frac{(l+m)!}{(l-m)!} \tag{23}$$

for the Legendre functions may also be found in Section 10.6.

## SPHERICAL HARMONICS

The functions

$$\boxed{Y_l^m(\theta, \phi) = P_l^{|m|}(\cos\theta)\, e^{im\phi}}$$

are the spherical harmonics. Their indices range over

$$-l \leq m \leq l, \quad 0 \leq l < \infty.$$

They are the eigenfunctions of the problem (3), (9). Equation (3) is the equation for the eigenfunctions of the Laplacian on the spherical surface. They are complete:

**Theorem 1.**   Every function on the surface of a sphere $\{r = a\}$ (specifically, every function whose square is integrable) can be expanded in a series of the spherical harmonics $Y_l^m(\theta, \phi)$.

Ignoring the constant coefficients, which are arbitrary anyway, the first few spherical harmonics are as follows:

| $l$ | $m$ | | | |
|---|---|---|---|---|
| 0 | 0 | 1 | | |
| 1 | 0 | $\cos\theta = \dfrac{z}{r}$ | | |
| 1 | ±1 | $\sin\theta\cos\phi = \dfrac{x}{r}$ | *and* | $\sin\theta\sin\phi = \dfrac{y}{r}$ |
| 2 | 0 | $3\cos^2\theta - 1 = \dfrac{3z^2 - r^2}{r^2}$ | | |
| 2 | ±1 | $\sin\theta\cos\theta\cos\phi = \dfrac{xz}{r^2}$ | *and* | $\dfrac{yz}{r^2}$ |
| 2 | ±2 | $\sin^2\theta\cos 2\phi = \dfrac{x^2 - y^2}{r^2}$ | *and* | $\dfrac{xy}{r^2}$ |

See Section 10.6 for the first few associated Legendre functions, from which this table is derived.

## Example 2.

*Solve the Dirichlet problem*

$$\boxed{\begin{aligned} \Delta u &= 0 \quad \text{in the ball } D \\ u &= g \quad \text{on bdy } D \end{aligned}}$$

by the method of separation of variables. (This is the three-dimensional analog of the two-dimensional problem in a disk that we did in Section 6.3.) When we separate variables, we get precisely (2) and (3) except that $\lambda = 0$. Because the $\lambda$ term is missing, the $R$ equation (2) is exactly of Euler type and so has the solution

$$R(r) = r^\alpha \qquad \text{where} \quad \alpha(\alpha - 1) + 2\alpha - \gamma = 0, \quad \text{or} \quad \alpha^2 + \alpha - \gamma = 0.$$

Equation (3) together with the boundary conditions (9) has already been solved: $Y = Y_l^m(\theta, \phi)$ with $\gamma = l(l + 1)$. Therefore,

$$0 = \alpha^2 + \alpha - l(l + 1) = (\alpha - l)(\alpha + l + 1).$$

We reject the negative root $\alpha = -l - 1$, which would lead to a singularity at the origin. So $\alpha = l$. Therefore, we have the separated solutions

$$r^l \cdot P_l^m(\cos\theta) \cdot e^{im\phi}. \tag{24}$$

The complete solution is

$$u = \sum_{l=0}^{\infty} \sum_{m=-l}^{l} A_{lm} \, r^l \, P_l^m(\cos\theta) \, e^{im\phi}. \tag{25}$$

The coefficients are determined by the expansion of $g(\theta, \phi)$ in spherical harmonics.    □

It is a remarkable fact that the *solid spherical harmonics* (24) are polynomials(!) in the cartesian coordinates $x$, $y$, $z$. To prove this, we use the fact, mentioned above, that the associated Legendre functions have the form $(\sqrt{1-s^2})^m \, p(s)$ where $p(s)$ is a polynomial and $m$ is an integer. Therefore, the solid harmonics (24) have the form

$$r^l \cdot \sin^m\theta \cdot p(\cos\theta) \cdot e^{im\phi}$$

for some polynomial $p$ of degree $l - m$. It is an even polynomial if $l - m$ is even, and an odd polynomial if $l - m$ is odd. So we can write (24) as

$$(r \sin\theta \, e^{i\phi})^m \cdot r^{l-m} p\left(\frac{z}{r}\right). \tag{26}$$

In either case only even powers of $r$ appear in the last factor $r^{l-m} p(z/r)$, so that it is a polynomial in the variables $z$ and $r^2$. Therefore, the solid spherical harmonic (26) is the polynomial $(x + iy)^m$ multiplied by a polynomial in $z$ and $x^2 + y^2 + z^2$. Therefore, (24) is a polynomial in $x$, $y$, and $z$.

## EXERCISES

1.  Calculate the normalizing constants for the spherical harmonics using the appropriate facts about the Legendre functions.

2.  Verify the first six entries in the table of spherical harmonics.

3.  Show that the spherical harmonics satisfy $Y_l^m = (-1)^m \, \overline{Y_l^{-m}}$.

4.  Solve the wave equation in the ball $\{r < a\}$ of radius $a$, with the conditions $\partial u / \partial r = 0$ on $\{r = a\}$,

    $$u = z = r\cos\theta \quad \text{when } t = 0, \quad \text{and} \quad u_t \equiv 0 \quad \text{when } t = 0.$$

5.  Solve the diffusion equation in the ball of radius $a$, with $u = B$ on the boundary and $u = C$ when $t = 0$, where $B$ and $C$ are constants. (*Hint:* The answer is radial.)

6.  ("*A Recipe for Eggs Fourier*," by J. Goldstein) Consider an egg to be a homogeneous ball of radius $\pi$ centimeters. Initially, at 20°C, it is placed in a pot of boiling water (at 100°C). How long does it take for the center to reach 50°C? Assume that the diffusion constant is $k = 6 \times 10^{-3}$ cm$^2$/sec. (*Hint:* The temperature is a function of $r$ and $t$. Approximate $u(0, t)$ by the first term of the expansion.)

7. (a)  Consider the diffusion equation in the ball of radius $a$, with $\partial u/\partial r = B$ on the boundary and $u = C$ when $t = 0$, where $B$ and $C$ are constants. Find the nondecaying terms in the expansion of the solution. (*Hint:* The answer is radial.)

   (b)  Find the decaying terms, including a simple equation satisfied by the eigenvalues.

8. (a)  Let $B$ be the ball $\{x^2 + y^2 + z^2 < a^2\}$. Find all the *radial* eigenfunctions of $-\Delta$ in $B$ with the Neumann BCs. By "radial" we mean "depending only on the distance $r$ to the origin." [*Hint:* A simple method is to let $v(r) = ru(r)$.]

   (b)  Find a simple explicit formula for the eigenvalues.

   (c)  Write the solution of $u_t = k \, \Delta u$ in $B$, $u_r = 0$ on bdy $B$, $u(\mathbf{x}, 0) = \phi(r)$ as an infinite series, including the formulas for the coefficients.

   (d)  In part (c), why does $u(\mathbf{x}, t)$ depend only on $r$ and $t$?

9. Solve the diffusion equation in the ball $\{x^2 + y^2 + z^2 < a^2\}$ with $u = 0$ on the boundary and a radial initial condition $u(\mathbf{x}, 0) = \phi(r)$, where $r^2 = x^2 + y^2 + z^2$. (*Hint:* See the hint for Exercise 8(a).)

10. Find the harmonic function in the *exterior* $\{r > a\}$ of a sphere that satisfies the boundary condition $\partial u/\partial r = -\cos\theta$ on $r = a$ and which is bounded at infinity.

11. Find the harmonic function in the half-ball $\{x^2 + y^2 + z^2 < a^2, z > 0\}$ with the BC $u = f(z)$ on the hemisphere $\{z = (a^2 - x^2 - y^2)^{1/2}\}$ and the BC $u \equiv 0$ on the disk $\{z = 0, x^2 + y^2 < a^2\}$. Include the formulas for the coefficients. (*Hint:* Use spherical coordinates and extend the solution to be odd across the $xy$ plane.)

12. A substance diffuses in infinite space with initial concentration $\phi(r) = 1$ for $r < a$, and $\phi(r) = 0$ for $r > a$. Find a formula for the concentration at later times. (*Hint:* It is radial. You can substitute $v = ru$ to get a problem on a half-line.)

13. Repeat Exercise 12 by computer using the methods of Section 8.2.

## 10.4 NODES

Let $v(\mathbf{x})$ be an eigenfunction of the laplacian,

$$-\Delta v = \lambda v \quad \text{in } D, \tag{1}$$

together with one of the standard boundary conditions. Its *nodal set* $\mathcal{N}$ is defined simply as the set of points $\mathbf{x} \in D$ where $v(\mathbf{x}) = 0$. By definition boundary points are not in the nodal set.

For example, in one dimension with Dirichlet's condition, we have $v_n(x) = \sin(n\pi x/l)$ on the interval $0 < x < l$. This eigenfunction has a node each of the $n - 1$ times it crosses the axis in the open interval $(0, l)$. $\mathcal{N}$ consists of these $n - 1$ points.

The nodal set is important because it allows us to visualize the sets where $v(\mathbf{x})$ is positive or negative. These sets are divided by the nodal set. In one, two, and three dimensions the nodal set consists of points, curves, and surfaces, respectively.

Here is an interpretation of the nodes in terms of waves. As we know, the function

$$u(\mathbf{x}, t) = (A \cos \sqrt{\lambda} ct + B \sin \sqrt{\lambda} ct)\, v(\mathbf{x}), \qquad (2)$$

for any $A$ and $B$, solves the wave equation $u_{tt} = c^2 \Delta u$ in $D$ with the same boundary conditions as $v(\mathbf{x})$. The nodal set is stationary. That is, *the points* $\mathbf{x}$ *in* $\mathcal{N}$ *do not move at all* because, at such a point, $u(\mathbf{x}, t) = 0$ for all $t$. So, for instance, when a guitar player puts his finger on a string, he eliminates certain notes. If he puts his finger exactly in the middle, he eliminates all the frequencies $n\pi ct/l$ with *odd* $n$, because only for even $n$ does the eigenfunction vanish in the middle, $v_n(l/2) = 0$. The nodal sets of ancient Chinese bells form interesting patterns which precisely explain their acoustic tones (see [Sh]).

## Example 1. The Square

In two dimensions the nodal set can be a lot more interesting than in one. Consider the Dirichlet problem in a square $D = \{0 < x < \pi, 0 < y < \pi\}$. Just as in Section 10.1,

$$v_{nm}(x, y) = \sin nx \sin my \quad \text{and} \quad \lambda_{nm} = n^2 + m^2. \qquad (3)$$

The four smallest eigenvalues are as follows:

| $\lambda$ | $v(x, y)$ |
|---|---|
| 2 | $A \sin x \sin y$ |
| 5 | $A \sin 2x \sin y + B \sin x \sin 2y$ |
| 8 | $A \sin 2x \sin 2y$ |
| 10 | $A \sin 3x \sin y + B \sin x \sin 3y$ |

The eigenvalues $\lambda = 5$ and $\lambda = 10$ are double. Because the eigenvalues are $m^2 + n^2$, the multiplicity problem reduces to the question: *In how many ways can a given integer $\lambda$ be written as the sum of two squares?*

The nodal lines for the eigenfunctions $\sin nx \sin my$ are simply line segments parallel to the coordinate axes. However, in the case of multiple eigenvalues many other nodal curves can occur. The zeros of the eigenfunction $A \sin mx \sin ny + B \sin nx \sin my$ for a square are an example.

In Figure 1 are drawn some pictures of nodal curves in cases of multiplicity where the eigenfunctions (3) are denoted by $u_{nm}$. □

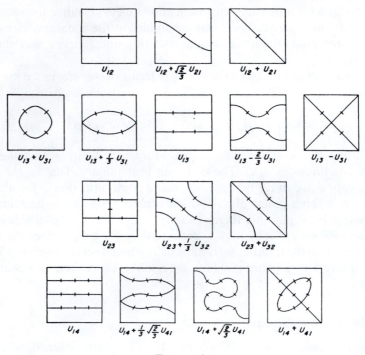

**Figure 1**

## Example 2. The Ball

Consider the ball $D = \{r < a\}$, with the eigenfunctions

$$v_{lmj}(r, \theta, \phi) = r^{-\frac{1}{2}} J_{l+\frac{1}{2}}(r\sqrt{\lambda_{lj}}) P_l^m(\cos\theta)(A\cos m\phi + B\sin m\phi). \quad (4)$$

Its nodal set is a union of the following kinds of surfaces.

- (i)  Spheres inside $D$, which correspond to zeros of the Bessel function.
- (ii)  Vertical planes $\phi = $ constant.
- (iii)  Horizontal planes $\theta = $ constant.

There are $j - 1$ spheres, $m$ vertical planes, and $l - m$ horizontal planes. The vertical and horizontal planes intersect any sphere (with its center at the origin) in great circles through the poles (of constant longitude) and horizontal circles (of constant latitude). They divide the sphere into regions called *tessera* on each of which the eigenfunction is of one sign. For details, see [St] or [TS]. □

How many regions can the nodal set divide a general domain $D$ into? The following result delineates the possibilities. (We assume that $D$ is a "connected" set.)

**Theorem 1.**

(i) The first eigenfunction $v_1(\mathbf{x})$ (the one corresponding to the smallest eigenvalue $\lambda_1$) cannot have any nodes.

(ii) For $n \geq 2$, the $n$th eigenfunction $v_n(\mathbf{x})$ (corresponding to the $n$th eigenvalue, counting multiplicity) divides the domain $D$ into at least two and at most $n$ pieces.

For instance, in one dimension, there are $n - 1$ nodal points that divide the interval $(0, l)$ into exactly $n$ pieces. In the case of the square illustrated above, the eigenvalues are 2, 5, 5, 8, 10, 10, .... The fifth and sixth eigenvalues on this list are $l = 10$, whose eigenfunctions divide the square into two or three or four pieces.

It is easy to see why the nodes of $v_n(\mathbf{x})$ divide $D$ into *at least* two pieces if $n \geq 2$. Indeed, by (i) we have $v_1(\mathbf{x}) \neq 0$ for all $\mathbf{x} \in D$. Since it is continuous and $D$ is connected, we may assume that $v_1(\mathbf{x}) > 0$ for all $\mathbf{x} \in D$. But we know that $v_n(\mathbf{x})$ is orthogonal to $v_1(\mathbf{x})$:

$$\iiint_D v_n(\mathbf{x})\, v_1(\mathbf{x})\, d\mathbf{x} = 0. \tag{5}$$

Therefore, $v_n(\mathbf{x})$ cannot be everywhere of one sign. So $v_n(\mathbf{x})$ must be somewhere positive and somewhere negative. By continuity these points must be separated by the nodal set where $v_n(\mathbf{x})$ is zero. The other statements of the theorem will be proven in Exercises 11.6.7 and 11.6.8.

## EXERCISES

1. For the Dirichlet problem in a square whose eigenfunctions are given by (3), list the nine smallest distinct eigenvalues. What are their multiplicities?

2. Sketch the nodal set of the eigenfunction

   $$v(x, y) = \sin 3x \sin y + \sin x \sin 3y \text{ in the square } (0, \pi)^2.$$

   (*Hint:* Use formulas for $\sin 3x$ and $\sin 3y$ together with factorization to rewrite it as $v(x, y) = 2 \sin x \sin y \, (3 - 2 \sin^2 x - 2 \sin^2 y)$.)

3. Small changes can alter the nature of the nodal set drastically. Use a computer program to show that the eigenfunction $\sin 12x \sin y + \sin x \sin 12y$ has a nodal set that divides the square $(0, \pi)^2$ into 12 subregions, but that the eigenfunction $\sin 12x \sin y + v \sin x \sin 12y$, for $v$ near 1, has a nodal set that divides it into only two subregions. Show that the nodal sets are as sketched in Figure 2.

4. Read about the nodal patterns of ancient Chinese bells in [Sh].

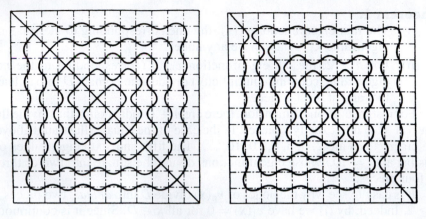

**Figure 2**

## 10.5  BESSEL FUNCTIONS

We have seen that in problems with circular and spherical symmetry the radial parts of the eigenfunctions satisfy *Bessel's differential equation*

$$\frac{d^2u}{dz^2} + \frac{1}{z}\frac{du}{dz} + \left(1 - \frac{s^2}{z^2}\right)u = 0. \tag{1}$$

The purpose of this section is to summarize the most important properties of the solutions of Bessel's equation. Some of the proofs are omitted. There are whole volumes on Bessel functions, such as [Bo].

In Section 10.2 we solved Bessel's ODE in the case that the order $s$ is an integer. Now we allow $s$ to be any real number. We write a prospective solution as

$$u(z) = z^{\alpha}\sum_{k=0}^{\infty}a_k z^k \qquad (a_0 \neq 0). \tag{2}$$

We get, exactly as in Section 10.2, the value (two values, actually) of the exponent $\alpha$ and the values of all the coefficients:

$$\alpha = \pm s, \quad a_k = 0 \text{ for } k \text{ odd,}$$

$$a_k = -\frac{a_{k-2}}{(\alpha + k)^2 - s^2} \quad (k = 2, 4, 6, \ldots). \tag{3}$$

### BESSEL FUNCTION

The Bessel function is defined as that particular solution of (1) with the exponent $\alpha = +s$ and with the leading coefficient $a_0 = [2^s\Gamma(s + 1)]^{-1}$. Here $\Gamma$ denotes the gamma function (see Section A.5). Actually, the choice of $a_0$

is quite arbitrary; we have simply made the standard choice. The gamma function has the property that $\Gamma(s + 1) = s!$ if $s$ is an integer. For any real $s$,

$$\Gamma(s + 1) = s\Gamma(s) = \cdots = s(s - 1) \cdots (s - n)\Gamma(s - n).$$

Therefore, if we write $k = 2j$, it follows from (3) that

$$a_{2j} = (-1)^j [2^{2j+s}\Gamma(j + 1)\Gamma(j + s + 1)]^{-1}.$$

So from (2) the Bessel function is

$$J_s(z) = \sum_{j=0}^{\infty} \frac{(-1)^j}{\Gamma(j + 1)\Gamma(j + s + 1)} \left(\frac{z}{2}\right)^{2j+s}. \tag{4}$$

This series sums up to a bona fide solution of (1) for any real $s$, except if $s$ is a negative integer. [In the latter case, the series is not defined because $\Gamma(s)$ is infinite for negative integers $s$.]

For a given value of $s$, not an integer, the functions $J_s(z)$ and $J_{-s}(z)$ provide a pair of linearly independent solutions of (1). Hence *all* the solutions of (1) are $AJ_s(z) + BJ_{-s}(z)$ for arbitrary constants $A$ and $B$. Note, however, that $J_{-s}(z)$ is infinite at $z = 0$.

## ZEROS

The zeros of $J_s(z)$ are the roots of the equation $J_s(z) = 0$. It can be shown that there are an infinite number of them: $0 < z_1 < z_2 < \cdots$. Each one is a simple zero, meaning that $J'_s(z_j) \neq 0$. Between any two zeros of $J_s$ is a zero of $J_{s+1}$, and vice versa. We therefore say that the zeros of $J_s$ and $J_{s+1}$ separate each other. In fact, the zeros of any two linearly independent solutions of (1) separate each other. The first zeros of $J_0(z)$ are $z = 2.405, 5.520, 8.654, 11.79, \ldots$. (see Figure 10.2.1).

## ASYMPTOTIC BEHAVIOR

It can be shown that as $z \to \infty$ the Bessel function has the form

$$J_s(z) = \sqrt{\frac{2}{\pi z}} \cos\left(z - \frac{s\pi}{2} - \frac{\pi}{4}\right) + O(z^{-3/2}). \tag{5}$$

(Precisely, this means that $[J_s(z) - \sqrt{2/\pi z} \cos(z - s\pi/2 - \pi/4)]z^{3/2}$ is bounded as $z \to \infty$.) Thus it looks just like a damped cosine.

## RECURSION RELATIONS

These are identities that relate the Bessel functions for different values of $s$. Three of them are

$$J_{s\pm1}(z) = \frac{s}{z}J_s(z) \mp J_s'(z) \tag{6}$$

and

$$J_{s-1}(z) + J_{s+1}(z) = \frac{2s}{z}J_s(z). \tag{7}$$

The first pair (6) follows easily from the series expansion (4), while (7) follows from (6) (see Exercise 3).

## NORMALIZING CONSTANTS

In Sections 10.2 and 10.3 we used the values of the definite integrals

$$\int_0^a [J_s(z)]^2 \, z \, dz = \tfrac{1}{2}a^2[J_s'(a)]^2 + \tfrac{1}{2}(a^2 - s^2)[J_s(a)]^2. \tag{8}$$

To prove (8), we observe that $J_s(z)$ satisfies Bessel's equation $(zu')' + z^{-1}(z^2 - s^2)u = 0$. Multiplying the equation by $2zu'$, we get the identity

$$[(zu')^2 + (z^2 - s^2)u^2]' = 2zu^2.$$

Integrating from 0 to $a$, we deduce that

$$2\int_0^a zu^2 \, dz = (zu')^2 + (z^2 - s^2)u^2 \Big|_0^a = a^2u'(a)^2 + (a^2 - s^2)u(a)^2$$

since $u(0) = 0$ for $s \neq 0$. This proves (8).

In particular, if $\beta a$ is a zero of $J_s(z)$, then (8) simplifies to

$$\int_0^a [J_s(\beta r)]^2 r \, dr = \tfrac{1}{2}a^2[J_s'(\beta a)]^2 = \tfrac{1}{2}a^2[J_{s\pm1}(\beta a)]^2 \tag{9}$$

where the last expression comes from (6).

## BESSEL FUNCTIONS OF HALF-INTEGER ORDER

By this we mean $J_s(z)$ with $s = n + \frac{1}{2}$, where $n$ is an integer. (This is the case we encountered in Section 10.3.) In this case the change of variables $u = z^{-1/2}v$ is highly effective. In fact, Bessel's equation (1) then is converted to the equation

$$v'' + \left(1 - \frac{s^2 - \frac{1}{4}}{z^2}\right)v = 0 \tag{10}$$

(see Exercise 4).

The simplest case is $s = \frac{1}{2}$, in which case the equation is just $v'' + v = 0$, with the solutions $v = A \cos z + B \sin z$. So $u(z) = A \cos z / \sqrt{z} + B \sin z / \sqrt{z}$. The Bessel function $J_{1/2}(z)$ is the solution that is finite at $z = 0$ (with an appropriate choice of constant $B$). Thus

$$J_{1/2}(z) = \sqrt{\frac{2}{\pi z}} \sin z. \tag{11}$$

Similarly, $J_{-1/2}(z) = \sqrt{2/\pi z} \cos z$. The recursion relations (6) now provide explicit formulas for all the half-integers; namely,

$$J_{n+\frac{1}{2}}(z) = (-1)^n \sqrt{\frac{2}{\pi}} \, z^{n+\frac{1}{2}} \left( z^{-1} \frac{d}{dz} \right)^n \frac{\sin z}{z}. \tag{12}$$

## OTHER SOLUTIONS OF BESSEL'S EQUATION

For $s$ not an integer, the *Neumann function* is defined as

$$N_s(z) = \frac{\cos \pi s}{\sin \pi s} J_s(z) - \frac{1}{\sin \pi s} J_{-s}(z). \tag{13}$$

(This is unrelated to the Neumann function of Exercise 7.4.21.) Of course, this is just another particular solution of (1). One can show that as $z \to \infty$ this solution satisfies

$$N_s(z) = \sqrt{\frac{2}{\pi z}} \sin \left( z - \frac{s \pi}{2} - \frac{\pi}{4} \right) + O(z^{-3/2}). \tag{14}$$

Still another pair of solutions is the pair of *Hankel functions*:

$$\begin{aligned} H_s^{\pm}(z) &= J_s(z) \pm i N_s(z) \\ &= \sqrt{\frac{2}{\pi z}} \exp[\pm i(z - s\pi/2 - \pi/4)] + O(z^{-3/2}). \end{aligned} \tag{15}$$

The property that distinguishes the Neumann and Hankel functions from the other solutions is their special behavior for large $z$.

## BESSEL FUNCTIONS OF INTEGER ORDER $s = n$

As we saw above, this is the case when we have found only one $(J_n(z))$ of the solutions of Bessel's equation (1). A second linearly independent solution for

$s = n$ is the Neumann function, which is defined as

$$N_n(z) = \lim_{s \to n} N_s(z) = \pi^{-1} \lim_{s \to n} \frac{\partial}{\partial s} [J_s(z) - (-1)^n J_{-s}(z)]$$

$$= \frac{2}{\pi} J_n(z) \log \left(\frac{z}{2}\right) + \sum_{k=-n}^{\infty} a_k z^k \tag{16}$$

for some coefficients $a_k$. We omit the derivations of these formulas.

## TWO IDENTITIES

An interesting pair of identities involving integer-order Bessel functions is

$$e^{iz \sin \theta} = \sum_{n=-\infty}^{\infty} e^{in\theta} J_n(z) \tag{17}$$

and

$$J_n(z) = \frac{1}{\pi} \int_0^{\pi} \cos (z \sin \theta - n\theta) \, d\theta. \tag{18}$$

To prove (17) and (18), let's begin by changing notation, replacing $z$ by $r$. Let's think of $r$ and $\theta$ as polar coordinates in the $xy$ plane. The function $e^{ir \sin \theta} = e^{iy}$ is periodic in $\theta$, so that we can expand it in a complex Fourier series

$$e^{ir \sin \theta} = \sum_{n=-\infty}^{\infty} g_n(r) e^{in\theta}.$$

Its coefficients are

$$g_n(r) = \frac{1}{2\pi} \int_{-\pi}^{\pi} e^{ir \sin \theta - in\theta} \, d\theta = \frac{1}{\pi} \int_0^{\pi} \cos(r \sin \theta - n\theta) \, d\theta.$$

*So it is enough to prove that $g_n(r) = J_n(r)$.*

Integrating by parts twice, we get

$$2\pi n^2 g_n(r) = n^2 \int_{-\pi}^{\pi} e^{ir \sin \theta} \cdot e^{-in\theta} \, d\theta$$

$$= nr \int_{-\pi}^{\pi} e^{ir \sin \theta} \cos \theta \cdot e^{-in\theta} \, d\theta$$

$$= \int_{-\pi}^{\pi} e^{ir \sin \theta - in\theta} \cdot (r^2 \cos^2 \theta + ir \sin \theta) \, d\theta. \tag{19}$$

Upon differentiating this integral, using Section A.3, we find that

$$r^2 g_n'' + r g_n' + (r^2 - n^2) g_n$$

$$= \frac{1}{2\pi} \int_{-\pi}^{\pi} e^{ir\sin\theta - in\theta}(-r^2 \sin^2\theta + ir\sin\theta + r^2 - n^2) \, d\theta. \quad (19a)$$

Due to $-r^2 \sin^2\theta + r^2 = r^2 \cos^2\theta$ and identity (19), the last integral equals

$$\frac{1}{2\pi} \int_{-\pi}^{\pi} e^{ir\sin\theta - in\theta}(n^2 - n^2) \, d\theta = 0.$$

So by (19a), $g_n(r)$ satisfies Bessel's equation of order $n$. Which solution is it? Since $g_n(0)$ is finite, $g_n(r) = A_n J_n(r)$ for some constant $A_n$. Now the $k$th derivative is

$$g_n^{(k)}(0) = \frac{i^k}{2\pi} \int_{-\pi}^{\pi} e^{-in\theta} \sin^k \theta \, d\theta.$$

Hence $0 = g_n(0) = g_n'(0) = \cdots = g_n^{(n-1)}(0)$ and $g_n^{(n)}(0) = 2^{-n}$ as we can see by writing the sine as exponentials. Since also $J_n^{(n)}(0) = 2^{-n}$, we deduce that $g_n(r) \equiv J_n(r)$. This completes the proof of (17) and (18).

## EXERCISES

1. Show that

$$J_0(z) = 1 - \left(\frac{z}{2}\right)^2 + \frac{1}{(2!)^2}\left(\frac{z}{2}\right)^4 - \frac{1}{(3!)^2}\left(\frac{z}{2}\right)^6 + \cdots$$

and

$$J_1(z) = -J_0'(z) = \frac{z}{2} - \frac{1}{2!}\left(\frac{z}{2}\right)^3 + \frac{3}{(3!)^2}\left(\frac{z}{2}\right)^5 + \cdots .$$

2. Write simple formulas for $J_{3/2}$ and $J_{-3/2}$.

3. Derive the recursion relations (6) and (7).

4. Show that the substitution $u = z^{-1/2} v$ converts Bessel's equation into (10).

5. Show that if $u$ satisfies Bessel's equation, then $v = z^\alpha u(\lambda z^\beta)$ satisfies the differential equation

$$v'' + \frac{1 - 2\alpha}{z} v' + \left[(\lambda\beta z^{\beta-1})^2 - \frac{s^2\beta^2 - \alpha^2}{z^2}\right] v = 0.$$

6.  Use (11) and the recursion relations to compute $J_{3/2}$ and $J_{5/2}$. Verify (12) in these cases.

7.  Find all the solutions $u(x)$ of the ODE $xu'' - u' + xu = 0$. (*Hint:* Substitute $u = xv$.)

8.  Show that $H_{1/2}^{\pm}(z) = \sqrt{2/\pi z}\, e^{\pm i(z-\pi/2)}$ exactly!

9.  (a)  Show that $u(r,\, t) = e^{i\omega t} H_s^{\pm}(\omega r/c)$ solves the three-dimensional wave equation.
    (b)  Show that it has the asymptotic form $(1 + i)\sqrt{c/\pi\omega r}\, e^{i\omega(t \pm r/c)}$. The plus sign gives us an incoming wave (an approximate function of $t + r/c$), the minus sign an outgoing wave (an approximate function of $t - r/c$).

10. Prove that the three definitions of the Neumann function of integer order given by (16) are equivalent.

11. Fill in the details in the derivation of (17) and (18).

12. Show that $\cos(x \sin \theta) = J_0(x) + 2\sum_{k=1}^{\infty} J_{2k}(x) \cos 2k\theta$.

13. Substitute $t = e^{i\theta}$ in (17) to get the famous identity

$$e^{(1/2)z(t-1/t)} = \sum_{n=-\infty}^{\infty} J_n(z)t^n.$$

The function on the left is called the *generating function* for the Bessel functions of integer order. It is the function whose power series expansion in $t$ has the integer Bessel functions as coefficients.

14. Solve the equation $-u_{xx} - u_{yy} + k^2 u = 0$ in the disk $\{x^2 + y^2 < a^2\}$ with $u \equiv 1$ on the boundary circle. Write your answer in terms of the Bessel functions $J_s(iz)$ of imaginary argument.

15. Solve the equation $-u_{xx} - u_{yy} + k^2 u = 0$ in the *exterior* $\{x^2 + y^2 > a^2\}$ of the disk with $u \equiv 1$ on the boundary circle and $u(x,\, y)$ bounded at infinity. Write your answer in terms of the Hankel functions $H_s(iz)$ of imaginary argument.

16. Solve the equation $-u_{xx} - u_{yy} - u_{zz} + k^2 u = 0$ in the ball $\{x^2 + y^2 + z^2 < a^2\}$, with $u \equiv 1$ on the boundary sphere. Write your answer in terms of elementary functions.

17. Solve the equation $-u_{xx} - u_{yy} - u_{zz} + k^2 u = 0$ in the *exterior* $\{x^2 + y^2 + z^2 > a^2\}$ of the ball with $u \equiv 1$ on the boundary sphere and $u(x,\, y,\, z)$ bounded at infinity. Write your answer in terms of elementary functions.

18. Find an equation for the eigenvalues and find the eigenfunctions of $-\Delta$ in the disk $\{x^2 + y^2 < a^2\}$ with the Robin BC $\partial v/\partial r + hv = 0$ on the circle, where $h$ is a constant.

19. Find an equation for the eigenvalues and find the eigenfunctions of $-\Delta$ in the annulus $\{a^2 < x^2 + y^2 < b^2\}$ with Dirichlet BCs on both circles.

# 10.6   LEGENDRE FUNCTIONS

In problems with spherical symmetry, as in Section 10.3, we encountered *Legendre's differential equation*

$$[(1 - z^2)u']' + \gamma u = 0, \tag{1}$$

where $\gamma = l(l + 1)$ for some integer $l \geq 0$. The purpose of this section is to summarize the most important properties of its polynomial solutions. Some of the proofs are omitted. For more details, see [Sa] or [MF].

## LEGENDRE POLYNOMIAL

The ODE (1) has "singular points" where $1 - z^2 = 0$. Thus $z = \pm 1$. (See Section A.4 for a brief discussion of singular points.) The equation (1) is easily solved by a power series:

$$u(z) = \sum_{k=0}^{\infty} a_k z^k.$$

Upon substituting the power series into (1), we get

$$\sum_{k=0}^{\infty} k(k - 1)a_k z^{k-2} - \sum_{k=0}^{\infty} (k^2 + k - \gamma)a_k z^k = 0.$$

We replace $k - 2$ by $k$ in the first sum. The coefficients of like powers of $z$ must match, so that

$$a_{k+2} = a_k \frac{k(k + 1) - \gamma}{(k + 2)(k + 1)} \qquad (k = 0, 1, 2, \ldots). \tag{2}$$

Both $a_0$ and $a_1$ are arbitrary constants. Since $\gamma = l(l + 1)$ for some integer $l$, we see from (2) that $a_{l+2} = a_{l+4} = \cdots = 0$.

Thus we always get at least one solution that has only a finite number of nonzero coefficients $a_l, a_{l-2}, \ldots$, that is, a polynomial. It is the Legendre polynomial $P_l(z)$. If $l$ is even, $P_l(z)$ has only even powers. If $l$ is odd, $P_l(z)$ has only odd powers. The integer $l$ is the degree of $P_l(z)$. Making a certain choice of the first coefficient $a_0$ or $a_1$, it follows from (2) that the Legendre polynomials are

$$P_l(z) = \frac{1}{2^l} \sum_{j=0}^{m} \frac{(-1)^j}{j!} \frac{(2l - 2j)!}{(l - 2j)!(l - j)!} z^{l-2j} \tag{3}$$

where $m = l/2$ if $l$ is even, and $m = (l - 1)/2$ if $l$ is odd. The first six Legendre polynomials are as follows:

| $l$ | $P_l(z)$ |
| --- | --- |
| 0 | 1 |
| 1 | $z$ |
| 2 | $\frac{1}{2}(3z^2 - 1)$ |
| 3 | $\frac{1}{2}(5z^3 - 3z)$ |
| 4 | $\frac{1}{8}(35z^4 - 30z^2 + 3)$ |
| 5 | $\frac{1}{8}(63z^5 - 70z^3 + 15z)$ |

The $P_l(z)$ satisfy the *recursion relation*

$$(l + 1)P_{l+1}(z) - (2l + 1)z P_l(z) + l P_{l-1}(z) = 0 \tag{4}$$

(see Exercise 1). As mentioned in Section 10.3, they satisfy the *orthogonality* relation

$$\int_{-1}^{1} P_l(z) P_{l'}(z) \, dz = 0 \qquad \text{for } l \neq l'. \tag{5}$$

In case $\gamma$ is not the product $l(l + 1)$ of two successive integers, it turns out that there is no polynomial solution $u \not\equiv 0$ of (1). Both linearly independent solutions can be expressed as power series as in Section A.4 in powers of $(z - 1)$. Only one of these solutions is singular at $z = 1$. The nonsingular solution is called a *Legendre function*.

## NORMALIZING CONSTANTS

The normalizing constants are

$$\int_{-1}^{1} [P_l(z)]^2 \, dz = \frac{2}{2l + 1}. \tag{6}$$

Let's prove (6). Taking the inner product of (4) with $P_{l-1}(z)$ and using the orthogonality (5), we get

$$(2l + 1)(z P_l, P_{l-1}) = l(P_{l-1}, P_{l-1}).$$

Replacing $l$ by $l - 1$ in (4) and taking the inner product with $P_l$, we get

$$(2l - 1)(z P_{l-1}, P_l) = l(P_l, P_l).$$

Combining these two identities, we get

$$(2l + 1)(P_l, P_l) = (2l - 1)(P_{l-1}, P_{l-1})$$

valid for $l = 2, 3, \ldots$ and also for $l = 1$. Hence

$$
\begin{aligned}
(P_l, P_l) &= \frac{2(l - 1) + 1}{2l + 1}(P_{l-1}, P_{l-1}) \\
&= \frac{2(l - 2) + 1}{2l + 1}(P_{l-2}, P_{l-2}) = \cdots \\
&= \frac{3}{2l + 1}(P_1, P_1) = \frac{1}{2l + 1}(P_0, P_0) = \frac{2}{2l + 1}.
\end{aligned}
$$

## RODRIGUES' FORMULA

The formula of Rodrigues expresses the Legendre polynomials explicitly. It is

$$P_l(z) = \frac{1}{2^l l!} \frac{d^l}{dz^l}(z^2 - 1)^l. \tag{7}$$

It follows that $P_l(1) = 1$. The proof is left for Exercise 2.

## ZEROS

Inside the interval $-1 < z < 1$, $P_l(z)$ has exactly $l$ zeros. Furthermore, its $k$th derivative $d^k P_l / dz^k$ has exactly $l - k$ zeros for $1 \le k \le l$. None of these derivatives vanishes at either endpoint.

To prove these statements, we'll use Rodrigues' formula. The polynomial $Q(z) = (z^2 - 1)^l$ has no zeros in the interval $(-1, 1)$ except at the endpoints $\pm 1$. By Rolle's theorem from elementary calculus, its first derivative $Q'$ has at least one interior zero as well as zeros at the two endpoints. Similarly, its second derivative $Q''$ has at least two interior zeros separated by the zero of $Q'$, its third derivative $Q'''$ has at least three separated by the two of $Q''$, and so on. Thus $P_l$, which is essentially the $l$th derivative $Q^{(l)}$ of $Q$, has at least $l$ interior zeros. Because it is a polynomial of degree $l$, it must have exactly $l$ zeros. But now the game changes because $Q^{(l)}$ no longer vanishes at the endpoints $\pm 1$. Because there are only $l - 1$ subintervals between the zeros of $Q^{(l)}$, its derivative $Q^{(l+1)}$ is only guaranteed to have $l - 1$ zeros. Because $Q^{(l+1)}$, which by (7) is almost the same as $P_l'$, is a polynomial of degree $l - 1$, it must have exactly $l - 1$ zeros. Similarly, $Q^{(l+2)}$, which is essentially $P_l''$, has exactly $l - 2$ zeros, and so on.

## GENERATING FUNCTION

The identity

$$(1 - 2tz + t^2)^{-1/2} = \sum_{l=0}^{\infty} P_l(z)t^l \tag{8}$$

is valid for $|z| < 1$ and $|t| < 1$. The function on the left side of (8) is called the generating function for the Legendre polynomials because all of them appear as the coefficients in the power series expansion.

To prove (8), we note that the left side $g(t, z)$ of (8) is an analytic function of $t$ for $|t| < 1$, which precisely means that it has a power series expansion

$$g(t, z) = \sum_{l=0}^{\infty} Q_l(z)t^l \tag{9}$$

with some coefficients $Q_l(z)$. (The reader who has not yet studied analytic functions can simply expand the function $g(t, z)$ using the binomial expansion in powers of $2tz - t^2$ and then expand the powers and rearrange the terms to get (9).)

On the other hand, explicit differentiation of $g(t, z)$ shows that it satisfies the PDE

$$[(1 - z^2)g_z]_z + t[tg]_{tt} = 0.$$

(Check it.) Plugging the expansion (9) into this PDE, we find that

$$\sum_{l=0}^{\infty} [(1 - z^2)Q_l'(z)]'t^l + \sum_{l=0}^{\infty} l(l + 1)Q_l(z)t^l = 0.$$

Because the coefficients must match, we must have

$$[(1 - z^2)Q_l'(z)]' + l(l + 1)Q_l(z) = 0.$$

So $Q_l$ satisfies Legendre's differential equation! On the other hand, putting $z = 1$ in the definition of $g(t, z)$, we have

$$g(t, 1) = (1 - 2t + t^2)^{-1/2} = (1 - t)^{-1} = \sum_{l=0}^{\infty} t^l,$$

so that $Q_l(1) = 1$. This determines *which* solution of Legendre's equation $Q_l$ is, namely $Q_l \equiv P_l$. This proves (8).

## ASSOCIATED LEGENDRE FUNCTIONS

The associated Legendre equation is

$$[(1 - z^2)u']' + \left(\gamma - \frac{m^2}{1 - z^2}\right)u = 0, \tag{10}$$

where $\gamma = l(l + 1)$ and $m \le l$, $m$ and $l$ being integers.

We define the associated Legendre functions as

$$P_l^m(z) = (1 - z^2)^{m/2} \frac{d^m}{dz^m} P_l(z), \tag{11}$$

where the integer $m$ is a superscript. Let's show that (11) is indeed a solution of (10). In fact, $v = (d^m/dz^m)P_l(z)$ has to satisfy the $m$-times-differentiated Legendre equation

$$(1 - z^2)v'' - 2(m + 1)zv' + [\gamma - m(m + 1)]v = 0.$$

Substituting $v(z) = (1 - z^2)^{-m/2} w(z)$, we get the equation

$$[(1 - z^2)w']' + \left(\gamma - \frac{m^2}{1 - z^2}\right)w = 0,$$

which is precisely (10).

The orthogonality property of the associated Legendre functions has already been stated in Section 10.3; namely, $P_l^m$ and $P_{l'}^m$ are orthogonal on the interval $(-1, 1)$. The normalizing constants are

$$\int_{-1}^1 [P_l^m(z)]^2 \, dz = \frac{2(l + m)!}{(2l + 1)(l - m)!}. \tag{12}$$

Rodrigues' formula for the associated functions follows immediately from the $m = 0$ case (7); its statement is left to the reader.

## EXERCISES

1. Show that the Legendre polynomials satisfy the recursion relation (4).
2. (a) Prove Rodrigues' formula (7).
   (b) Deduce that $P_l(1) = 1$.
3. Show that $P_{2n}(0) = (-1)^n (2n)!/2^{2n}(n!)^2$.
4. Show that $\int_{-1}^1 x^2 P_l(x) \, dx = 0$ for $l \geq 3$.
5. Let $f(x) = x$ for $0 \leq x < 1$, and $f(x) = 0$ for $-1 < x \leq 0$. Find the coefficients $a_l$ in the expansion $f(x) = \sum_{l=0}^\infty a_l P_l(x)$ of $f(x)$ in terms of Legendre polynomials in the interval $(-1, 1)$.
6. Find the harmonic function in the ball $\{x^2 + y^2 + z^2 < a^2\}$ with $u = \cos^2 \theta$ on the boundary.
7. Find the harmonic function in the ball $\{x^2 + y^2 + z^2 < a^2\}$ with the boundary condition $u = A$ on the top hemisphere $\{x^2 + y^2 + z^2 = a^2, z > 0\}$ and with $u = B$ on the bottom hemisphere $\{x^2 + y^2 + z^2 = a^2, z < 0\}$, where $A$ and $B$ are constants.
8. Solve the diffusion equation in the solid cone $\{x^2 + y^2 + z^2 < a^2, \theta < \alpha\}$ with $u = 0$ on the whole boundary and with general initial conditions. [*Hint:* Separate variables and write the solution as a series with terms of the separated form $T(t)R(r)q(\phi)p(\cos \theta)$. Show that $p(s)$ satisfies the

associated Legendre equation. Expand $p(s)$ in powers of $(s - 1)$. In terms of such a function, write the equations that determine the eigenvalues.]

## 10.7 ANGULAR MOMENTUM IN QUANTUM MECHANICS

This section is a follow-up to Section 9.5. We consider Schrödinger's equation with a radial potential $V(r)$, where $r = |\mathbf{x}| = (x^2 + y^2 + z^2)^{1/2}$. Thus

$$i u_t = -\tfrac{1}{2} \Delta u + V(r) u \tag{1}$$

in all of space with the "boundary condition"

$$\iiint |u(\mathbf{x}, t)|^2 \, d\mathbf{x} < \infty. \tag{2}$$

We separate out the time variable to get $u(\mathbf{x}, t) = v(\mathbf{x})e^{-i\lambda t/2}$, where

$$-\Delta v + 2V(r)v = \lambda v.$$

Recall that $\lambda$ is interpreted as the energy.

Next we separate out the radial variable $v = R(r)Y(\theta, \phi)$ to get

$$\lambda r^2 - 2r^2 V(r) + \frac{r^2 R_{rr} + 2r R_r}{R} + \frac{1}{Y} \left\{ \frac{1}{\sin^2 \theta} Y_{\phi\phi} + \frac{1}{\sin \theta} [(\sin \theta) Y_\theta]_\theta \right\} = 0.$$

Because the variable $r$ occurs only in the first three terms, the radial equation is

$$R_{rr} + \frac{2}{r} R_r + \left[ \lambda - 2V(r) - \frac{\gamma}{r^2} \right] R = 0, \tag{3}$$

whereas the $Y$ equation is precisely (10.3.3). Therefore, $Y(\theta, \phi)$ is a spherical harmonic:

$$Y(\theta, \phi) = Y_l^m(\theta, \phi) = P_l^m(\cos \theta) \, e^{im\phi} \tag{4}$$

for $|m| \leq l$ (see Section 10.3). Furthermore, $\gamma = l(l + 1)$. The index $l$ is called the *orbital quantum number*, and $m$ is the *magnetic quantum number*.

In quantum mechanics the *angular momentum operator* is defined as the cross product

$$\mathbf{L} = -i\mathbf{x} \times \nabla = -i \begin{pmatrix} y\partial_z - z\partial_y \\ z\partial_x - x\partial_z \\ x\partial_y - y\partial_x \end{pmatrix} = \begin{pmatrix} L_x \\ L_y \\ L_z \end{pmatrix}. \tag{5}$$

This operator $\mathbf{L}$ has the spherical harmonics as its eigenfunctions, as we now explain. In spherical coordinates $\mathbf{L}$ takes the form (Exercise 1)

$$
\begin{cases}
L_x = i(\cot\theta \cos\phi \, \partial_\phi + \sin\phi \, \partial_\theta) \\
L_y = i(\cot\theta \sin\phi \, \partial_\phi - \cos\phi \, \partial_\theta) \\
L_z = -i\partial_\phi.
\end{cases}
\tag{6}
$$

Therefore,

$$
\begin{aligned}
|\mathbf{L}|^2 &= L_x^2 + L_y^2 + L_z^2 \\
&= -\frac{1}{\sin^2\theta}\frac{\partial^2}{\partial\phi^2} - \frac{1}{\sin\theta}\frac{\partial}{\partial\theta}\sin\theta\frac{\partial}{\partial\theta},
\end{aligned}
\tag{7}
$$

which is exactly the angular part of the negative laplacian operator! So from (10.3.3) we have the equation

$$
|\mathbf{L}|^2\big(Y_l^m\big) = l(l+1)\,Y_l^m.
\tag{8}
$$

Because of the explicit form of $L_z$, we also have the equation

$$
L_z\big(Y_l^m\big) = mY_l^m.
\tag{9}
$$

The effect of the operators $L_x$ and $L_y$ on the spherical harmonics is not as simple. If we denote

$$
L_\pm = L_x \pm iL_y,
$$

known as the *raising and lowering operators*, then

$$
\begin{aligned}
L_+\big(Y_l^m\big) &= [(l-m)(l+m+1)]^{1/2}\,Y_l^{m+1} \\
L_-\big(Y_l^m\big) &= [(l+m)(l-m+1)]^{1/2}\,Y_l^{m-1}.
\end{aligned}
\tag{10}
$$

In quantum mechanics there cannot be a pure rotation about a single axis. Indeed, let's choose coordinates so that the axis of rotation is the $z$ axis. A pure rotation about the $z$ axis would mean $\phi$ dependence but no $\theta$ dependence. But such a spherical harmonic would mean that $m \neq 0$ and $l = 0$. This cannot happen because $|m| \leq l$. The physicist's explanation is that in quantum mechanics you cannot define an axis of rotation except in an averaged way. There will always remain some probability of a rotation off the axis.

## HYDROGEN ATOM

The radial equation (3) can be solved analytically only for certain potentials $V(r)$. We take up the case again of the *hydrogen atom* where $V(r) = -1/r$. Thus, after separating the time variable, we have

$$
R_{rr} + \frac{2}{r}R_r + \left[\lambda + \frac{2}{r} - \frac{l(l+1)}{r^2}\right]R = 0.
\tag{11}
$$

The case when $l = 0$ has already been discussed in Section 9.5. As in Section 9.5, the boundary conditions are $R(0)$ finite and $R(\infty) = 0$. We shall follow exactly the same method as in that section.

Consider only the case $\lambda < 0$ and let $\beta = \sqrt{-\lambda}$. Let $w(r) = e^{\beta r} R(r)$. Then

$$w_{rr} + 2\left(\frac{1}{r} - \beta\right)w_r + \left[\frac{2(1 - \beta)}{r} - \frac{l(l + 1)}{r^2}\right]w = 0. \tag{12}$$

We look for solutions in the form of a power series $w(r) = \sum_{k=0}^{\infty} a_k r^k$. Thus

$$\sum k(k - 1)a_k r^{k-2} + 2\sum k a_k r^{k-2} - 2\beta \sum k a_k r^{k-1}$$
$$+ 2(1 - \beta)\sum a_k r^{k-1} - l(l + 1)\sum a_k r^{k-2} = 0.$$

Switching the dummy index $k$ to $k - 1$ in the third and fourth sums, we get

$$\sum_{k=0}^{\infty} [k(k - 1) + 2k - l(l + 1)]a_k r^{k-2}$$

$$+ \sum_{k=1}^{\infty} [-2\beta(k - 1) + (2 - 2\beta)]a_{k-1} r^{k-2} = 0.$$

That is,

$$\sum_{k=0}^{\infty} [k(k + 1) - l(l + 1)]a_k r^{k-2} + \sum_{k=1}^{\infty} 2(1 - k\beta)a_{k-1} r^{k-2} = 0. \tag{13}$$

Thus we are led to the recursion relations

$$
\begin{aligned}
k = 0: \quad & l(l + 1)a_0 = 0 \\
k = 1: \quad & [2 - l(l + 1)]a_1 = -2(1 - \beta)a_0 \\
k = 2: \quad & [6 - l(l + 1)]a_2 = -2(1 - 2\beta)a_1 \\
k = 3: \quad & [10 - l(l + 1)]a_3 = -2(1 - 3\beta)a_2,
\end{aligned}
$$

and in general

$$[k(k + 1) - l(l + 1)]a_k = -2(1 - k\beta)a_{k-1}. \tag{14}$$

We know that $l$ has to be a nonnegative integer. Thus $a_0 = 0$ if $l \neq 0$. In fact, it is clear from (14) that every coefficient has to vanish until the $l$th one:

$$a_0 = a_1 = \cdots a_{l-1} = 0.$$

But then $a_l$ is completely arbitrary. Once $a_l$ is chosen, the succeeding coefficients are determined by it. There will be a polynomial solution only if the coefficients terminate. This will happen whenever $\beta = 1/n$ with $n$ an integer greater than $l$. Thus the eigenvalues are

$$\lambda = -\frac{1}{n^2} \tag{15}$$

(as in Section 9.5). The index $n$ is called the *principal quantum number*.
Thus for each integer $l$, $0 \leq l < n$, we have eigenfunctions of the form

$$v_{nlm}(r, \theta, \phi) = e^{-r/n} L_n^l(r) \cdot Y_l^m(\theta, \phi). \tag{16}$$

These are the wave functions of the hydrogen atom. We have used the (non-standard) notation $w(r) = L_n^l(r)$ for the polynomial. The ODE (11) is a relative of the *associated Laguerre equation*. The polynomials $L_n^l(r)$ have the form

$$L(r) = \sum_{k=l}^{n-1} a_k r^k = a_l r^l + \cdots + a_{n-1} r^{n-1}. \tag{17}$$

[The *associated Laguerre polynomials* are $r^{-l} L(r)$.] The eigenfunctions satisfy the PDE

$$-\Delta v_{nlm} - \frac{2}{r} v_{nlm} = -\frac{1}{n^2} v_{nlm} \tag{18}$$

for $n$, $l$, $m$ integers with $0 \leq |m| \leq l \leq n - 1$.

We conclude that the separated solutions of the full Schrödinger equation for the hydrogen atom are

$$e^{it/2n^2} \cdot e^{-r/n} \cdot L_n^l(r) \cdot Y_l^m(\theta, \phi). \tag{19}$$

We should beware however that, as in Section 9.5, the eigenfunctions $v_{nlm}(\mathbf{x})$ are *not* complete among the functions of three variables.

We have seen that the eigenvalue $\lambda_n = -1/n^2$ has *many* eigenfunctions, one corresponding to each $m$ and $l$ such that $0 \leq l < n$ and $|m| \leq l$. Thus there are $2l + 1$ eigenfunctions for each $l$ and altogether there are $\sum_{l=0}^{n-1} (2l + 1) = n^2$ eigenfunctions for $\lambda_n$. (In physics, the values of $l$ are traditionally denoted by the letters $s(l = 0)$, $p(l = 1)$, $d(l = 2)$, and $f(l = 3)$, a legacy from the early observations of spectral lines. For a reference, see [Ed].)

## EXERCISES

1. Show that in spherical coordinates the angular momentum operator $\mathbf{L}$ is given by (6).

2. Prove the identity $L_x L_y - L_y L_x = iL_z$ and the two similar identities formed by cyclically permuting the subscripts.

3. (a) Write down explicitly the eigenfunction of the PDE (18) in the case $n = 1$.
   (b) What are the four eigenfunctions with $n = 2$?
   (c) What are the nine eigenfunctions with $n = 3$?

4.  Show that if $\beta$ is not the reciprocal of an integer, the ODE (11) has no solution that vanishes as $r \to \infty$.

5.  (a)  Write Schrödinger's equation in *two* dimensions in polar coordinates $iu_t = -\frac{1}{2}(u_{rr} + u_r/r + u_{\theta\theta}/r^2) + V(r)u$, with a radial potential $V(r)$. Find the separated eigenfunctions $u = T(t)R(r)\Theta(\theta)$, leaving $R$ in the form of a solution of an ODE.

    (b)  Assume that $V(r) = \frac{1}{2}r^2$. Substitute $\rho = r^2$ and $R(r) = e^{-\rho/2}\rho^{-n/2}L(\rho)$ to show that $L$ satisfies the Laguerre ODE

    $$L_{\rho\rho} + \left[-1 + \frac{\nu + 1}{\rho}\right]L_\rho + \frac{\mu}{\rho}L = 0$$

    for some constants $\nu$ and $\mu$.

# 11

# GENERAL EIGENVALUE PROBLEMS

The eigenvalues are the most important quantities in PDE problems. Only for special domains, as in Sections 10.2 and 10.3, is it possible to get explicit formulas for the eigenvalues of the laplacian. Can anything be said about the eigenvalues of the laplacian for a domain of arbitrary shape? The answer is yes and provides the subject of this chapter.

We first show that the eigenvalues always minimize the energy subject to certain constraints. We use this idea in Section 2 to derive a practical method for computing the eigenvalues. In Section 3 we prove the completeness of the eigenfunctions. In Section 4 we consider more general eigenvalue problems, including the Sturm–Liouville problems. Then we deduce some consequences of completeness. Finally, in Section 6 we study the size of the $n$th eigenvalue for $n$ large.

## 11.1 THE EIGENVALUES ARE MINIMA OF THE POTENTIAL ENERGY

The basic eigenvalue problem with Dirichlet boundary conditions is

$$-\Delta u = \lambda u \text{ in } D, \quad u = 0 \text{ on bdy } D, \tag{1}$$

where $D$ is an *arbitrary* domain (open set) in 3-space that has a piecewise-smooth boundary. In this chapter we denote the eigenvalues by

$$0 < \lambda_1 \le \lambda_2 \le \lambda_3 \le \cdots \le \lambda_n \le \cdots, \tag{2}$$

**299**

repeating each one according to its multiplicity. (We know they are all positive by Section 10.1.) Each domain $D$ has its own particular sequence of eigenvalues.

As usual, we let

$$(f, g) = \iiint_D f(\mathbf{x})\overline{g(\mathbf{x})}\, d\mathbf{x} \quad \text{and} \quad \|f\| = (f, f)^{1/2} = \left( \iiint_D |f(\mathbf{x})|^2\, d\mathbf{x} \right)^{1/2}.$$

Although we'll work in three dimensions, everything we'll say is also valid in two dimensions (or any number of dimensions for that matter). We have already seen in Section 7.1 (Dirichlet principle) that the function that minimizes the energy and satisfies an inhomogeneous boundary condition is the harmonic function. The eigenvalue problem as well is equivalent to a minimum problem for the energy, as we shall now show.

It may be helpful to pause for a moment to think about minima in ordinary calculus. If $E(u)$ is a function of one variable and $u$ is a point where it has a minimum, then the derivative is zero at that point: $E'(u) = 0$. The method of Lagrange multipliers states the following. If $E(\mathbf{u})$ is a scalar function of a vector variable that has a minimum among all $\mathbf{u}$ which satisfy a constraint $F(\mathbf{u}) = 0$, then $\mathbf{u}$ satisfies the equation $\nabla E(\mathbf{u}) = \lambda \nabla F(\mathbf{u})$ for some constant scalar $\lambda$. The constant $\lambda$ is called the *Lagrange multiplier*.

Now consider the *minimum problem:*

$$m = \min \left\{ \frac{\|\nabla w\|^2}{\|w\|^2} : w = 0 \text{ on bdy } D, w \neq 0 \right\} \tag{MP}$$

(where $w(\mathbf{x})$ is a $C^2$ function). This notation means that we should find the smallest possible value of the quotient.

$$Q = \frac{\|\nabla w\|^2}{\|w\|^2} \tag{3}$$

among all functions $w(\mathbf{x})$ that vanish on the boundary but are not identically zero in $D$. This $Q$ is called the *Rayleigh quotient*.

What do we mean by a "solution" of (MP)? We mean a $C^2$ function $u(\mathbf{x})$, not the zero function, such that $u = 0$ on bdy $D$ and such that

$$\frac{\|\nabla u\|^2}{\|u\|^2} \leq \frac{\|\nabla w\|^2}{\|w\|^2} \tag{4}$$

for all $w$ with $w = 0$ on bdy $D$ and $w \neq 0$. Notice that if $u(\mathbf{x})$ is a solution of (MP), so is $Cu(\mathbf{x})$ for any constant $C \neq 0$.

**Theorem 1. Minimum Principle for the First Eigenvalue**   Assume that $u(\mathbf{x})$ is a solution of (MP). Then the value of the minimum equals the first (smallest) eigenvalue $\lambda_1$ of (1) and $u(\mathbf{x})$ is its eigenfunction.

That is,

$$\lambda_1 = m = \min\left\{\frac{\|\nabla w\|^2}{\|w\|^2}\right\} \quad \text{and} \quad -\Delta u = \lambda_1 u \quad \text{in } D. \tag{5}$$

The moral is that "the first eigenvalue is the minimum of the energy." This is a true fact in most physical systems. The first eigenfunction $u(\mathbf{x})$ is called the *ground state*. It is the state of lowest energy.

**Proof.**   *By a* trial function, *we shall mean any $C^2$ function $w(\mathbf{x})$ such that $w = 0$ on bdy $D$ and $w \not\equiv 0$.* Let $m$ be the minimum value of the Rayleigh quotient among all trial functions. Clearly, the constant $m$ is nonnegative. Let $u(\mathbf{x})$ be a solution of (MP). By assumption, we have

$$m = \frac{\iiint |\nabla u|^2\, d\mathbf{x}}{\iiint |u|^2\, d\mathbf{x}} \leq \frac{\iiint |\nabla w|^2\, d\mathbf{x}}{\iiint |w|^2\, d\mathbf{x}}$$

for all trial functions $w(\mathbf{x})$. Let's abbreviate, writing $\int$ instead of $\iiint d\mathbf{x}$. Let $v(\mathbf{x})$ be *any other* trial function and let $w(\mathbf{x}) = u(\mathbf{x}) + \epsilon v(\mathbf{x})$ where $\epsilon$ is *any constant.* Then

$$f(\epsilon) = \frac{\int |\nabla(u + \epsilon v)|^2}{\int |u + \epsilon v|^2} \tag{6}$$

has a minimum at $\epsilon = 0$. By ordinary calculus, $f'(0) = 0$. Expanding both squares in (6), we have

$$f(\epsilon) = \frac{\int (|\nabla u|^2 + 2\epsilon \nabla u \cdot \nabla v + \epsilon^2 |\nabla v|^2)}{\int (u^2 + 2\epsilon uv + \epsilon^2 v^2)}.$$

So the derivative is easy to compute:

$$0 = f'(0) = \frac{(\int u^2)(2 \int \nabla u \cdot \nabla v) - (\int |\nabla u|^2)(2 \int uv)}{(\int u^2)^2}.$$

Hence

$$\int \nabla u \cdot \nabla v = \frac{\int |\nabla u|^2}{\int u^2} \int uv = m \int uv. \tag{7}$$

By Green's first identity (G1 in Chapter 7) and the boundary condition $v = 0$, we can rewrite (7) as

$$\iiint (\Delta u + mu)(v)\, d\mathbf{x} = 0.$$

(Check it.) It is valid for all trial functions $v(\mathbf{x})$. Since $v(\mathbf{x})$ can be chosen in an arbitrary way inside $D$, we deduce just as at the end of Section 7.1 that $\Delta u + mu = 0$ in $D$. Therefore, the minimum value $m$ of $Q$ is an eigenvalue of $-\Delta$ and $u(\mathbf{x})$ is its eigenfunction!

To show that $m$ is the *smallest* eigenvalue of $-\Delta$, let $-\Delta v_j = \lambda_j v_j$, where $\lambda_j$ is any eigenvalue at all. By the definition of $m$ as the minimum of $Q$ and by (G1), we have

$$m \le \frac{\int |\nabla v_j|^2}{\int v_j^2} = \frac{\int (-\Delta v_j)(v_j)}{\int v_j^2}$$

$$= \frac{\int (\lambda_j v_j) v_j}{\int v_j^2} = \lambda_j. \tag{8}$$

Therefore, $m$ is smaller than any other eigenvalue. This completes the proof of Theorem 1. □

A formulation equivalent to (5) is that

$$\lambda_1 = \min \iiint |\nabla w|^2 \, d\mathbf{x} \tag{9}$$

subject to the conditions $w \in C^2$, $w = 0$ on bdy $D$, and $\iiint w^2 d\mathbf{x} = 1$ (see Exercise 2). Thus $\lambda_1$ *minimizes the potential energy* subject to these conditions.

### Example 1.

The minimum of $\int_0^1 [w'(x)]^2 dx$, among all functions such that $\int_0^1 w^2 \, dx = 1$ and $w(0) = w(1) = 0$, is the first eigenvalue of $-d^2/dx^2$ with Dirichlet BCs and therefore equals exactly $\pi^2$. (Who would have imagined that this minimum had anything to do with the number $\pi$?) A direct derivation (from scratch) of this minimum is quite tricky; see Exercise 3. □

### THE OTHER EIGENVALUES

All the other eigenvalues listed in (2) are minima as well, but with additional constraints, as we now demonstrate.

**Theorem 2. Minimum Principle for the $n$th Eigenvalue** Suppose that $\lambda_1, \ldots, \lambda_{n-1}$ are already known, with the eigenfunctions $v_1(\mathbf{x}), \ldots, v_{n-1}(\mathbf{x})$, respectively. Then

$$\lambda_n = \min \left\{ \frac{\|\nabla w\|^2}{\|w\|^2} : w \not\equiv 0, w = 0 \text{ on bdy } D, w \in C^2, \right.$$
$$\left. 0 = (w, v_1) = (w, v_2) = \cdots = (w, v_{n-1}) \right\}, \tag{MP$_n$}$$

assuming that the minimum exists. Furthermore, the minimizing function is the $n$th eigenfunction $v_n(\mathbf{x})$.

This is the same as the minimum problem (MP) except for the additional constraints of being orthogonal to all of the earlier eigenfunctions. Because there are more constraints now, the minimum value for $(MP)_n$ must be higher than the minimum value for $(MP)_{n-1}$, which is higher than the minimum value for (MP). In fact, Theorem 2 says that you get exactly the next eigenvalue $\lambda_n \geq \lambda_{n-1}$.

**Proof of Theorem 2.** Let $u(\mathbf{x})$ denote the minimizing function for $(MP)_n$, which exists by assumption. Let $m^*$ denote this new minimum value, so that $m^*$ is the value of the Rayleigh quotient at $u(\mathbf{x})$. Thus $u = 0$ on bdy $D$, $u$ is orthogonal to $v_1, \ldots, v_{n-1}$, and the quotient $Q$ is smaller for $u$ than for any other function $w$ that satisfies the conditions in $(MP)_n$. As in the proof of Theorem 1, we substitute $w = u + \epsilon v$, where $v$ satisfies the same constraints. Exactly as before,

$$\iiint (\Delta u + m^* u) \, v \, d\mathbf{x} = 0 \tag{10}$$

for all trial functions $v$ that are orthogonal to $v_1, \ldots, v_{n-1}$. Furthermore, by Green's second formula (G2) (see Section 7.2),

$$\iiint (\Delta u + m^* u) \, v_j \, d\mathbf{x} = \iiint u \, (\Delta v_j + m^* v_j) \, d\mathbf{x}$$
$$= (m^* - \lambda_j) \iiint u v_j \, d\mathbf{x} = 0 \tag{11}$$

because $u$ is orthogonal to $v_j$ for $j = 1, \ldots, n - 1$.

Now let $h(\mathbf{x})$ be an arbitrary trial function (a $C^2$ function such that $h = 0$ on bdy $D$ and $h \not\equiv 0$). Let

$$v(\mathbf{x}) = h(\mathbf{x}) - \sum_{k=1}^{n-1} c_k v_k(\mathbf{x}), \quad \text{where } c_k = \frac{(h, v_k)}{(v_k, v_k)}. \tag{12}$$

Then $(v, v_j) = 0$ for $j = 1, \ldots, n - 1$. (Check it!) This function $v$ is the "part" of $h$ that is orthogonal to each of the $n - 1$ functions $v_1, \ldots, v_{n-1}$. So $v$ satisfies all the required constraints. Therefore, (10) is valid for this $v$. So a linear combination of (10) and (11) yields

$$\iiint (\Delta u + m^* u) \, h \, d\mathbf{x} = 0. \tag{13}$$

This is true for all trial functions $h$. Therefore, we again deduce the equation $-\Delta u = m^* u$, which means that $m^*$ is an eigenvalue. It is left as an exercise to show that $m^* = \lambda_n$. □

The *existence* of the minima (MP) and $(MP)_n$ is a delicate mathematical issue that we have avoided. The early proofs in the mid-nineteenth century had serious gaps and it took some 50 years to fill them! In fact, there are domains $D$ with rough boundaries for which (MP) does not have any $C^2$ solution at all. For further information, see [Ga] or [CH].

## EXERCISES

1.  Let $f(x)$ be a function such that $f(0) = f(3) = 0$, $\int_0^3 [f(x)]^2\, dx = 1$, and $\int_0^3 [f'(x)]^2\, dx = 1$. Find such a function if you can. If it cannot be found, explain why not.

2.  Show that Theorem 1 can be reformulated as (9).

3.  Construct a direct (but unmotivated) proof of Example 1 as follows, *without using any knowledge of eigenvalues.* Let $w(x)$ be a $C^2$ function such that $w(0) = w(1) = 0$.
    (a)  Expand $\int_0^1 [w'(x) - \pi w(x)\cot(\pi x)]^2\, dx$ and integrate the cross-term by parts.
    (b)  Show that $w^2(x)\cot \pi x \to 0$ as $x \to 0$ or 1.
    (c)  Deduce that

    $$\int_0^1 [w'(x)]^2\, dx - \pi^2 \int_0^1 [w(x)]^2\, dx$$
    $$= \int_0^1 [w'(x) - \pi w(x)\cot \pi x]^2\, dx \geq 0.$$

    (d)  Show that if $w(x) = \sin \pi x$, then part (c) is an equality and therefore the minimum of Example 1 is $\pi^2$.

4.  In the proof of Theorem 2 we showed that $-\Delta u = m^* u$. Show that $m^* = \lambda_n$.

5.  (a)  Show that the lowest eigenvalue $\lambda_1$ of $-\Delta$ with the *Robin* boundary condition $\partial u/\partial n + a(\mathbf{x})u = 0$ is given by

    $$\lambda_1 = \min \left\{ \frac{\iiint_D |\nabla w|^2\, d\mathbf{x} + \iint_{\text{bdy} D} aw^2\, dS}{\iiint_D w^2\, d\mathbf{x}} \right\}$$

    among all the $C^2$ functions $w(\mathbf{x})$ for which $w \not\equiv 0$.
    (b)  Show that $\lambda_1$ increases as $a(\mathbf{x})$ increases.

## 11.2   COMPUTATION OF EIGENVALUES

The eigenvalues, particularly the smallest one, are important in many engineering applications. The most useful technique for computing them is based on their characterization as minima of the energy.

Beginning with the first eigenvalue, we know from (11.1.5) that

$$\lambda_1 \leq \frac{\|\nabla w\|^2}{\|w\|^2} \tag{1}$$

for all $w$ vanishing on the boundary. A supremely clever choice of the trial function $w$ would give an equality (namely, $w = v_1$), but since we don't know $v_1$ in advance, we should be satisfied with a moderately clever choice that might provide a moderately good approximation.

## Example 1.

Consider finding the first eigenvalue of $-u'' = \lambda u$ on the interval $0 \leq x \leq l$ with the Dirichlet BC $u(0) = u(l) = 0$, assuming that we didn't already know the correct answer $\lambda_1 = \pi^2/l^2$. Take as a simple trial function the quadratic $w(x) = x(l - x)$. It clearly satisfies the boundary conditions. The Rayleigh quotient is

$$\frac{\|w'\|^2}{\|w\|^2} = \frac{\int_0^l (l - 2x)^2 \, dx}{\int_0^l x^2(l - x)^2 \, dx} = \frac{10}{l^2}. \tag{2}$$

This is an amazingly good approximation to the exact value $\lambda_1 = \pi^2/l^2 = 9.87/l^2$. □

Further examples are found in the exercises. The method does not usually work as well as in this simple example because it is too difficult to guess a good trial function, but it can always be improved by using more trial functions, as we shall now show.

## RAYLEIGH–RITZ APPROXIMATION

Let $w_1, \ldots, w_n$ be $n$ arbitrary trial functions ($C^2$ functions that vanish on bdy $D$). Let

$$a_{jk} = (\nabla w_j, \nabla w_k) = \iiint_D \nabla w_j \cdot \nabla w_k \, d\mathbf{x}$$

and

$$b_{jk} = (w_j, w_k) = \iiint_D w_j \cdot w_k \, d\mathbf{x}.$$

Let $A$ be the $n \times n$ symmetric matrix $(a_{jk})$ and $B$ the $n \times n$ symmetric matrix $(b_{jk})$. Then *the roots of the polynomial equation*

$$\boxed{\text{determinant of } (A - \lambda B) = 0} \tag{3}$$

*are approximations to the first $n$ eigenvalues $\lambda_1, \ldots, \lambda_n$.*

**Example 2.**

Consider the radial vibrations of a circular membrane (disk) of radius 1:

$$-\Delta u = -u_{rr} - \frac{1}{r}u_r = \lambda u \quad (0 < r < 1), \quad u = 0 \text{ at } r = 1. \tag{4}$$

Then $-(ru_r)_r = \lambda ru$ and the Rayleigh quotient is

$$Q = \frac{\iint |\nabla u|^2 \, d\mathbf{x}}{\iint u^2 \, d\mathbf{x}} = \frac{\int_0^1 ru_r^2 \, dr}{\int_0^1 ru^2 \, dr}. \tag{5}$$

The trial functions are required to satisfy the boundary conditions $u_r(0) = 0 = u(1)$. A simple choice of a pair of them is $1 - r^2$ and $(1 - r^2)^2$. Then

$$\frac{A}{2\pi} = \begin{pmatrix} \int_0^1 4r^2 r \, dr & \int_0^1 8r^2(1 - r^2)r \, dr \\ \int_0^1 8r^2(1 - r^2)r \, dr & \int_0^1 16r^2(1 - r^2)^2 r \, dr \end{pmatrix} = \begin{pmatrix} 1 & \frac{2}{3} \\ \frac{2}{3} & \frac{2}{3} \end{pmatrix}$$

and

$$\frac{B}{2\pi} = \begin{pmatrix} \int_0^1 (1 - r^2)^2 r \, dr & \int_0^1 (1 - r^2)^3 r \, dr \\ \int_0^1 (1 - r^2)^3 r \, dr & \int_0^1 (1 - r^2)^4 r \, dr \end{pmatrix} = \begin{pmatrix} \frac{1}{6} & \frac{1}{8} \\ \frac{1}{8} & \frac{1}{10} \end{pmatrix}.$$

Hence a calculation of the determinant (3) yields the quadratic equation

$$\frac{1}{(2\pi)^2} \det(A - \lambda B) = \left(1 - \frac{\lambda}{6}\right)\left(\frac{2}{3} - \frac{\lambda}{10}\right) - \left(\frac{2}{3} - \frac{\lambda}{8}\right)^2$$

or $\lambda^2/960 - 2\lambda/45 + 2/9 = 0$. The eigenvalues are approximately equal to the roots of this quadratic. They are $\lambda_1 \sim 5.784$ and $\lambda_2 \sim 36.9$. The true eigenvalues are the squares of the roots of the Bessel function $J_o$ (as we know from Section 10.2), the first two of which are $\lambda_1 \sim 5.783$ and $\lambda_2 \sim 30.5$. The first is amazingly accurate; the second is a poor approximation. To do a better job with $\lambda_2$, we could use a different pair of trial functions or else use three trial functions. $\square$

**Informal Derivation of the RRA**    Let $w_1, \ldots, w_n$ be arbitrary linearly independent trial functions. As an *approximation* to the true minimum problem $(MP)_n$, let's impose the additional condition that $w(\mathbf{x})$ is a linear combination of $w_1(\mathbf{x}), \ldots, w_n(\mathbf{x})$. We can then write

$$w(\mathbf{x}) = \sum_{k=1}^{n} c_k w_k(\mathbf{x}). \tag{6}$$

If we were extremely clever, then $w(\mathbf{x})$ would itself be an eigenfunction. Then $-\Delta w = \lambda w$, where $\lambda$ is both the eigenvalue and the value of the Rayleigh quotient. Then we would have $(\nabla w, \nabla w_j) = \lambda(w, w_j)$. Into the last equation we substitute (6) and obtain

$$\sum_k a_{jk} c_k = \lambda \sum_k b_{jk} c_k. \tag{7}$$

In matrix notation we can write (7) as $A\mathbf{c} = \lambda B\mathbf{c}$, where $\mathbf{c} \in \mathbb{R}^n$. Since $w \neq 0$, we have $\mathbf{c} \neq \mathbf{0}$. This means that $A - \lambda B$ would be a singular matrix, or in other words that the determinant of $(A - \lambda B)$ would be zero. We are not so supremely clever; nevertheless, we use the determinant as our method of approximating the eigenvalues.

## MINIMAX PRINCIPLE

What we really want is an *exact* formula rather than just an approximation. To motivate it, let's denote the roots of the polynomial equation (3) by

$$\lambda_1^* \leq \cdots \leq \lambda_n^*.$$

Let's just consider the biggest one, $\lambda_n^*$. It's easy to see by linear algebra that

$$\lambda_n^* = \max_{\mathbf{c} \neq 0} \frac{A\mathbf{c} \cdot \mathbf{c}}{B\mathbf{c} \cdot \mathbf{c}}. \tag{8}$$

(see Exercise 9). Therefore,

$$\lambda_n^* = \max \frac{\sum a_{jk} c_j c_k}{\sum b_{jk} c_j c_k}$$

$$= \max \frac{\left( \nabla \left( \sum c_j w_j \right), \nabla \left( \sum c_k w_k \right) \right)}{\left( \sum c_j w_j, \sum c_k w_k \right)},$$

where the maxima are taken over all (nonzero) $n$-tuples $c_1, \ldots, c_n$. Therefore,

$$\lambda_n^* = \max \left\{ \frac{\|\Delta w\|^2}{\|w\|^2}, \text{ taken over all functions } w \right.$$
$$\left. \text{that are linear combinations of } w_1, \ldots, w_n \right\}. \tag{9}$$

Formula (9) leads to the minimax principle, which states that *the smallest possible value of $\lambda_n^*$ is the $n$th eigenvalue $\lambda_n$*. Thus $\lambda_n$ is a minimum of a maximum!

**Theorem 1. Minimax Principle** If $w_1, \ldots, w_n$ is an arbitrary set of trial functions, define $\lambda_n^*$ by formula (9). Then the $n$th eigenvalue is

$$\lambda_n = \min \lambda_n^* \tag{10}$$

where the minimum is taken over *all* possible choices of $n$ trial functions $w_1, \ldots, w_n$.

**Proof.**    We begin by fixing any choice of $n$ trial functions $w_1, \ldots, w_n$. Then we choose $n$ constants $c_1, \ldots, c_n$ (not all zero), so that the linear combination $w(\mathbf{x}) \equiv \sum_{j=1}^{n} c_j w_j(\mathbf{x})$ is orthogonal to the first $(n-1)$ eigenfunctions $v_1(\mathbf{x}), \ldots, v_{n-1}(\mathbf{x})$. That is,

$$(w, v_k) = \sum_{j=1}^{n} c_j (w_j, v_k) = 0, \quad k = 1, \ldots, n - 1. \tag{11}$$

This can surely be accomplished because the constants $c_1, \ldots, c_n$ are $n$ unknowns that need only satisfy the $(n-1)$ linear equations (11). Because there are fewer linear equations than unknowns, there is always a nonzero solution.

By the minimum principle for the $n$th eigenvalue (in Section 11.1), we have $\lambda_n \leq \|\nabla w\|^2 / \|w\|^2$. On the other hand, because the maximum in (9) is taken over all possible linear combinations, we have $\|\nabla w\|^2 / \|w\|^2 \leq \lambda_n^*$. Thus

$$\lambda_n \leq \frac{\|\nabla w\|^2}{\|w\|^2} \leq \lambda_n^*. \tag{12}$$

Inequality (12) is true for *all* choices of $n$ trial functions $w_1, \ldots, w_n$. Therefore, $\lambda_n \leq \min \lambda_n^*$, where the minimum is taken over all such choices. This proves half of (10).

To show that the two numbers in (10) are equal, we'll exhibit a special choice of the trial functions $w_1, \ldots, w_n$. In fact, we choose them to be the first $n$ eigenfunctions: $w_1 = v_1, \ldots, w_n = v_n$. We may assume they are normalized: $\|v_j\| = 1$. With *this* choice of the $w$'s, we have

$$\lambda_n^* = \max_{c_1, \ldots, c_n} \frac{\|\nabla(\sum c_j v_j)\|^2}{\|\sum c_j v_j\|^2}. \tag{13}$$

Using Green's first formula again as in (11.1.8), the last quotient equals $\sum c_j^2 \lambda_j / \sum c_j^2$, which is at most $\sum c_j^2 \lambda_n / \sum c_j^2 = \lambda_n$. Thus (13) implies that $\lambda_n^* \leq \lambda_n$ for this *particular* choice of $n$ trial functions. But we already showed in (12) the opposite inequality $\lambda_n \leq \lambda_n^*$ for *all* choices of trial functions. So

$$\lambda_n = \lambda_n^* \text{ for our particular choice.} \tag{14}$$

Therefore $\lambda_n$ *equals* the minimum of $\lambda_n^*$ over *all* possible choices of $n$ trial functions. This is the minimax principle.    □

The Rayleigh–Ritz method is reminiscent of the finite element method of Section 8.5. The finite element method is distinguished from it by its use of piecewise-linear trial functions or other very simple functions defined "in pieces."

## EXERCISES

1. For the eigenvalue problem $-u'' = \lambda u$ in the interval $(0, 1)$, with $u(0) = u(1) = 0$, choose the pair of trial functions $x - x^2$ and $x^2 - x^3$ and compute the Rayleigh–Ritz approximations to the first two eigenvalues. Compare with the exact values.

2. Do the same with the pair of trial functions $w_1(x) = x - x^2$ and $w_2(x) = $ the piecewise-linear function such that $w_2(0) = 0$, $w_2(\frac{1}{4}) = 1$, $w_2(\frac{3}{4}) = -1$, $w_2(1) = 0$. Compare with the answer to Exercise 1.

3. Consider $-\Delta$ in the square $(0, \pi)^2$ with Dirichlet BCs. Compute the Rayleigh quotient with the trial function $xy(\pi - x)(\pi - y)$. Compare with the first eigenvalue.

4. Consider $-\Delta$ in the ball $\{r^2 = x^2 + y^2 + z^2 < 1\}$ with Dirichlet BCs.
   (a) Compute the Rayleigh quotient with the trial function $(1 - r)$. Compare with the first eigenvalue.
   (b) Repeat using the trial function $\cos \frac{1}{2}\pi r$.

5. For the eigenvalue problem $-u'' = \lambda u$ in the interval $(0, 1)$, but with the *mixed* boundary conditions $u'(0) = u(1) = 0$, choose the trial function $1 - x^2$ and compute the Rayleigh quotient. Compare with the first eigenvalue.

6. For the eigenvalue problem $-u'' = \lambda u$ in the interval $(0, 1)$, but with the *mixed* boundary conditions $u'(0) = u(1) = 0$, choose the pair of trial functions $1 - x^2$ and $1 - x^3$.
   (a) Compute the $2 \times 2$ matrices $A$ and $B$.
   (b) Solve the polynomial equation (3).
   (c) How close are the roots to the first two eigenvalues?

7. (a) What is the exact value of the first eigenvalue of $-\Delta$ in the unit disk with Dirichlet BCs?
   (b) Compute the Rayleigh quotient of the function $1 - r$. Compare with the exact value.

8. Estimate the first eigenvalue of $-\Delta$ with Dirichlet BCs in the triangle $\{x + y < 1, x > 0, y > 0\}$:
   (a) Using a Rayleigh quotient with the trial function $xy(1 - x - y)$.
   (b) Using the Rayleigh–Ritz approximation with two trial functions of your own choosing.

9. (a) Show that if $A$ is a real symmetric matrix and $\lambda_n^*$ is its largest eigenvalue, then
$$\lambda_n^* = \max_{\mathbf{c} \neq 0} \frac{A\mathbf{c} \cdot \mathbf{c}}{|\mathbf{c}|^2}.$$
   (b) If $B$ is another real symmetric matrix, $B$ is positive definite, and $\lambda_n^*$ is the largest root of the polynomial equation $\det(A - \lambda B) = 0$, show that
$$\lambda_n^* = \max_{\mathbf{c} \neq 0} \frac{A\mathbf{c} \cdot \mathbf{c}}{B\mathbf{c} \cdot \mathbf{c}}.$$
   [Hint for (b): Use the fact that $B$ has a unique square root that is positive definite symmetric.]

## 11.3   COMPLETENESS

In this section we'll begin with a discussion of the Neumann boundary condition. Then we'll prove, in a very simple way, the completeness of the eigenfunctions *for general domains*.

## THE NEUMANN BOUNDARY CONDITION

This means that $\partial u/\partial n = 0$ on bdy $D$. We denote the eigenvalues by $\tilde{\lambda}_j$ and the eigenfunctions by $\tilde{v}_j(\mathbf{x})$. Thus

$$-\Delta \tilde{v}_j(\mathbf{x}) = \tilde{\lambda}_j \tilde{v}_j(\mathbf{x}) \quad \text{in } D$$

$$\frac{\partial \tilde{v}_j}{\partial n} = 0 \quad \text{on bdy } D. \tag{1}$$

We number them in ascending order:

$$\boxed{0 = \tilde{\lambda}_1 < \tilde{\lambda}_2 \leq \tilde{\lambda}_3 \leq \cdots \leq \tilde{\lambda}_3 \leq \cdots.}$$

(This is not the same numbering as in Section 4.2.) The first eigenfunction $\tilde{v}_1(\mathbf{x})$ is a constant.

**Theorem 1.**   For the Neumann condition we define a "trial function" as *any* $C^2$ function $w(\mathbf{x})$ such that $w \not\equiv 0$. With this new definition of a trial function, each of the preceding characterizations of the eigenvalues [the minimum principle, (11.1.5) and (MP)$_n$, the Rayleigh–Ritz approximation, and the minimax principle (11.2.10)] is valid.

Thus there is no condition on the trial functions at the boundary. The trial functions are *not* supposed to satisfy the Neumann or any other boundary condition. For this reason the Neumann condition is sometimes called the *free condition*, as distinguished from the "fixed" Dirichlet condition. (This terminology also makes sense for physical reasons.)

**Proof.**   We repeat the various steps of the preceding proofs. There are differences only where the boundary comes into play. The proof of Theorem 11.1.1 ends up with (11.1.7):

$$\iiint\limits_{D} (-\nabla u \cdot \nabla v + muv)\, d\mathbf{x} = 0,$$

which now is valid for *all* $C^2$ functions $v(\mathbf{x})$ with no restriction on bdy $D$. This can be rewritten by Green's first formula (G1) as

$$\iiint\limits_{D} (\Delta u + mu)\, v\, d\mathbf{x} = \iint\limits_{bdyD} \frac{\partial u}{\partial n} v\, dS. \tag{2}$$

In (2) we first choose $v(\mathbf{x})$ arbitrarily *inside D* and $v = 0$ on bdy $D$. As before, we still get $\Delta u + mu = 0$ *inside D*.

Thus the left side of (2) vanishes for every trial function $v(\mathbf{x})$. So

$$\iint\limits_{bdy D} \frac{\partial u}{\partial n} v \, dS = 0 \tag{3}$$

for *all trial functions* $v$. But the new kind of trial function can be chosen to be an arbitrary function on bdy $D$. Let's choose $v = \partial u/\partial n$ on bdy $D$. Then (3) implies that the integral of $(\partial u/\partial n)^2$ vanishes. By the first vanishing theorem of Section A.1, we deduce that $\partial u/\partial n = 0$ on bdy $D$. This is the Neumann condition!

The same sort of modification is made in Theorem 11.1.2. The proof of the minimax principle is completely unchanged, except for the broader definition of what a trial function is. □

## COMPLETENESS

**Theorem 2.**  For the Dirichlet boundary condition, the eigenfunctions are complete in the $L^2$ sense. The same is true for the Neumann condition.

This theorem means the following. Let $\lambda_n$ be all the eigenvalues and $v_n(\mathbf{x})$ the corresponding eigenfunctions. The eigenfunctions may be assumed to be mutually orthogonal. Let $f(\mathbf{x})$ be any $L^2$ function in $D$, that is, $\|f\| < \infty$. Let $c_n = (f, v_n)/(v_n, v_n)$. Then

$$\left\| f - \sum_{n=1}^{N} c_n v_n \right\|^2 = \iiint\limits_{D} \left| f(\mathbf{x}) - \sum_{n=1}^{N} c_n v_n(\mathbf{x}) \right|^2 d\mathbf{x} \to 0 \text{ as } N \to \infty. \tag{4}$$

The proof of Theorem 2, given below, depends upon: (i) the existence of the minima discussed in Section 11.1, and (ii) the fact (which will be derived in Section 11.6) that the sequence of eigenvalues $\lambda_n$ tends to $+\infty$ as $n \to \infty$.

**Proof.**  We will prove (4) only in the special case that $f(\mathbf{x})$ is a trial function. In advanced texts it is shown that (4) is also valid for any $L^2$ function $f(\mathbf{x})$. Let's begin with the Dirichlet case. Then $f(\mathbf{x})$ is a $C^2$ function that vanishes on bdy $D$. Let's denote the remainder after the subtraction of $N$ terms of the Fourier expansion of $f$ by

$$r_N(\mathbf{x}) = f(\mathbf{x}) - \sum_{n=1}^{N} c_n v_n(\mathbf{x}). \tag{5}$$

By the orthogonality,

$$\left(r_N, v_j\right) = \left(f - \sum_{n=1}^{N} c_n v_n, v_j\right)$$

$$= (f, v_j) - \sum_{n=1}^{N} c_n(v_n, v_j)$$

$$= (f, v_j) - c_j(v_j, v_j) = 0$$

for $j = 1, 2, \ldots, N$. Thus the remainder is a trial function that satisfies the constraints of (MP). Hence

$$\lambda_{N+1} = \min_{w} \frac{\|\nabla w\|^2}{\|w\|^2} \leq \frac{\|\nabla r_N\|^2}{\|r_N\|^2} \tag{6}$$

by Theorem 11.1.2.

Next we expand

$$\|\nabla r_N\|^2 = \int \left|\nabla\left[f - \sum_{n=1}^{N} c_n v_n\right]\right|^2$$

$$= \int \left(|\nabla f|^2 - 2\sum_n c_n \nabla f \cdot \nabla v_n + \sum_{n,m} c_n c_m \nabla v_n \cdot \nabla v_m\right), \tag{7}$$

using the shorthand notation for the integrals. By Green's first identity and the boundary condition ($f = v_n = 0$ on bdy $D$), the middle term in (7) contains the integral

$$\int \nabla f \cdot \nabla v_n = -\int f \Delta v_n = \lambda_n \int f v_n \tag{8}$$

while the last term in (7) contains the integral

$$\int \nabla v_n \cdot \nabla v_m = -\int v_n \Delta v_m = \delta_{mn} \lambda_n \int v_n^2, \tag{9}$$

where $\delta_{mn}$ equals zero for $m \neq n$ and equals one for $m = n$. So the expression (7) simplifies to

$$\|\nabla r_N\|^2 = \int |\nabla f|^2 - 2\sum_n c_n \lambda_n(f, v_n) + \sum_{n,m} \delta_{mn} c_n^2 \lambda_n(v_n, v_n). \tag{10}$$

Recalling the definition of $c_n$, the last two sums combine to produce the identity

$$\|\nabla r_N\|^2 = \int |\nabla f|^2 - \sum_n c_n^2 \lambda_n(v_n, v_n).$$

All the sums run from 0 to $N$. If we throw away the last sum, we get the inequality

$$\|\nabla r_N\|^2 \leq \int |\nabla f|^2 = \|\nabla f\|^2. \tag{11}$$

If we combine (11) and (6), we get

$$\|r_N\|^2 \leq \frac{\|\nabla r_N\|^2}{\lambda_N} \leq \frac{\|\nabla f\|^2}{\lambda_N}, \tag{12}$$

recalling that $\lambda_N > 0$.

We will show in Section 11.6 that $\lambda_N \to \infty$. So the right side of (12) tends to zero since the numerator is fixed. Therefore, $\|r_N\| \to 0$ as $N \to \infty$. This proves (4) in case $f(\mathbf{x})$ is a trial function and the boundary condition is Dirichlet.

In the Neumann case, too, we assume that $f(\mathbf{x})$ is a trial function. We use the notation $\tilde{\lambda}_j$ and $\tilde{v}_j(\mathbf{x})$ for the eigenvalues and eigenfunctions. Then the proof is the same as above (with tildes). Notice that (8) and (9) still are valid because the eigenfunctions $\tilde{v}_j(\mathbf{x})$ satisfy the Neumann BCs. The only other difference from the preceding proof occurs as a consequence of the fact that the first eigenvalue vanishes: $\tilde{\lambda}_1 = 0$. However, $\tilde{\lambda}_2 > 0$ (see Exercise 1). So, when writing the inequality (12), we simply assume that $N \geq 2$.  □

Some consequences of completeness are discussed in Section 11.5. In Section 11.6 it is proved that $\lambda_N \to \infty$.

## EXERCISES

1. We well know that the smallest eigenvalue for the Neumann BCs is $\tilde{\lambda}_1 = 0$ (with the constant eigenfunction). Show that $\tilde{\lambda}_2 > 0$. This is the same as saying that zero is a simple eigenvalue, that is, of multiplicity 1.

2. Let $f(\mathbf{x})$ be a function in $D$ and $g(\mathbf{x})$ a function on bdy $D$. Consider the minimum of the functional

$$\frac{1}{2} \iiint_D |\nabla w|^2 \, d\mathbf{x} - \iiint_D fw \, d\mathbf{x} - \iint_{\text{bdy } D} gw \, dS$$

among all $C^2$ functions $w(\mathbf{x})$. Assume that $\iiint_D f \, d\mathbf{x} + \iint_{\text{bdy } D} g \, dS = 0$. Show that a solution of this minimum problem leads to a solution of the Neumann problem

$$-\Delta u = f \quad \text{in } D, \qquad \frac{\partial u}{\partial n} = g \quad \text{on bdy } D.$$

3. Let $g(\mathbf{x})$ be a function on bdy $D$. Consider the minimum of the functional

$$\frac{1}{2} \iiint_D |\nabla w|^2 \, d\mathbf{x} - \iiint_D fw \, d\mathbf{x}$$

among all $C^2$ functions $w(\mathbf{x})$ for which $w = g$ on bdy $D$. Show that a solution of this minimum problem leads to a solution of the Dirichlet problem

$$-\Delta u = f \quad \text{in } D, \qquad u = g \quad \text{on bdy } D.$$

## 11.4   SYMMETRIC DIFFERENTIAL OPERATORS

Our purpose here is to indicate how extensively all the theory of this chapter (and much of this book!) can be generalized to allow for variable coefficients. Variable coefficients correspond to inhomogeneities in the physical medium.
   We consider the PDE

$$\boxed{-\nabla \cdot (p\nabla u) + qu = \lambda m u} \tag{1}$$

in a domain $D$ where the coefficients $p$, $q$, and $m$ are functions of $\mathbf{x}$. More explicitly, the equation is

$$-\sum \frac{\partial}{\partial x_j}\left(p(\mathbf{x})\frac{\partial u}{\partial x_j}\right) + q(\mathbf{x})u(\mathbf{x}) = \lambda m(\mathbf{x})u(\mathbf{x}) \tag{1'}$$

if we write $\mathbf{x} = (x, y, z) = (x_1, x_2, x_3)$ and let the sum run from 1 to 3. We assume that $p(\mathbf{x})$ is a $C^1$ function, that $q(\mathbf{x})$ and $m(\mathbf{x})$ are continuous functions, and that $p(\mathbf{x}) > 0$ and $m(\mathbf{x}) > 0$. By the *Dirichlet eigenvalue problem* we mean the search for a solution $u(\mathbf{x})$ of (1) in a domain $D$ for some $\lambda$ that satisfies

$$u = 0 \quad \text{on bdy } D. \tag{2}$$

**Theorem 1.**   All the previous results (Theorems 11.1.1, 11.1.2, 11.2.1, 11.3.2, and the Rayleigh–Ritz approximation) are valid for the Dirichlet eigenvalue problem (1)-(2) with only the following two changes:
   (i)   The Rayleigh quotient is now defined as

$$Q = \frac{\iiint_D [p(\mathbf{x})|\nabla w(\mathbf{x})|^2 + q(\mathbf{x})[w(\mathbf{x})]^2]\, d\mathbf{x}}{\iiint_D m(\mathbf{x})[w(\mathbf{x})]^2\, d\mathbf{x}}. \tag{3}$$

   (ii)   The inner product is now defined as

$$(f, g) = \iiint_D m(\mathbf{x})f(\mathbf{x})\overline{g(\mathbf{x})}\, d\mathbf{x}. \tag{4}$$

The trial functions $w(\mathbf{x})$ are the same as before, namely all the $C^2$ functions that satisfy (2).

The *Neumann eigenvalue problem* consists of the PDE (1) together with the boundary condition $\partial u / \partial n = 0$. As before, we get the same results with the trial functions free to be arbitrary $C^2$ functions.

The *Robin eigenvalue problem* is the PDE (1) together with the boundary condition

$$\frac{\partial u}{\partial n} + a(\mathbf{x})u(\mathbf{x}) = 0 \qquad \text{on bdy } D, \tag{5}$$

where $a(\mathbf{x})$ is a given continuous function. The Rayleigh quotient in the Robin case is

$$Q = \frac{\iiint_D [p|\nabla w|^2 + qw^2] \, d\mathbf{x} + \iint_{\text{bdy}D} aw^2 \, dS}{\iiint_D mw^2 \, d\mathbf{x}}. \tag{6}$$

The trial functions in this case are free. Everything else is the same.

## Example 1.

The hydrogen atom. In Section 9.5 we studied the eigenvalue problem $-\Delta v - (2/r)v = \lambda v$. In that case, the potential was $q(\mathbf{x}) = -2/|\mathbf{x}|$, and $m(\mathbf{x}) = p(\mathbf{x}) \equiv 1$. □

## Example 2.

Heat flow in an inhomogeneous medium (Section 1.3, Example 5) leads to the eigenvalue problem $\nabla \cdot (\kappa \nabla v) = \lambda c\rho v$, where $\kappa, c, \rho$ are functions of $\mathbf{x}$. □

## STURM–LIOUVILLE PROBLEMS

This is the name given to the *one-dimensional* case of equation (1) together with a set of "symmetric" boundary conditions. Thus $x$ is a single variable, running over an interval $D = [a, b]$. The differential equation in the interval $D$ is

$$\boxed{-(pu')' + qu = \lambda mu \qquad \text{for } a < x < b.} \tag{7}$$

What does it mean for the boundary conditions for (7) (at $x = a$ and $x = b$) to be symmetric? The idea is the same as in Section 5.3 except for the modification due to the variable coefficients. Namely, a set of boundary conditions [as in (5.3.4)] is called *symmetric* for the ODE (7) if

$$(fg' - f'g)p \Big|_{x=a}^{x=b} = 0. \tag{8}$$

Thus the only difference from Section 5.3 is the presence of the function $p(\mathbf{x})$ in (8).

In most texts the Sturm–Liouville problem is treated by methods of ODEs. Those methods are less sophisticated but more explicit than the ones we have used. The reader is referred to [CL] or [CH] for such a discussion. The main results in those references are the same as ours: that the eigenvalues $\lambda_n$ tend to $+\infty$ and that the eigenfunctions are complete.

## Example 3.

When studying Dirichlet's problem for an annulus in Section 6.4, we separated the variables to get the radial equation

$$r^2 R'' + r R' - \lambda R = 0 \qquad \text{or} \qquad (r R')' = \lambda r^{-1} R$$

for $a < r < b$ with boundary conditions $R(a) = R(b) = 0$. In this case $m(r) = r^{-1}$ and $p(r) = r$. □

**Singular Sturm–Liouville Problems** One often encounters Sturm–Liouville problems of one of the following types:

(i) The coefficient $p(\mathbf{x})$ vanishes at one or both of the endpoints $x = a$ or $x = b$.

(ii) One or more of the coefficients $p(\mathbf{x})$, $q(\mathbf{x})$, or $m(\mathbf{x})$ becomes infinite at $a$ or $b$.

(iii) One of the endpoints is itself infinite: $a = -\infty$ or $b = +\infty$.

*If either (i) or (ii) or (iii) occurs, the Sturm–Liouville problem is called singular.* The boundary condition at a singular endpoint is usually modified from the usual types. This is easiest to understand by looking at some examples, several of which we have already encountered in this book.

## Example 4.

When studying the vibrations of a drumhead in Section 10.2, we encountered Bessel's equation,

$$(r R')' - \frac{n^2}{r} R = \lambda r R \quad \text{in } [0, a]$$

with $R(0)$ finite and $R(a) = 0$. This example is singular because $p(r) = r$ vanishes at the left endpoint. The BC at the left endpoint is a finiteness condition. □

## Example 5.

When studying the solid vibrations of a ball in Section 10.3, the $\theta$-part of the spherical harmonic satisfied the ODE

$$-(\sin\theta\, p'(\theta))' + \frac{m^2}{\sin\theta} p(\theta) = \gamma \sin\theta\; p(\theta) \quad \text{in } (0, \pi)$$

with $p(0)$ and $p(\pi)$ finite. Notice that $\sin\theta > 0$ in $0 < \theta < \pi$, but $\sin\theta = 0$ at both endpoints, so this problem is singular at both ends. It is singular at both ends due to both (i) and (ii). At both of these ends the BC is a finiteness condition.    □

### Example 6.

When studying quantum mechanics in Section 10.7, the radial part of the wave function satisfied the ODE

$$R'' + \frac{2}{r}R' + \left[\lambda - 2V(r) - \frac{\gamma}{r^2}\right]R = 0$$

or

$$-(r^2R')' + [2r^2V(r) + \gamma]\, R = \lambda r^2 R \quad \text{in } 0 < r < \infty.$$

Either way one writes the equation, it is singular at both ends. At $r = 0$ we require a finiteness condition, while at $r = \infty$ we require a vanishing condition.    □

### EXERCISES

1.  (a)  Find the exact eigenvalues and eigenfunctions of the problem

$$-(xu')' = \frac{\lambda}{x}u \quad \text{in the interval } 1 < x < b,$$

with the boundary conditions $u(1) = u(b) = 0$.
    (b)  What is the orthogonality condition?
        (*Hint:* It is convenient to change the variable $x = e^s$.)

2.  Repeat Exercise 1 with the boundary conditions $u'(1) = 0$ and $u'(b) + hu(b) = 0$. Find the eigenvalues graphically.

3.  Consider the eigenvalue problem $-v'' - xv = \lambda v$ in $(0, \pi)$ with $v(0) = v(\pi) = 0$.
    (a)  Compute the Rayleigh quotient for this equation for $w(x) = \sin x$.
    (b)  Find the first eigenvalue exactly, by looking up Airy functions in a reference. Compare with part (a).

4.  Solve the problem $(x^2u_x)_x + x^2u_{yy} = 0$ with the conditions $u(1, y) = u(2, y) \equiv 0$ and $u(x, 1) = u(x, -1) = f(x)$, where $f(x)$ is a given function.

5.  Consider the PDE $u_{tt} = (1/x)(xu_x)_x$ in the interval $0 < x < l$. The boundary conditions are $|u(0)| < \infty$ and $u(l, t) = \cos\omega t$. Find the eigenfunction expansion for the solution.

6.  Show that the Robin BCs are symmetric for any Sturm–Liouville operator.

7.  Consider the operator $v \mapsto (pv^{(m)})^{(m)}$, where the superscript denotes the $m$th derivative, in an interval with the boundary conditions $v = v' = v'' = \ldots = v^{(m-1)} = 0$ at both ends. Show that its eigenvalues are real.

## 11.5    COMPLETENESS AND SEPARATION OF VARIABLES

In this section we show how completeness (Section 11.3) allows us to (i) solve an inhomogeneous elliptic problem and (ii) fully justify the separation of variables technique.

### INHOMOGENEOUS ELLIPTIC PROBLEM

We begin with the solution of the elliptic PDE

$$-\nabla \cdot (p(\mathbf{x})\nabla u) + q(\mathbf{x})\, u(\mathbf{x}) = a\, m(\mathbf{x})u(\mathbf{x}) + f(\mathbf{x}) \qquad \text{in } D \qquad (1)$$

with homogeneous Dirichlet, Neumann, or Robin boundary conditions, where $p, q, m$ satisfy the conditions of Section 11.4, $f$ is a given real function, and $a$ is a given constant.

### Theorem 1.

(a) If *a is not an eigenvalue* (of the corresponding problem with $f \equiv 0$), then there exists a unique solution for all functions $f(\mathbf{x})$ [such that $\iiint f^2(1/m)\, d\mathbf{x} < \infty$].

(b) If *a is an eigenvalue* of the homogeneous problem, then *either* there is no solution at all *or* there are an infinite number of solutions, depending on the function $f(\mathbf{x})$.

**Proof.**    Theorem 1 is sometimes called the *Fredholm alternative*. To prove it, let's first take the case (a) when $a$ is not an eigenvalue. Call the eigenvalues $\lambda_n$ and the eigenfunctions $v_n(\mathbf{x})$ $(n = 1, 2, \dots )$. Let $\delta$ be the distance between the number $a$ and the nearest eigenvalue $\lambda_n$. Let's write $\int$ for $\iiint_D$. By a *solution* we mean a function $u(\mathbf{x})$ that satisfies the PDE (1) and one of the three standard boundary conditions. By the completeness, we know that $u(\mathbf{x})$ can be expanded in a series of the $v_n(\mathbf{x})$ as

$$u(\mathbf{x}) = \sum_{n=1}^{\infty} \frac{(u, v_n)}{(v_n, v_n)} v_n(\mathbf{x}), \qquad (2)$$

where the inner product is $(f, g) = \int_D fgm\, d\mathbf{x}$ and the infinite series converges in the $L^2$ sense.

Let's multiply the equation (1) by $v_n$ and integrate over $D$ to get

$$-\int \nabla \cdot (p\nabla u)\, v_n\, d\mathbf{x} + \int quv_n\, d\mathbf{x} = a \int muv_n\, d\mathbf{x} + \int fv_n\, d\mathbf{x}. \qquad (3)$$

Applying the divergence theorem (Green's second identity, essentially) to the first term and using the fact that both $u$ and $v_n$ satisfy the boundary conditions, the first term of (3) becomes $-\int u\nabla \cdot (p\nabla v_n)\, d\mathbf{x}$. Equation (3) simplifies greatly if the PDE satisfied by $v_n$ is taken into account. It reduces to

$$(\lambda_n - a) \int m u v_n \, d\mathbf{x} = \int f v_n \, d\mathbf{x}. \tag{4}$$

That is, $(u, v_n) = (\lambda_n - a)^{-1} \int f v_n \, d\mathbf{x}$. Substituting this into (2), we get

$$u(\mathbf{x}) = \sum_{n=1}^{\infty} \frac{\int f v_n \, d\mathbf{x}}{(\lambda_n - a) \int v_n^2 \, m d\mathbf{x}} \, v_n(\mathbf{x}). \tag{5}$$

This is an explicit formula for the solution, in the form of an infinite series, in terms of $f(\mathbf{x})$ and the eigenfunctions and eigenvalues.

Why does the series (5) converge in the $L^2$ sense? Because $|\lambda_n - a| \geq \delta$, we have

$$|u(x)| \leq \sum_{n=1}^{\infty} \frac{|\int f v_n \, d\mathbf{x}|}{\delta(v_n, v_n)} \, |v_n(\mathbf{x})|.$$

We may assume that $\|v_n\| = 1$. By Schwarz's inequality (Exercise 5.5.2) and Parseval's equality (5.4.19) it follows that

$$\|u\|^2 \leq \frac{1}{\delta^2} \sum_{n=1}^{\infty} \left| \int f v_n \, d\mathbf{x} \right|^2 \leq \frac{1}{\delta^2} \int f^2 \frac{1}{m} \, d\mathbf{x} < \infty. \tag{6}$$

This proves part (a) of Theorem 1.

As for part (b), on the other hand, suppose that $a = \lambda_N$ for some $N$. Then the same identity (4) with $n = N$ implies that $\int f v_N \, d\mathbf{x} = 0$. Therefore, in case $\int f v_N \, d\mathbf{x} \neq 0$, there cannot exist a solution. In case $\int f v_N \, d\mathbf{x} = 0$, there exist an infinite number of solutions! They are

$$u(\mathbf{x}) = \sum_{n \neq N} \frac{\int_D f v_n \, d\mathbf{x}}{(\lambda_n - \lambda_N)(v_n, v_n)} v_n(\mathbf{x}) + C v_N(\mathbf{x}) \tag{7}$$

for any constant $C$.   $\square$

## SEPARATION OF VARIABLES, REVISITED

The simplest case is the *separation of the time variable*. Let's consider our standard problem

$$u_t = k \, \Delta u \quad \text{in } D, \qquad u(\mathbf{x}, 0) = \phi(\mathbf{x}), \qquad u = 0 \quad \text{on bdy } D. \tag{8}$$

We could also consider the Neumann or Robin boundary conditions. Let's denote the eigenvalues and eigenfunctions by $-\Delta v_n = \lambda_n v_n$ in $D$, $v_n$ satisfying the boundary condition. The eigenfunctions are complete. If we assume that

$\int_D (|u|^2 + |\nabla u|^2)\,dx < \infty$ and we also make certain differentiability assumptions, we shall show that the usual expansion

$$u(\mathbf{x}, t) = \sum_{n=1}^{\infty} A_n e^{-\lambda_n k t}\, v_n(\mathbf{x}) \tag{9}$$

must hold.

To prove it, we simply expand $u(\mathbf{x}, t)$, for each $t$, as a series in the complete set of eigenfunctions,

$$u(\mathbf{x}, t) = \sum_{n=1}^{\infty} a_n(t) v_n(\mathbf{x}). \tag{10}$$

By completeness there are some coefficients $a_n(t)$, but so far they are unknown. Assuming differentiability (sufficient to differentiate the series term-by-term as in Section A.2), the PDE (8) takes the explicit form

$$\sum \left(\frac{da_n}{dt}\right) v_n = k \sum a_n \Delta v_n = -k \sum a_n \lambda_n v_n. \tag{11}$$

Because the coefficients in (11) must match, we get $da_n/dt = -\lambda_n k a_n$. This is a simple ODE with the solution $a_n(t) = A_n e^{-\lambda_n k t}$, which is exactly (9) as we wanted.

In Section 12.1 we will introduce a more general concept of differentiability with which it will be exceedingly straightforward to justify the passage from (10) to (11).    □

Now we justify the *separation of the space variables*. For simplicity, let's take $D$ to be a rectangle and write $D = D_1 \times D_2$, where $D_1$ is the $x$ interval and $D_2$ is the $y$ interval. (It will be clear that more general geometry is possible, such as annuli in polar coordinates, and so on.) We write $\mathbf{x} = (x, y)$ where $x \in D_1$, $y \in D_2$. We write $\Delta = \partial_{xx} + \partial_{yy}$ and consider any one of the three standard boundary conditions. Let the operator $-\partial_{xx}$ have the real eigenfunctions $v_n(x)$ with eigenvalues $\alpha_n$, and let the operator $-\partial_{yy}$ have $w_n(y)$ with $\beta_n$.

**Theorem 2.**    The set of products $\{v_n(x) w_m(y) : n = 1, 2, \ldots; m = 1, 2, \ldots\}$ is a *complete* set of eigenfunctions for $-\Delta$ in $D$ with the given boundary conditions.

**Proof.**    We note that each product $v_n(x) w_m(y)$ is an eigenfunction because

$$-\Delta(v_n w_m) = (-\partial_{xx} v_n) w_m + v_n(-\partial_{yy} w_m) = (\alpha_n + \beta_m) v_n w_m. \tag{12}$$

The eigenvalue is $\alpha_n + \beta_m$. The eigenfunctions are mutually orthogonal since

$$\iint v_n w_m v_{n'} w_{m'}\,dx\,dy = \left(\int v_n v_{n'}\,dx\right) \cdot \left(\int w_m w_{m'}\,dy\right) = 0$$

if either $n \neq n'$ or $m \neq m'$.

By Section 11.3, the eigenfunctions for all three problems (in $D$, in $D_1$, and in $D_2$) are complete. Among the eigenfunctions of $-\Delta$ in the rectangle $D$ are the products $v_n w_m$. Suppose now that there were an eigenfunction $u(x, y)$ in the rectangle, other than these products. Then, for some $\lambda$, $-\Delta u = \lambda u$ in $D$ and $u$ would satisfy the boundary conditions. If $\lambda$ were different from every one of the sums $\alpha_n + \beta_m$, then we would know (from Section 10.1) that $u$ is orthogonal to all the products $v_n w_m$. Hence

$$0 = (u, v_n w_m) = \int \left[ \int u(x, y) v_n(x)\, dx \right] w_m(y)\, dy. \tag{13}$$

So, by the completeness of the $w_m$,

$$\int u(x, y) v_n(x)\, dx = 0 \qquad \text{for all } y. \tag{14}$$

By the completeness of the $v_n$, (14) would imply that $u(x, y) = 0$ for all $x, y$. So $u(x, y)$ wasn't an eigenfunction after all.

One possibility remains, namely, that $\lambda = \alpha_n + \beta_m$ for certain $n$ and $m$. This could be true for one pair $m, n$ or several such pairs. If $\lambda$ were such a sum, we would consider the difference

$$\psi(x, y) = u(x, y) - \sum c_{nm} v_n(x) w_m(y), \tag{15}$$

where the sum is over all the $n, m$ pairs for which $\lambda = \alpha_n + \beta_m$ and where $c_{nm} = (u, v_n w_m)/\|v_n w_m\|^2$. The function $\psi$ defined by (15) is constructed so as to be orthogonal to *all* the products $v_n w_m$, for both $\alpha_n + \beta_m = \lambda$ and $\alpha_n + \beta_m \neq \lambda$. It follows by the same reasoning as above that $\psi(x, y) \equiv 0$. Hence $u(x, y) = \sum c_{nm} v_n(x) w_m(y)$, summed over $\alpha_n + \beta_m = \lambda$. That is, $u$ was *not* a new eigenfunction at all, but just a linear combination of those old products $v_n w_m$ which have the same eigenvalue $\lambda$. This completes the proof of Theorem 2.  $\square$

## EXERCISES

1. Verify that all the functions (7) are solutions of (1) if $a$ is an eigenvalue $\lambda_N$ and if $\int f v_N\, dx = 0$. Why does the series in (7) converge?

2. Use the completeness to show that the solutions of the wave equation in any domain with a standard set of BCs satisfy the usual expansion $u(\mathbf{x}, t) = \sum_{n=1}^{\infty} [A_n \cos(\sqrt{\lambda_n}\, ct) + B_n \sin(\sqrt{\lambda_n}\, ct)] v_n(\mathbf{x})$. In particular, show that the series converges in the $L^2$ sense.

3. Provide the details of the proof that $\psi(x, y)$, defined by (15), is identically zero.

## 11.6  ASYMPTOTICS OF THE EIGENVALUES

The main purpose of this section is to show that $\lambda_n \to +\infty$. In fact, we'll show exactly *how fast* the eigenvalues go to infinity. For the case of the Dirichlet boundary condition, the precise result is as follows.

**Theorem 1.**   For a *two*-dimensional problem $-\Delta u = \lambda u$ in any plane domain $D$ with $u = 0$ on bdy $D$, the eigenvalues satisfy the limit relation

$$\lim_{n\to\infty} \frac{\lambda_n}{n} = \frac{4\pi}{A}, \tag{1}$$

where $A$ is the *area* of $D$.

For a *three*-dimensional problem in any solid domain, the relation is

$$\lim_{n\to\infty} \frac{\lambda_n^{3/2}}{n} = \frac{6\pi^2}{V}, \tag{2}$$

where $V$ is the *volume* of $D$.

## Example 1.  The Interval

Let's compare Theorem 1 with the *one*-dimensional case where $\lambda_n = n^2\pi^2/l^2$. In that case,

$$\lim_{n\to\infty} \frac{\lambda_n^{1/2}}{n} = \frac{\pi}{l}, \tag{3}$$

where $l$ is the length of the interval! The same result (3) was also derived for the one-dimensional Neumann condition in Section 4.2 and the Robin conditions in Section 4.3.   □

## Example 2.  The Rectangle

Here the domain is $D = \{0 < x < a, 0 < y < b\}$ in the plane. We showed explicitly in Section 10.1 that

$$\lambda_n = \frac{l^2\pi^2}{a^2} + \frac{m^2\pi^2}{b^2} \tag{4}$$

with the eigenfunction $\sin(l\pi x/a) \cdot \sin(m\pi y/b)$. Because the eigenvalues are naturally numbered using a pair of integer indices, it is difficult to see the relationship between (4) and (1). For this purpose it is convenient to introduce the *enumeration function*

$$N(\lambda) \equiv \textit{the number of eigenvalues that do not exceed } \lambda. \tag{5}$$

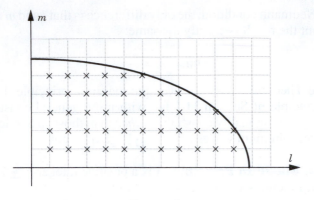

**Figure 1**

If the eigenvalues are written in increasing order as in (11.1.2), then $N(\lambda_n) = n$. Now we can express $N(\lambda)$ another way using (4). Namely, $N(\lambda)$ is the number of integer lattice points $(l, m)$ that are contained within the quarter-ellipse

$$\frac{l^2}{a^2} + \frac{m^2}{b^2} \le \frac{\lambda}{\pi^2} \qquad (l > 0, m > 0) \qquad (6)$$

in the $(l, m)$ plane (see Figure 1). Each such lattice point is the upper-right corner of a square lying within the quarter ellipse. Therefore, $N(\lambda)$ is at most the area of this quarter ellipse:

$$N(\lambda) \le \frac{\lambda ab}{4\pi}. \qquad (7)$$

For large $\lambda$, $N(\lambda)$ and this area may differ by approximately the length of the perimeter, which is of the order $\sqrt{\lambda}$. Precisely,

$$\frac{\lambda ab}{4\pi} - C\sqrt{\lambda} \le N(\lambda) \le \frac{\lambda ab}{4\pi} \qquad (8)$$

for some constant $C$. Substituting $\lambda = \lambda_n$ and $N(\lambda) = n$, (8) takes the form

$$\frac{\lambda_n ab}{4\pi} - C\sqrt{\lambda_n} \le n \le \frac{\lambda_n ab}{4\pi}, \qquad (9)$$

where the constant $C$ does not depend on $n$. Therefore, upon dividing by $n$, we deduce that

$$\lim_{n \to \infty} \frac{\lambda_n}{n} = \frac{4\pi}{ab}, \qquad (10)$$

which is Theorem 1 for a rectangle. $\qquad\qquad\qquad\qquad\qquad\qquad\square$

For the Neumann condition, the only difference is that $l$ and $m$ are allowed to be zero, but the result is exactly the same:

$$\lim_{n\to\infty} \frac{\tilde{\lambda}_n}{n} = \frac{4\pi}{ab}. \tag{11}$$

To prove Theorem 1, we will need the maximin principle. It is like the minimum principle of Section 11.1 but with more general constraints. The idea is that *any* orthogonality constraints *other* than those in Section 11.1 will lead to smaller minimum values of the Rayleigh quotient.

**Theorem 2. Maximin Principle**    Fix a positive integer $n \geq 2$. Fix $n - 1$ arbitrary trial functions $y_1(\mathbf{x}), \ldots, y_{n-1}(\mathbf{x})$. Let

$$\lambda_{n*} = \min \frac{\|\nabla w\|^2}{\|w\|^2} \tag{12}$$

among all trial functions $w$ that are orthogonal to $y_1, \ldots, y_{n-1}$. Then

$$\lambda_n = \max \lambda_{n*} \tag{13}$$

over *all* choices of the $n - 1$ piecewise continuous functions $y_1, \ldots, y_{n-1}$.

A function defined on a domain $D$ is called *piecewise continuous* if $D$ can be subdivided into a finite number of subdomains $D_i$ so that, for each $i$, the restriction of $f$ to $D_i$ has a continuous extension to the closure $\overline{D}_i$. The idea is that the function can have jump discontinuities on some curves inside $D$ but is otherwise continuous. This generalizes the one-dimensional concept in Section 5.4.

**Proof.**    Fix an arbitrary choice of $y_1, \ldots, y_{n-1}$. Let $w(\mathbf{x}) = \sum_{j=1}^{n} c_j v_j(\mathbf{x})$ be a linear combination of the first $n$ eigenfunctions which is chosen to be orthogonal to $y_1, \ldots, y_{n-1}$. That is, the constants $c_1, \ldots, c_n$ are chosen to satisfy the linear system

$$0 = \left( \sum_{j=1}^{n} c_j v_j, y_k \right) = \sum_{j=1}^{n} (v_j, y_k) c_j \quad \text{(for } k = 1, \ldots, n - 1).$$

Being a system of only $n - 1$ equations in $n$ unknowns, it has a solution $c_1, \ldots, c_n$, not all of which constants are zero. Then, by definition (12) of $\lambda_{n*}$,

$$\lambda_{n*} \leq \frac{\|\nabla w\|^2}{\|w\|^2} = \frac{\sum_{j,k} c_j c_k (-\Delta v_j, v_k)}{\sum_{j,k} c_j c_k (v_j, v_k)}$$

$$= \frac{\sum_{j=1}^{n} \lambda_j c_j^2}{\sum_{j=1}^{n} c_j^2} \leq \frac{\sum_{j=1}^{n} \lambda_n c_j^2}{\sum_{j=1}^{n} c_j^2} = \lambda_n, \tag{14}$$

where we've again taken $\|v_j\| = 1$. This inequality (14) is true for every choice of $y_1, \ldots, y_{n-1}$. Hence, max $\lambda_{n*} \leq \lambda_n$. This proves half of (13).

To demonstrate the equality in (13), we need only exhibit a special choice of $y_1, \ldots, y_{n-1}$ for which $\lambda_{n*} = \lambda_n$. Our special choice is the first $n - 1$ eigenfunctions: $y_1 = v_1, \ldots, y_{n-1} = v_{n-1}$. By the minimum principle (MP)$_n$ for the $n$th eigenvalue in Section 11.1, we know that

$$\lambda_{n*} = \lambda_n \qquad \text{for this choice.} \tag{15}$$

The maximin principle (13) follows directly from (14) and (15).    $\square$

The same maximin principle is also valid for the *Neumann boundary condition* if we use the "free" trial functions that don't satisfy any boundary condition. Let's denote the Neumann eigenvalues by $\tilde{\lambda}_j$. Now we shall simultaneously consider the Neumann and Dirichlet cases.

**Theorem 3.**   $\tilde{\lambda}_j \leq \lambda_j$ for all $j = 1, 2, \ldots$.

**Proof.**   Let's begin with the first eigenvalues. By Theorems 11.1.1 and 11.3.1, both $\tilde{\lambda}_1$ and $\lambda_1$ are expressed as the same minimum of the Rayleigh quotient except that the trial functions for $\lambda_1$ satisfy one extra constraint (namely, that $w = 0$ on bdy $D$). Having one less constraint, $\tilde{\lambda}_1$ has a greater chance of being small. Thus $\tilde{\lambda}_1 \leq \lambda_1$.

Now let $n \geq 2$. For the same reason of having one extra constraint, we have

$$\tilde{\lambda}_{n*} \leq \lambda_{n*}. \tag{16}$$

We take the maximum of both sides of (16) over all choices of piecewise continuous functions $y_1, \ldots, y_{n-1}$. By the maximin principle of this section (Theorem 2 and its Neumann analog), we have

$$\tilde{\lambda}_n = \max \tilde{\lambda}_{n*} \leq \max \lambda_{n*} = \lambda_n.$$

Actually it is true that $\tilde{\lambda}_{n+1} \leq \lambda_n$ but we omit the proof [Fi].    $\square$

**Example 3.**

For the interval $(0, l)$ in one dimension, the eigenvalues are $\lambda_n = n^2\pi^2/l^2$ and $\tilde{\lambda}_n = (n - 1)^2\pi^2/l^2$ (using our present notation with $n$ running from 1 to $\infty$). It is obvious that $\tilde{\lambda}_n < \lambda_n$.    $\square$

The general principle illustrated by Theorem 3 is that

*any additional constraint will increase the value of the maximin.*   (17)

In particular, we can use this principle as follows to prove the monotonicity of the eigenvalues with respect to the domain.

**Theorem 4.**   If the domain is enlarged, each Dirichlet eigenvalue is decreased.

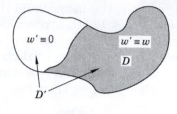

**Figure 2**

That is, if one domain $D$ is contained in another domain $D'$, then $\lambda_n \geq \lambda'_n$, where we use primes on eigenvalues to refer to the larger domain $D'$ (see Figure 2).

**Proof.**    In the Dirichlet case, consider the maximin expression (13) for $D$. If $w(\mathbf{x})$ is any trial function in $D$, we define $w(\mathbf{x})$ in all of $D'$ by setting it equal to zero outside $D$; that is,

$$w'(\mathbf{x}) = \begin{cases} w(\mathbf{x}) & \text{for } \mathbf{x} \text{ in } D \\ 0 & \text{for } \mathbf{x} \text{ in } D' \text{ but } \mathbf{x} \text{ not in } D. \end{cases} \tag{18}$$

Thus every trial function in $D$ corresponds to a trial function in $D'$ (but not conversely). So, compared to the trial functions for $D'$, the trial functions for $D$ have the extra constraint of vanishing in the rest of $D'$. By the general principle (17), the maximin for $D$ is larger than the maximin for $D'$. It follows that $\lambda_n \geq \lambda'_n$, as we wanted to prove. But we should beware that we are avoiding the difficulty that by extending the function to be zero, it is most likely no longer a $C^2$ function and therefore not a trial function. The good thing about the extended function $w'(\mathbf{x})$ is that it still is continuous. For a rigorous justification of this point, see [CH] or [Ga].    □

## SUBDOMAINS

Our next step in establishing Theorem 1 is to *divide the general domain $D$ into a finite number of subdomains $D_1, \ldots, D_m$* by introducing inside $D$ a system of smooth surfaces $S_1, S_2, \ldots$ (see Figure 3). Let $D$ have Dirichlet eigenvalues $\lambda_1 \leq \lambda_2 \leq \ldots$ and Neumann eigenvalues $\tilde{\lambda}_1 \leq \tilde{\lambda}_2 \leq \ldots$. Each of the subdomains $D_1, \ldots, D_m$ has its own collection of eigenvalues. We combine

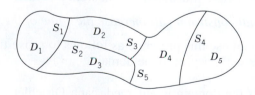

**Figure 3**

*all* of the Dirichlet eigenvalues of *all* of the subdomains $D_1, \ldots, D_m$ into a single increasing sequence $\mu_1 \leq \mu_2 \leq \cdots$. We combine *all* of their Neumann eigenvalues into another single increasing sequence $\tilde{\mu}_1 \leq \tilde{\mu}_2 \leq \ldots$.

By the maximin principle, each of these numbers can be obtained as the maximum over piecewise continuous functions $y_1, \ldots, y_{n-1}$ of the minimum over trial functions $w$ orthogonal to $y_1, \ldots, y_{n-1}$. As discussed above, although each $\mu_n$ is a Dirichlet eigenvalue of a single one of the subdomains, the trial functions can be defined in all of $D$ simply by making them vanish in the other subdomains. They will be continuous but not $C^2$ in the whole domain $D$. Thus each of the competing trial functions for $\mu_n$ has the extra restriction, compared with the trial functions for $\lambda_n$ for $D$, of vanishing on the internal boundaries. It follows from the general principle (17) that

$$\lambda_n \leq \mu_n \quad \text{for each } n = 1, 2, \ldots. \tag{20}$$

On the other hand, the trial functions defining $\tilde{\lambda}_n$ for the Neumann problem in $D$ are arbitrary $C^2$ functions. As above, we can characterize $\tilde{\mu}_n$ as

$$\tilde{\mu}_n = \max \tilde{\mu}_{n*} \qquad \tilde{\mu}_{n*} = \min \frac{\sum\limits_i \iint_{D_i} |\nabla w|^2 \, dx}{\|w\|^2}, \tag{21}$$

where the competing trial functions are arbitrary on each subdomain and orthogonal to $y_1, \ldots, y_{n-1}$. But these trial functions *are allowed to be discontinuous on the internal boundaries*, so they comprise a significantly more extensive class than the trial functions for $\tilde{\lambda}_n$, which are required to be continuous in $D$. Therefore, by (17) we have $\tilde{\mu}_n \leq \tilde{\lambda}_n$ for each $n$. Combining this fact with Theorem 3 and (20), we have proven the following inequalities.

**Theorem 5.**

$$\boxed{\tilde{\mu}_n \leq \tilde{\lambda}_n \leq \lambda_n \leq \mu_n.}$$

**Example 4.**

Let $D$ be the union of a finite number of rectangles $D = D_1 \cup D_2 \cup \ldots$ in the plane as in Figure 4. Each particular $\mu_n$ corresponds to one of these rectangles, say $D_p$ (where $p$ depends on $n$). Let $A(D_p)$ denote the

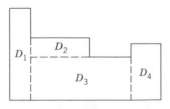

**Figure 4**

area of $D_p$. Let $M(\lambda)$ be the enumeration function for the sequence $\mu_1$, $\mu_2, \ldots$ defined above:

$$M(\lambda) \equiv \text{the number of } \mu_1, \mu_2, \ldots \text{ that do not exceed } \lambda. \qquad (22)$$

Then, adding up the integer lattice points which are located within the quarter ellipses, we get

$$\lim_{\lambda \to \infty} \frac{M(\lambda)}{\lambda} = \sum_p \frac{A(D_p)}{4\pi} = \frac{A(D)}{4\pi}, \qquad (23)$$

as for the case of a single rectangle. Since $M(\mu_n) = n$, the reciprocal of (23) takes the form

$$\lim_{n \to \infty} \frac{\mu_n}{n} = \frac{4\pi}{A(D)}. \qquad (24)$$

Similarly,

$$\lim_{n \to \infty} \frac{\tilde{\mu}_n}{n} = \frac{4\pi}{A(D)}. \qquad (25)$$

By Theorem 5 it follows that all the limits are equal: $\lim \lambda_n / n = \lim \tilde{\lambda}_n / n = 4\pi / A(D)$. This proves Theorem 1 for unions of rectangles. $\qquad \square$

   Now an arbitrary plane domain $D$ can be approximated by finite unions of rectangles just as in the construction of a double integral (and as in Section 8.4). With the help of Theorem 5, it is possible to prove Theorem 1. The details are omitted but the proof may be found in Sec. VI.4, vol. 1 of [CH].

## THREE DIMENSIONS

The *three-dimensional case* works the same way. We limit ourselves, however, to the basic example.

## Example 5. The Rectangular Box

Let $D = \{0 < x < a, 0 < y < b, 0 < z < c\}$. As in Example 2, the enumeration function $N(\lambda)$ is approximately the volume of the ellipsoid

$$\frac{l^2}{a^2} + \frac{m^2}{b^2} + \frac{k^2}{c^2} \leq \frac{\lambda}{\pi^2}$$

in the first octant. Thus for large $\lambda$

$$N(\lambda) \sim \frac{1}{8} \frac{4\pi}{3} \frac{a\lambda^{1/2}}{\pi} \frac{b\lambda^{1/2}}{\pi} \frac{c\lambda^{1/2}}{\pi} = \lambda^{3/2} \frac{abc}{6\pi^2} \qquad (26)$$

and the same for the Neumann case. Substituting $\lambda = \lambda_n$ and $N(\lambda) = n$, we deduce that

$$\lim_{n\to\infty} \frac{\lambda_n^{3/2}}{n} = \frac{6\pi^2}{abc} = \lim_{n\to\infty} \frac{\tilde{\lambda}_n^{3/2}}{n}. \tag{27}$$

For the union of a finite number of boxes of volume $V(D)$, we deduce that

$$\lim_{n\to\infty} \frac{\lambda_n^{3/2}}{n} = \frac{6\pi^2}{V(D)} = \lim_{n\to\infty} \frac{\tilde{\lambda}_n^{3/2}}{n}.$$

Then a general domain is approximated by unions of boxes.    □

For the very general case of a *symmetric differential operator* as (11.4.1), the statement of the theorem is modified (in three dimensions, say) to read

$$\lim_{n\to\infty} \frac{\lambda_n^{3/2}}{n} = \lim_{n\to\infty} \frac{\tilde{\lambda}_n^{3/2}}{n}$$
$$= \frac{6\pi^2}{\iiint_D [m(\mathbf{x})/p(\mathbf{x})]^{3/2} d\mathbf{x}}. \tag{28}$$

## EXERCISES

1. Prove that (9) implies (10).

2. Explain how it is possible that $\lambda_2$ is *both* a maximin *and* a minimax.

3. For $-\Delta$ in the interior $D = \{x^2 + y^2/4 < 1\}$ of an ellipse with Dirichlet BCs use the monotonicity of the eigenvalues with respect to the domain to find estimates for the first two eigenvalues. Inscribe or circumscribe rectangles or circles, for which we already know the exact values.
   (a) Find upper bounds.
   (b) Find lower bounds.

4. In the proof of Theorem 1 for an arbitrary domain $D$, one must approximate $D$ by unions of rectangles. This is a delicate limiting procedure. Outline the main steps required to carry out the proof.

5. Use the surface area of an ellipsoid to write the inequalities that make (26) a more precise statement.

6. For a symmetric differential operator in three dimensions as in (11.4.1), explain why Theorem 1 should be modified to be (28).

7. Consider the Dirichlet BCs in a domain $D$. Show that the first eigenfunction $v_1(\mathbf{x})$ *vanishes at no point of $D$* by the following method.
   (a) Suppose on the contrary that $v_1(\mathbf{x}) = 0$ at some point in $D$. Show that both $D^+ = \{\mathbf{x} \in D: v_1(\mathbf{x}) > 0\}$ and $D^- = \{\mathbf{x} \in D: v_1(\mathbf{x}) < 0\}$ are nonempty. (*Hint:* Use the maximum principle in Exercise 7.4.25.)
   (b) Let $v^+(\mathbf{x}) = v_1(\mathbf{x})$ for $\mathbf{x} \in D^+$ and $v^+(\mathbf{x}) = 0$ for $\mathbf{x} \in D^-$. Let $v^- = v_1 - v^+$. Notice that $|v_1| = v^+ - v^-$. Noting that $v_1 = 0$ on

bdy $D$, we may deduce that $\nabla v^+ = \nabla v_1$ in $D^+$, and $\nabla v^+ = 0$ outside $D^+$. Similarly for $\nabla v^-$. Show that the Rayleigh quotient $Q$ for the function $|v_1|$ is equal to $\lambda_1$. Therefore, both $v_1$ and $|v_1|$ are eigenfunctions with the eigenvalue $\lambda_1$.

(c)  Use the maximum principle on $|v_1|$ to show that $v_1 > 0$ in all of $D$ or $v_1 < 0$ in all of $D$.

(d)  Deduce that $\lambda_1$ is a simple eigenvalue (*Hint:* If $u(x)$ were another eigenfunction with eigenvalue $\lambda_1$, let $w$ be the component of $u$ orthogonal to $v_1$. Applying part (c) to $w$, we know that $w > 0$ or $w < 0$ or $w \equiv 0$ in $D$. Conclude that $w \equiv 0$ in $D$.)

8.  Show that the nodes of the $n$th eigenfunction $v_n(\mathbf{x})$ divide the domain $D$ into *at most $n$* pieces, assuming (for simplicity) that the eigenvalues are distinct, by the following method. Assume Dirichlet BCs.

(a)  Suppose on the contrary that $\{\mathbf{x} \in D : v_n(\mathbf{x}) \neq 0\}$ is the disjoint union of at least $n + 1$ components $D_1 \cup D_2 \cup \ldots \cup D_{n+1}$. Let $w_j(\mathbf{x}) = v_n(\mathbf{x})$ for $\mathbf{x} \in D_j$, and $w_j(\mathbf{x}) = 0$ elsewhere. You may assume that $\nabla w_j(\mathbf{x}) = \nabla v_n(\mathbf{x})$ for $\mathbf{x} \in D_j$, and $\nabla w_j(\mathbf{x}) = 0$ elsewhere. Show that the Rayleigh quotient for $w_j$ equals $\lambda_n$.

(b)  Show that the Rayleigh quotient for any linear combination $w = c_1 w_1 + \cdots + c_{n+1} w_{n+1}$ also equals $\lambda_n$.

(c)  Let $y_1, \ldots, y_n$ be any trial functions. Choose the $n + 1$ coefficients $c_j$ so that $w$ is orthogonal to each of $y_1, \ldots, y_n$. Use the maximin principle to deduce that $\lambda_{n+1*} \leq \|\nabla w\|^2 / \|w\|^2 = \lambda_n$. Hence deduce that $\lambda_{n+1} = \lambda_n$, which contradicts our assumption.

9.  Fill in the details of the proof of Theorem 5.

# 12

# DISTRIBUTIONS AND TRANSFORMS

The purpose of this chapter is to introduce two important techniques, both of which shed new light on PDE problems. The first is the theory of distributions, which permits a succinct and elegant interpretation of Green's functions. The second technique is that of the Fourier transformation and its cousin, the Laplace transformation. They permit a new and independent approach to many of the problems already encountered in this book. Some of the examples in this chapter require a knowledge of Chapters 7 or 9.

## 12.1 DISTRIBUTIONS

In several critical places in this book we've encountered *approximate delta functions*. See the diffusion kernel in Section 2.4 and the Dirichlet kernel in Section 5.5. They look roughly as shown in Figure 1. What exactly is a delta function? It is supposed to be infinite at $x = 0$ and zero at all $x \neq 0$. It should have integral 1: $\int_{-\infty}^{\infty} \delta(x)\, dx = 1$. Of course, this is impossible. However, the concept in physics is clear and simple. The delta function $\delta(x)$ is supposed to represent a point mass, that is, a particle of unit mass located at the origin. It is the idealization that it is located at a mathematical point that is the source of the logical difficulty. So how should we make sense of it? Certainly it's not a function. It's a more general object, called a distribution. A function is a rule that assigns numbers to numbers. A distribution is a *rule* (or *transformation* or *functional*) that assigns numbers to *functions*.

**Definition.** The *delta function* is the rule that assigns the number $\phi(0)$ to the *function* $\phi(x)$.

To give a proper definition, we need to say what kinds of $\phi(x)$ are used. A *test function* $\phi(x)$ is a real $C^\infty$ function (a function all of whose derivatives exist) that vanishes outside a finite interval. Thus $\phi \colon \mathbb{R} \to \mathbb{R}$ is defined and

**331**

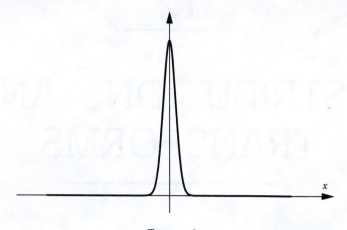

**Figure 1**

differentiable for all $-\infty < x < \infty$ and $\phi(x) \equiv 0$ for $x$ large (near $+ \infty$) and for $x$ small (near $- \infty$) (see Figure 2). Let $\mathcal{D}$ denote the collection of all test functions.

**Definition.** A *distribution f* is a functional (a rule): $\mathcal{D} \mapsto \mathbb{R}$ which is *linear* and *continuous* in the following sense. If $\phi \in \mathcal{D}$ is a test function, then we denote the corresponding real number by $(f, \phi)$.

By *linearity* we mean that

$$(f, a\phi + b\psi) = a(f, \phi) + b(f, \psi) \tag{1}$$

for all constants $a$, $b$ and all test functions $\phi$, $\psi$.

By *continuity* we mean the following. If $\{\phi_n\}$ is a sequence of test functions that vanish outside a common interval and converge uniformly to a test function $\phi$, and if all their derivatives do as well, then

$$(f, \phi_n) \to (f, \phi) \qquad \text{as } n \to \infty. \tag{2}$$

We'll sometimes specify a distribution $f$ by writing $\phi \mapsto (f, \phi)$.

## Example 1.

According to the first definition above, the delta function is the distribution $\phi \mapsto \phi(0)$. It is denoted by $\delta$. In integration theory it is called

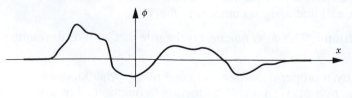

**Figure 2**

the unit point mass. It is a distribution because both (1) and (2) are true. (Why?) □

## Example 2.

The functional $\phi \mapsto \phi''(5)$ is a distribution. It is linear because $(a\phi + b\psi)''(5) = a\phi''(5) + b\psi''(5)$. It is a continuous functional because obviously $\phi_n''(5) \to \phi''(5)$ as $n \to \infty$ if $\phi_n \to \phi, \phi_n' \to \phi'$ and $\phi_n'' \to \phi''$ uniformly. □

## Example 3.

Let $f(x)$ be any ordinary (integrable) function. It corresponds to the distribution

$$\phi \mapsto \int_{-\infty}^{\infty} f(x)\, \phi(x)\, dx. \tag{3}$$

See Exercise 1 for the verification that this is a distribution. The function $f(x)$ is thereby considered to be a distribution. □

Because of Example 3, it is common to use the notation (*but it is only a notation*)

$$\int_{-\infty}^{\infty} \delta(x)\, \phi(x)\, dx = \phi(0) \tag{4}$$

and to speak of the delta function as if it were a true function.

## CONVERGENCE OF DISTRIBUTIONS

If $f_N$ is a sequence of distributions and $f$ is another distribution, we say that $f_N$ *converges weakly to* $f$ if

$$(f_N, \phi) \to (f, \phi) \qquad \text{as } N \to \infty \tag{5}$$

for all test functions $\phi$.

## Example 4.

The source function for the *diffusion equation* on the whole line is $S(x, t) = 1/\sqrt{4\pi kt}\, e^{-x^2/4kt}$ for $t > 0$. We proved in Section 3.5 that

$$\int_{-\infty}^{\infty} S(x, t)\phi(x)\, dx \to \phi(0) \qquad \text{as } t \to 0. \tag{6}$$

Because for each $t$ we may consider the function $S(x, t)$ as a distribution as in Example 3, this means that

$$\boxed{S(x, t) \to \delta(x) \qquad \text{weakly as } t \to 0.} \tag{7}$$

□

**Example 5.**

Let $K_N(\theta)$ be the *Dirichlet kernel* from Section 5.5. It is given by the formulas

$$K_N(\theta) = 1 + 2\sum_{n=1}^{N}\cos n\theta = \frac{\sin\left[\left(N + \frac{1}{2}\right)\theta\right]}{\sin\frac{1}{2}\theta}. \tag{8}$$

We proved that

$$\int_{-\pi}^{\pi} K_N(\theta)\,\phi(\theta)\,d\theta \to 2\pi\phi(0) \qquad \text{as } N \to \infty \tag{9}$$

for any periodic $C^1$ function $\phi(x)$. In fact, it can be verified that the periodicity is not required (because we are dealing only with the case $x = 0$ of Section 5.5). Therefore,

$$\boxed{K_N(\theta) \to 2\pi\,\delta(\theta) \quad \text{weakly as } N \to \infty \text{ in the interval } (-\pi, \pi).} \tag{10}$$

$\square$

## DERIVATIVE OF A DISTRIBUTION

The derivative of a distribution always exists and is another distribution. To motivate the definition, let $f(x)$ be any $C^1$ function and $\phi(x)$ any test function. Integration by parts shows that

$$\int_{-\infty}^{\infty} f'(x)\phi(x)\,dx = -\int_{-\infty}^{\infty} f(x)\phi'(x)\,dx \tag{11}$$

since $\phi(x) = 0$ for large $|x|$.

**Definition.**   For any distribution $f$, we define its *derivative* $f'$ by the formula

$$\boxed{(f', \phi) = -(f, \phi') \qquad \text{for all test functions } \phi.} \tag{12}$$

That $f'$ satisfies the linearity and continuity properties is left to Exercise 2. Most of the ordinary rules of differentiation are valid for distributions. It is easy to see that if $f_N \to f$ weakly, then $f'_N \to f'$ weakly. (Why?)

## Example 6.

Directly from the definition we see that the *derivatives of the delta function* are

$$(\delta', \phi) = -(\delta, \phi') = -\phi'(0) \tag{13}$$

$$(\delta'', \phi) = -(\delta', \phi') = +(\delta, \phi'') = +\phi''(0), \quad \text{etc.} \tag{14}$$

$\square$

### Example 7.

The *Heaviside function* (or step function) is defined by $H(x) = 1$ for $x > 0$, and $H(x) = 0$ for $x < 0$. For any test function, $(H', \phi) = -(H, \phi') = -\int_{-\infty}^{\infty} H(x)\phi'(x)\, dx = -\int_{0}^{\infty} \phi'(x)\, dx = -\phi(x)|_{0}^{\infty} = \phi(0)$. Thus

$$\boxed{H' = \delta.} \tag{15}$$

The *plus function* $p(x) = x^{+}$ is defined as $p(x) = x$ for $x \geq 0$, and $p(x) = 0$ for $x \leq 0$. Then $p' = H$ and $p'' = \delta$. $\qquad\square$

If $I$ is an open interval, a distribution $f$ in $I$ is a functional defined as before but where the test functions vanish outside a closed interval $\subset I$, and the required continuity is valid for sequences $\{\phi_n\}$ that vanish outside a common closed interval $\subset I$.

### Example 8.

We know from Chapter 5 and the comparison test in Section A.2 that

$$|x| = \frac{\pi}{2} - \sum_{n \text{ odd}} \frac{4}{n^2 \pi} \cos nx \quad \text{uniformly in } [-\pi, \pi]. \tag{16}$$

In particular, if we multiply this series by a test function and integrate, we see that (16) converges weakly as a series of distributions in $(-\pi, \pi)$. Therefore, the differentiated series also converges weakly. This means that

$$\sum_{n \text{ odd}} \frac{4}{n\pi} \sin nx = \begin{cases} 1 & \text{for } 0 < x < \pi \\ -1 & \text{for } -\pi < x < 0. \end{cases} = 2H(x) - 1 \tag{17}$$

Actually, (17) is true pointwise, as we showed in Chapter 5. We can keep differentiating weakly as often as we please. If we differentiate (17) once more and divide by 2, we get

$$\boxed{\sum_{n \text{ odd}} \frac{2}{\pi} \cos nx = \delta(x) \quad \text{in } (-\pi, \pi)} \tag{18}$$

What (18) actually means is that

$$\sum_{n \text{ odd}} \int_{-\pi}^{\pi} \phi(x) \cos nx \, dx = \frac{\pi \phi(0)}{2} \tag{19}$$

for all $C^{\infty}$ functions $\phi(x)$ that vanish near $\pm\pi$. $\qquad\square$

## Example 9.

Consider the complex series $\sum_{n=-\infty}^{\infty} e^{inx}$. Clearly, it diverges (e.g., at $x = 0$). However, the Dirichlet kernel (8) can be written as its partial sum

$$K_N(x) = \sum_{n=-N}^{N} e^{inx}. \tag{20}$$

From (10) we have $K_N \to 2\pi\delta$ as $N \to \infty$. Thus from (20) we get

$$\sum_{n=-\infty}^{\infty} e^{inx} = 2\pi\delta(x) \qquad \text{in the weak sense in } (-\pi, \pi). \tag{21}$$

$\square$

## DISTRIBUTIONS IN THREE DIMENSIONS

A test function $\phi(\mathbf{x}) = \phi(x, y, z)$ is a real $C^\infty$ function that vanishes outside some ball. $\mathscr{D}$ denotes the set of all test functions of $\mathbf{x}$. Then the definition of a *distribution* is identical to the one-dimensional case except we replace 'common interval' by 'common ball.'

The "delta function" $\delta$ is defined as the functional $\phi \mapsto \phi(\mathbf{0})$. Its partial derivative $\partial\delta/\partial z$ is defined as the functional $\phi \mapsto -(\partial\phi/\partial z)(\mathbf{0})$. If $f(\mathbf{x})$ is any ordinary integrable function, it is considered to be the same as the distribution $\phi \mapsto \int_{-\infty}^{\infty} \int_{-\infty}^{\infty} \int_{-\infty}^{\infty} f(\mathbf{x})\phi(\mathbf{x}) \, d\mathbf{x}$.

## Example 10.

Let's reconsider the plucked string. The initial condition is $\phi(x) = (b - b|x|/a)^+$, where $()^+$ denotes the plus function of Example 7. The solution is

$$u(x, t) = \frac{1}{2}\phi(x + ct) + \frac{1}{2}\phi(x - ct)$$

$$= \frac{1}{2}\left(b - \frac{b|x + ct|}{a}\right)^+ + \frac{1}{2}\left(b - \frac{b|x - ct|}{a}\right)^+.$$

By the chain rule, we calculate its various derivatives as

$$u_x = -\frac{b}{2a}H(x + ct) - \frac{b}{2a}H(x - ct),$$

$$u_{xx} = -\frac{b}{2a}[\delta(x + ct) + \delta(x - ct)],$$

$$u_t = -\frac{bc}{2a}H(x + ct) + \frac{bc}{2a}H(x - ct),$$

$$u_{tt} = -\frac{bc^2}{2a}[\delta(x + ct) + \delta(x - ct)].$$

Therefore $u_{tt} = c^2 u_{xx}$ and $u$ is called a "weak" solution of the wave equation. In general, a *weak solution* of the wave equation is a distribution $u$ for which

$$(u, \phi_{tt} - c^2\phi_{xx}) = 0$$

for all test functions $\phi(x, t)$. □

## Example 11.

Let $S$ denote the sphere $\{|\mathbf{x}| = a\}$. Then the distribution $\phi \mapsto \iint_S \phi \, dS$ is denoted $\delta(|\mathbf{x}| - a)$. This notation makes sense because formally

$$\iiint \delta(|\mathbf{x}| - a)\phi(\mathbf{x}) \, d\mathbf{x} = \int_0^\infty \int_0^{2\pi} \int_0^\pi \phi(\mathbf{x}) \sin\theta \, d\theta \, d\psi \, \delta(r - a) r^2 dr$$

$$= a^2 \int_0^{2\pi} \int_0^\pi \phi(\mathbf{x}) \sin\theta \, d\theta \, d\psi$$

$$= \iint_S \phi \, dS.$$ □

## Example 12.

Let $C$ be a smooth curve in space. Then the line integral over $C$ defines the distribution $\phi \mapsto \int_C \phi \, ds$, where $ds$ denotes the arc length. □

## EXERCISES

1. Verify directly from the definition that $\phi \mapsto \int_{-\infty}^\infty f(x)\phi(x) \, dx$ is a distribution if $f(x)$ is any function that is integrable on each bounded set.

2. Let $f$ be any distribution. Verify that the functional $f'$ defined by $(f', \phi) = -(f, \phi')$ satisfies the linearity and continuity properties and therefore is another distribution.

3. Verify that the derivative is a linear operator on the vector space of distributions.

4. Denoting $p(x) = x^+$, show that $p' = H$ and $p'' = \delta$.

5. Verify, directly from the definition of a distribution, that the discontinuous function $u(x, t) = H(x - ct)$ is a weak solution of the wave equation.

6. Use Chapter 5 directly to prove (19) for all $C^1$ functions $\phi(x)$ that vanish near $\pm\pi$.

7. Let a sequence of $L^2$ functions $f_n(x)$ converge to a function $f(x)$ in the mean-square sense. Show that it also converges weakly in the sense of distributions.

8. (a) Show that the product $\delta(x)\delta(y)\delta(z)$ makes sense as a three-dimensional distribution.

(b)  Show that $\delta(\mathbf{x}) = \delta(x)\delta(y)\delta(z)$, where the first delta function is the three-dimensional one.

9.  Show that no sense can be made of the square $[\delta(x)]^2$ as a distribution.

10.  Verify that Example 11 is a distribution.

11.  Verify that Example 12 is a distribution.

12.  Let $\chi_a(x) = 1/2a$ for $-a < x < a$, and $\chi_a(x) = 0$ for $|x| > a$. Show that $\chi_a \to \delta$ weakly as $a \to 0$.

## 12.2  GREEN'S FUNCTIONS, REVISITED

Here we reinterpret the Green's functions and source functions for the most important PDEs.

### LAPLACE OPERATOR

We saw in Section 6.1 that $1/r$ is a harmonic function in three dimensions except at the origin, where $r = |\mathbf{x}|$. Let $\phi(\mathbf{x})$ be a test function. By Exercise 7.2.2 we have the identity

$$\phi(\mathbf{0}) = - \iiint \frac{1}{r} \Delta\phi(\mathbf{x}) \frac{d\mathbf{x}}{4\pi}.$$

This means precisely that

$$\Delta\left(-\frac{1}{4\pi r}\right) = \delta(\mathbf{x}) \tag{1}$$

in three dimensions. Because $\delta(\mathbf{x})$ vanishes except at the origin, formula (1) explains why $1/r$ is a harmonic function away from the origin and it explains exactly how it differs from being harmonic at the origin.

Consider now the Dirichlet problem for the Poisson equation,

$$\Delta u = f \quad \text{in } D, \qquad u = 0 \quad \text{on bdy } D.$$

Its solution is

$$u(\mathbf{x}_0) = \iiint_D G(\mathbf{x}, \mathbf{x}_0) f(\mathbf{x}) \, d\mathbf{x} \tag{2}$$

from Theorem 7.3.2, where $G(\mathbf{x}, \mathbf{x}_0)$ is the Green's function. Now fix the point $\mathbf{x}_0 \in D$. The left side of (2) can be written as

$$u(\mathbf{x}_0) = \iiint_D \delta(\mathbf{x} - \mathbf{x}_0) u(\mathbf{x}) \, d\mathbf{x}.$$

We assume that $u(\mathbf{x})$ is an arbitrary test function whose support (closure of the set where $u \neq 0$) is a bounded subset of $D$. The right side of (2) is

$$u(\mathbf{x}_0) = \iiint_D G(\mathbf{x}, \mathbf{x}_0) \, \Delta u(\mathbf{x}) \, d\mathbf{x} = \iiint_D \Delta G(\mathbf{x}, \mathbf{x}_0) \, u(\mathbf{x}) \, d\mathbf{x},$$

where $\Delta G$ is understood in the sense of distributions. Because $u(\mathbf{x})$ can be an arbitrary test function in $D$, we deduce that

$$\boxed{\Delta G(\mathbf{x}, \mathbf{x}_0) = \delta(\mathbf{x} - \mathbf{x}_0) \qquad \text{in } D.} \tag{3}$$

This is the best way to understand the Green's function. As we saw in Section 7.3, the function $G(\mathbf{x}, \mathbf{x}_0) + (4\pi |\mathbf{x} - \mathbf{x}_0|)^{-1}$ is harmonic in the whole of the domain $D$, including at $\mathbf{x}_0$. Thus

$$\Delta G = -\Delta \frac{1}{4\pi |\mathbf{x} - \mathbf{x}_0|} = \delta(\mathbf{x} - \mathbf{x}_0) \qquad \text{in } D,$$

which is the same result as (3). $G(\mathbf{x}, \mathbf{x}_0)$ is the unique distribution that satisfies the PDE (3) and the boundary condition

$$\boxed{G = 0 \quad \text{for } \mathbf{x} \in \text{bdy } D.} \tag{4}$$

$G(\mathbf{x}, \mathbf{x}_0)$ may be interpreted as the steady-state temperature distribution of an object $D$ that is held at zero temperature on the boundary due to a unit source of heat at the point $\mathbf{x}_0$.

## DIFFUSION EQUATION

Consider the one-dimensional diffusion equation on the whole line. As we saw in Example 4 of Section 12.1, the source function solves the problem

$$\boxed{S_t = k \, \Delta S \quad (-\infty < x < \infty, \quad 0 < t < \infty), \qquad S(x, 0) = \delta(x).} \tag{5}$$

It is a function for $t > 0$ that becomes a distribution as $t \searrow 0$.

Let $R(x, t) = S(x - x_0, t - t_0)$ for $t > t_0$ and let $R(x, t) \equiv 0$ for $t < t_0$. Then $R$ satisfies the inhomogeneous diffusion equation

$$\boxed{R_t - k \, \Delta R = \delta(x - x_0) \, \delta(t - t_0) \quad \text{for } -\infty < x < \infty, \, -\infty < t < \infty} \tag{6}$$

(see Exercise 7). The same interpretations are true in any dimension by Section 9.4.

## WAVE EQUATION

The source function for the wave equation is the solution of the problem

$$
\begin{aligned}
S_{tt} &= c^2 \Delta S \qquad (-\infty < x, y, z < \infty, -\infty < t < \infty) \\
S(\mathbf{x}, 0) &= 0 \qquad S_t(\mathbf{x}, 0) = \delta(\mathbf{x}).
\end{aligned}
\tag{7}
$$

It is called the *Riemann function*. To find a formula for it, let $\psi(\mathbf{x})$ be any test function and let

$$
u(\mathbf{x}, t) = \int S(\mathbf{x} - \mathbf{y}, t)\, \psi(\mathbf{y})\, d\mathbf{y}. \tag{8}
$$

Then $u(\mathbf{x}, t)$ satisfies the wave equation with initial data $u(\mathbf{x}, 0) \equiv 0$ and $u_t(\mathbf{x}, 0) = \psi(\mathbf{x})$.

Now in *one* dimension (8) must take the form

$$
\int_{-\infty}^{\infty} S(x - y, t)\, \psi(y)\, dy = u(x, t) = \frac{1}{2c} \int_{x-ct}^{x+ct} \psi(y)\, dy
$$

for $t \geq 0$ by Section 2.1. Therefore, $S(x - y, t)$ equals either $\frac{1}{2}c$ or 0, depending on whether $y - x$ is in the interval $(-ct, +ct)$ or not. Replacing $x - y$ by $x$, we conclude that

$$
S(x, t) = \begin{cases} \dfrac{1}{2c} & \text{for } |x| < ct \\[2mm] 0 & \text{for } |x| > ct. \end{cases}
$$

Using the Heaviside function, we can rewrite it as

$$
S(x, t) = \frac{1}{2c} H(c^2 t^2 - x^2) \operatorname{sgn}(t) \qquad \text{for } c^2 t^2 \neq x^2, \tag{9}
$$

where we now permit $-\infty < t < \infty$ and $\operatorname{sgn}(t)$ denotes the sign of $t$. Notice that although the Riemann function is a function, it does have a jump discontinuity along the characteristics that issue from the origin. This is an example of the propagation of singularities discussed in Section 9.3.

In *three* dimensions we derived the formula

$$
\iiint S(\mathbf{x} - \mathbf{y}, t)\, \psi(\mathbf{y})\, d\mathbf{y} = u(\mathbf{x}, t) = \frac{1}{4\pi c^2 t} \iint_{|\mathbf{x}-\mathbf{y}|=ct} \psi(\mathbf{y})\, dS_y
$$

$$
= \frac{1}{4\pi c^2 t} \iiint \delta(ct - |\mathbf{x} - \mathbf{y}|)\, \psi(\mathbf{y})\, d\mathbf{y}
$$

for $t \geq 0$, with a minor shift of notation from Section 9.2. Therefore, $S(\mathbf{x}, t) = 1/(4\pi c^2 t)\, \delta(ct - |\mathbf{x}|)$ for $t \geq 0$. By Exercise 8 this can be rewritten as

$$S(\mathbf{x}, t) = \frac{1}{2\pi c}\delta(c^2 t^2 - |\mathbf{x}|^2)\,\mathrm{sgn}(t). \tag{10}$$

Formula (10) is valid for negative as well as positive $t$. Like (9), it is written in relativistic form. It is a distribution that vanishes except on the light cone, where it is a delta function. The fact that it vanishes *inside* the cone is exactly equivalent to Huygens's principle.

Some explanation is in order for expressions like those in (10). If $f$ is a distribution and $w = \psi(r)$ is a function with $\psi'(r) \neq 0$, the meaning of the distribution $g(r) = f[\psi(r)]$ is

$$(g, \phi) = \left( f \; , \; \frac{\phi[\psi^{-1}(w)]}{|\psi'[\psi^{-1}(w)]|} \right) \tag{11}$$

for all test functions $\phi$. This definition is motivated by the change of variables $w = \psi(r)$ for ordinary integrals

$$\int f[\psi(r)]\,\phi(r)\,dr = \int f(w)\,\phi(r)\frac{dw}{|\psi'(r)|}.$$

For instance, in (10) we have $f = \delta$ and $w = c^2 t^2 - r^2$ with $r > 0$, so that

$$(\delta(c^2 t^2 - r^2), \phi(r)) = \left( \delta(w), \frac{\phi(\sqrt{c^2 t^2 - w})}{2(\sqrt{c^2 t^2 - w})} \right) = \frac{1}{2c|t|}\phi(c|t|).$$

In *two* dimensions the formula is, for $t > 0$,

$$S(\mathbf{x}, t) = \begin{cases} \dfrac{1}{2\pi c}(c^2 t^2 - |\mathbf{x}|^2)^{-1/2} & \text{for } |\mathbf{x}| < ct \\ 0 & \text{for } |\mathbf{x}| > ct \end{cases} \tag{12}$$

(see Exercise 9). In this case the Riemann function is a certain smooth function inside the light cone (depending only on the relativistic quantity $c^2 t^2 - |\mathbf{x}|^2$) with a singularity on the cone. It becomes infinite as the cone is approached from the inside.

## BOUNDARY AND INITIAL CONDITIONS

Consider a diffusion process inside a bounded region $D$ in three dimensions with Dirichlet boundary conditions. The source function is defined as the solution of the problem

$$S_t = k\Delta S \qquad \text{for } \mathbf{x} \in D$$
$$S = 0 \qquad \text{for } \mathbf{x} \in \text{bdy } D \qquad (13)$$
$$S = \delta(\mathbf{x} - \mathbf{x}_0) \qquad \text{for } t = 0.$$

We denote it by $S(\mathbf{x}, \mathbf{x}_0, t)$. Let $u(\mathbf{x}, t)$ denote the solution of the same problem but with the initial function $\phi(\mathbf{x})$. Let $\lambda_n$ and $X_n(\mathbf{x})$ denote the eigenvalues and (normalized) eigenfunctions for the domain $D$, as in Chapter 11. Then

$$u(\mathbf{x}, t) = \sum_{n=1}^{\infty} c_n e^{-\lambda_n kt} X_n(\mathbf{x})$$

$$= \sum_{n=1}^{\infty} \left[ \iiint_D \phi(\mathbf{y}) X_n(\mathbf{y}) \, d\mathbf{y} \right] e^{-\lambda_n kt} X_n(\mathbf{x})$$

$$= \iiint_D \left[ \sum_{n=1}^{\infty} e^{-\lambda_n kt} X_n(\mathbf{x}) X_n(\mathbf{y}) \right] \phi(\mathbf{y}) \, d\mathbf{y},$$

assuming that the switch of summation and integration is justified. Therefore, we have the formula

$$S(\mathbf{x}, \mathbf{x}_0, t) = \sum_{n=1}^{\infty} e^{-\lambda_n kt} X_n(\mathbf{x}) X_n(\mathbf{x}_0). \qquad (14)$$

However, the convergence of this series is a delicate question that we do not pursue.

## EXERCISES

1. Give an interpretation of $G(\mathbf{x}, \mathbf{x}_0)$ as a stationary wave or as the steady-state diffusion of a substance.

2. An infinite string, at rest for $t < 0$, receives an instantaneous transverse blow at $t = 0$ which imparts an initial velocity of $V \delta(x - x_0)$, where $V$ is a constant. Find the position of the string for $t > 0$.

3. A semi-infinite string ($0 < x < \infty$), at rest for $t < 0$ and held at $u = 0$ at the end, receives an instantaneous transverse blow at $t = 0$ which imparts an initial velocity of $V \delta(x - x_0)$, where $V$ is a constant and $x_0 > 0$. Find the position of the string for $t > 0$.

4. Let $S(x, t)$ be the source function (Riemann function) for the one-dimensional wave equation. Calculate $\partial S/\partial t$ and find the PDE and initial conditions that it satisfies.

5. A force acting only at the origin leads to the wave equation $u_{tt} = c^2 \Delta u + \delta(\mathbf{x})f(t)$ with vanishing initial conditions. Find the solution.

6. Find the formula for the general solution of the inhomogeneous wave equation in terms of the source function $S(\mathbf{x}, t)$.

7. Let $R(x, t) = S(x - x_0, t - t_0)$ for $t > t_0$ and let $R(x, t) \equiv 0$ for $t < t_0$. Let $R(x, t_0)$ remain undefined. Verify that $R$ satisfies the inhomogeneous diffusion equation

$$R_t - k\,\Delta R = \delta(x - x_0)\delta(t - t_0).$$

8. (a) Prove that $\delta(a^2 - r^2) = \delta(a - r)/2a$ for $a > 0$ and $r > 0$.
   (b) Deduce that the three-dimensional Riemann function for the wave equation for $t > 0$ is

$$S(\mathbf{x}, t) = \frac{1}{2\pi c}\delta(c^2 t^2 - |\mathbf{x}|^2).$$

9. Derive the formula (12) for the Riemann function of the wave equation in two dimensions.

10. Consider an applied force $f(t)$ that acts only on the $z$ axis and is independent of $z$, which leads to the wave equation

$$u_{tt} = c^2(u_{xx} + u_{yy}) + \delta(x, y)f(t)$$

with vanishing initial conditions. Find the solution.

11. For any $a \neq b$, derive the identity

$$\delta[(\lambda - a)(\lambda - b)] = \frac{1}{|a - b|}[\delta(\lambda - a) + \delta(\lambda - b)].$$

12. A rectangular plate $\{0 \leq x \leq a, 0 \leq y \leq b\}$ initially has a hot spot at its center so that its initial temperature distribution is $u(x, y, 0) = M\delta(x - \frac{a}{2}, y - \frac{b}{2})$. Its edges are maintained at zero temperature. Let $k$ be the diffusion constant. Find the temperature at any later time in the form of a series.

13. Calculate the distribution $\Delta(\log r)$ in two dimensions.

## 12.3 FOURIER TRANSFORMS

Just as problems on finite intervals lead to Fourier series, problems on the whole line $(-\infty, \infty)$ lead to Fourier integrals. To understand this relationship, consider a function $f(x)$ defined on the interval $(-l, l)$. Its Fourier series, in complex notation, is

$$f(x) = \sum_{n=-\infty}^{\infty} c_n e^{in\pi x/l}$$

where the coefficients are

$$c_n = \frac{1}{2l} \int_{-l}^{l} f(y) e^{-in\pi y/l} dy.$$

(As usual, complex notation is more efficient.) The Fourier integral comes from letting $l \to \infty$. However, this limit is one of the trickiest in all of mathematics because the interval grows simultaneously as the terms change. If we write $k = n\pi/l$, and substitute the coefficients into the series, we get

$$f(x) = \frac{1}{2\pi} \sum_{n=-\infty}^{\infty} \left[ \int_{-l}^{l} f(y) e^{-iky} dy \right] e^{ikx} \frac{\pi}{l}. \tag{1}$$

As $l \to \infty$, the interval expands to the whole line and the points $k$ get closer together. In the limit we should expect $k$ to become a continuous variable, and the sum to become an integral. The distance between two successive $k$'s is $\Delta k = \pi/l$, which we may think of as becoming $dk$ in the limit. Therefore, we *expect* the result

$$f(x) = \frac{1}{2\pi} \int_{-\infty}^{\infty} \left[ \int_{-\infty}^{\infty} f(y) e^{-iky} dy \right] e^{ikx} dk. \tag{2}$$

This is in fact correct, although we shall not provide a rigorous proof (see, e.g., [Fo]). It is a continuous version of the completeness property of Fourier series.

Another way to state the identity (2) is

$$\boxed{f(x) = \int_{-\infty}^{\infty} F(k) e^{ikx} \frac{dk}{2\pi}} \tag{3}$$

where

$$\boxed{F(k) = \int_{-\infty}^{\infty} f(x) e^{-ikx} dx.} \tag{4}$$

$F(k)$ is called the *Fourier transform* of $f(x)$. Notice that the relationship is almost reversible: $f(x)$ is almost the Fourier transform of $F(k)$, the only difference being the minus sign in the exponent and the $2\pi$ factor. The variables $x$ and $k$ play dual roles; $k$ is called the frequency variable.

Following is a table of some important transforms.

|  | $f(x)$ | $F(k)$ |  |
| --- | --- | --- | --- |
| Delta function | $\delta(x)$ | $1$ | (5) |
| Square pulse | $H(a - \|x\|)$ | $\dfrac{2}{k}\sin ak$ | (6) |
| Exponential | $e^{-a\|x\|}$ | $\dfrac{2a}{a^2 + k^2}$ $\ (a > 0)$ | (7) |
| Heaviside function | $H(x)$ | $\pi\,\delta(k) + \dfrac{1}{ik}$ | (8) |
| Sign | $H(x) - H(-x)$ | $\dfrac{2}{ik}$ | (9) |
| Constant | $1$ | $2\pi\,\delta(k)$ | (10) |
| Gaussian | $e^{-x^2/2}$ | $\sqrt{2\pi}\,e^{-k^2/2}$ | (11) |

Some of these are left to the exercises. Some of them require reinterpretation in terms of distribution theory. The last one (11) is particularly interesting: The transform of a gaussian is again a gaussian! To derive (11), we complete the square in the exponent:

$$F(k) = \int_{-\infty}^{\infty} e^{-x^2/2} e^{-ikx}\,dx = \int_{-\infty}^{\infty} e^{-(x+ik)^2/2}\,dx \cdot e^{+i^2k^2/2}$$

$$= \int_{-\infty}^{\infty} e^{-y^2/2}\,dy \cdot e^{-k^2/2} = \sqrt{2\pi}\,e^{-k^2/2}$$

where $y = x + ik$. This change of variables is not really fair because $ik$ is complex, but it can be justified as a "shift of contours," as is done in any complex analysis course. The last step uses the formula from Exercise 2.4.7.

Another example, which we will need later, is the following one (which can be found on page 406 of [MOS]):

The transform of $\frac{1}{2} J_0(\sqrt{1 - x^2})H(1 - x^2)$ is $\dfrac{\sin\sqrt{k^2 + 1}}{\sqrt{k^2 + 1}}$ (12)

where $J_0$ is the Bessel function of order zero.

## PROPERTIES OF THE FOURIER TRANSFORM

Let $F(k)$ be the transform of $f(x)$ and let $G(k)$ be the transform of $g(x)$. Then we have the following table:

|       | Function | Transform |
|-------|----------|-----------|
| (i)   | $\dfrac{df}{dx}$ | $ikF(k)$ |
| (ii)  | $xf(x)$ | $i\dfrac{dF}{dk}$ |
| (iii) | $f(x-a)$ | $e^{-iak}F(k)$ |
| (iv)  | $e^{iax}f(x)$ | $F(k-a)$ |
| (v)   | $af(x)+bg(x)$ | $aF(k)+bG(k)$ |
| (vi)  | $f(ax)$ | $\dfrac{1}{\lvert a \rvert}F\left(\dfrac{k}{a}\right)$   $(a \neq 0)$ |

### Example 1.

By (iii) the transform of $\delta(x-a)$ is $e^{-iak}$ times the transform of $\delta(x)$. Therefore,

$$\text{the transform of} \quad \tfrac{1}{2}\delta(x+a) + \tfrac{1}{2}\delta(x-a) \quad \text{is} \quad \cos ak. \tag{13}$$

$\square$

Another important property is Parseval's equality (also called *Plancherel's theorem* in this context), which states that

$$\int_{-\infty}^{\infty} |f(x)|^2 dx = \int_{-\infty}^{\infty} |F(k)|^2 \frac{dk}{2\pi}. \tag{14}$$

If one of these integrals is finite, so is the other. Also,

$$\int_{-\infty}^{\infty} f(x)\overline{g(x)}\, dx = \int_{-\infty}^{\infty} F(k)\overline{G(k)}\frac{dk}{2\pi}. \tag{15}$$

## THE HEISENBERG UNCERTAINTY PRINCIPLE

In quantum mechanics $k$ is called the momentum variable and $x$ the position variable. The wave functions $f(x)$ are always normalized, so that $\int_{-\infty}^{\infty} |f(x)|^2 \, dx = 1$. To be precise, let $f(x)$ be a test function. The expected value of the square of the position is $\overline{x}^2 = \int_{-\infty}^{\infty} |xf(x)|^2 \, dx$. The expected value of the square of the momentum is $\overline{k}^2 = \int_{-\infty}^{\infty} |kF(k)|^2 \, dk/2\pi$. The uncertainty principle asserts that

$$\boxed{\overline{x} \cdot \overline{k} \geq \frac{1}{2}.} \tag{16}$$

Thus $\bar{x}$ and $\bar{k}$ can't both be too close to zero. This principle has the interpretation that you can't precisely determine that both the position and the momentum are at the origin. In other words, if you make a very precise measurement of position, then momentum can't be measured with much precision at the same time. Actually, by our choice of units we have omitted the very small Planck's constant, which should appear on the right side of (16).

**Proof of (16).**   By Schwarz's inequality (see Exercise 5.5.2), we have

$$\left| \int_{-\infty}^{\infty} x f(x) f'(x)\, dx \right| \le \left[ \int_{-\infty}^{\infty} |x f(x)|^2\, dx \right]^{1/2} \left[ \int_{-\infty}^{\infty} |f'(x)|^2\, dx \right]^{1/2}. \quad (17)$$

By the definitions of $\bar{x}$ and $\bar{k}$, by property (i) of Fourier transforms, and by Parseval's equality the right side of (17) equals

$$\bar{x} \left[ \int_{-\infty}^{\infty} |ik F(k)|^2 \frac{dk}{2\pi} \right]^{1/2} = \bar{x}\,\bar{k}. \quad (18)$$

On the other hand, integrating the left side of (17) by parts, we get

$$\int_{-\infty}^{\infty} x f(x) f'(x)\, dx = \tfrac{1}{2} x [f(x)]^2 \Big|_{-\infty}^{\infty} - \int_{-\infty}^{\infty} \tfrac{1}{2} [f(x)]^2\, dx = 0 - \tfrac{1}{2} \quad (19)$$

since $f(x)$ is normalized. Therefore, (17) takes the form $\tfrac{1}{2} \le \bar{x}\,\bar{k}$, which is (16).   $\square$

## CONVOLUTION

A useful concept is the convolution of two functions. If $f(x)$ and $g(x)$ are two functions of a real variable, their convolution (written $f * g$) is defined to be

$$(f * g)(x) = \int_{-\infty}^{\infty} f(x - y) g(y)\, dy.$$

We have seen many PDE formulas that are convolutions, such as formula (2.4.6). Its most interesting property is its relationship with the Fourier transform.

If the Fourier transform of $f(x)$ is $F(k)$ and that of $g(x)$ is $G(k)$, the Fourier transform of the convolution $(f * g)(x)$ is the *product* $F(k)G(k)$. To prove this, just observe that the Fourier transform of $f * g$ is

$$\int (f * g)(x) e^{-ikx}\, dx = \iint f(x - y) g(y)\, dy\, e^{-ikx}\, dx.$$

If we switch the order of integration and then substitute $z = x - y$ in the inner $x$ integral, we get

$$\iint f(z) e^{-ik(y+z)}\, dz\, g(y)\, dy = \int f(z) e^{-ikz}\, dz \cdot \int g(y) e^{-iky}\, dy$$
$$= F(k) \cdot G(k).$$

The convolution also plays a prominent role in probability theory.

## THREE DIMENSIONS

In three dimensions the Fourier transform is defined as

$$F(\mathbf{k}) = \int_{-\infty}^{\infty} \int_{-\infty}^{\infty} \int_{-\infty}^{\infty} f(\mathbf{x}) e^{-i\mathbf{k}\cdot\mathbf{x}} \, d\mathbf{x},$$

where $\mathbf{x} = (x, y, z)$, $\mathbf{k} = (k_1, k_2, k_3)$, and $\mathbf{k} \cdot \mathbf{x} = xk_1 + yk_2 + zk_3$. Then one recovers $f(\mathbf{x})$ from the formula

$$f(\mathbf{x}) = \int_{-\infty}^{\infty} \int_{-\infty}^{\infty} \int_{-\infty}^{\infty} F(\mathbf{k}) \, e^{+i\mathbf{k}\cdot\mathbf{x}} \frac{d\mathbf{k}}{(2\pi)^3}.$$

## EXERCISES

1. Verify each entry in the table of Fourier transforms. (Use (15) as needed.)
2. Verify each entry in the table of properties of Fourier transforms.
3. Show that

$$\frac{1}{2\pi^2 cr} \int_0^{\infty} \sin kct \, \sin kr \, dk = \frac{1}{8\pi^2 cr} \int_{-\infty}^{\infty} [e^{ik(ct-r)} - e^{ik(ct+r)}] \, dk$$

$$= \frac{1}{4\pi cr} [\delta(ct - r) - \delta(ct + r)].$$

4. Prove the following properties of the convolution.
   (a) $f * g = g * f$.
   (b) $(f * g)' = f' * g = f * g'$, where $'$ denotes the derivative in one variable.
   (c) $f * (g * h) = (f * g) * h$.
5. (a) Show that $\delta * f = f$ for any distribution $f$, where $\delta$ is the delta function.
   (b) Show that $\delta' * f = f'$ for any distribution $f$, where $'$ is the derivative.
6. Let $f(x)$ be a continuous function defined for $-\infty < x < \infty$ such that its Fourier transform $F(k)$ satisfies

$$F(k) = 0 \qquad \text{for } |k| > \pi.$$

   Such a function is said to be *band-limited*.
   (a) Show that

$$f(x) = \sum_{n=-\infty}^{\infty} f(n) \frac{\sin[\pi(x - n)]}{\pi(x - n)}.$$

   Thus $f(x)$ is completely determined by its values at the integers! We say that $f(x)$ is *sampled* at the integers.
   (b) Let $F(k) = 1$ in the interval $(-\pi, \pi)$ and $F(k) = 0$ outside this interval. Calculate both sides of (a) directly to verify that they are equal.
   (*Hints:* (a) Write $f(x)$ in terms of $F(k)$. Notice that $f(n)$ is the $n$th Fourier coefficient of $F(k)$ on $[-\pi, \pi]$. Deduce that $F(k) = \Sigma f(n)e^{-ink}$

in $[-\pi, \pi]$. Substitute this back into $f(x)$, and then interchange the integral with the series.)

7.  (a)  Let $f(x)$ be a continuous function on the line $(-\infty, \infty)$ that vanishes for large $|x|$. Show that the function

$$g(x) = \sum_{n=-\infty}^{\infty} f(x + 2\pi n)$$

is periodic with period $2\pi$.

(b)  Show that the Fourier coefficients $c_m$ of $g(x)$ on the interval $(-\pi, \pi)$ are $F(m)/2\pi$, where $F(k)$ is the Fourier transform of $f(x)$.

(c)  In the Fourier *series* of $g(x)$ on $(-\pi, \pi)$, let $x = 0$ to obtain the *Poisson summation formula*

$$\sum_{n=-\infty}^{\infty} f(2\pi n) = \sum_{n=-\infty}^{\infty} \frac{1}{2\pi} F(n).$$

8.  Let $\chi_a(x)$ be the function in Exercise 12.1.12. Compute its Fourier transform $\hat{\chi}_a(k)$. Use it to show that $\hat{\chi}_a \to 1$ weakly as $a \to 0$.

9.  Use Fourier transforms to solve the ODE $-u_{xx} + a^2 u = \delta$, where $\delta = \delta(x)$ is the delta function.

## 12.4  SOURCE FUNCTIONS

In this section we show how useful the Fourier transform can be in finding the source function of a PDE *from scratch*.

### DIFFUSION

The source function is properly defined as the unique solution of the problem

$$S_t = S_{xx} \quad (-\infty < x < \infty, \quad 0 < t < \infty), \qquad S(x, 0) = \delta(x) \quad (1)$$

where we have taken the diffusion constant to be 1. Let's assume no knowledge at all about the form of $S(x, t)$. We only assume it has a Fourier transform as a distribution in $x$, for each $t$. Call its transform

$$\hat{S}(k, t) = \int_{-\infty}^{\infty} S(x, t) e^{-ikx} \, dx.$$

(Here $k$ denotes the frequency variable, not the diffusion constant.) By property (i) of Fourier transforms, the PDE takes the form

$$\frac{\partial \hat{S}}{\partial t} = (ik)^2 \hat{S} = -k^2 \hat{S}, \qquad \hat{S}(k, 0) = 1. \quad (2)$$

For each $k$ this is an ODE that is easy to solve. The solution is

$$\hat{S}(k, t) = e^{-k^2 t}. \quad (3)$$

All we have to do is find the function with this transform. The variable $t$ is now fixed. Our table shows that the transform of $f(x) = e^{-(1/2)x^2}/\sqrt{2\pi}$ is $F(k) = e^{-(1/2)k^2}$. By property (vi), the transform of $e^{-(1/2)a^2x^2}/\sqrt{2\pi}$ is $(1/a)e^{-(1/2)k^2/a^2}$ for any $a > 0$. Choosing $a = 1/\sqrt{2t}$, we find that the transform of $e^{-x^2/4t}/\sqrt{2\pi}$ is $\sqrt{2t}\,e^{-k^2t}$. Therefore, the transform of $1/\sqrt{4\pi t}\,e^{-x^2/4t}$ is $e^{-k^2t}$, so that $S(x, t) = 1/\sqrt{4\pi t}\,e^{-x^2/4t}$. This result agrees with Section 2.4.

## WAVES

By definition, the source function for the *one*-dimensional wave equation satisfies

$$S_{tt} = c^2 S_{xx}, \qquad S(x, 0) = 0 \qquad S_t(x, 0) = \delta(x). \tag{4}$$

The same method used for diffusions now leads to

$$\frac{\partial^2 \hat{S}}{\partial t^2} = -c^2 k^2 \hat{S}, \qquad \hat{S}(k, 0) = 0, \qquad \frac{\partial \hat{S}}{\partial t}(k, 0) = 1. \tag{5}$$

This ODE has the solution

$$\hat{S}(k, t) = \frac{1}{kc} \sin kct = \frac{e^{ikct} - e^{-ikct}}{2ikc}. \tag{6}$$

Therefore,

$$S(x, t) = \int_{-\infty}^{\infty} \frac{e^{ik(x+ct)} - e^{ik(x-ct)}}{4\pi i kc} \, dk. \tag{7}$$

Now, according to our table, the transform of $\operatorname{sgn}(x) \equiv H(x) - H(-x)$ is $2/ik$. By property (iii) of Fourier transforms, the transform of $\operatorname{sgn}(x + a)/4c$ is $e^{iak}/2ikc$. Therefore, from either (6) or (7) for $t > 0$,

$$S(x, t) = \frac{\operatorname{sgn}(x + ct) - \operatorname{sgn}(x - ct)}{4c} = \begin{cases} (1 - 1)/4c = 0 & \text{for } |x| > ct > 0 \\ (1 + 1)/4c = 1/2c & \text{for } |x| < ct, \end{cases}$$

$$\boxed{S(x, t) = \frac{H(c^2t^2 - x^2)}{2c}.} \tag{8}$$

In *three* dimensions the source function has a (three-dimensional) Fourier transform $\hat{S}(\mathbf{k}, t)$ which satisfies

$$\frac{\partial^2 \hat{S}}{\partial t^2} = -c^2(k_1^2 + k_2^2 + k_3^2)\hat{S}, \qquad \hat{S}(\mathbf{k}, 0) = 0, \qquad \frac{\partial \hat{S}}{\partial t}(\mathbf{k}, 0) = 1,$$

where $\mathbf{k} = (k_1, k_2, k_3)$. Letting $k^2 = |\mathbf{k}|^2 = (k_1^2 + k_2^2 + k_3^2)$, the solution of the ODE is (6) again. Therefore,

$$S(\mathbf{x}, t) = \iiint \frac{1}{kc} \sin kct \; e^{i\mathbf{k} \cdot \mathbf{x}} \frac{d\mathbf{k}}{8\pi^3}. \tag{9}$$

This integral can be calculated conveniently using spherical coordinates in the **k** variables. We choose the polar axis (the "z-axis") to be along the **x** direction and denote the spherical coordinates by $k, \theta,$ and $\phi$. Also let $r = |\mathbf{x}|$. Then $\mathbf{k} \cdot \mathbf{x} = kr \cos\theta$ and

$$S(\mathbf{x}, t) = \int_0^{2\pi} \int_0^{\pi} \int_0^{\infty} (kc)^{-1} \sin kct \, e^{ikr \cos\theta} k^2 \sin\theta \, \frac{dk \, d\theta \, d\phi}{8\pi^3}. \tag{10}$$

The $\phi$ and $\theta$ integrals can be exactly evaluated to get

$$\frac{1}{2\pi^2 cr} \int_0^{\infty} \sin kct \sin kr \, dk.$$

Writing in terms of complex exponentials and switching $k$ to $-k$ in some terms, we get (by Exercise 12.3.3)

$$\frac{1}{8\pi^2 cr} \int_{-\infty}^{\infty} [e^{ik(ct-r)} - e^{ik(ct+r)}] \, dk = \frac{1}{4\pi cr} [\delta(ct - r) - \delta(ct + r)]. \tag{11}$$

Notice how the characteristic variables show up again! For $t > 0$, we have $ct + r > 0$, so that $\delta(ct + r) = 0$. Therefore,

$$\boxed{S(\mathbf{x}, t) = \frac{1}{4\pi cr} \delta(ct - r) = \frac{1}{4\pi c^2 t} \delta(ct - r)} \tag{12}$$

for $t > 0$, in agreement with our previous answer.

## LAPLACE'S EQUATION IN A HALF-PLANE

We use the Fourier transform to rework the problem of Section 7.4,

$$\begin{aligned} u_{xx} + u_{yy} &= 0 & &\text{in the half-plane } y > 0, \\ u(x, 0) &= \delta(x) & &\text{on the line } y = 0. \end{aligned} \tag{13}$$

We cannot transform the $y$ variable, but can transform $x$ because it runs from $-\infty$ to $\infty$. Let

$$U(k, y) = \int_{-\infty}^{\infty} e^{-ikx} u(x, y) \, dx \tag{14}$$

be the Fourier transform. Then $U$ satisfies the ODE

$$-k^2 U + U_{yy} = 0 \quad \text{for } y > 0, \qquad U(k, 0) = 1. \tag{15}$$

The solutions of the ODE are $e^{\pm yk}$. We must reject a positive exponent because $U$ would grow exponentially as $|k| \to \infty$ and would not have a Fourier transform. So $U(k, y) = e^{-y|k|}$. Therefore,

$$u(x, y) = \int_{-\infty}^{\infty} e^{ikx} e^{-y|k|} \frac{dk}{2\pi}. \tag{16}$$

This improper integral clearly converges for $y > 0$. It is split into two parts and integrated directly as

$$
u(x, y) = \left. \frac{1}{2\pi(ix - y)} e^{ikx-ky} \right|_0^\infty + \left. \frac{1}{2\pi(ix + y)} e^{ikx+ky} \right|_{-\infty}^0
$$

$$
= \frac{1}{2\pi}\left( \frac{1}{y - ix} + \frac{1}{y + ix} \right) = \frac{y}{\pi(x^2 + y^2)},
$$

(17)

in agreement with Exercise 7.4.6.

## EXERCISES

1. Use the Fourier transform directly to solve the heat equation with a convection term, namely, $u_t = \kappa u_{xx} + \mu u_x$ for $-\infty < x < \infty$, with an initial condition $u(x, 0) = \phi(x)$, assuming that $u(x, t)$ is bounded and $\kappa > 0$.

2. Use the Fourier transform in the $x$ variable to find the harmonic function in the half-plane $\{y > 0\}$ that satisfies the Neumann condition $\partial u/\partial y = h(x)$ on $\{y = 0\}$.

3. Use the Fourier transform to find the bounded solution of the equation $-\Delta u + m^2 u = \delta(\mathbf{x})$ in free three-dimensional space with $m > 0$.

4. If $p(x)$ is a polynomial and $f(x)$ is any continuous function on the interval $[a, b]$, show that $g(x) = \int_a^b p(x - s)f(s)\,ds$ is also a polynomial.

5. In the three-dimensional half-space $\{(x, y, z): z > 0\}$, solve the Laplace equation with $u(x, y, 0) = \delta(x, y)$, where $\delta$ denotes the delta function, as follows.
   (a) Show that
   $$
   u(x, y, z) = \int_{-\infty}^\infty \int_{-\infty}^\infty e^{ikx+ily} e^{-z\sqrt{k^2+l^2}} \frac{dk\,dl}{4\pi^2}.
   $$
   (b) Letting $\rho = \sqrt{k^2 + l^2}$, $r = \sqrt{x^2 + y^2}$, and $\theta$ be the angle between $(x, y)$ and $(k, l)$, so that $xk + yl = \rho r \cos\theta$, show that
   $$
   u(x, y, z) = \int_0^{2\pi} \int_0^\infty e^{i\rho r\cos\theta} e^{-z\rho} \rho\,d\rho \frac{d\theta}{4\pi^2}.
   $$
   (c) Carry out the integral with respect to $\rho$ and then use an extensive table of integrals to evaluate the $\theta$ integral.

6. Use the Fourier transform to solve $u_{xx} + u_{yy} = 0$ in the infinite strip $\{0 < y < 1, -\infty < x < \infty\}$, together with the conditions $u(x, 0) = 0$ and $u(x, 1) = f(x)$.

## 12.5 LAPLACE TRANSFORM TECHNIQUES

The Laplace transform is a close relative of the Fourier transform. In this section, however, we apply it to the *time* rather than the space variable. It allows us to solve some PDE problems in a very simple way.

For a function $f(t)$ we define its *Laplace transform* as

$$F(s) = \int_0^\infty f(t)\, e^{-st}\, dt.$$

(1)

(The only essential differences from the Fourier transform are that the exponential is real and that the variable is the time.) For instance, the Laplace transform of the function $f(t) \equiv 1$ is $F(s) = \int_0^\infty 1 \cdot e^{-st}\, dt = 1/s$ for $s > 0$. If $f(t)$ is any bounded function, then $F(s)$ is defined for $s > 0$.

Following is a table of Laplace transforms.

| $f(t)$ | $F(s)$ | |
|---|---|---|
| $e^{at}$ | $\dfrac{1}{s-a}$ | (2) |
| $\cos \omega t$ | $\dfrac{s}{s^2 + \omega^2}$ | (3) |
| $\sin \omega t$ | $\dfrac{\omega}{s^2 + \omega^2}$ | (4) |
| $\cosh at$ | $\dfrac{s}{s^2 - a^2}$ | (5) |
| $\sinh at$ | $\dfrac{a}{s^2 - a^2}$ | (6) |
| $t^k$ | $\dfrac{k!}{s^{k+1}}$ | (7) |
| $H(t - b)$ | $\dfrac{1}{s} e^{-bs}$ | (8) |
| $\delta(t - b)$ | $e^{-bs}$ | (9) |
| $a(4\pi t^3)^{-1/2} e^{-a^2/4t}$ | $e^{-a\sqrt{s}}$ | (10) |
| $(\pi t)^{-1/2} e^{-a^2/4t}$ | $\dfrac{1}{\sqrt{s}} e^{-a\sqrt{s}}$ | (11) |
| $1 - \mathscr{E}rf\dfrac{a}{\sqrt{4t}}$ | $\dfrac{1}{s} e^{-a\sqrt{s}}$ | (12) |

Here are some *properties* of the Laplace transform. Let $F(s)$ and $G(s)$ be the Laplace transforms of $f(t)$ and $g(t)$, respectively. Then we have the following table of properties.

|       | Function | Transform |
|-------|----------|-----------|
| (i)   | $af(t) + bg(t)$ | $aF(s) + bG(s)$ |
| (ii)  | $\dfrac{df}{dt}$ | $sF(s) - f(0)$ |
| (iii) | $\dfrac{d^2 f}{dt^2}$ | $s^2 F(s) - sf(0) - f'(0)$ |
| (iv)  | $e^{bt} f(t)$ | $F(s - b)$ |
| (v)   | $\dfrac{f(t)}{t}$ | $\displaystyle\int_s^\infty F(s')\,ds'$ |
| (vi)  | $tf(t)$ | $-\dfrac{dF}{ds}$ |
| (vii) | $H(t - b)f(t - b)$ | $e^{-bs} F(s)$ |
| (viii)| $f(ct)$ | $\dfrac{1}{c} F\left(\dfrac{s}{c}\right)$ |
| (ix)  | $\displaystyle\int_0^t g(t - t')f(t')\,dt'$ | $F(s)G(s)$ |

The last property says that the transform of the "convolution" is the product of the transforms.

## Example 1.

By (4) and (v), the transform of $(\sin t)/t$ is

$$\int_s^\infty \frac{ds'}{s'^2 + 1} = \frac{\pi}{2} - \tan^{-1} s = \tan^{-1} \frac{1}{s}. \qquad \square$$

Complex integration together with residue calculus is a useful technique for computing Laplace transforms, but it goes beyond the scope of this book. We limit ourselves to writing down the *inversion formula*

$$f(t) = \int_{\alpha - i\infty}^{\alpha + i\infty} e^{st} F(s) \frac{ds}{2\pi i}. \tag{13}$$

This is an integral over the vertical line $s = \alpha + i\beta$ in the complex plane, where $-\infty < \beta < \infty$. For further information on Laplace transforms, see [We], for instance.

## Example 2.

Here is an efficient way of solving the simple ODE

$$u_{tt} + \omega^2 u = f(t) \qquad \text{with } u(0) = u'(0) = 0.$$

By the properties on page 354, the Laplace transform $U(s)$ satisfies

$$s^2 U(s) + \omega^2 U(s) = F(s).$$

Hence $U(s) = F(s)/(s^2 + \omega^2)$. Now the Laplace transform of $\sin \omega t$ is $\omega/(s^2 + \omega^2)$. So we use the last property (ix) in the table to get the solution $u(t)$ expressed as the convolution

$$u(t) = \int_0^t \frac{1}{\omega} \sin[\omega(t - t')] f(t') \, dt'. \qquad \square$$

Now we illustrate the applications of the Laplace transform to PDEs for some one-dimensional problems. Although each of them can be solved in other ways, the Laplace transform provides an easy alternative method. It is a particularly useful technique to handle an inhomogeneous boundary condition, providing an alternative approach to the problems of Section 5.6. We begin with a really simple inhomogeneous example.

### Example 3.

Solve the diffusion equation $u_t = ku_{xx}$ in $(0, l)$, with the conditions

$$u(0, t) = u(l, t) = 1, \qquad u(x, 0) = 1 + \sin \frac{\pi x}{l}.$$

Using properties (i) and (ii) and noting that the partials with respect to $x$ commute with the transforms with respect to $t$, the Laplace transform $U(x, s)$ satisfies

$$sU(x, s) - u(x, 0) = kU_{xx}(x, s).$$

The boundary conditions become $U(0, s) = U(l, s) = 1/s$, using for instance (2) with $a = 0$. So we have an ODE in the variable $x$ together with some boundary conditions. The solution is easily seen to be

$$U(x, s) = \frac{1}{s} + \frac{1}{s + k\pi^2/l^2} \sin \frac{\pi x}{l}.$$

(Check it!) As a function of $s$, this expression has the form $s^{-1} + b(s - a)^{-1}$ where $b = \sin(\pi x/l)$ and $a = -k\pi^2/l^2$. Therefore, the first entry (2) in our table of Laplace transforms yields the answer

$$u(x, t) = 1 + be^{at} = 1 + e^{-k\pi^2 t/l^2} \sin \frac{\pi x}{l}. \qquad \square$$

### Example 4.

Solve the wave equation $u_{tt} = c^2 u_{xx}$ for $0 < x < \infty$, with the conditions

$$u(0, t) = f(t), \qquad u(x, 0) = u_t(x, 0) \equiv 0.$$

For $x \to +\infty$ we assume that $u(x, t) \to 0$. Because the initial conditions vanish, the Laplace transform satisfies

$$s^2 U = c^2 U_{xx}, \qquad U(0, s) = F(s).$$

Solving this ODE, we get

$$U(x, s) = a(s)e^{-sx/c} + b(s)e^{+sx/c},$$

where $a(s)$ and $b(s)$ are to be determined. From the assumed property of $u$, we expect that $U(x, s) \to 0$ as $x \to +\infty$. Therefore $b(s) \equiv 0$. Hence $U(x, s) = F(s)e^{-sx/c}$. Now we use property (vii) directly to get the answer

$$u(x, t) = H\left(t - \frac{x}{c}\right) f\left(t - \frac{x}{c}\right).$$

For another method, see (3.4.19).    □

## Example 5.

Solve the diffusion equation $u_t = ku_{xx}$ for $0 < x < \infty$, with the conditions

$$u(0, t) = f(t), \qquad u(x, 0) \equiv 0$$

with $u(x, t) \to 0$ as $x \to +\infty$. The Laplace transform satisfies

$$sU = kU_{xx}, \qquad U(0, s) = F(s), \qquad U(+\infty) = 0.$$

Its solution is $U(x, s) = F(s)e^{-x\sqrt{s/k}}$ because the positive exponent is not allowed. By (10), the function $e^{-x\sqrt{s/k}}$ is the Laplace transform of $E(x, t) = [x/(2\sqrt{k\pi}\, t^{3/2})]\, e^{-x^2/4kt}$. The convolution property (ix) then states that

$$u(x, t) = \int_0^t E(t - t')f(t')\, dt'$$

$$= \frac{x}{2\sqrt{k\pi}} \int_0^t \frac{1}{(t - t')^{3/2}} e^{-x^2/4k(t-t')} f(t')\, dt'.$$

This is the solution formula.

For instance, suppose that $f(t) \equiv 1$. Then the last integral can be simplified by substituting $p = x[4k(t - t')]^{-1/2}$ to obtain the solution

$$u(x, t) = \frac{2}{\sqrt{\pi}} \int_{x/\sqrt{4kt}}^{\infty} e^{-p^2}\, dp = 1 - \mathscr{E}rf \frac{x}{\sqrt{4kt}}.$$

For another method, see Exercise 3.3.2.    □

## EXERCISES

1. Verify the entries (2)–(9) in the table of Laplace transforms.
2. Verify each entry in the table of properties of the Laplace transform.
3. Find $f(t)$ if its Laplace transform is $F(s) = 1/[s(s^2 + 1)]$.

4. Show that the Laplace transform of $t^k$ is $\Gamma(k+1)/s^{k+1}$ for any $k > -1$, where $\Gamma(p)$ is the gamma function. (*Hint:* Use property (viii) of the Laplace transform.)

5. Use the Laplace transform to solve $u_{tt} = c^2 u_{xx}$ for $0 < x < l$, $u(0, t) = u(l, t) = 0$, $u(x, 0) = \sin(\pi x/l)$, and $u_t(x, 0) = -\sin(\pi x/l)$.

6. Use the Laplace transform to solve

$$u_{tt} = c^2 u_{xx} + \cos \omega t \sin \pi x \qquad \text{for } 0 < x < 1$$

$$u(0, t) = u(1, t) = u(x, 0) = u_t(x, 0) = 0.$$

Assume that $\omega > 0$ and be careful of the case $\omega = c\pi$. Check your answer by direct differentiation.

7. Use the Laplace transform to solve $u_t = k u_{xx}$ in $(0, l)$, with $u_x(0, t) = 0$, $u_x(l, t) = 0$, and $u(x, 0) = 1 + \cos(2\pi x/l)$.

# 13

# PDE PROBLEMS FROM PHYSICS

This chapter contains five independent sections. Section 13.1 requires knowledge of Section 9.2, Section 13.2 requires only Chapter 1, part of Section 13.3 requires Section 10.3, Section 13.4 requires Section 9.5, and part of Section 13.5 requires Chapter 12.

## 13.1 ELECTROMAGNETISM

Electromagnetism describes the effects of charged particles on each other. Charged particles create an electric field $\mathbf{E}$, whereas moving ones also create a magnetic field $\mathbf{B}$. These fields are vector fields, that is, vector functions of space and time: $\mathbf{E}(\mathbf{x}, t)$ and $\mathbf{B}(\mathbf{x}, t)$, where $\mathbf{x} = (x, y, z)$. Maxwell proposed that these two functions are universally governed by the equations

$$\text{(I)} \quad \frac{\partial \mathbf{E}}{\partial t} = c\nabla \times \mathbf{B} \qquad \text{(III)} \quad \nabla \cdot \mathbf{E} = 0 \tag{1}$$

$$\text{(II)} \quad \frac{\partial \mathbf{B}}{\partial t} = -c\nabla \times \mathbf{E} \qquad \text{(IV)} \quad \nabla \cdot \mathbf{B} = 0,$$

where $c$ is the speed of light, at least in a vacuum. That is, these equations exactly govern the propagation of electromagnetic radiation such as light, radio waves, and so on, in the absence of any interference. The Maxwell equations are the subject of this section. Notice that there are two vector equations and two scalar equations.

In the presence of interference, the physics is governed instead by the *inhomogeneous* Maxwell equations

$$\text{(I)} \quad \frac{\partial \mathbf{E}}{\partial t} = c\nabla \times \mathbf{B} - 4\pi \mathbf{J} \qquad \text{(III)} \quad \nabla \cdot \mathbf{E} = 4\pi \rho \tag{2}$$

$$\text{(II)} \quad \frac{\partial \mathbf{B}}{\partial t} = -c\nabla \times \mathbf{E} \qquad \text{(IV)} \quad \nabla \cdot \mathbf{B} = 0,$$

where $\rho(\mathbf{x}, t)$ is the charge density and $\mathbf{J}(\mathbf{x}, t)$ is the current density. The equations (2) imply the *continuity equation* $\partial\rho/\partial t = \nabla \cdot \mathbf{J}$ (see Exercise 1).

Special cases of the Maxwell equations are well known by themselves. (II) is called Faraday's law. (III) is called Coulomb's law. In case $\mathbf{E}$ does not depend on time, (I) reduces to $c\nabla \times \mathbf{B} = 4\pi\mathbf{J}$, which is Ampère's law. The trouble with Ampère's law is that it requires $\nabla \cdot \mathbf{J} = 0$. So if $\nabla \cdot \mathbf{J} \neq 0$, Maxwell proposed adding the term $\partial\mathbf{E}/\partial t$ for the sake of mathematical consistency, thereby coming up with his complete set of equations (2).

As we know, inhomogeneous linear equations can be solved once the homogeneous ones are. So we concentrate on solving the homogeneous equations (1). To (1) must be adjoined some initial conditions. They are

$$\mathbf{E}(\mathbf{x}, 0) = \mathbf{E}^0(\mathbf{x}) \qquad \mathbf{B}(\mathbf{x}, 0) = \mathbf{B}^0(\mathbf{x}). \tag{3}$$

The two vector fields $\mathbf{E}^0(\mathbf{x})$ and $\mathbf{B}^0(\mathbf{x})$ are arbitrary except for the obvious restriction that $\nabla \cdot \mathbf{E}^0 = \nabla \cdot \mathbf{B}^0 = 0$, which comes from (III) and (IV).

Our main goal is to solve (1) and (3) in all of 3-space, that is, without boundary conditions. This can be done very simply by reduction to the wave equation, already solved in Chapter 9.

## SOLUTION OF (1),(3)

First notice that $\mathbf{E}$ satisfies the wave equation. In fact, from (I) and then (II), we have

$$\frac{\partial^2\mathbf{E}}{\partial t^2} = \frac{\partial}{\partial t}(c\nabla \times \mathbf{B}) = c\nabla \times \frac{\partial\mathbf{B}}{\partial t} = c\nabla \times (-c\nabla \times \mathbf{E}).$$

By a standard vector identity, this equals $c^2(\Delta\mathbf{E} - \nabla(\nabla \cdot \mathbf{E})) = c^2\Delta\mathbf{E}$, because $\nabla \cdot \mathbf{E} = 0$ in the equations (1). Thus

$$\frac{\partial^2\mathbf{E}}{\partial t^2} = c^2\Delta\mathbf{E}, \tag{4}$$

which means that each component of $\mathbf{E} = (E_1, E_2, E_3)$ satisfies the ordinary wave equation. Similarly for the magnetic field:

$$\frac{\partial^2\mathbf{B}}{\partial t^2} = c^2\Delta\mathbf{B}. \tag{5}$$

Now $\mathbf{E}$ satisfies the initial conditions

$$\mathbf{E}(\mathbf{x}, 0) = \mathbf{E}^0(\mathbf{x}) \qquad \text{and} \qquad \frac{\partial\mathbf{E}}{\partial t}(\mathbf{x}, 0) = c\nabla \times \mathbf{B}^0(\mathbf{x}). \tag{6}$$

Similarly, $\mathbf{B}$ satisfies the initial conditions

$$\mathbf{B}(\mathbf{x}, 0) = \mathbf{B}^0(\mathbf{x}) \qquad \text{and} \qquad \frac{\partial\mathbf{B}}{\partial t}(\mathbf{x}, 0) = -c\nabla \times \mathbf{E}^0(\mathbf{x}). \tag{7}$$

We now show that any solution of (4)–(7) also satisfies our problem (1),(3).

Assuming (4)–(7), it is obvious that the initial conditions (3) are satisfied, so we just need to check (1). Let's begin with equation (III). Let $u = \nabla \cdot \mathbf{E}$. We know that $(\partial^2/\partial t^2 - \Delta)u = \nabla \cdot (\partial^2/\partial t^2 - \Delta)\mathbf{E} = 0$. So the scalar $u(\mathbf{x}, t)$ satisfies the wave equation and the initial conditions $u(\mathbf{x}, 0) = \nabla \cdot \mathbf{E}^0 = 0$ (by assumption) and $\partial u/\partial t(\mathbf{x}, 0) = \nabla \cdot (c\nabla \times \mathbf{B}^0) = 0$ (because the divergence of the curl of any vector field is zero). By the uniqueness of solutions of the wave equation, $u(\mathbf{x}, t) \equiv 0$. This proves (III).

How about (I)? Let $\mathbf{F} = \partial \mathbf{E}/\partial t - c\nabla \times \mathbf{B}$. Then $(\partial^2/\partial t^2 - \Delta)\mathbf{F} = 0$ because both $\mathbf{E}$ and $\mathbf{B}$ satisfy the wave equation. The initial conditions of $\mathbf{F}(\mathbf{x}, t)$ are $\mathbf{F}(\mathbf{x}, 0) = c\nabla \times \mathbf{B}^0 - c\nabla \times \mathbf{B}^0 = 0$ and

$$\frac{\partial \mathbf{F}}{\partial t}(\mathbf{x}, 0) = \frac{\partial^2 \mathbf{E}}{\partial t^2} - c\nabla \times \left.\frac{\partial \mathbf{B}}{\partial t}\right|_{t=0}$$

$$= c^2 \Delta \mathbf{E}^0 - c\nabla \times (-c\nabla \times \mathbf{E}^0)$$

$$= c^2 \nabla(\nabla \cdot \mathbf{E}^0) = 0$$

by yet another vector identity. Therefore, $\mathbf{F}(\mathbf{x}, t) \equiv 0$. This proves (I). Equations (II) and (IV) are left as an exercise.

Now we may solve (4) and (6) for $\mathbf{E}$ separately. Each of the three components of $\mathbf{E}(\mathbf{x}, t)$ separately satisfies the ordinary wave equation with an initial condition. So we may apply the formula (9.2.3) to get

$$\mathbf{E}(\mathbf{x}_0, t_0) = \frac{1}{4\pi c^2 t_0} \iint_S c\nabla \times \mathbf{B}^0 \, dS + \frac{\partial}{\partial t_0} \frac{1}{4\pi c^2 t_0} \iint_S \mathbf{E}^0 \, dS,$$

where $S = \{|\mathbf{x} - \mathbf{x}_0| = ct_0\}$ is the sphere of center $\mathbf{x}_0$ and radius $ct_0$. We carry out the time derivative in the last term by using spherical coordinates. We get

$$\mathbf{E}(\mathbf{x}_0, t_0) = \frac{1}{4\pi ct_0} \iint_S \left( \nabla \times \mathbf{B}^0 + \frac{1}{ct_0}\mathbf{E}^0 + \frac{\partial \mathbf{E}^0}{\partial r} \right) dS, \qquad (8)$$

where $r = |\mathbf{x} - \mathbf{x}_0|$. Similarly, we get

$$\mathbf{B}(\mathbf{x}_0, t_0) = \frac{1}{4\pi ct_0} \iint_S \left( -\nabla \times \mathbf{E}^0 + \frac{1}{ct_0}\mathbf{B}^0 + \frac{\partial \mathbf{B}^0}{\partial r} \right) dS. \qquad (9)$$

Formulas (8) and (9) are the solution of (1),(3). For further discussion, see [Ja] or [Fd].

## EXERCISES

1. Derive the continuity equation $\partial\rho/\partial t + \nabla \cdot \mathbf{J} = 0$ from the inhomogeneous Maxwell equations.

2. Derive the equations of electrostatics from the Maxwell equations by assuming that $\partial\mathbf{E}/\partial t = \partial\mathbf{B}/\partial t \equiv 0$.

3. From $\nabla \cdot \mathbf{B} = 0$ it follows that there exists a vector function $\mathbf{A}$ such that $\nabla \times \mathbf{A} = \mathbf{B}$. This is a well-known fact in vector analysis; see [EP], [Kr], [Sg1].
   (a) Show from Maxwell's equations that there also exists a scalar function $u$ such that $-\nabla u = \mathbf{E} + c^{-1}\partial\mathbf{A}/\partial t$.
   (b) Deduce from (2) that

   $$-c^{-1}\nabla \cdot \frac{\partial\mathbf{A}}{\partial t} - \Delta u = 4\pi\rho$$

   and    $$\frac{1}{c^2}\frac{\partial^2\mathbf{A}}{\partial t^2} - \Delta\mathbf{A} + \nabla\left(\nabla \cdot \mathbf{A} + c^{-1}\frac{\partial u}{\partial t}\right) = \frac{4\pi}{c}\mathbf{J}.$$

   (c) Show that if $\mathbf{A}$ is replaced by $\mathbf{A} + \nabla\lambda$ and $u$ by $u - (1/c)\partial\lambda/\partial t$, then the equations in parts (a) and (b) are still valid for the new $\mathbf{A}$ and the new $u$. This property is called *gauge invariance*.
   (d) Show that the scalar function $\lambda$ may be chosen so that the new $\mathbf{A}$ and the new $u$ satisfy $\nabla \cdot \mathbf{A} + c^{-1}\partial u/\partial t = 0$.
   (e) Conclude that the new potentials satisfy

   $$\frac{1}{c^2}\frac{\partial^2 u}{\partial t^2} - \Delta u = 4\pi\rho \qquad \text{and} \qquad \frac{1}{c^2}\frac{\partial^2\mathbf{A}}{\partial t^2} - \Delta\mathbf{A} = \frac{4\pi}{c}\mathbf{J}.$$

   $\mathbf{A}$ is called the *vector potential* and $u$ the *scalar potential*. The equations in part (e) are inhomogeneous wave equations. The transformation in part (c) is the simplest example of a *gauge transformation*.

4. Show that each component of $\mathbf{E}$ and of $\mathbf{B}$ satisfies the wave equation.

5. Derive carefully the formulas (8) and (9) for the solution of Maxwell's equations.

6. Prove that (II) and (IV) follow from the solution formulas (8)-(9).

7. Prove that (3) follows directly from (8)-(9).

8. Solve the inhomogeneous Maxwell equations.

## 13.2  FLUIDS AND ACOUSTICS

We shall model a fluid (gas or liquid) by its velocity field. That is, $\mathbf{v}(\mathbf{x}, t)$ is the velocity of the fluid at the point $\mathbf{x}$ at the time $t$. Another basic quantity is the mass density $\rho(\mathbf{x}, t)$, which is a scalar. This will lead us to the eulerian form of the fluid equations.

## FLUIDS

The first equation is merely the *conservation of mass*. Take any region $D$. The amount of mass within $D$ at time $t$ is $M(t) = \iiint_D \rho \, d\mathbf{x}$. Fluid can exit the region only through the boundary. The rate of exiting in the unit outward normal direction $\mathbf{n}$ at a point is $\rho\mathbf{v} \cdot \mathbf{n}$. Thus

$$\iiint_D \frac{\partial \rho}{\partial t} \, d\mathbf{x} = \frac{d}{dt} \iiint_D \rho \, d\mathbf{x} = -\iint_{\text{bdy } D} \rho\mathbf{v} \cdot \mathbf{n} \, dS. \tag{1}$$

The minus sign indicates that the mass within $D$ is decreasing if the fluid is escaping from $D$. By the divergence theorem, the last expression is $-\iiint_D \nabla \cdot (\rho\mathbf{v}) \, d\mathbf{x}$. Because this result is valid for all regions $D$, we may apply the second vanishing theorem in Section A.1 to deduce that

$$\boxed{\frac{\partial \rho}{\partial t} = -\nabla \cdot (\rho\mathbf{v}).} \tag{2}$$

This is the *equation of continuity*.

Next we balance the forces on the portion of fluid in $D$. This is Newton's law of motion or the *conservation of momentum*. Balancing the momentum in the same way that we balanced the mass, we get

$$\frac{d}{dt} \iiint_D \rho v_i \, d\mathbf{x} + \iint_{\text{bdy } D} \rho v_i \mathbf{v} \cdot \mathbf{n} \, dS + \iint_{\text{bdy } D} p n_i \, dS = \iiint_D \rho F_i \, d\mathbf{x}, \tag{3}$$

where $p(\mathbf{x}, t)$ is the pressure and $\mathbf{F}(\mathbf{x}, t)$ is the totality of "external" forces at the point $\mathbf{x}$. The index $i$ runs over the three components. The first term in (3) is the rate of change of momentum, the second term is the flux of momentum across the boundary, the third term is the net pressure at the boundary, and the fourth term is the net external force. Applying the divergence theorem to the second and third terms, we get

$$\iiint_D \left[ \frac{\partial(\rho v_i)}{\partial t} + \nabla \cdot (\rho v_i \mathbf{v}) + \frac{\partial p}{\partial x_i} - \rho F_i \right] d\mathbf{x} = 0. \tag{4}$$

Because $D$ is arbitrary, the last integrand vanishes. Carrying out the derivatives in this integrand, we get

$$\rho \left[ \frac{\partial v_i}{\partial t} + \mathbf{v} \cdot \nabla v_i \right] + v_i \left[ \frac{\partial \rho}{\partial t} + \nabla \cdot (\rho\mathbf{v}) \right] = -\frac{\partial p}{\partial x_i} + \rho F_i.$$

But the second term in brackets vanishes because of (2). So we end up with the *equation of motion*

$$\frac{\partial \mathbf{v}}{\partial t} + (\mathbf{v} \cdot \nabla)\mathbf{v} = \mathbf{F} - \frac{1}{\rho}\nabla p. \tag{5}$$

Finally, we need an equation for the pressure $p(\mathbf{x}, t)$. This usually takes the form

$$p(\mathbf{x}, t) = f(\rho(\mathbf{x}, t)), \tag{6}$$

where $f$ is some empirically determined, increasing function. This is the *equation of state*. For a gas the equation of state is often taken to be $p = c\rho^{\gamma}$, where $c$ and $\gamma$ are constants. In this case the entropy is a constant and the fluid flow is called *adiabatic*.

The fluid equations are the equations of continuity, motion, and state. They form a system of five scalar equations for the five scalar unknowns $\rho$, $\mathbf{v}$, and $p$. In contrast to Maxwell's equations, they are highly nonlinear and therefore very difficult to analyze. We all know how turbulent a fluid can become, and this turbulence is a consequence of the nonlinear character of the equations.

The equation of motion (5) was first derived by Leonhard Euler in 1752. In case the fluid is viscous, there are internal forces of the form $\mathbf{F} = \nu\,\nabla\mathbf{v}$, where $\nu > 0$ represents the strength of the viscosity. This viscous equation, together with (2), was derived in 1821 and is called the Navier-Stokes equation (NS). The mathematical properties of the Euler and the NS equations are related to the turbulence of the fluid and are still only partly understood. Figure 1 is a photograph of a jet becoming increasingly turbulent from left to right. The

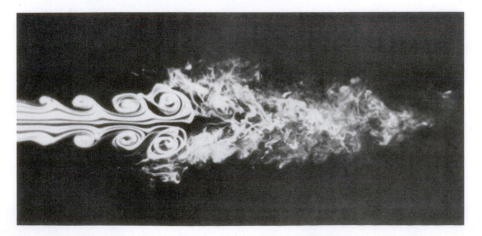

**Figure 1** A turbulent jet (Photograph by Robert Drubka and Hassan Nagib)

well-posedness of NS is one of the five celebrated Millenium Problems (see http://www.claymath.org/millennium).

## ACOUSTICS

Now let us consider the propagation of sound in a gas. Assume that there are no external forces ($\mathbf{F} = \mathbf{0}$). Sound is the result of vibrations in air which under normal circumstances are fairly small. The waves are longitudinal because the individual molecules move in the same direction as the waves propagate. The smallness of the vibrations will lead us to the *linear* wave equation. In the air we know that equations (2), (5), and (6) are valid, so that

$$\frac{\partial \rho}{\partial t} + \nabla \cdot (\rho \mathbf{v}) = 0 \qquad \text{and} \qquad \frac{\partial \mathbf{v}}{\partial t} + (\mathbf{v} \cdot \nabla)\mathbf{v} = -\frac{1}{\rho}\nabla(f(\rho)). \qquad (7)$$

Quiet air has a constant density $\rho = \rho_0$ and a vanishing velocity $\mathbf{v} = \mathbf{0}$. We assume that the vibrations that disturb the air are so small that

$$\rho(\mathbf{x}, t) = \rho_0 + O(\epsilon) \qquad \text{and} \qquad \mathbf{v} = O(\epsilon), \qquad (8)$$

where $O(\epsilon)$ is a small quantity. We write the equation of continuity as

$$\frac{\partial \rho}{\partial t} + \rho_0 \nabla \cdot \mathbf{v} = -\nabla \cdot ((\rho - \rho_0)\mathbf{v}).$$

Expanding the functions $1/\rho$ and $f'(\rho)$ in Taylor series in powers of $\rho - \rho_0$, the equation of motion (7) becomes

$$\frac{\partial \mathbf{v}}{\partial t} + (\mathbf{v} \cdot \nabla)\mathbf{v} = -\left[ \frac{1}{\rho_0} - \frac{1}{\rho_0^2}(\rho - \rho_0) + O(\rho - \rho_0)^2 \right]$$

$$\times \left[ f'(\rho_0) + f''(\rho_0)(\rho - \rho_0) + O(\rho - \rho_0)^2 \right]\nabla(\rho - \rho_0).$$

We assume that $\rho - \rho_0$ and $\mathbf{v}$ and their first derivatives are small to the order $O(\epsilon)$. Dropping all the terms of order $O(\epsilon)^2$, we therefore get the approximate equations

$$\frac{\partial \rho}{\partial t} + \rho_0 \nabla \cdot \mathbf{v} = 0, \qquad \frac{\partial \mathbf{v}}{\partial t} = -\frac{f'(\rho_0)}{\rho_0}\nabla \rho. \qquad (9)$$

These are the *linearized equations of acoustics.*

It follows from (9) that

$$\frac{\partial^2 \rho}{\partial t^2} = -\rho_0 \nabla \cdot \frac{\partial \mathbf{v}}{\partial t}$$

$$= -\rho_0 \nabla \cdot \left( -\frac{f'(\rho_0)}{\rho_0}\nabla \rho \right)$$

$$= f'(\rho_0)\, \Delta \rho,$$

so that $\rho$ satisfies the *wave equation* with the wave speed

$$c_0 = \sqrt{f'(\rho_0)}.$$

As for the velocity **v**, its curl satisfies

$$\frac{\partial}{\partial t} \nabla \times \mathbf{v} = \nabla \times \frac{\partial \mathbf{v}}{\partial t} = -\frac{f'(\rho_0)}{\rho_0} \nabla \times \nabla \rho = \mathbf{0}.$$

$\nabla \times \mathbf{v}$ is called the *vorticity*. We assume now that $\nabla \times \mathbf{v} = \mathbf{0}$ at $t = 0$, which means that the motion of the air is *irrotational* initially. Then the vorticity vanishes ($\nabla \times \mathbf{v} = \mathbf{0}$) for all $t$. This implies that

$$\frac{\partial}{\partial x_i} \frac{\partial v_j}{\partial x_j} = \frac{\partial}{\partial x_j} \frac{\partial v_i}{\partial x_j}$$

so that $\nabla(\nabla \cdot \mathbf{v}) = \Delta \mathbf{v}$. It follows that

$$\frac{\partial^2 \mathbf{v}}{\partial t^2} = -\frac{f'(\rho_0)}{\rho_0} \nabla \frac{\partial \rho}{\partial t}$$

$$= -\frac{f'(\rho_0)}{\rho_0} \nabla(-\rho_0 \nabla \cdot \mathbf{v})$$

$$= f'(\rho_0) \nabla(\nabla \cdot \mathbf{v}) = f'(\rho_0) \Delta \mathbf{v}.$$

*Thus both $\rho$ and all three components of* **v** *satisfy the wave equation with the same wave speed $c_0$.* Naturally, $c_0$ is the speed of sound and $c_0^2$ is the derivative of the pressure with respect to the density.

### Example

In air at normal atmospheric pressure we have approximately $p = f(\rho) = p_0(\rho/\rho_0)^{7/5}$, where $p_0 = 1.033\,\text{kg/cm}^2$ and $\rho_0 = 0.001293\,\text{g/cm}^3$. Hence $c_0 = \sqrt{(1.4)p_0/\rho_0} = 336\,\text{m/s}$.    □

For more information on fluids, see [Me], and on acoustics, see [MI] for instance.

### EXERCISES

1. Assuming in (5) that $\mathbf{F} = \mathbf{0}$, and that **v** is a gradient ($\mathbf{v} = \nabla\phi$), which means that the flow is irrotational and unforced, show that $\int dp/\rho + \partial\phi/\partial t + \frac{1}{2}|\nabla\phi|^2 = \text{constant}$. (*Hint:* Into (5) substitute $\mathbf{v} = \nabla\phi$ and $p = f(\rho)$.) This is called Bernoulli's Law.

2. In particular, in a steady flow, show that low pressure corresponds to high velocity. (*Hint:* Set $\partial\phi/\partial t = 0$.)

## 13.3   SCATTERING

A scattering, or diffraction, problem consists of an incoming wave, an interaction, and an outgoing wave. Because the interaction itself may have very complicated effects, we focus our attention on its consequence, the incoming $\rightarrow$ outgoing process, known as the *scattering process*.

### INHOMOGENEOUS STRING

As our first example, we take an infinite vibrating string made of two different materials so that its density is $\rho(x) = \rho_1$ for $x < 0$ and $\rho_2$ for $x > 0$. A wave would travel along the left half string at speed $c_1 = \sqrt{T/\rho_1}$ and along the right half at speed $c_2 = \sqrt{T/\rho_2}$. Thus any wave along the string would satisfy

$$u_{tt} = c^2(x)u_{xx} \qquad \text{where } c(x) = \begin{cases} c_1 & \text{for } x < 0. \\ c_2 & \text{for } x > 0. \end{cases} \qquad (1)$$

Let a wave $u(x, t) = f(x - c_1 t)$ come in from the left, where $f(s) = 0$ for $s > 0$. This is the *incoming, or incident, wave* (see Figure 1). What eventually happens to this wave?

We know from Section 2.1 that

$$u(x, t) = \begin{cases} F(x - c_1 t) + G(x + c_1 t) & \text{for } x < 0 \text{ and all } t \\ H(x - c_2 t) + K(x + c_2 t) & \text{for } x > 0 \text{ and all } t. \end{cases} \qquad (2)$$

Even though $c(x)$ is discontinuous at $x = 0$, the physics requires that $u(x, t)$ and $u_x(x, t)$ are continuous everywhere. (Why?) The initial conditions are

$$u(x, 0) = f(x) \qquad \text{and} \qquad u_t(x, 0) = -c_1 f'(x). \qquad (3)$$

(Why?) In particular, the initial data are zero for $x > 0$. Combining (2) and (3), we can derive

$$u(x, t) = f(x - c_1 t) + \frac{c_2 - c_1}{c_2 + c_1} f(-c_1 t - x) \qquad \text{for } x < 0 \qquad (4)$$

$x = 0$

**Figure 1**

and

$$u(x, t) = \frac{2c_2}{c_2 + c_1} f\left(\frac{c_1}{c_2}(x - c_2 t)\right) \qquad \text{for } x > 0. \tag{5}$$

(see Exercise 1).

This result is interpreted as follows. The first term in (4) is the incoming wave. The last term in (4) is the *reflected wave*, which travels to the left at speed $c_1$ with a reflection coefficient $(c_2 - c_1)/(c_2 + c_1)$. The expression (5) is the *transmitted wave*, which travels to the right at speed $c_2$ with a transmission coefficient $2c_2/(c_2 + c_1)$.

In this example the medium undergoes an abrupt change at a single point. In a more general situation we could study the effects of any impurities or *inhomogeneities* in the medium. For instance, we could consider the equation

$$u_{tt} - \nabla \cdot (p \nabla u) + qu = 0, \tag{6}$$

where $\mathbf{x} = (x, y, z)$ and the functions $p = p(\mathbf{x})$ and $q = q(\mathbf{x})$ represent inhomogeneities. We would assume that $p(\mathbf{x}) \to c^2 > 0$ and $q(\mathbf{x}) \to 0$ as $|\mathbf{x}| \to \infty$, meaning that the inhomogeneities are "localized". Another kind of scatterer is a rigid body. This is our next example.

## SCATTERING OF A PLANE WAVE BY A SPHERE

Let the sphere be $\{|\mathbf{x}| = R\}$, on which we assume Dirichlet boundary conditions. Let the incident wave be $Ae^{i(\omega t - kz)}$ with constants $A$, $\omega$, and $k$ that satisfy $\omega^2 = c^2 k^2$. This is a solution of the wave equation traveling along the $z$ axis at speed $c$, called a *traveling plane wave*. The scattering problem is to solve

$$\begin{array}{ll} u_{tt} - c^2 \Delta u = 0 & \text{in } |\mathbf{x}| > R \\ u = 0 & \text{on } |\mathbf{x}| = R \\ u(x, t) \sim Ae^{i(\omega t - kz)} & \text{as } t \to -\infty. \end{array} \tag{7}$$

Because we expect the time behavior of the solution to be $e^{i\omega t}$, we look for a solution of the form

$$u(x, t) = Ae^{i(\omega t - kz)} + e^{i\omega t} v(\mathbf{x}). \tag{8}$$

This requires the reflected wave $v(\mathbf{x})$ to satisfy the problem

$$\begin{array}{ll} \omega^2 v + c^2 \Delta v = 0 & \text{in } |\mathbf{x}| > R \\ v = -Ae^{-ikz} & \text{on } |\mathbf{x}| = R \end{array} \tag{9}$$

$v$ satisfies an "outgoing radiation condition" at $\infty$.

The *radiation condition* ought to mean that $e^{i\omega t} v(\mathbf{x})$ has no incoming part: it should be purely outgoing. For instance, the spherical wave $e^{i\omega t} e^{\pm ikr}/r$ is incoming with the $+$ sign and outgoing with the $-$ sign (for positive $k$ and $\omega$). For the outgoing one, $rv$ is bounded and $r(\partial v/\partial r + ikv) \to 0$ as $r \to \infty$. Thus the problem to be satisfied by $v(\mathbf{x})$ is

$$\Delta v + k^2 v = 0 \qquad \text{in } |\mathbf{x}| > R$$

$$v = -Ae^{-ikz} \qquad \text{on } |\mathbf{x}| = R \tag{10}$$

$$rv \text{ bounded and } r\left(\frac{\partial v}{\partial r} + ikv\right) \to 0 \quad \text{as } r \to \infty.$$

To solve (10), we shall make use of the methods of Chapter 10. It is natural to use spherical coordinates $(r, \theta, \phi)$. Then $v(\mathbf{x})$ will not depend on $\phi$ due to the symmetry of the problem. As in Section 10.3, we expand $v(\mathbf{x})$ in spherical harmonics. Because $\Delta v = k^2 v$, the expansion has the form

$$v(r, \theta) = \sum a_l \, R_{l+\frac{1}{2}}(kr) \, P_l(\cos\theta), \tag{11}$$

where $R_{l+\frac{1}{2}}$ is related to Bessel's equation, see (10.3.2), and $P_l$ is a Legendre polynomial. Because $v$ is independent of the angle $\phi$, the associated index $m$ is zero.

The outgoing radiation condition determines the asymptotic behavior as $r \to \infty$. We look at Section 10.5 for the facts we need. From (10.5.15) with $s = l + \frac{1}{2}$ and $z = kr$, we have

$$H_s^-(kr) \sim \sqrt{\frac{2}{\pi kr}} e^{-i(kr - s\pi/2 - \pi/4)} \qquad \text{as } r \to \infty.$$

Among all the solutions of Bessel's equation, this is the one we are looking for. Thus $R_{l+\frac{1}{2}}(kr) = H_{l+\frac{1}{2}}^-(kr)/\sqrt{kr}$.

The coefficients in (11) are determined by the boundary condition in (10). Putting $r = R$, we require

$$-Ae^{-ikz} = -Ae^{-ikR\cos\theta} = \sum_{l=0}^{\infty} a_l \frac{H_{l+\frac{1}{2}}^-(kR)}{\sqrt{kR}} P_l(\cos\theta). \tag{12}$$

In order to find the $a_l$, we shall use a three-dimensional version of the identity (10.5.17), as follows.

Notice that $e^{-iz} = e^{-ir\cos\theta}$ obviously solves the equation $\Delta w + w = 0$. So it has an expansion in spherical harmonics, as in Section 10.3, of the form

$$e^{-ir\cos\theta} = \sum_{l=0}^{\infty} b_l \frac{1}{\sqrt{r}} J_{l+\frac{1}{2}}(r) P_l(\cos\theta) \tag{13}$$

for some coefficients $b_l$. By the orthogonality of the Legendre polynomials and by the values of their normalizing constants from Section 10.6, we deduce that

$$b_l \frac{1}{\sqrt{r}} J_{l+\frac{1}{2}}(r) = \frac{2l+1}{2} \int_{-1}^{1} e^{-irs} P_l(s)\, ds. \tag{14}$$

(See Exercise 3).

Let's compare the asymptotic behavior as $r \to \infty$ of both sides of (14). The right side of (14) is integrated by parts twice to get

$$\left(l + \tfrac{1}{2}\right) \left[ \frac{i}{r} e^{-irs} P_l(s) \Big|_{-1}^{1} - \left(\frac{i}{r}\right)^2 e^{-irs} P_l'(s) \Big|_{-1}^{1} + \left(\frac{i}{r}\right)^2 \int_{-1}^{1} e^{-irs} P_l''(s)\, ds \right].$$

Of these three terms the dominant one is the first one, since it has a $1/r$ factor instead of $1/r^2$. So the right side of (14) is

$$\left(l + \tfrac{1}{2}\right) \frac{i}{r} \left[ e^{-ir} P_l(1) - e^{ir} P_l(-1) \right] + O\left(\frac{1}{r^2}\right)$$

$$= \frac{2}{r}(-i)^l \left(l + \tfrac{1}{2}\right) \sin\left(r - \frac{l\pi}{2}\right) + O\left(\frac{1}{r^2}\right) \tag{15}$$

by Exercise 4. On the other hand, using (10.5.5), the left side of (14) is asymptotically

$$b_l \frac{1}{\sqrt{r}} \sqrt{\frac{2}{\pi r}} \cos\left[ r - \left(l + \tfrac{1}{2}\right)\frac{\pi}{2} - \frac{\pi}{4} \right] = b_l \sqrt{\frac{2}{\pi}} \frac{1}{r} \sin\left(r - \frac{l\pi}{2}\right). \tag{16}$$

Comparing (15) and (16), we deduce that

$$b_l = \sqrt{2\pi}(-i)^l \left(l + \tfrac{1}{2}\right).$$

Putting this result into (13), we have derived the expansion

$$e^{-ir\cos\theta} = \sqrt{2\pi} \sum_{l=0}^{\infty} (-i)^l \left(l + \tfrac{1}{2}\right) \frac{1}{\sqrt{r}} J_{l+\frac{1}{2}}(r) P_l(\cos\theta). \tag{17}$$

We multiply (17) by $-A$ and replace $r$ by $kR$. Then this expansion must be in agreement with (12), so that

$$a_l \frac{H_{l+\frac{1}{2}}^{-}(kR)}{\sqrt{kR}} = -A\sqrt{2\pi}(-i)^l (l + \tfrac{1}{2}) \frac{J_{l+\frac{1}{2}}(kR)}{\sqrt{kR}}.$$

This is the formula for the coefficients in (11). Thus we have proven the following theorem.

**Theorem 1.** The scattering of the plane wave $Ae^{ik(ct-z)}$ by the sphere $|\mathbf{x}| = R$ with Dirichlet boundary conditions leads to the complete solution

$$u(\mathbf{x}, t) = Ae^{ik(ct-z)} - Ae^{ikct}\sqrt{\frac{2\pi}{kr}}\sum_{l=0}^{\infty}(-i)^l\left(l + \frac{1}{2}\right)$$

$$\times \frac{J_{l+\frac{1}{2}}(kR)}{H_{l+\frac{1}{2}}^{-}(kR)}H_{l+\frac{1}{2}}^{-}(kr)P_l(\cos\theta).$$

For more on scattering, see [MF], [AJS], or [AS].

## EXERCISES

1. Derive (4) and (5) from (2) and (3).
2. A point mass $M$ is attached to an infinite homogeneous string at the origin by means of a spring. This leads to the wave equation (with a speed $c$) with the jump conditions

$$T[u_x(0+, t) - u_x(0-, t)] = ku(0-, t) + Mu_{tt}(0-, t)$$
$$= ku(0+, t) + Mu_{tt}(0+, t).$$

   Find the reflected and transmitted waves if a wave $f(x - ct)$ is initially traveling from the left (i.e., $f(x - ct) = 0$ for $x \geq 0, t \leq 0$). For simplicity, take $c = T = k = M = 1 = c_1 = c_2$.
3. Use the orthogonality of the Legendre polynomials to derive (14).
4. Derive (15).
5. Repeat the problem of scattering by a sphere for the case of Neumann boundary conditions. (This could be acoustic scattering off a rigid ball.).
6. Do the problem of scattering by an infinitely long cylinder with Dirichlet conditions. (*Hint:* See Section 10.5)
7. Solve the problem of scattering of a point source off a plane:

$$\Delta v + k^2 v = \delta(x^2 + y^2 + (z - a)^2) \text{ in } z > 0, \qquad v = 0 \qquad \text{on } z = 0$$

   where $a > 0$. What is the "reflected" or "scattered" wave? (*Hint:* First solve the equation in all space without a BC. Then use the method of reflection as in Section 7.4.)

## 13.4   CONTINUOUS SPECTRUM

In quantum mechanics a variety of phenomena are described by the Schrödinger equation

$$iu_t = -\Delta u + V(\mathbf{x})u$$

with a real potential $V(\mathbf{x})$. By separating variables $v(\mathbf{x}, t) = e^{-i\lambda t}\psi(\mathbf{x})$, we get

$$-\Delta\psi + V(\mathbf{x})\psi = \lambda\psi. \tag{1}$$

We say that $\lambda$ is an eigenvalue if (1) has a nonzero solution with $\iiint |\psi|^2\,d\mathbf{x} < \infty$. The set of all the eigenvalues is called the *discrete spectrum* (or point spectrum). The eigenfunction $\psi(\mathbf{x})$ is called a *bound state*. In this section we shall assume that the potential $V(\mathbf{x}) \to 0$ at a certain sufficiently rapid rate as $|\mathbf{x}| \to \infty$. Then the solutions of (1) ought to behave like the solutions of $-\Delta\psi = \lambda\psi$ as $|\mathbf{x}| \to \infty$.

The simpler equation $-\Delta\psi = \lambda\psi$ has the harmonic plane wave solutions $\psi(\mathbf{x}, \mathbf{k}) = e^{-i\mathbf{x}\cdot\mathbf{k}}$ for $|\mathbf{k}|^2 = \lambda$. In Chapter 12 we analyzed problems like this using Fourier transforms. However, the plane waves that we are now considering are not eigenfunctions because $\iiint |\psi|^2\,d\mathbf{x} = \infty$. Thus we say that the set of all positive numbers $\lambda$ ($0 < \lambda < \infty$) comprise the *continuous spectrum* of the operator $-\Delta$. It is because of the continuous spectrum that the usual Fourier expansion has to be replaced by the Fourier integral.

The problem (1) with a potential also has a continuous spectrum. This means that for $\lambda = |\mathbf{k}|^2 > 0$, there is a solution $f(\mathbf{x}, \mathbf{k})$ of (1) such that

$$f(\mathbf{x}, \mathbf{k}) \sim e^{-i\mathbf{k}\cdot\mathbf{x}} \text{ as } |\mathbf{x}| \to \infty.$$

Problem (1) may or may not have bound states as well.

For the *hydrogen atom* where $V(\mathbf{x}) = c/r$ ($c = $ constant), the discrete spectrum (corresponding to the bound states) was found in Sections 9.5 and 10.7 to be $\{-1/n^2 : n \text{ is a positive integer}\}$. The continuous spectrum turns out to be the whole interval $[0, +\infty)$. The whole spectrum, continuous and discrete, is sketched in Figure 1. The potential $c/r$ for the hydrogen atom has a singularity at the origin and does not tend to zero particularly rapidly as $r \to \infty$. For a nicer potential that is smooth and tends to zero rapidly (exponentially fast, say, or faster than a certain power) as $r \to \infty$, it has been proven that the continuous spectrum is $[0, \infty)$ and the discrete spectrum consists of a *finite* number of negative values $\lambda_N \leq \cdots \leq \lambda_1 < 0$ where $N \geq 0$. See Vol. IV, p. 98 of [RS], or p. 117 of [Dd], or [AS].

*Now we study the continuous spectrum from the point of view of scattering theory.* For simplicity we take the one-dimensional case, writing (1) in the form

$$-\psi_{xx} + V(x)\psi = k^2\psi \qquad (-\infty < x < \infty), \tag{2}$$

where $\lambda = k^2 > 0$ is in the continuous spectrum. The role of the scatterer is played by the potential $V(x)$, which we assume satisfies

$$\int_{-\infty}^{\infty} (1 + x^2)\,V(x)\,dx < \infty. \tag{3}$$

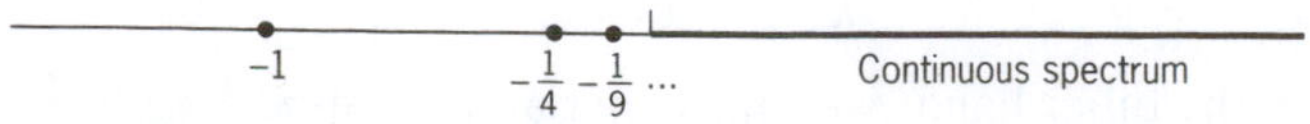

Figure 1

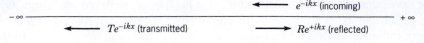

**Figure 2**

If $e^{-ikx}$ is a wave coming in from $x = +\infty$, there will be a reflected wave going back to $+\infty$ and a transmitted wave going through to $-\infty$ (see Figure 2):

$$\psi(x) \sim e^{-ikx} + Re^{+ikx} \qquad \text{as } x \to +\infty$$

$$\psi(x) \sim Te^{-ikx} \qquad \text{as } x \to -\infty. \tag{4}$$

The reflection coefficient is $R$, which depends on $k$. So does the transmission coefficient $T$. It can be proved, under condition (3), that for each $k$ there is a unique solution with these properties (4) (see [AS] or [Dd]).

**Theorem 1.**

$$\boxed{|R|^2 + |T|^2 = 1.}$$

This theorem means that you add $R$ and $T$ in a pythagorean way. It follows, obviously, that both $|R|$ and $|T|$ are $\leq 1$.

**Proof.**   Let $f(x)$ and $g(x)$ be the solutions of (2) such that

$$f(x) \sim e^{-ikx} \text{ as } x \to -\infty \qquad \text{and} \qquad g(x) \sim e^{+ikx} \text{ as } x \to +\infty. \tag{5}$$

(More precisely, $e^{ikx} f(x) \to 0$ as $x \to -\infty$, and so on. It can be shown that $f(x)$ and $g(x)$ exist and are unique. Their complex conjugates $\bar{f}(x)$ and $\bar{g}(x)$ are solutions as well (why?) and satisfy

$$\bar{f}(x) \sim e^{+ikx} \text{ as } x \to -\infty \qquad \text{and} \qquad \bar{g}(x) \sim e^{-ikx} \text{ as } x \to +\infty. \tag{6}$$

Now $g$ and $\bar{g}$ are linearly independent because they satisfy linearly independent conditions at $+\infty$. So *every* solution of the ODE (2) can be expressed in terms of these two. Thus

$$f(x) = ag(x) + b\bar{g}(x) \tag{7}$$

for some complex constants $a$ and $b$. (All these functions and constants depend on $k$.) The wronskian, defined as $W(g, \bar{g}) = g\bar{g}_x - g_x\bar{g}$, must be independent of $x$ because

$$\frac{dW}{dx} = g\bar{g}_{xx} + g_x\bar{g}_x - g_x\bar{g}_x - g_{xx}\bar{g} = g \cdot \bar{g}_{xx} - g_{xx} \cdot \bar{g}$$
$$= g \cdot (V(x) - k^2)\bar{g} - (V(x) - k^2)g \cdot \bar{g} = 0$$

from (2). On the other hand, we know that as $x \to +\infty$,

$$W(g, \bar{g}) \sim (e^{ikx})(-ike^{-ikx}) - (ike^{ikx})(e^{-ikx}) \sim -2ik.$$

Therefore, $W(g, \bar{g}) \equiv -2ik$ for all $x$. Similarly, $W(f, \bar{f}) \equiv +2ik$ for all $x$. Using (7) and the fact that $W(g, g) = 0 = W(\bar{g}, \bar{g})$, we have

$$W(f, \bar{f}) = W(ag + b\bar{g}, \bar{a}\bar{g} + \bar{b}g) = a\bar{a}W(g, \bar{g}) + b\bar{b}W(\bar{g}, g)$$
$$= (|a|^2 - |b|^2)W(g, \bar{g}).$$

Therefore, $2ik = (|a|^2 - |b|^2)(-2ik)$, which means that

$$|a|^2 - |b|^2 = -1. \tag{8}$$

We return to $\psi(x)$, which is defined by the conditions (4). Comparing (4) to (6), we must have

$$\psi(x) = \bar{g}(x) + Rg(x) \qquad \text{(from the conditions at } +\infty)$$

and

$$\psi(x) = Tf(x) \qquad \text{(from the conditions at } -\infty).$$

Hence $Tf(x) = \bar{g}(x) + Rg(x)$. From (7) it follows that $a = R/T$ and $b = 1/T$. Putting this into (8), we deduce that $|R/T|^2 - |1/T|^2 = -1$, or

$$|R|^2 + |T|^2 = 1.$$

For further information on the continuous spectrum, see [RS] or [AS].

## EXERCISES

1. Find all the eigenvalues (discrete spectrum) of

$$-\psi'' + q\psi = \lambda\psi \quad \text{with } \psi(-x) = \psi(x) \quad (-\infty < x < \infty)$$

where $q(x) = -Q$ for $|x| < 1$, and $q(x) = 0$ outside this interval. The depth $Q$ is a positive constant. This is the *square well potential*. (*Hint:* First show that $-Q < \lambda < 0$. The eigenfunctions and their first derivatives should be continuous.)

2. Find all the eigenvalues (discrete spectrum) of

$$-\psi'' + q\psi = \lambda\psi \qquad (-\infty < x < \infty)$$

for the potential $q(x) = -Q\delta(x)$, a positive constant times the delta function. (*Hint:* The eigenfunctions in this case are merely continuous.)

## 13.5 EQUATIONS OF ELEMENTARY PARTICLES

In the last 50 years various PDE models of hyperbolic type have played central roles in our understanding of the elementary particles: electrons, protons, neutrons, mesons, quarks, and so on. Here we shall describe some of these equations. For deeper study, see [MF] or [Bl].

One of the simplest is the *Klein–Gordon equation*

$$u_{tt} - c^2 \Delta u + m^2 u = 0 \tag{1}$$

in three dimensions, where $m$ is the mass of the particle. It is of hyperbolic type (see Section 1.6) and in fact its only difference from the wave equation is the last term. Its solution is given at the end of this section. Much more difficult is the *nonlinear Klein–Gordon equation*

$$u_{tt} - c^2 \Delta u + m^2 u + gu^3 = 0, \tag{2}$$

where $g$ is constant, which is a model for mesons.

## DIRAC EQUATION

The Dirac equation for the electron was devised to be a "square root" of the Klein–Gordon equation. It is

$$c^{-1}\gamma^0 \frac{\partial u}{\partial t} + \gamma^1 \frac{\partial u}{\partial x} + \gamma^2 \frac{\partial u}{\partial y} + \gamma^3 \frac{\partial u}{\partial z} + imu = 0, \tag{3}$$

where $m$ is the mass, $i = \sqrt{-1}$, and $\gamma^0, \gamma^1, \gamma^2,$ and $\gamma^3$ are certain $4 \times 4$ *matrices*. The coefficient matrices are

$$\gamma^0 = \begin{pmatrix} 1 & & & \\ & 1 & & \\ & & -1 & \\ & & & -1 \end{pmatrix} \qquad \gamma^j = \begin{pmatrix} 0 & 0 & & -\sigma^j \\ 0 & 0 & & \\ & -\sigma^j & 0 & 0 \\ -\sigma^j & & 0 & 0 \end{pmatrix}.$$

for $j = 1, 2, 3$, where

$$\sigma^1 = \begin{pmatrix} 0 & 1 \\ 1 & 0 \end{pmatrix} \qquad \sigma^2 = \begin{pmatrix} 0 & -i \\ i & 0 \end{pmatrix} \qquad \sigma^3 = \begin{pmatrix} 1 & 0 \\ 0^. & -1 \end{pmatrix}$$

are the $2 \times 2$ Pauli matrices. The solution $u(x, y, z, t)$ is a four-dimensional complex vector at each point in space-time. The coefficient matrices have the following properties:

$$(\gamma^0)^* = \gamma^0, \quad (\gamma^j)^* = \gamma^j \quad \text{for } j = 1, 2, 3, \tag{4}$$

where $*$ is the conjugate transpose,

$$(\gamma^0)^2 = I, \qquad (\gamma^j)^2 = -I \quad \text{for } j = 1, 2, 3 \tag{5}$$

and

$$\gamma^\alpha \gamma^\beta + \gamma^\beta \gamma^\alpha = 0 \qquad \text{for } \alpha \neq \beta, \quad \alpha, \beta = 0, 1, 2, 3. \tag{6}$$

Dirac is a square root of Klein–Gordon in the sense that

$$\left( \frac{1}{c}\gamma^0 \frac{\partial}{\partial t} + \gamma^1 \frac{\partial}{\partial x} + \gamma^2 \frac{\partial}{\partial y} + \gamma^3 \frac{\partial}{\partial z} + im \right)^2 = \frac{1}{c^2}\frac{\partial^2}{\partial t^2} - \Delta + m^2$$

as operators (see Exercise 2).

The famous equations of quantum electrodynamics (QED) form a system that combines Dirac's and Maxwell's equations with nonlinear coupling terms. We will not write them down here. They describe the interaction between electrons (governed by Dirac's equation) and photons (governed by Maxwell's equations). The predictions of the theory agree with experiments to 12 figures, making it the most accurate theory in all of physics.

## GAUGE THEORY

This is the central idea by which physicists today are attempting to unify the four fundamental forces of nature (gravitational, electromagnetic, weak, and strong). On the PDE level it is based on the *Yang–Mills equations*. They are just like the Maxwell equations except that each of the components has (three-dimensional) vector values. They are, putting $c = 1$ for simplicity,

$$\text{(I)} \quad \begin{cases} D_0 \mathbf{B}_1 = D_3 \mathbf{E}_2 - D_2 \mathbf{E}_3 \\ D_0 \mathbf{B}_2 = D_1 \mathbf{E}_3 - D_3 \mathbf{E}_1 \\ D_0 \mathbf{B}_3 = D_2 \mathbf{E}_1 - D_1 \mathbf{E}_2 \end{cases} \qquad \text{(II)} \quad \begin{cases} D_0 \mathbf{E}_1 = D_2 \mathbf{B}_3 - D_3 \mathbf{B}_2 \\ D_0 \mathbf{E}_2 = D_3 \mathbf{B}_1 - D_1 \mathbf{B}_3 \\ D_0 \mathbf{E}_3 = D_1 \mathbf{B}_2 - D_2 \mathbf{B}_1 \end{cases}$$

$$\text{(III)} \quad D_1 \mathbf{B}_1 + D_2 \mathbf{B}_2 + D_3 \mathbf{B}_3 = 0 \qquad \text{(IV)} \quad D_1 \mathbf{E}_1 + D_2 \mathbf{E}_2 + D_3 \mathbf{E}_3 = 0.$$

But here the operators $D_1, D_2$, and $D_3$ are *nonlinear* versions of the ordinary partial derivatives. These *covariant derivative* operators are defined as

$$D_0 U = \frac{\partial \mathbf{U}}{\partial t} - \mathbf{A}_0 \times \mathbf{U}$$

$$D_k U = \frac{\partial \mathbf{U}}{\partial x_k} + \mathbf{A}_k \times \mathbf{U} \tag{8}$$

for $k = 1, 2, 3$ and $\mathbf{x} = (x_1, x_2, x_3) = (x, y, z)$. The unknowns in the Yang–Mills equations are the 10 variables

$$\mathbf{E}_1, \mathbf{E}_2, \mathbf{E}_3, \mathbf{B}_1, \mathbf{B}_2, \mathbf{B}_3, \mathbf{A}_0, \mathbf{A}_1, \mathbf{A}_2, \mathbf{A}_3.$$

Finally, to (I)–(IV) are adjoined the equations

$$\text{(V)} \quad \mathbf{E}_k = \frac{\partial \mathbf{A}_0}{\partial x_k} + \frac{\partial \mathbf{A}_k}{\partial t} + \mathbf{A}_k \times \mathbf{A}_0$$

and

$$\text{(VI)} \quad \mathbf{B}_1 = \frac{\partial \mathbf{A}_2}{\partial x_3} - \frac{\partial \mathbf{A}_3}{\partial x_2} + \mathbf{A}_3 \times \mathbf{A}_2$$

and similar equations for $\mathbf{B}_2$ and $\mathbf{B}_3$ (with the indices permuted cyclically). The total *energy* for the Yang–Mills equations is

$$\mathscr{E} = \tfrac{1}{2} \iiint (|\mathbf{E}_1|^2 + |\mathbf{E}_2|^2 + |\mathbf{E}_3|^2 + |\mathbf{B}_1|^2 + |\mathbf{B}_2|^2 + |\mathbf{B}_3|^2) \, d\mathbf{x} \tag{9}$$

and the total *momentum* is

$$\mathcal{P}_1 = \iiint (\mathbf{B}_2 \cdot \mathbf{E}_3 - \mathbf{B}_3 \cdot \mathbf{E}_2)\,dx, \quad \text{etc.} \quad \text{(cyclically).} \quad (10)$$

These quantities are invariants (see Exercise 3).

An equivalent way to write the Yang–Mills equations is to consider each of the 10 dependent variables not as a vector but as a skew-hermitian complex $2 \times 2$ matrix with zero trace. Thus each of $\mathbf{E}_1, \ldots, \mathbf{A}_3$ becomes such a matrix. Such a matrix has three real components, so that it is equivalent to a vector; see Exercise 11. Then everywhere replace each vector product like $\mathbf{A} \times \mathbf{B}$ by $AB - BA$ and each scalar product like $\mathbf{A} \cdot \mathbf{B}$ by *trace* $(AB^*)$.

What is special about the Yang–Mills equations is their *gauge invariance*. Namely, let $G(\mathbf{x}, t)$ be any unitary $2 \times 2$ matrix function with determinant $= 1$. If $(E_k, B_k, A_k, A_0)$ (where $k = 1, 2, 3$) is any solution of (I)–(VI) with values considered as matrices, then so is $(E'_k, B'_k, A'_k, A'_0)$, where

$$E'_k = G^{-1}E_kG, \qquad\qquad B'_k = G^{-1}B_kG,$$

$$A'_k = G^{-1}A_kG + G^{-1}\frac{\partial G}{\partial x_k}, \qquad A'_0 = G^{-1}A_0G - G^{-1}\frac{\partial G}{\partial t} \quad (11)$$

(where $k = 1, 2, 3$). The products are ordinary matrix multiplications. This invariance has powerful consequences that lie at the heart of the idea from physics.

We have described the equations of gauge theory with the *gauge group* $\mathcal{G}$ being the group of unitary $2 \times 2$ matrices with determinant $= 1$. More general gauge theories use larger gauge groups.

## SOLUTION OF THE KLEIN–GORDON EQUATION

First we solve it in *one dimension*:

$$u_{tt} - c^2 u_{xx} + m^2 u = 0 \qquad (-\infty < x < \infty)$$

$$u(x, 0) = \phi(x) \qquad u_t(x, 0) = \psi(x), \qquad\qquad (12)$$

where $c$ and $m$ are positive constants. We use the method of Fourier transforms as in Section 12.4. The source function has the Fourier transform $\hat{S}(k, t)$, where

$$\frac{\partial^2 \hat{S}}{\partial t^2} = -c^2 k^2 \hat{S} - m^2 \hat{S}, \qquad \hat{S}(k, 0) = 0, \qquad \frac{\partial \hat{S}}{\partial t}(k, 0) = 1.$$

This ODE has the solution

$$\hat{S}(k, t) = \frac{\sin \left[t \sqrt{c^2 k^2 + m^2}\right]}{\sqrt{c^2 k^2 + m^2}}, \qquad\qquad (13)$$

so that

$$S(x, t) = \int_{-\infty}^{\infty} \frac{\sin [t\sqrt{c^2k^2 + m^2}]}{\sqrt{c^2k^2 + m^2}} e^{ikx} \frac{dk}{2\pi}.$$

Fortunately, this is (almost) an entry in our table of Fourier transforms in Section 12.3. From Exercise 5 it follows that

$$S(x, t) = \frac{1}{2c} J_0 \left( m\sqrt{t^2 - \frac{x^2}{c^2}} \right) \qquad \text{for } |x| < ct \qquad (14)$$

and $S(x, t) = 0$ for $|x| > ct \geq 0$. Thus the source function has the same jump discontinuity on the light cone as the wave equation. In fact, as $m \to 0$ it converges to the source function for the wave equation (12.2.9).

In *three dimensions* the same method, using three-dimensional Fourier transforms, leads to the formula

$$S(\mathbf{x}, t) = \int_{-\infty}^{\infty} \int_{-\infty}^{\infty} \int_{-\infty}^{\infty} \frac{\sin [t\sqrt{c^2k^2 + m^2}]}{\sqrt{c^2k^2 + m^2}} e^{i\mathbf{k}\cdot\mathbf{x}} \frac{d\mathbf{k}}{8\pi^3}, \qquad (15)$$

where $k^2 = |\mathbf{k}|^2$. Now we must use spherical coordinates. We let $\theta$ denote the angle between $\mathbf{k}$ and $\mathbf{x}$, and let $r = |\mathbf{x}|$. Then (15) takes the form

$$S(\mathbf{x}, t) = \int_0^{2\pi} \int_0^{\pi} \int_0^{\infty} \frac{\sin [t\sqrt{c^2k^2 + m^2}]}{\sqrt{c^2k^2 + m^2}} e^{ikr \cos\theta} \frac{k^2 \sin\theta \, dk \, d\theta \, d\phi}{8\pi^3}. \qquad (16)$$

The $\phi$ and $\theta$ integrals are easily integrated out to get

$$S(\mathbf{x}, t) = \frac{1}{2\pi^2 r} \int_0^{\infty} \frac{\sin [t\sqrt{c^2k^2 + m^2}]}{\sqrt{c^2k^2 + m^2}} k \sin kr \, dk. \qquad (17)$$

Now we can write $k \sin kr = \partial(-\cos kr)/\partial r$, pull the $\partial/\partial r$ outside the integral, and use the fact that the integrand is an even function of $k$, to get

$$S(\mathbf{x}, t) = -\frac{1}{4\pi^2 r} \frac{\partial}{\partial r} \int_{-\infty}^{\infty} \frac{\sin [t\sqrt{c^2k^2 + m^2}]}{\sqrt{c^2k^2 + m^2}} e^{ikr} \, dk. \qquad (18)$$

Using Exercise 6, we get

$$S(\mathbf{x}, t) = -\frac{1}{4\pi cr}\frac{\partial}{\partial r}\left[H\left(t^2 - \frac{r^2}{c^2}\right)J_0\left(m\sqrt{t^2 - \frac{r^2}{c^2}}\right)\right]. \tag{19}$$

Carrying out the derivative and using the identity $J_0' = -J_1$ and $J_0(0) = 1$, we get

$$S(\mathbf{x}, t) = \frac{1}{2\pi c}\delta(c^2t^2 - r^2)$$
$$\tag{20}$$
$$- mH(c^2t^2 - r^2)\frac{J_1[(m/c)\sqrt{c^2t^2 - r^2}]}{4\pi c^2\sqrt{c^2t^2 - r^2}}.$$

This means that the source function for the Klein–Gordon equation in three dimensions is a delta function on the light cone plus a Bessel function inside the cone. If $m = 0$, the formula reduces to the wave equation case.

## EXERCISES

1. Prove properties (4), (5), and (6) of the Dirac matrices.

2. Prove that the Dirac operator is a square root of the Klein–Gordon operator.

3. For the Yang–Mills equations, show that the energy $\mathcal{E}$ and the momentum $\mathcal{P}$ are invariants. Assume the solutions vanish sufficiently fast at infinity.

4. Prove the gauge invariance of the Yang–Mills equations. See Exercise 11 on page 379.

5. Use (12.3.12) in the table of Fourier transforms to carry out the last step in the derivation of the formula (14) for the one-dimensional Klein–Gordon equation.

6. Fill in the details of the derivation of (20) in three dimensions.

7. Rederive the solution of the one-dimensional Klein–Gordon equation by the "method of descent" as follows. Calling it $u(x, t)$, define $v(x, y, t) = e^{imy/c}u(x, t)$. Show that $v$ satisfies the two-dimensional wave equation. Use the formula from Chapter 9 to solve for $v(0, 0, t)$, assuming that $\phi(x) \equiv 0$. Transform it to

$$u(0, t) = \int_{|x|<ct}\int_{|y|<\mu} e^{imy}\frac{1}{\sqrt{\mu^2 - y^2}}dy\,\psi(x)\frac{dx}{2\pi},$$

where $\mu = \sqrt{t^2 - x^2}$. From a table of definite integrals, the inner integral equals $\pi J_0(m\mu)$.

8. The *telegraph equation* or *dissipative wave equation* is

$$u_{tt} - c^2 \Delta u + v u_t = 0,$$

where $v > 0$ is the coefficient of dissipation. Show that the energy is decreasing:

$$\frac{d\mathscr{E}}{dt} = -v \iiint u_t^2 \, d\mathbf{x} \leq 0.$$

9. Solve the telegraph equation with $v = 1$ in one dimension as follows. Substituting $u(x, t) = e^{-t/2} v(x, t)$, show that $u_{tt} - c^2 u_{xx} - \frac{1}{4} u = 0$. This is the Klein–Gordon equation with imaginary mass $m = i/2$. Deduce the source function for the telegraph equation.

10. Solve the equation $u_{xy} + u = 0$ in the quarter plane $Q = \{x > 0, y > 0\}$ with the boundary conditions $u(x, 0) = u(0, y) = 1$. (Method 1: Reduce it to the Klein-Gordon equation by a rotation of $\pi/4$. Method 2: Look for a solution of the form $u(x, y) = f(xy)$ and show that $f$ satisfies an ODE that is almost Bessel's equation.)

11. Let $A$ be a skew-hermitian complex $2 \times 2$ matrix with trace $= 0$. Show that $A$ has the form

$$A = i \begin{pmatrix} \alpha_1 & \alpha_2 + i\alpha_3 \\ \alpha_2 - i\alpha_3 & -\alpha_1 \end{pmatrix}$$

with three real components $\alpha_1, \alpha_2, \alpha_3$. For any such matrix $A$, let $\mathbf{A}$ be the real vector $\mathbf{A} = [\alpha_1, \alpha_2, \alpha_3]$. Show that the vector corresponding to the matrix $\frac{1}{2}(AB - BA)$ is exactly $\mathbf{A} \times \mathbf{B}$.

# 14

# NONLINEAR PDES

With nonlinear equations the superposition principle ceases to hold. Therefore, the method of eigenfunctions and the transform methods cannot be used. New phenomena occur, such as shocks and solitons. We now pursue five such topics. The latter part of Section 14.1 requires knowledge of Section 12.1, Section 14.2 of Section 13.4, Section 14.3 of Section 7.1, Section 14.4 merely of Chapter 4, and Section 14.5 of Section 13.2.

## 14.1   SHOCK WAVES

Shock waves occur in explosions, traffic flow, glacier waves, airplanes breaking the sound barrier, and so on. They are modeled by nonlinear hyperbolic PDEs. The simplest type is the first-order equation

$$u_t + a(u)u_x = 0. \tag{1}$$

A system of two nonlinear equations of a similar type is

$$\rho_t + (\rho v)_x = 0 \qquad \text{and} \qquad v_t + vv_x + \rho^{-1}f(\rho)_x = 0,$$

which we encountered in Section 13.2. In this section we limit ourselves to discussing the single equation (1).

However, we shall begin with a review of first-order linear equations, Section 1.2. Consider the equation

$$u_t + a(x, t)\, u_x = 0. \tag{2}$$

[If one encounters the more general equation $b(x, t)u_t + a(x, t)u_x = 0$, one would first divide by $b(x, t)$.] Consider the characteristic curves, which are defined as the solutions of the ODE, $dx/dt = a(x, t)$. Every point $(x_0, t_0)$ in the $(x, t)$ plane has a unique characteristic curve passing through it, because the ODE can be uniquely solved with the initial condition $x(t_0) = x_0$. Call this solution $x = x(t; x_0, t_0)$.

Now, along such a curve (parametrized by $t$) we calculate

$$0 = u_x a + u_t = \frac{\partial u}{\partial x}\frac{dx}{dt} + \frac{\partial u}{\partial t} = \frac{d}{dt}[u(x(t), t)] = \frac{du}{dt}.$$

So the derivative of $u$ along the curve must vanish. Thus $u(x(t), t)$ is constant along each such curve. That is, $u(x(t), t) = u(x_0, t_0)$. If we draw these curves in the $xy$ plane, any differentiable function $u(x, t)$ that is constant on each characteristic curve is a solution of the PDE (2).

## Example 1.

Let's solve the PDE

$$u_t + e^{x+t}u_x = 0. \tag{3}$$

The characteristic equation is $dx/dt = e^{x+t}$. This ODE separates as $e^{-x}dx = e^t dt$. Its solutions are $e^{-x} = -e^t + C$ where $C$ is an arbitrary constant, or $x = -\log(C - e^t)$. So the general solution of (3) is

$$u(x, t) = f(C) = f(e^{-x} + e^t) \tag{4}$$

where $f$ is an arbitrary differentiable function of one variable.  □

## Example 2.

Let's solve (3) with the initial condition $u(x, 0) = \phi(x)$. Using the preceding formula, we must have $\phi(x) = u(x, 0) = f(e^{-x} + 1)$. We find $f$ in terms of $\phi$ by substituting $s = e^{-x} + 1$ or $x = -\log(s - 1)$ to get $f(s) = \phi[-\log(s - 1)]$.

So the solution is

$$u(x, t) = \phi[-\log(e^{-x} + e^t - 1)].$$

For instance, if we want to solve (3) with $u(x, 0) = x^3$, then

$$u(x, t) = -[\log(e^{-x} + e^t - 1)]^3. \tag{5}$$

□

We begin our discussion of nonlinear equations with a very specific but typical example.

## Example 3.

The equation

$$u_t + uu_x = 0 \tag{6}$$

is the simplest form of the basic equation of fluids (13.2.5). It is nonlinear and therefore a lot subtler than the linear equations of Section 1.2. Nevertheless, we use the geometric method. The characteristic curves for (6) are the curves that are given by solutions of the ODE

$$\frac{dx}{dt} = u(x, t). \tag{7}$$

Because the PDE (6) is nonlinear, this characteristic equation (7) depends on the unknown function $u(x, t)$ itself! Each solution $u(x, t)$ of (6) will give a different set of characteristics. By the existence and uniqueness theorem for ODEs (see Section A.4), there is a unique curve passing through any given point $(x_0, t_0)$. So if we could lay our hands on a solution, it would provide us with a family of curves perfectly filling out the $xt$ plane without intersections.

At first, because we don't know a solution $u(x, t)$, it seems we could know nothing about such a characteristic curve $(x(t), t)$. But notice that $u$ is a constant on it:

$$\frac{d}{dt}[u(x(t), t)] = u_t + \frac{dx}{dt}u_x = u_t + uu_x = 0 \quad (!) \tag{8}$$

by the chain rule. The solution $u(x(t), t)$ is a constant on each such curve, even though we still don't know what the curve is. Hence it is also true that $dx/dt = u(x(t), t) = $ constant. From these observations we deduce three principles.

($\alpha$) *Each characteristic curve is a straight line.* So each solution $u(x, t)$ has a family of straight lines (of various slopes) as its characteristics.

($\beta$) *The solution is constant on each such line.*

($\gamma$) *The slope of each such line is equal to the value of $u(x, t)$ on it.*

Suppose now that we ask for a solution of the PDE that satisfies the initial condition

$$\boxed{u(x, 0) = \phi(x).} \tag{9}$$

That is, we specify the solution on the line $t = 0$. Then, by ($\gamma$), the characteristic line that passes through $(x_0, 0)$ must have slope $\phi(x_0)$. Similarly, the characteristic line through $(x_1, 0)$ must have slope $\phi(x_1)$. If the two lines intersect (see Figure 1), we're in trouble. For $u = \phi(x_0)$ on one line and $u = \phi(x_1)$ on the other line, so that $\phi(x_0) = \phi(x_1)$, which is impossible because they have different slopes!

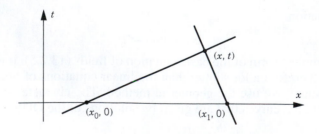

Figure 1

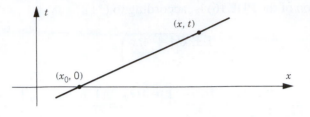

**Figure 2**

There are three conceivable ways out of this quandary. One is to avoid any such intersection of characteristic lines. This will happen for $t \geq 0$ whenever the function $\phi(x)$ is increasing for all $x$. A second way is to extend the notion of solution to allow discontinuities. This leads to the theory of shock waves. A third way is to simply admit that the solution usually exists only near the initial line $t = 0$ and that away from this line it may break down in some unknown manner.

We can write a formula for the solution of (6), where it exists, as follows. Consider the characteristic line passing through $(x_0, 0)$ and $(x, t)$ (see Figure 2). Its slope is

$$\frac{x - x_0}{t - 0} = \frac{dx}{dt} = u(x, t) = u(x_0, 0) = \phi(x_0),$$

so that

$$\boxed{x - x_0 = t\phi(x_0).} \tag{10}$$

Equation (10) gives $x_0$ implicitly as a function of $(x, t)$. Then

$$\boxed{u(x, t) = \phi(x_0(x, t))} \tag{11}$$

is the solution. [The implicit form of (10) is related to the geometric intersection problem discussed above.]  □

## Example 4.

Let us continue Example 3 by making a particular choice of the initial function $\phi(x)$. Let $\phi(x) = x^2$. Then (10) takes the form

$$x - x_0 = t x_0^2 \qquad \text{or} \qquad t x_0^2 + x_0 - x = 0.$$

We solve this quadratic equation for $x_0$ explicitly as

$$x_0 = \frac{-1 \pm \sqrt{1 + 4tx}}{2t} \qquad \text{for } t \neq 0.$$

The solution of the PDE (6) is, according to (11),

$$u(x, t) = \phi(x_0) = \left( \frac{-1 \pm \sqrt{1 + 4tx}}{2t} \right)^2$$

$$= \frac{1 \mp 2\sqrt{1 + 4tx} + (1 + 4tx)}{4t^2} = \frac{1 + 2tx \mp \sqrt{1 + 4tx}}{2t^2}$$

for $t \neq 0$. This formula is supposed to be the solution of the problem

$$u_t + u u_x = 0, \qquad u(x, 0) = x^2,$$

but it is not defined along the line $t = 0$. So we require that

$$x^2 = u(x, 0) = \lim_{t \to 0} \frac{1 + 2tx \mp \sqrt{1 + 4tx}}{2t^2}.$$

With the plus sign this expression has no limit (it is $2/0 = \infty$), so it can't be a solution. With the minus sign, however, there is some hope because the limit is $0/0$. We use L'Hôpital's rule twice (with $x$ constant) to calculate the limit as

$$\lim \frac{2x - 2x(1 + 4tx)^{-1/2}}{4t} = \lim \frac{4x^2(1 + 4tx)^{-3/2}}{4} = x^2,$$

as it should be. Therefore, the solution is

$$u(x, t) = \frac{1 + 2tx - \sqrt{1 + 4tx}}{2t^2} \qquad \text{for } t \neq 0. \qquad (12)$$

This is the formula for the unique (continuous) solution. It is a solution, however, only in the region $1 + 4tx \geq 0$, which is the region between the two branches of the hyperbola $tx = -\frac{1}{4}$.   □

Let's return now to the general equation

$$u_t + a(u) u_x = 0 \qquad (1)$$

The characteristic curves for (1) are the solutions of the ODE

$$\frac{dx}{dt} = a(u(x, t)). \qquad (13)$$

Calling such a curve $x = x(t)$, we observe that

$$\frac{d}{dt} u(x(t), t) = u_x \frac{dx}{dt} + u_t = u_x a(u) - a(u)u_x = 0.$$

Therefore, *the characteristics are straight lines and the solution is constant along them.*

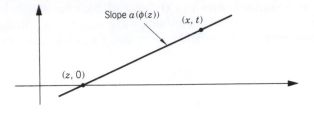

**Figure 3**

Thus we can solve equation (1) with the initial condition $u(x, 0) = \phi(x)$, provided that the characteristics do not intersect. The characteristic line through $(x, t)$ and $(z, 0)$ has the "slope" (see Figure 3)

$$\frac{x - z}{t - 0} = \frac{dx}{dt} = a(u(x, t)) = a(u(z, 0)) = a(\phi(z)).$$

Hence $x - z = t\,a(\phi(z))$. This gives $z$ implicitly as a function of $x$ and $t$. Writing $z = z(x, t)$, the solution of (1) is given by the formula

$$u(x, t) = u(z, 0) = \phi(z) = \phi(z(x, t)). \tag{14}$$

In case no pair of characteristic lines intersect in the half-plane $t > 0$, there exists a solution $u(x, t)$ throughout this half-plane. This can happen only if the slope is increasing as a function of the intercept:

$$a(\phi(z)) \leq a(\phi(w)) \qquad \text{for } z \leq w.$$

In other words, the lines spread out for $t > 0$. Such a solution is called an *expansive wave* or *rarefaction wave*.

In the general case, however, some characteristics will cross and the solution will exist only in the region up to the time of crossing. What does it look like near that time? The wave speed is $a(u)$. Because it depends on $u$, some parts of the wave move faster than others. At any "compressive" part of the wave, a movie of the solution will typically look as in Figure 4, where the "crest" moves faster and the wave is said to "*break*." The bigger $u$ is, the faster the wave moves, so the bigger, faster part of the wave overtakes the smaller, slower part of the wave. This leads to a triple-valued "solution." This is the situation when a water wave breaks on the beach, as well as at the shock front of an explosion.

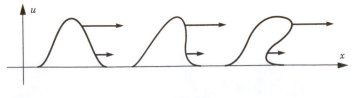

**Figure 4**

When the solution ceases to exist due to compression, what happens? Mathematically speaking, where a shock wave occurs, the solution $u(x, t)$ has a jump discontinuity. This usually occurs along a curve in the $xt$ plane.

## DISCONTINUOUS SOLUTIONS

As we are dealing with a solution of a PDE that is not even continuous, let alone differentiable, what is the meaning of the PDE? In Section 12.1 we discussed a way that very general "functions" could be solutions of PDEs. Therefore, for equation (1) we ask that it be valid *in the sense of distributions*. Let $A'(u) = a(u)$. The equation (1) can be written as $u_t + A(u)_x = 0$. For it to be valid in the sense of distributions means precisely that

$$\int_0^\infty \int_{-\infty}^\infty [u\psi_t + A(u)\psi_x]\,dx\,dt = 0 \tag{15}$$

for all test functions $\psi(x, t)$ defined in the half-plane. (A test function is a $C^\infty$ function in the $xt$ plane that is zero outside a bounded set.) A solution of this type is called a "weak" solution.

Suppose now that the jump discontinuity, called a *shock*, occurs along the curve $x = \xi(t)$ (see Figure 5). Because it is a jump, the limits $u^+(t) = u(x+, t)$ and $u^-(t) = u(x-, t)$ from the right and the left exist. We assume that the solution is smooth elsewhere. The speed of the shock is $s(t) = d\xi/dt$, which is the reciprocal of the slope in Figure 5. Now we split the inner integral in (15) into the piece from $-\infty$ to $\xi(t)$ and the piece from $\xi(t)$ to $+\infty$. On each piece separately (where the function is $C^1$) we apply the divergence theorem. We obtain

$$\int_0^\infty \int_{-\infty}^{\xi(t)} [-u_t\psi - A(u)_x\psi]\,dx\,dt - \int_{x=\xi(t)} [u^+\psi n_t + A(u^+)\psi n_x]\,dl$$

$$+ \int_0^\infty \int_{\xi(t)}^{+\infty} [-u_t\psi - A(u)_x\psi]\,dx\,dt$$

$$+ \int_{x=\xi(t)} [u^-\psi n_t + A(u^-)\psi n_x]\,dl = 0, \tag{16}$$

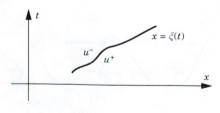

**Figure 5**

where $(n_x, n_t)$ denotes the unit normal vector to the shock curve which points to the left. But $u_t + A(u)_x = 0$ in the ordinary sense in both regions separately, so that the two double integrals in (16) vanish. Hence

$$\int_{x=\xi(t)} [u^+\psi n_t + A(u^+)\psi n_x]\, dl = \int_{x=\xi(t)} [u^-\psi n_t + A(u^-)\psi n_x]\, dl.$$

Because $\psi(x, t)$ is arbitrary, we can cancel it and we get the result that

$$\boxed{\frac{A(u^+) - A(u^-)}{u^+ - u^-} = -\frac{n_t}{n_x} = s(t).} \tag{17}$$

This is the *Rankine–Hugoniot formula* for the speed of the shock wave.

To summarize, a *shock wave* is a function with jump discontinuities along a finite number of curves on each of which (17) holds and off of which the PDE (1) holds.

### Example 5.

Let $a(u) = u$ and $\phi_1(x) = 1$ for $x > 0$, and $\phi_1(x) = 0$ for $x < 0$. Then $A(u) = \frac{1}{2}u^2$ and $a(\phi_1(x))$ is an increasing function of $x$. A continuous solution is the expansion wave $u_1(x, t) = 0$ for $x \le 0$, $x/t$ for $0 \le x \le t$, and 1 for $x \ge t$. It is a solution of the PDE for $t \ge 0$ because $(x/t)_t + (x/t)(x/t)_x = -x/t^2 + (x/t)(1/t) = 0$. The characteristics are sketched in Figure 6. □

### Example 6.

Let $a(u) = u$ and $\phi_2(x) = 0$ for $x > 0$, and $\phi_2(x) = 1$ for $x < 0$. Then $a(\phi_2(x))$ is a decreasing function of $x$ and there is no continuous solution. But there is a shock wave solution, namely $u_2(x, t) = 0$ for $x < st$, and $u_2(x, t) = 1$ for $x > st$. It has a discontinuity along the line $x = st$.

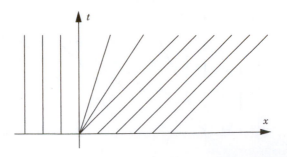

**Figure 6**

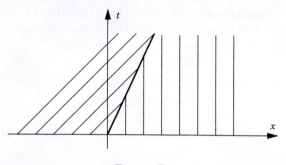

**Figure 7**

What is $s$? By (17),

$$s = \frac{A(u^+) - A(u^-)}{u^+ - u^-} = \frac{\frac{1}{2}0^2 - \frac{1}{2}1^2}{0 - 1} = \frac{1}{2}.$$

Therefore, the solution is as sketched in Figure 7. $\qquad\square$

## Example 7.

By extending the concept of a solution to allow jumps, we have introduced the possibility that the solution is not unique. Indeed, consider the same initial data as in Example 5, and the solution $u_3(x, t) = 0$ for $x < t/2$, and $u_3(x, t) = 1$ for $x > t/2$. This is obviously a solution for $x \neq t/2$ and furthermore it satisfies (17) because $s = \frac{1}{2} = (\frac{1}{2}1^2 - \frac{1}{2}0^2)/(1 - 0)$. Therefore there are at least two solutions with the initial condition $\phi_1(x)$ (see Figure 8).

Which one is physically correct? We could argue that the continuous one is preferred. But there may be other situations where neither one is continuous. Here both mathematicians and physicists are guided by the concept of entropy in gas dynamics. It requires that the wave speed just behind the shock is greater than the wave speed just ahead of it; that

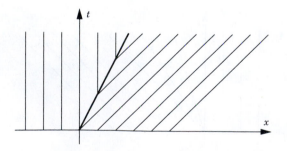

**Figure 8**

is, the wave behind the shock is "catching up" to the wave ahead of it. Mathematically, this means that on a shock curve we have

$$a(u^-) > s > a(u^+).$$    (18)

This is the *entropy criterion* for a solution. Notice that (18) is satisfied for Example 6 but not for Example 7. Therefore, Example 7 is rejected. Finally, the definition of a shock wave is complete. *Along its curves of discontinuity it must satisfy both* (17) *and* (18). For further discussion of shocks, see [Wh] or [Sm].    □

## EXERCISES

1.  (a)  Use direct differentiation to check that (4) solves (3).
    (b)  Check directly that (5) solves the initial condition $u(x, 0) = x^3$.

2.  Solve $(1 + t)u_t + xu_x = 0$. Then solve it with the initial condition $u(x, 0) = x^5$ for $t > 0$.

3.  Solve the nonlinear equation $u_t + uu_x = 0$ with the auxiliary condition $u(x, 0) = x$. Sketch some of the characteristic lines.

4.  Sketch some typical characteristic lines for Example 4.

5.  Solve $u_t + u^2 u_x = 0$ with $u(x, 0) = 2 + x$.

6.  Verify by differentiation that the formula (12) provides a solution of the differential equation (6).

7.  Solve $xu_t + uu_x = 0$ with $u(x, 0) = x$. (*Hint:* Change variables $x \mapsto x^2$.)

8.  Show that a smooth solution of the problem $u_t + uu_x = 0$ with $u(x, 0) = \cos \pi x$ must satisfy the equation $u = \cos[\pi(x - ut)]$. Show that $u$ ceases to exist (as a single-valued continuous function) when $t = 1/\pi$. (*Hint:* Graph $\cos^{-1} u$ versus $\pi(x - ut)$ as functions of $u$.)

9.  Check by direct differentiation that the formula $u(x, t) = \phi(z)$, where $z$ is given implicitly by $x - z = t \, a(\phi(z))$, does indeed provide a solution of the PDE (1).

10. Solve $u_t + uu_x = 0$ with the initial condition $u(x, 0) = 1$ for $x \le 0$, $1 - x$ for $0 \le x \le 1$, and 0 for $x \ge 1$. Solve it for all $t \ge 0$, allowing for a shock wave. Find exactly where the shock is and show that it satisfies the entropy condition. Sketch the characteristics.

11. Show that (15) is equivalent to the statement

$$\frac{d}{dt} \int_a^b u(x, t) \, dx + A(u(b, t)) - A(u(a, t)) = 0 \qquad \text{for all } a, b.$$

12. Solve $u_t + uu_x = 1$ with $u(x, 0) = x$.

## 14.2   SOLITONS

A soliton is a localized traveling wave solution of a nonlinear PDE that is remarkably stable. One PDE that has such a solution is the *Korteweg–deVries (KdV) equation*,

$$u_t + u_{xxx} + 6uu_x = 0 \qquad (-\infty < x < \infty). \tag{1}$$

(The "six" is of no special significance.) It has been known for a century that this equation describes water waves along a canal. The soliton was first observed by J. S. Russell in 1834, who wrote:

> I was observing the motion of a boat which was rapidly drawn along a narrow channel by a pair of horses, when the boat suddenly stopped—not so the mass of water in the channel which it had put in motion; it accumulated round the prow of the vessel in a state of violent agitation, then suddenly leaving it behind, rolled forward with great velocity, assuming the form of a large solitary elevation, a rounded, smooth and well-defined heap of water, which continued its course along the channel apparently without change of form or diminution of speed. I followed it on horseback, and overtook it still rolling on at a rate of some eight or nine miles an hour, preserving its original figure some thirty feet long and a foot to a foot and a half in height. Its height gradually diminished, and after a chase of one or two miles I lost it in the windings of the channel.

The same equation has also come up in the theory of plasmas and several other branches of physics.

Associated with equation (1) are the three fundamental quantities

$$\text{mass} = \int_{-\infty}^{\infty} u \, dx$$

$$\text{momentum} = \int_{-\infty}^{\infty} u^2 \, dx$$

$$\text{energy} = \int_{-\infty}^{\infty} \left( \tfrac{1}{2} u_x^2 - u^3 \right) dx,$$

each of which is a constant of motion (an invariant) (see Exercise 1). In fact, it has been discovered that there are an *infinite* number of other invariants involving higher derivatives.

Let's look for a *traveling wave solution* of (1), that is,

$$u(x, t) = f(x - ct).$$

We get the ODE $- cf' + f''' + 6ff' = 0$. Integrating it once leads to $-cf + f'' + 3f^2 = a$, where $a$ is a constant. Multiplying it by $2f'$ and integrating

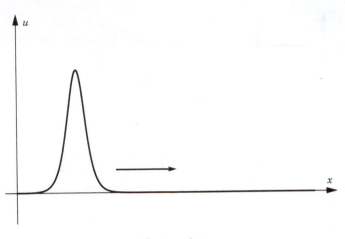

**Figure 1**

again leads to

$$-cf^2 + (f')^2 + 2f^3 = 2af + b, \qquad (2)$$

where $b$ is another constant.

We are looking for a solution that is solitary like Russell's, meaning that away from the heap of water, there is no elevation. That is, $f(x), f'(x)$, and $f''(x)$ should tend to zero as $x \to \pm\infty$. Then we must have $a = b = 0$, so that $-cf^2 + (f')^2 + 2f^3 = 0$. The solution of this first-order ODE is

$$f(x) = \tfrac{1}{2}c \ \text{sech}^2\!\left[\tfrac{1}{2}\sqrt{c}(x - x_0)\right], \qquad (3)$$

where $x_0$ is the constant of integration and sech is the hyperbolic secant, $\text{sech} \, x = 2/(e^x + e^{-x})$ (see Exercise 3). It decays to zero exponentially as $x \to \pm\infty$ (see Figure 1).

With this function $f$, $u(x, t) = f(x - ct)$ is the *soliton*. It travels to the right at speed $c$. Its amplitude is $c/2$. There is a soliton for every $c > 0$. It is tall, thin, and fast if $c$ is large, and is short, fat, and slow if $c$ is small.

The remarkable stability of the soliton, discovered only in the 1960s by computer experimentation by M. Kruskal and N. Zabusky, can be described as follows. If we start with two solitons, the faster one will overtake the slower one and, after a complicated nonlinear interaction, the two solitons will emerge *unscathed* as they move to the right, except for a slight delay. In fact, it was observed from the computer output that every solution of (1), with *any* initial function $u(x, 0) = \phi(x)$, seems to decompose as $t \to +\infty$ into a finite number of solitons (of various speeds $c$) plus a *dispersive tail* which gradually disappears (see Figure 2).

This kind of behavior is expected for linear problems because each eigenfunction evolves separately as in Section 4.1, but that it could happen for a nonlinear problem was a complete surprise at the time. This special behavior

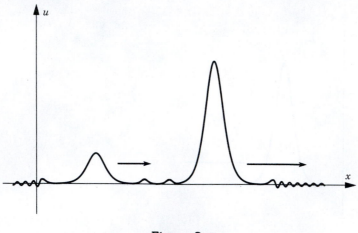

Figure 2

induced physicists to use the soliton as a mathematical model of a stable elementary particle.

## INVERSE SCATTERING

The explanation of soliton stability lies in the *inverse scattering* analysis, which shows that the highly nonlinear equation (1) has in fact a close but complicated relationship with a *linear* equation! The associated linear equation is just Schrödinger's:

$$-\psi_{xx} - u\psi = \lambda\psi \qquad (4)$$

with the parameter $\lambda$ and the "potential function" $u = u(x, t)$. In (4), the time variable $t$ is to be regarded merely as another parameter, so that we are dealing with a family of potentials that are functions of $x$ depending on the parameter $t$.

We know from Section 13.4 that (in general) there are some eigenvalues $\lambda$ for (4). That is, there are expected to be solutions of (4) with $\int_{-\infty}^{\infty} |\psi|^2 dx < \infty$ and $\lambda$ is expected to depend on $t$. If $\lambda(t)$ is an eigenvalue and $\psi(x, t)$ its eigenfunction that satisfies equation (4), let's substitute

$$u = -\lambda - \frac{\psi_{xx}}{\psi} \qquad (5)$$

into (1) to get

$$\lambda_t \psi^2 + (\psi h_x - \psi_x h)_x = 0, \qquad (6)$$

where

$$h = \psi_t - 2(-u + 2\lambda)\psi_x - u_x\psi$$

(see Exercise 5). We may normalize so that $\int_{-\infty}^{\infty} |\psi|^2 dx = 1$. Integrating (6), we then get $\lambda_t = 0$, so that $\lambda$ is a constant. Thus we have found the key relationship between (1) and (4). *If u solves (1) and $\lambda$ is an eigenvalue of (4), then $\lambda$ does not depend on t.*

Thus each eigenvalue provides another constant of the motion. Let's denote all the discrete spectrum of (4) by

$$\lambda_N \leq \lambda_{N-1} \leq \cdots \leq \lambda_1 < 0,$$

where $\lambda_n = -\kappa_n^2$ has the eigenfunction $\psi_n(x, t)$. The limiting behavior

$$\psi_n(x, t) \sim c_n(t)\, e^{-\kappa_n x} \qquad \text{as } x \to +\infty$$

can be proven from (4) (see [AS] or [Ne]). The $c_n(t)$ are called the *normalizing constants*.

As we know, a complete analysis of (4) involves the continuous spectrum as well. So let $\lambda = k^2 > 0$ be a fixed positive number. From Section 13.4 we know that there is a solution of (4) with the asymptotic behavior

$$
\begin{array}{ll}
\psi(x) \sim e^{-ikx} + R e^{+ikx} & \text{as } x \to +\infty \\[4pt]
\psi(x) \sim T e^{-ikx} & \text{as } x \to -\infty.
\end{array}
\tag{7}
$$

Here the reflection coefficient $R$ and the transmission coefficient $T$ may depend on both $k$ and the "parameter" $t$.

**Theorem 1.** $\partial T/\partial t \equiv 0$, $\partial R/\partial t = 8ik^3 R$, and $dc_n/dt = 4\kappa_n^3 c_n$.

It follows immediately from Theorem 1 that

$$
T(k, t) = T(k, 0), \qquad R(k, t) = R(k, 0)e^{8ik^3 t}, \qquad c_n(t) = c_n(0)e^{4\kappa_n^3 t}.
$$

**Proof.**   Identity (6) is also true for any $\lambda > 0$. But $\lambda$ is a fixed constant. So (6) says that $\psi h_x - \psi_x h$ depends only on $t$. So $\psi h_x - \psi_x h$ equals its asymptotic value as $x \to -\infty$, which is the sum of several terms. From the formula for $h$, the fact that $u(x, t) \to 0$, and from the asymptotic expression (7) for $\psi(x, t)$, all of these terms cancel (see Exercise 6). Hence $\psi h_x - \psi_x h \equiv 0$.

Furthermore, dividing the last result by $\psi^2$, it follows that the quotient $h/\psi$ is a function of $t$ only. So $h/\psi$ also equals its asymptotic value as $x \to +\infty$. That is,

$$
\begin{aligned}
\frac{h}{\psi} &= \frac{\psi_t - 2(-u + 2\lambda)\psi_x - u_x \psi}{\psi} \\[8pt]
&\sim \frac{R_t e^{ikx} - 4\lambda(-ike^{-ikx} + Rike^{ikx}) - 0}{e^{-ikx} + R e^{ikx}} \\[8pt]
&= \frac{[R_t - 4ik\lambda R]e^{ikx} + [4ik\lambda]e^{-ikx}}{R e^{ikx} + e^{-ikx}}.
\end{aligned}
$$

For this expression to be independent of $x$, the numerator and denominator must be linearly dependent as functions of $x$. Therefore,

$$\frac{R_t - 4ik\lambda R}{R} = \frac{4ik\lambda}{1}.$$

Thus $R_t = 8ik\lambda R = 8ik^3 R$, proving the second part of the theorem. The first and third parts are left for Exercise 8.

## SOLUTION OF THE KdV

The scattering theory of the associated Schrödinger equation leads to the resolution of the KdV equation with an initial condition $u(x, 0) = \phi(x)$. Schematically, the method is

$$\phi(x) \rightarrow \text{scattering data at time } 0$$
$$\rightarrow \text{scattering data at time } t \tag{8}$$
$$\rightarrow u(x, t).$$

By the *scattering data* we mean the reflection and transmission coefficients, the eigenvalues, and the normalizing constants. The first arrow is the *direct scattering problem*, finding the scattering data of a given potential. The second one is trivial due to the theorem above. The third one is the *inverse scattering problem*, finding the potential with given scattering data. The third step is the difficult one. There is a complicated procedure, the Gelfand–Levitan method, to carry it out. It turns out that the transmission coefficient $T(t)$ is not needed, and the required scattering data is just $\{R, \kappa_n, c_n\}$. All three steps in (8) have unique solutions, leading to the unique solution of (1) with the given initial condition.

## Example 1.

Suppose that we take $u(x, 0)$ to be the initial data of a single soliton, say $u(x, 0) = 2 \operatorname{sech}^2 x$. The unique solution, of course, is $u(x, t) = 2 \operatorname{sech}^2(x - 4t)$, which is (3) with $c = 4$. It turns out that in this case $R(k, t) \equiv 0$ and there is exactly one negative eigenvalue of the Schrödinger operator (4) (see Exercise 9). □

All of the cases with vanishing reflection coefficient can be explicitly computed. If there are $N$ negative eigenvalues of (4), one obtains a complicated but explicit solution of (1) which as $t \rightarrow \pm\infty$ resolves into $N$ distinct solitons. This solution is called an *N-soliton*.

## Example 2.

Let $\phi(x) = 6 \operatorname{sech}^2 x$. Then it turns out that $R(k, t) \equiv 0$, $N = 2$, $\lambda_1 = -1$, and $\lambda_2 = -4$. The solution of (1) is the 2-soliton, given by

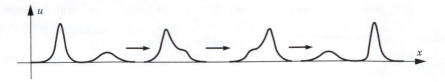

**Figure 3**

the formula

$$u(x, t) = \frac{12[3 + 4\cosh(2x - 8t) + \cosh(4x - 64t)]}{[3\cosh(x - 28t) + \cosh(3x - 36t)]^2}.$$

A movie of it is shown in Figure 3. It looks asymptotically like a pair of single solitons of amplitudes 8 and 2 which are interacting. There is a *phase shift*, which means they are slightly shifted from what they would have been if they were not interacting.    □

There are a number of other "soliton equations," including the cubic Schrödinger equation

$$iu_t + u_{xx} + |u|^2 u = 0,$$

the (ridiculously named) sine-Gordon equation

$$u_{tt} - u_{xx} + \sin u = 0,$$

and the Kadomstev–Petviashvili equation

$$(u_t + u_{xxx} + 6uu_x)_x + 3u_{yy} = 0.$$

Each such equation arises in a variety of problems in physics. Soliton equations are, however, very special; almost any change in the nonlinear term will spoil it. For instance, the equation $u_t + u_{xxx} + u^p u_x = 0$ is a soliton equation only if $p = 2$ or 3. For more information, see [AS], [Dd] or [Ne].

## EXERCISES

1. Show that the mass, momentum, and energy are constants of motion (invariants) for the KdV equation by direct differentiation with respect to time.

2. Show that $\int (xu - 3tu^2)\, dx$ is also an invariant.

3. Derive the formula (3) for the soliton.

4. Show that there also exist *periodic* traveling wave solutions of KdV, as follows. Let $P(f) = -2f^3 + cf^2 + 2af + b$.
   (a) By solving the ODE $(f')^2 = P(f)$, find an implicit formula defining $f(x)$.
   (b) Show that each simple zero of $P(f)$ corresponds to a minimum or maximum of $f(x)$.

(c)   Show that one can choose $a$ and $b$ in such a way that the cubic polynomial $P(f)$ has three real zeros $f_1 < f_2 < f_3$. Show that $P(f) > 0$ for $f_2 < f < f_3$ and for $f < f_1$.

(d)   Show that there is a periodic solution $f(x)$ with $\max_x f(x) = f_3$ and $\min_x f(x) = f_2$.

(e)   Look up what an elliptic integral is (in [MOS], for instance) and transform the formula in (a) to an *elliptic integral of the first kind*.

5.   (*difficult*) Verify the key identity (6).

6.   Use the asymptotic formula (7) to prove that $\psi h_x - \psi_x h \equiv 0$.

7.   Carry out the following alternative proof that an eigenvalue $\lambda$ in equation (4) is independent of $t$. Differentiate (4) with respect to $t$, multiply the result by $\psi$, integrate over $x$, integrate by parts, substitute $\partial u/\partial t$ from (1), and simplify the result.

8.   Prove the rest of Theorem 1.

9.   Consider the single soliton $2 \operatorname{sech}^2 x$ and its associated eigenvalue problem $\psi_{xx} + (2 \operatorname{sech}^2 x + \lambda)\psi = 0$.

(a)   Substitute $s = \tanh x$ to convert it to

$$[(1 - s^2)\psi']' + \left(2 + \frac{\lambda}{1 - s^2}\right)\psi = 0 \qquad \text{where } ' = \frac{d}{ds}.$$

This is the associated Legendre equation of degree one (see Section 10.6).

(b)   Show that a solution that vanishes at $x = \pm\infty$ ($s = \pm 1$) is $\lambda = -1$ and $\psi = \frac{1}{2}\sqrt{1 - s^2} = \frac{1}{2} \operatorname{sech} x$. This is the unique bound state of this Schrödinger equation!

(c)   (*difficult*) By analyzing the solutions of the associated Legendre equation for $\lambda = k^2 > 0$, show that $R(k) \equiv 0$ and $T(k) = (ik - 1)/(ik + 1)$.

10.   The linearized KdV equation is $u_t + u_{xxx} = 0$. Solve it by Fourier transformation with the initial condition $u(x, 0) = \phi(x)$, assuming appropriate decay as $x \to \pm\infty$. Write the answer in terms of the Airy function

$$A(\xi) = \int_{-\infty}^{\infty} e^{ik\xi + ik^3/3} \frac{dk}{2\pi}$$

assuming this integral makes sense.

11.   Show that the KdV equation is invariant under the scaling transformation: $x \mapsto kx, t \mapsto k^3 t, u \mapsto k^{-2}u$ for any $k > 0$.

12.   Use the idea in Exercise 11 together with the ideas in Section 2.4 to show that the linearized KdV equation $u_t + u_{xxx} = 0$, $u(0) = \phi$ must have a solution of the form $u(x, t) = (k/t^{1/3}) \int_{-\infty}^{\infty} B((x - y)/t^{1/3}) \phi(y) \, dy$.

13.  For the cubic Schrödinger equation $iu_t + u_{xx} + |u|^2 u = 0$, show that $Q = \int |u|^2 \, dx$ and $2E = \int (u_x^2 - \frac{1}{2}|u|^4) \, dx$ are constants of motion (independent of time).

## 14.3  CALCULUS OF VARIATIONS

Dirichlet's principle (Section 7.1) is a problem in the calculus of variations. In general, the calculus of variations is concerned with minimizing or maximizing a quantity that depends on an arbitrary function. The Dirichlet principle asserts that one can find the state of *least* potential energy among all the functions in $D$ that satisfy a boundary condition on bdy $D$. This state is the harmonic function. A second example we have seen in Section 11.2 is the computation of the eigenvalues as minima of the energy.

### Example 1.

Here is another instance of a minimum. Consider the *refraction of light* as it passes through an inhomogeneous medium. The velocity of light $c(\mathbf{x})$ changes from point to point. Thus $|d\mathbf{x}/dt| = c(\mathbf{x})$ so that $dt = |d\mathbf{x}|/c(\mathbf{x})$. Writing $\mathbf{x} = (x, y)$ and assuming a light ray travels in the $xy$ plane along a curve $y = u(x)$, the time required for it to get from $x = a$ to $x = b$ is therefore

$$T = \int_a^b \frac{1}{c(x, u)} \sqrt{1 + \left(\frac{du}{dx}\right)^2} \, dx. \qquad (1)$$

Fermat's principle states that the path must *minimize* this time $T$. In a homogeneous medium the solution of this minimization problem is of course the straight line (see Exercise 1).  □

### Example 2.

Another famous problem is to find the *shape of a soap bubble* that spans a wire ring. The shape must be chosen so as to minimize the energy, and according to the physics of stretched membranes, the energy is proportional to the surface area. Thus, if we want to find the soap bubble bounded by a wire ring, we must minimize the surface area subject to the boundary condition given by the wire ring. The geometry is simplest when the surface can be written as the graph of a (single-valued) function $z = f(x, y)$ where $(x, y)$ lies over a plane region $D$ and the wire ring is a curve $\{(x, y) \in \text{bdy } D, z = h(x, y)\}$ lying over bdy $D$ (see Figure 1). The surface area is

$$A = \iint_D \sqrt{1 + u_x^2 + u_y^2} \, dx \, dy. \qquad (2)$$

We must *minimize* $A$ among all functions $u(x, y)$ in the closure $\overline{D}$ that satisfy $u(x, y) = h(x, y)$ on bdy $D$.  □

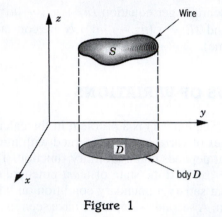

**Figure 1**

## Example 3.

Sometimes in the calculus of variations one does not minimize or maximize but one just looks for a *saddle point*. For instance, let's take the vibrating string (Section 1.3). Physicists define the *action* as the kinetic energy *minus* the potential energy, which in this case is

$$A[u] = \int_{t_1}^{t_2} \int_0^L \left[ \rho \left( \frac{\partial u}{\partial t} \right)^2 - T \left( \frac{\partial u}{\partial x} \right)^2 \right] dx \, dt \tag{3}$$

and assert that the action is "stationary." This means that the "derivative" of $A[u]$ is zero at a solution $u$. That is, $(d/d\epsilon)A(u + \epsilon v) = 0$ at $\epsilon = 0$. Explicit differentiation of $A(u + \epsilon v)$ with respect to $\epsilon$ leads to

$$\int_{t_1}^{t_2} \int_0^L \left( \rho \frac{\partial u}{\partial t} \frac{\partial v}{\partial t} - T \frac{\partial u}{\partial x} \frac{\partial v}{\partial x} \right) dx \, dt = 0$$

for all functions $v(x, t)$ that vanish on the space-time boundary. We can integrate by parts to get

$$-\int_{t_1}^{t_2} \int_0^L \left( \frac{1}{2} \rho \frac{\partial^2 u}{\partial t^2} - \frac{1}{2} T \frac{\partial^2 u}{\partial x^2} \right) (v) \, dx \, dt = 0.$$

Because $v$ is arbitrary inside $D$, we deduce that

$$\rho \frac{\partial^2 u}{\partial t^2} - T \frac{\partial^2 u}{\partial x^2} = 0 \quad \text{in } D,$$

which is the wave equation!    □

The particular examples above require us to maximize, minimize, or find the *stationary points* of a functional of the form

$$E[u] = \int_a^b F(x, u, u') \, dx \tag{4}$$

or

$$E[u] = \iint_D F(x, y, u, u_x, u_y) \, dx \, dy. \tag{5}$$

The basic idea of the calculus of variations is to set the first derivative equal to zero, just as in ordinary calculus.

Let's carry this out for the case of (4) assuming that $u(x)$ is specified at both ends. Let $v(x)$ be any function that vanishes at both ends. We denote $p = u'(x)$ and $F = F(x, u, p)$. We consider

$$g(\epsilon) = E[u + \epsilon v] = \int_a^b F(x, u + \epsilon v, u' + \epsilon v') \, dx$$

as a function of the single variable $\epsilon$. We set its derivative equal to zero at $\epsilon = 0$ to get

$$\int_a^b \left( \frac{\partial F}{\partial u} v + \frac{\partial F}{\partial p} v' \right) dx = 0.$$

Integrating by parts, we get

$$\int_a^b \left( \frac{\partial F}{\partial u} - \frac{d}{dx} \frac{\partial F}{\partial p} \right) (v) \, dx = 0.$$

Because $v(x)$ is an arbitrary function on $a < x < b$, we could choose it to be the first factor in the integral and therefore we deduce from the first vanishing theorem in Section A.1 that

$$\boxed{\frac{\partial F}{\partial u} = \frac{d}{dx} \frac{\partial F}{\partial p}.} \tag{6}$$

That is,

$$\frac{\partial F}{\partial u}(x, u(x), u'(x)) = \frac{d}{dx} \left[ \frac{\partial F}{\partial p}(x, u(x), u'(x)) \right].$$

This is the Euler-Lagrange equation for (4), an ODE that must be satisfied by $u(x)$.

For the two-dimensional problem (5), we denote $p = u_x, q = u_y$ and $F = F(x, y, u, p, q)$. The same procedure leads to

$$0 = \iint_D \left( \frac{\partial F}{\partial u} v + \frac{\partial F}{\partial p} v_x + \frac{\partial F}{\partial q} v_y \right) dx \, dy$$

$$= \iint_D \left( \frac{\partial F}{\partial u} - \frac{\partial}{\partial x} \frac{\partial F}{\partial p} - \frac{\partial}{\partial y} \frac{\partial F}{\partial q} \right) (v) \, dx \, dy.$$

In this case the Euler-Lagrange equation takes the form

$$\frac{\partial F}{\partial u} = \frac{\partial}{\partial x}\frac{\partial F}{\partial p} + \frac{\partial}{\partial y}\frac{\partial F}{\partial q},$$
(7)

a PDE to be satisfied by $u(x, y)$. In equation (7), $p = u_x(x, y)$ and $q = u_y(x, y)$.

Let's apply this to the soap bubble problem, Example 2 above. The integrand is $F(u_x, u_y) = \sqrt{1 + u_x^2 + u_y^2}$, which does not depend on $u$ but only on its first derivatives. Differentiating $F$, the Euler-Lagrange equation (7) takes the form

$$\left(\frac{u_x}{\sqrt{1 + u_x^2 + u_y^2}}\right)_x + \left(\frac{u_y}{\sqrt{1 + u_x^2 + u_y^2}}\right)_y = 0,$$
(8)

called the *minimal surface equation*. This PDE is nonlinear but nevertheless is reminiscent of the Laplace equation. It is an elliptic equation (see Exercise 5). It shares some properties with the Laplace equation but is much harder to solve. If the minimal surface equation is linearized around the zero solution, the square root is replaced by 1 and the Laplace equation is obtained.

For more on the calculus of variations, see [Ak] or [Ga].

## EXERCISES

1.  Use the calculus of variations to prove that the shortest path between two points is a straight line. (*Hint:* Minimize the integral (1) where $c(x, u) \equiv 1$.)

2.  Find the shortest curve in the $xy$ plane that joins the two given points $(0, a)$ and $(1, b)$ and that has a given area $A$ below it (above the $x$-axis and between $x = 0$ and $x = 1$); $a$ and $b$ are positive.

3.  Prove Snell's law of reflection: Given two points $P$ and $Q$ on one side of a plane $\Pi$, the shortest broken line path from $P$ to a point of $\Pi$ and thence to $Q$ is the unique path that makes equal angles with $\Pi$. (You may use ordinary calculus.)

4.  Find the curve $y = u(x)$ that makes the integral $\int_0^1 (u'^2 + xu)\, dx$ stationary subject to the constraints $u(0) = 0$ and $u(1) = 1$.

5.  (a)  Carry out the derivatives in the minimal surface equation (8) to write it as a second-order equation.
    (b)  Show that the equation is elliptic in the sense of Section 1.6.

6.  Find the minimal surfaces *of revolution*. (*Hint:* The area is $\int_a^b 2\pi y \sqrt{1 + y'^2}\, dx$.)

7.  Show that there are an infinite number of functions  that minimize

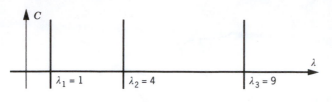

**Figure 1**

the integral

$$\int_0^2 (y')^2 (1 + y')^2 \, dx \quad \text{subject to } y(0) = 1 \text{ and } y(2) = 0.$$

They are continuous functions with piecewise continuous first derivatives.

8.  In the kinetic theory of gases, equilibrium is attained when the entropy is minimized subject to constant mass, momentum, and energy. Find the resulting distribution of particles $f(\mathbf{v})$ where $\mathbf{v}$ denotes the velocity. Here entropy $= \iiint f(\mathbf{v}) \log f(\mathbf{v}) \, d\mathbf{v} = H$, energy $= \iiint \frac{1}{2} |\mathbf{v}|^2 f(\mathbf{v}) \, d\mathbf{v} = E$, mass $= \iiint f(\mathbf{v}) \, d\mathbf{v} = 1$, and momentum $= \iiint \mathbf{v} f(\mathbf{v}) \, d\mathbf{v} = 0$.

9.  Repeat Exercise 8 without the energy constraint.

10. (a)  If the action is $A[u] = \iint \left( \frac{1}{2} u_x u_t + u_x^3 - \frac{1}{2} u_{xx}^2 \right) dx \, dt$, find the Euler-Lagrange equation.

    (b)  If $v = u_x$, show that $v$ satisfies the KdV equation.

11. If the action is $A[u] = \iint (u_{xx}^2 - u_t^2) \, dx \, dt$, show that the Euler-Lagrange equation is the *beam equation* $u_{tt} + u_{xxxx} = 0$, the equation for a stiff rod.

## 14.4   BIFURCATION THEORY

A bifurcation means a fork in the road. The road here is a path of solutions that depend on a parameter. Bifurcation theory means the study of how the solutions of differential equations depend on parameters and, in particular, the study of the forks in the road.

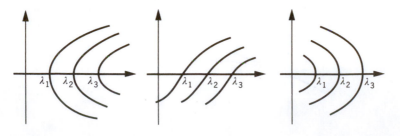

**Figure 2**

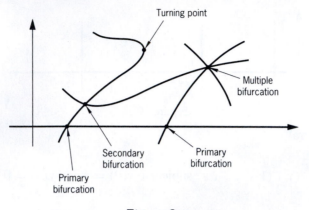

Figure 3

A trivial example is afforded by the eigenvalue problem $-u'' = \lambda u$ in $[0, \pi]$ with $u = 0$ at $x = 0, \pi$. The solutions are $\lambda = n^2$ with $u(x) = C \sin nx$ with the arbitrary parameter $C$. The picture of all the solutions in the $\lambda C$ plane is therefore drawn schematically in Figure 1. There is a horizontal line $(C = 0)$ and an infinite number of vertical lines $\lambda = n^2$, one at each eigenvalue. There is a "fork" at each eigenvalue on the $\lambda$ axis.

If nonlinear terms are present, they will distort these lines into curves. Typically they may look like one of the examples shown in Figure 2. There can also be secondary (or tertiary, etc.) bifurcations as shown in Figure 3.

## Example 1.

A rod is subjected to a compressive load $\lambda$ and bends as shown in Figure 4. (For instance, it could be a vertical yardstick resting on the ground with a weight at its upper end.) Let $u(x)$ be the angle of bending. Under reasonable assumptions about the rod, one can show that

$$\frac{d^2u}{dx^2} + \lambda \sin u = 0 \qquad \text{for } 0 \le x \le l \tag{1}$$

$$\frac{du}{dx}(0) = 0 = \frac{du}{dx}(l). \tag{2}$$

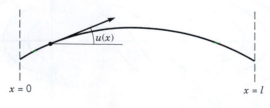

$x = 0$ $\qquad\qquad\qquad\qquad\qquad\qquad$ $x = l$

Figure 4

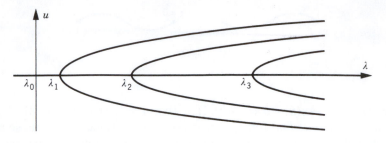

**Figure 5**

The ODE is nonlinear. The equilibrium state is $u \equiv 0$. The parameter is $\lambda$.

We can "linearize" it by replacing $\sin u = u - u^3/3! + \cdots$ simply by $u$. So the linearized problem is $v'' + \lambda v = 0$ with $v'(0) = 0 = v'(l)$. Its exact solutions are $\lambda_n = n^2 \pi^2 / l^2$ with $v_n(x) = C \cos(n \pi x / l)$ ($n = 0, 1, 2, \ldots$). A detailed analysis of the nonlinear problem (1)-(2) shows that each line is distorted to the right as in Figure 5.

This bifurcation diagram can be interpreted as follows. If the load is small, then $\lambda < \pi^2/l^2$ and merely a slight compression of the rod results. But as the load increases past the value $\pi^2/l^2$, the rod can buckle one way or the other (see Figure 6). These are the two branches of the first "pitchfork" in Figure 5. As the load gets still bigger, $\lambda > 4\pi^2/l^2$, there are two more theoretical possibilities as in Figure 7. Which states are most likely to occur for large loads? It can be shown that the two simple buckled states in Figure 6 will almost certainly occur because all the other states (the trivial state $u \equiv 0$ as well as the complicated states as in Figure 7) are *unstable*. That is, the states in Figure 7 can occur but a tiny perturbation of them will make them "pop" into one of the stable states of Figure 6. (With a very heavy load of course, the rod would break.) □

## Example 2.

We now consider the problem

$$u_{xx} + f(u) = 0 \qquad \text{for } -l \leq x \leq l \tag{3}$$

$$u(-l) = 0 = u(l), \tag{4}$$

$$x = 0 \qquad\qquad x = l \qquad x = 0 \qquad\qquad x = l$$

**Figure 6**

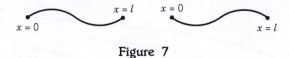

Figure 7

where $f(u) = -u(u - a)(u - b)$. Here $0 < a < \frac{1}{2}b$ (see Figure 8). This is a model of a *reaction-diffusion* problem. The nonlinear term represents a chemical reaction, the unknown $u$ is the concentration of a substance, and $u_{xx}$ is the diffusion term. The corresponding time-dependent problem is the nonlinear diffusion equation $u_t = u_{xx} + f(u)$.

Rather than being fixed, $l$ is regarded as a parameter. Of course, there is the trivial solution $u \equiv 0$. Are there any others? How do they depend on the parameter? To analyze the problem, we draw the *phase plane* picture for the ODE (3), letting $v = u'$ (see Figure 9). Note that $H(u, v) = \frac{1}{2}v^2 + F(u)$, where $F' = f$, is independent of $x$. We are look-ing for orbits that satisfy the boundary conditions, which means that the curves should begin and end on the $v$ axis; that is, we want $u = 0$ when $x = \pm l$. At $x = -l$, the curve passes through a point $(0, p)$. By the sym-metry ($v \to -v$), the curve will be symmetric with respect to the $u$ axis, and at $x = 0$ it will pass through a point $(\alpha(p), 0)$ on the $u$ axis. Thus $H(u, v) = F(\alpha(p))$, so that

$$\frac{du}{dx} = v = \sqrt{2}\sqrt{F(\alpha(p)) - F(u)}.$$

Solving this for $dx$ and integrating from $x = 0$ to $x = l$, we get

$$l = \frac{1}{\sqrt{2}} \int_0^{\alpha(p)} \frac{du}{\sqrt{F(\alpha(p)) - F(u)}}. \tag{5}$$

We would like to solve (5) for $p$ if $0 < p < A$, where $A$ is defined in Figure 9. Let's denote the right side of (5) by $\beta(p)$ so that (5) takes the form $l = \beta(p)$. We can show that the graph of $\beta(p)$ has a single minimum $\beta_0$ in $0 < p < A$ and that $\beta(p) \to \infty$ as $p \to 0$ or $A$ (see Figure 10). Thus if $l < \beta_0$, there is no solution except the trivial one. If $l > \beta_0$, there

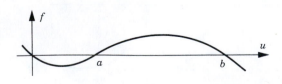

Figure 8

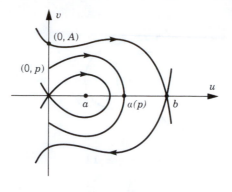

**Figure 9**

are exactly two nontrivial solutions $u_1(x)$ and $u_2(x)$, which in the phase plane look like Figure 11. □

## Example 3.

Finally, consider the time-dependent reaction-diffusion problem

$$u_t = u_{xx} + f(u) \quad \text{in } -l < x < l, \qquad u(-l, t) = 0 = u(l, t) \qquad (6)$$

with the same $f$ as in Example 2. Note that the solutions $u_0 \equiv 0$, $u_1(x)$, and $u_2(x)$ are stationary solutions of the parabolic PDE (6). It can be shown (see Section 24D in [Sm]) that $u_0 \equiv 0$ and $u_2(x)$ are stable, but that $u_1(x)$ is unstable. This means that if the initial condition $u(x, 0) = \phi(x)$ is near enough to 0 or $u_2(x)$, then the solution will always remain nearby in the future. The opposite is true of $u_1(x)$. No matter how near the function $\phi(x)$ is to the function $u_1(x)$, the solution of (6) may get away from $u_1(x)$ in the future; it might go to $u_0$ or to $u_2(x)$ as $t \to +\infty$. □

For further reading on bifurcation, see [Sm] or [IJ].

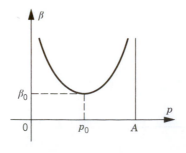

**Figure 10**

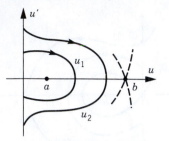

Figure 11

## EXERCISES

1.  (*difficult*) Derive the bifurcation diagram for Example 1 by integrating the ODE as explicitly as possible.

2.  Show that the curves in the phase plane (Figure 9) must be symmetric with respect to the $u$ axis.

3.  (*difficult*) Show that in Figure 10 the graph of $\beta(p)$ has a single minimum $\beta_0$ in $0 < p < A$ and that $\beta(p) \to \infty$ as $p \to 0$ or $A$.

4.  Show that the solutions of Example 2 look as pictured in Figure 11.

## 14.5   WATER WAVES

A famous problem in physics is to understand the shapes of water waves, for instance ocean waves. A water wave can be gentle and regular as in Figure 1, or it can be turbulent with a very complicated structure like a fractal or a random pattern. An extreme example of a water wave is a tsunami. A tsunami is regular in the open ocean but it has a lot of energy and becomes very turbulent as soon as it nears the shore.

Water can be regarded for practical purposes as an incompressible fluid without viscosity. In Section 13.2 we derived Euler's equations for the motion

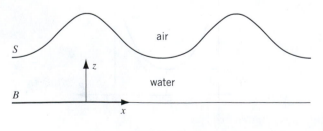

Figure 1

of a fluid:

$$\frac{\partial \rho}{\partial t} = -\nabla \cdot (\rho \mathbf{v}), \qquad \frac{\partial \mathbf{v}}{\partial t} + (\mathbf{v} \cdot \nabla)\mathbf{v} = \mathbf{F} - \frac{1}{\rho}\nabla p. \qquad (1)$$

Here $\rho(\mathbf{x}, t)$ is the density of the fluid, $\mathbf{v}(\mathbf{x}, t)$ is its velocity, and $p(\mathbf{x}, t)$ is its pressure at the position $\mathbf{x}$. $\mathbf{F}(\mathbf{x}, t)$ is a force acting on the water particles. We assume incompressibility, which means that $\rho$ is a constant. We also assume that the only force is gravity, so that $\mathbf{F}$ is the constant vector of magnitude $g$ pointing downwards. Therefore the basic equations reduce to

$$\nabla \cdot \mathbf{v} = 0, \qquad \frac{\partial \mathbf{v}}{\partial t} + (\mathbf{v} \cdot \nabla)\mathbf{v} = -\frac{1}{\rho}\nabla p + \begin{pmatrix} 0 \\ 0 \\ -g \end{pmatrix}. \qquad (2)$$

We also assume that the air above the water is quiescent. Let $P_0$ be the atmospheric pressure.

We assume the bottom $B$ is flat and the average water depth is $d$. The water particles cannot penetrate the bottom, so that $\mathbf{v}$ is tangent to $B$. On the water surface $S$, which is *unknown*, there are two conditions coming from the physics, assuming that there is no surface tension:

(i)   The water particles that start on $S$ remain on $S$.

(ii)  The water pressure from below is equal to the air pressure from above:
      $p = P_0$ on $S$.

## UNIDIRECTIONAL WATER WAVES

Now let's consider waves that move in only one horizontal direction $x$. They depend on $x$ and $z$ but are independent of $y$. Furthermore, there is no motion in the $y$ direction. Such a wave is called unidirectional. We write

$$\mathbf{x} = \begin{pmatrix} x \\ y \\ z \end{pmatrix}, \qquad \mathbf{v} = \begin{pmatrix} u \\ 0 \\ w \end{pmatrix}.$$

Thus the PDEs inside the water reduce to

$$\begin{cases} u_x + w_z = 0 \\ u_t + u u_x + w u_z = -p_x \\ w_t + u w_x + w w_z = -p_z - g \end{cases} \qquad (3)$$

if $\rho = 1$. Let the bottom and the water surface be given by

$$B = \{z = 0\}, \qquad S = \{z = \eta(x, t)\}.$$

The boundary conditions are

$$\begin{cases} w = 0 \text{ on } B \\ p = P_0 \text{ on } S \\ w = \eta_t + u\eta_x \text{ on } S. \end{cases} \tag{4}$$

The third equation in (4) comes from the consideration that if a water particle is given by $x = x(t)$, $z = z(t)$ and remains on the surface $S$, then $z = \eta(x, t)$ so that by differentiation, $dz/dt = \eta_t + \eta_x dx/dt$. Since $dx/dt = u$ and $dz/dt = w$, we get the third equation in (4). Equations (3) and (4) are an example of a *free boundary problem* because the boundary $S$ is one of the unknowns. Altogether, the unknowns are $u, w, p$, and $\eta$. It's a big system of equations. (It would have been even more complicated without the unidirectional assumption.)

Two auxiliary functions are useful to introduce. One is the *vorticity* $\nabla \times \mathbf{v}$. Because of the unidirectional assumption, its first two components vanish. Its third component is

$$\omega = w_x - u_z, \tag{5}$$

which we call the (scalar) *vorticity*. A second useful function is the *stream function*, defined by

$$\psi_x = -w \qquad \psi_z = u. \tag{6}$$

It exists because of the incompressibility, $u_x + w_z = 0$, by basic vector analysis (in a simply connected domain); see the end of Section A.1. It is determined up to an additive constant. By Exercise 1, $\omega_t + u\omega_x + w\omega_z = 0$. Note also that

$$\psi_{xx} + \psi_{zz} = -w_x + u_z = -\omega, \tag{7}$$

which is the Poisson equation (6.1.1) for $\psi$.

Let's now assume the water wave is periodic in the horizontal direction $x$:

$$u(x + L, z, t) = u(x, z, t), \quad w(x + L, z, t) = w(x, z, t),$$
$$\eta(x + L, t) = \eta(x, t).$$

The total energy of the wave (within a period) is defined as

$$\mathcal{E} = \iint\limits_{D_\eta} \left\{ \frac{1}{2}(u^2 + w^2) + gz \right\} dz \, dx \tag{8}$$

where $D_\eta$ is the region $\{0 < z < \eta(x, t), 0 < x < L\}$ occupied by the water. By Exercise 2, $\mathcal{E}$ is a constant. This is conservation of energy. The mass (within a period) is

$$m = \iint\limits_{D_\eta} dz \, dx \tag{9}$$

because the density $\rho = 1$.

Even with these assumptions, the problem (3)-(4) is still very difficult so we will content ourselves to study only the *traveling waves*. Ocean waves are generated by wind and have different speeds, wavelengths, and heights. When they move far past the influence of the generating winds, they sort themselves into groups with similar speeds and wavelengths, thereby forming a regular pattern of wave "trains". For instance, waves originating from Antarctic storms have been recorded close to the Alaskan coast. The wave trains are typically unidirectional, their motion being identical in any direction parallel to the crest line, as well as periodic and traveling at a constant speed.

A traveling wave has the form

$$u = u(x - ct, z), \quad w = w(x - ct, z), \quad p = p(x - ct, z), \quad \eta = \eta(x - ct)$$

where its speed $c$ is a constant. Then $w_t = -cw_x$. By Exercise 1,

$$(u - c)\omega_x + w\omega_z = 0. \tag{10}$$

Now we calculate

$$(u - c)\psi_x + w(\psi_z - c) = (u - c)(-w) + w(u - c) = 0.$$

The last two equations say that the plane vector $\begin{pmatrix} u - c \\ w \end{pmatrix}$ is orthogonal to both $\nabla \omega$ and $\nabla(\psi - cz)$. Because the gradients are normal vectors to the level curves in the $xz$ plane, the level curves of $\omega$ and $\psi - cz$ coincide! Thus one is a function of the other. Let's write

$$\omega = \gamma(\psi - cz). \tag{11}$$

The function $\gamma$ is arbitrary. Combining (7) and (11), we have

$$\psi_{xx} + \psi_{zz} = -\gamma(\psi - cz) \tag{12}$$

in the fluid domain $D_\eta$. This is a single scalar PDE for the single unknown $\psi$. The equation is elliptic because there are no derivatives in the nonlinear term.

We can completely eliminate $t$ from the problem simply by letting $X = x - ct$. Then $D_\eta = \{0 < z < \eta(X), 0 < X < L\}$ is independent of $t$. We can also eliminate $c$ by setting $\Psi = \psi - cz$. Its derivative along the water surface $S$ is

$$\frac{d}{dX}[\Psi(X, \eta(X))] = \frac{d}{dX}[\psi(X, \eta(X)) - c\eta(X)] = \psi_X + (\psi_z - c)\eta'(X)$$

$$= \psi_x + \psi_z\eta_x + \eta_t = -w + u\eta_x + \eta_t = 0.$$

So $\Psi$ is a constant on the surface! In view of (6), the constant is quite arbitrary, so we may take its value on $S$ to be 0. With the help of Exercise 3, we then have the following reduced problem for the traveling waves.

$$
\begin{cases}
\Psi_{XX} + \Psi_{zz} = -\gamma(\Psi) & \text{in } D_\eta \\
\Psi = 0 & \text{on } S \\
\frac{1}{2}|\nabla\Psi|^2 + gz = a = \text{constant} & \text{on } S \\
\Psi = b = \text{constant} & \text{on } B.
\end{cases}
\tag{13}
$$

This is a scalar PDE with a free boundary, a lot simpler than the original system (3)-(4) which had more unknowns. The PDE is the Laplace equation except for the nonlinear term. Of course, it still has the free surface $S$ and a nonlinear boundary condition on it. For some mathematical analysis on this problem, see [CS][Js].

## WATER WAVES AND KORTEWEG-DE VRIES

Now we shall show how, under appropriate conditions, water waves (not necessarily traveling or periodic ones) lead to the KdV equation (see Section 14.2). This time we simplify the problem by considering only waves of small amplitude, for which $S$ does not differ much from the horizontal. Of course, many realistic waves indeed have relatively small amplitudes. Thus we will now assume that $S = \{z = 1 + \epsilon\eta(x,t)\}$, where $\epsilon$ is a small parameter.

First, the basic equations (3)-(4) for water waves can be rescaled to lead to the equations

$$
\begin{cases}
u_x + w_z = 0 \\
u_t + \epsilon(uu_x + wu_z) = -p_x \\
w_t + \epsilon(uw_x + ww_z) = -\epsilon^{-1}p_z
\end{cases}
\tag{14}
$$

in $D = \{0 < z < 1 + \epsilon\eta(x,t)\}$, together with the boundary conditions

$$
\begin{cases}
w = 0 \text{ on } B \\
p = \eta \text{ on } S \\
w = \eta_t + \epsilon u\eta_x \text{ on } S.
\end{cases}
\tag{15}
$$

The system (14)-(15) arises from the weird rescaling

$$
x = \tilde{x}, \quad z = \delta^2\tilde{z}, \quad t = \delta^{-1}\tilde{t}, \quad u = \delta^5\tilde{u}, \quad w = \delta^7\tilde{w},
$$

$$
p = P_0 + \delta^2 - \delta^2 g\tilde{z} + \delta^6\tilde{p}, \quad \eta = \delta^2 + \delta^6\tilde{\eta}, \quad \epsilon = \delta^4.
$$

One gets (14)-(15) in the tilde-variables $\tilde{x}$, $\tilde{p}$, and so forth. See Exercise 5. A detailed explanation of this rescaling is given in [Js].

So far $\epsilon$ has been an arbitrary positive parameter but from now on we take it to be small. If, as a first really simple approximation, we set $\epsilon = 0$ in (14) and (15), then we get $p(x,z,t) = \eta(x,t)$ and $u_t = -\eta_x$, from which it follows that

$$
\eta_{tt} = \eta_{xx}.
\tag{16}
$$

See Exercise 6. By Section 2.1, $\eta$ is the sum of a function of $x + t$ and $x - t$. Let's assume $\eta(x, t) = f(x - t)$. This means we are taking a right-moving wave.

Therefore for small $\epsilon$ we are motivated to introduce the new variables

$$\xi = x - t, \qquad \tau = \epsilon t.$$

Introduction of $\tau$ means that we are looking at the behavior of the water wave at a large time because $t = \tau/\epsilon$ will be large if $\tau$ is of modest size and $\epsilon$ is small. Then the system (14)-(15) takes the new form

$$\begin{cases} u_\xi + w_z = 0 \\ -u_\xi + \epsilon(u_\tau + uu_\xi + wu_z) = -p_\xi \\ -\epsilon w_\xi + \epsilon^2(w_\tau + uw_\xi + ww_z) = -p_z \end{cases} \qquad (17)$$

in $D = \{0 < z < 1 + \epsilon\eta(x, t)\}$, together with the boundary conditions

$$\begin{cases} w = 0 \text{ on } B \\ p = \eta \text{ on } S \\ w = -\eta_\xi + \epsilon(\eta_\tau + u\eta_\xi) \text{ on } S \end{cases} \qquad (18)$$

where $B = \{z = 0\}$ is the bottom and $S = \{z = 1 + \epsilon\eta(x, t)\}$ is the water surface.

If we now consider small (but nonzero) $\epsilon$, then we ought to be able to expand all the functions in Taylor series in powers of $\epsilon$. If $\epsilon$ is very small, it is reasonable to take just the first two terms in each series:

$$u = u_0 + \epsilon u_1, \; w = w_0 + \epsilon w_1, \; p = p_0 + \epsilon p_1, \; \eta = \eta_0 + \epsilon\eta_1. \qquad (19)$$

Of course, this is not the exact solution but merely an approximation to it. We plug (19) into (17)-(18) and drop all terms of order $\epsilon^2$, thereby obtaining

$$\begin{cases} u_{0\xi} + \epsilon u_{1\xi} + w_{0z} + \epsilon w_{1z} = 0 \\ -u_{0\xi} - \epsilon u_{1\xi} + \epsilon(u_{0\tau} + u_0 u_{0\xi} + w_0 u_{0z}) = -p_{0\xi} - \epsilon p_{1\xi} \\ -\epsilon w_{0\xi} = -p_{0z} - \epsilon p_{1z}. \end{cases} \qquad (20)$$

For the boundary conditions at $\{z = 1 + \epsilon\eta\}$, we have to expand $w(x, 1 + \epsilon\eta) = w + \epsilon\eta w_z + O(\epsilon^2)$ and similarly for the other variables. Therefore

$$\begin{cases} w_0 + \epsilon w_1 = 0 \text{ on } \{z = 0\} \\ p_0 + \epsilon\eta_0 p_{0z} + \epsilon p_1 = \eta_0 + \epsilon\eta_1 \text{ on } \{z = 1\} \\ w_0 + \epsilon\eta_0 w_{0z} + \epsilon w_1 = -\eta_{0\xi} + \epsilon(-\eta_{1\xi} + \eta_{0\tau} + u_0\eta_{0\xi}) \text{ on } \{z = 1\}. \end{cases} \qquad (21)$$

Putting $\epsilon = 0$, we get

$$-u_{0\xi} = -p_{0\xi}, \quad p_{0z} = 0, \quad u_{0\xi} + w_{0z} = 0,$$
$$w_0 = 0 \text{ on } \{z = 0\}, \quad p_0 = \eta_0 \quad \text{and} \quad w_0 = -\eta_{0\xi} \text{ on } \{z = 1\}.$$

Therefore

$$p_0 = \eta_0 = u_0, \qquad w_0 = -z\eta_{0\xi}$$

for all $z$ (assuming that $u_0 = 0$ whenever $\eta_0 = 0$). Using this result and matching the coefficients of $\epsilon$ in (20)-(21), we get

$$\begin{cases} u_{1\xi} + w_{1z} = 0 \\ -u_{1\xi} + u_{0\tau} + u_0 u_{0\xi} = -p_{1\xi} \\ p_{1z} = w_{0\xi}. \end{cases} \tag{22}$$

$$\begin{cases} w_1 = 0 \ \text{ on } \{z = 0\} \\ p_1 = \eta_1 \ \text{ on}\{z = 1\} \\ w_1 + \eta_0 w_{0z} = -\eta_{1\xi} + \eta_{0\tau} + u_0\eta_{0\xi} \ \text{ on } \{z = 1\}. \end{cases} \tag{23}$$

Because $u_{0z} = p_{0z} = 0$ and $w_{0z} = -\eta_{0\xi}$, we get $p_{1zz} = w_{0\xi z} = w_{0z\xi} = -\eta_{0\xi\xi}$. Thus $p_1 = \frac{1}{2}(1 - z^2)\eta_{0\xi\xi} + \eta_1$ so that

$$w_{1z} = u_{1\xi} = -p_{1\xi} - u_{0\tau} - u_0 u_{0\xi} = -\eta_{1\xi} - \frac{1}{2}(1 - z^2)\eta_{0\xi\xi\xi} - \eta_{0\tau} - \eta_0\eta_{0\xi}$$

and

$$w_1 = -(\eta_{1\xi} + \eta_{0\tau} + \eta_0\eta_{0\xi} + \frac{1}{2}\eta_{0\xi\xi\xi})z + \frac{1}{6}z^3\eta_{0\xi\xi\xi}.$$

However, the surface condition is now expressed as $w_1 - \eta_0\eta_{0\xi} = -\eta_{1\xi} + \eta_{0\tau} + \eta_0\eta_{0\xi}$ on $\{z = 1\}$. Combining the last two equations on $\{z = 1\}$, we have

$$2\eta_{0\tau} + 3\eta_0\eta_{0\xi} + \frac{1}{3}\eta_{0\xi\xi\xi} = 0. \tag{24}$$

Up to a simple rescaling, (24) is the same as the KdV equation (14.2.1)! This is the reason that one can see solitons on water surfaces, as Russell did at the canal.

## EXERCISES

1.  Use differentiation to show that $\omega_t + u\omega_x + w\omega_z = 0$. (Vorticity equation)

2.  Show that $\mathcal{E}$ is independent of $t$ (Conservation of energy)
    [*Hint:* Let $q = \frac{1}{2}u^2 + \frac{1}{2}w^2 + p + gz$. Show that (3) can be written as $u_t + q_x = w\omega$, $w_t + q_z = -u\omega$. Deduce that
    $$\left(\tfrac{1}{2}u^2 + \tfrac{1}{2}w^2 + gz\right)_t + (uq)_x + (wq)_z = 0.$$
    Integrate over $z$, use the boundary conditions, and then integrate over $x$.]

3.  For a traveling wave, show that $\frac{1}{2}|\nabla\Psi|^2 + gz$ is a constant on $S$. (Bernoulli's condition) [*Hint:* Let $Q = \frac{1}{2}\Psi_X^2 + \frac{1}{2}\Psi_z^2 + gz + p + \Gamma(\Psi)$, where $\Gamma' = \gamma$. Use the equations to show that $Q$ is a constant throughout the fluid. Then deduce Bernoulli's condition on $S$.]

4. Consider the "flat" waves, solutions of (13) that depend on $z$ but not on $X$, and for which $\Psi_z > 0$. For these waves the water particles move only horizontally. Solve the resulting system in terms of an implicit equation. Show that there is one identity that must hold between the constants $a$, $b$, and the water height $d$.

5. Show how (14)-(15) follow from (3)-(4) by using the indicated scaling. [This problem is not difficult but requires plenty of patience!]

6. Give the details of how the wave equation (16) follows from the $\epsilon = 0$ form of (14)-(15). (*Hint:* The key observation is that $p$ and therefore $u_t$ are independent of $z$.)

7. Use a simple rescaling to change (24) into (14.2.1).

# APPENDIX

This is a brief guide to some of the concepts used in this book. It is designed to be used as a reference by the reader. For more details, see a book on advanced calculus such as [F1].

## A.1  CONTINUOUS AND DIFFERENTIABLE FUNCTIONS

We write $[b, c]$ = the closed interval $\{b \leq x \leq c\}$, $(b, c)$ = the open interval $\{b < x < c\}$, and we define similarly the half-open intervals $[b, c)$ and $(b, c]$.

The key concept in calculus is that of the limit. Let $f(x)$ be a (real) function of one variable. The function $f(x)$ *has the limit L as x approaches a* if for any number $\epsilon > 0$ (no matter how small) there exists a number $\delta > 0$ such that $0 < |x - a| < \delta$ implies that $|f(x) - L| < \epsilon$. We write $\lim_{x \to a} f(x) = L$. An equivalent definition is in terms of limits of sequences: For any sequence $x_n \to a$ it follows that $f(x_n) \to L$.

To have a limit, the function has to be defined on both sides of $a$. But it doesn't matter what the function is at $x = a$ itself; it need not even be defined there. For instance, even though the function $f(x) = (\sin x)/x$ is not defined at $x = 0$, its limit as $x \to 0$ exists (and equals 1).

We can also define the one-sided limits. The function $f(x)$ has the *limit L from the right* as $x$ approaches $a$ if for any number $\epsilon > 0$ (no matter how small) there exists a number $\delta > 0$ such that $0 < x - a < \delta$ implies that $|f(x) - L| < \epsilon$. The only difference from the ordinary (two-sided) limit is the removal of the absolute value, which implies that $x > a$. In this text the number $L$ is denoted by $f(x+)$.

The *limit from the left* is defined the same way except that $0 < a - x < \delta$. The value of the limit from the left is denoted by $f(x-)$. If both the limits from the right and from the left exist and if $f(x+) = f(x-)$, then the (ordinary) limit $\lim_{x \to a} f(x)$ exists and equals $f(x+) = f(x-)$.

The function $f(x)$ is *continuous at a point a* if the limit of $f(x)$ as $x \to a$ exists and equals $f(a)$. In particular, the function has to be defined at $a$. For instance, the function defined by $f(x) = (\sin x)/x$ for $x \neq 0$ and $f(0) = 1$ is a continuous function at every point. A function is *continuous in an interval* $b \leq x \leq c$ if it is continuous at each point of that interval. (At the endpoints only one-sided continuity is defined.) Intuitively, the graph of a continuous function can be drawn without lifting the pen from the page. Thus the graph may have corners but not jumps or skips.

**Intermediate Value Theorem.**   If $f(x)$ is continuous in a finite closed interval $[a, b]$, and $f(a) < p < f(b)$, then there exists at least one point $c$ in the interval such that $f(c) = p$.

**Theorem on the Maximum.**   If $f(x)$ is continuous in a finite closed interval $[a, b]$, then it has a maximum in that interval. That is, there is a point $m \in [a, b]$ such that $f(x) \leq f(m)$ for all $x \in [a, b]$. By applying this theorem to the function $-f(x)$, it follows that $f(x)$ also has a minimum.

**Vanishing Theorem.**   Let $f(x)$ be a continuous function in a finite closed interval $[a, b]$. Assume that $f(x) \geq 0$ in the interval and that $\int_a^b f(x)\, dx = 0$. Then $f(x)$ is identically zero.

**Proof.**   The idea of this theorem is that the graph $y = f(x)$ lies above the $x$-axis, but the area under it is zero. We provide a proof here because it is not a particularly standard fact but it is used repeatedly in this book. Suppose, on the contrary, that $f(c) > 0$ for some $c \in (a, b)$. Let $\epsilon = \frac{1}{2} f(c)$ in the definition of continuity at $c$. Thus there exists $\delta > 0$ such that $|x - c| < \delta$ implies that $f(x) > \frac{1}{2} f(c)$. Then the region under the graph encloses a rectangle of height $\frac{1}{2} f(c)$ and width $2\delta$, so that

$$\int_a^b f(x)\, dx \geq \int_{c-\delta}^{c+\delta} f(x)\, dx \geq \tfrac{1}{2} f(c) \cdot 2\delta = \delta f(c) > 0. \qquad (1)$$

This contradicts the assumption. So $f(x) = 0$ for all $a < x < b$. It follows from the continuity that $f(a) = f(b) = 0$ as well.     □

A function is said to have a *jump discontinuity* (or simply to have a *jump*) if both one-sided limits $f(x-)$ and $f(x+)$ exist (as finite numbers) but they are unequal. The number $f(x+) - f(x-)$ is called the value of the jump. Other kinds of discontinuities may occur as well.

A function $f(x)$ is called *piecewise continuous* on a finite closed interval $[a, b]$ if there are a finite number of points $a = a_0 \leq a_1 \leq \cdots \leq a_n = b$ such that $f(x)$ is continuous on each open subinterval $(a_{j-1}, a_j)$ and all the one-sided limits $f(a_j^-)$ for $1 \leq j \leq n$ and $f(a_j^+)$ for $0 \leq j \leq n - 1$ exist. Thus it has jumps at a finite number of points but is otherwise continuous. Every piecewise continuous function is integrable. See Section 5.4 for examples.

A function $f(x)$ is *differentiable at a point* $a$ if the limit of $[f(x) - f(a)]/(x - a)$ as $x \to a$ exists. The value of the limit is denoted by $f'(a)$ or $(df/dx)(a)$. A function is *differentiable in an open interval* $(b, c)$ if it is differentiable at each point $a$ in the interval. It is easy to show that any function which is differentiable at a point is continuous there.

## FUNCTIONS OF TWO OR MORE VARIABLES

Such functions are defined on domains in the space of independent variables. In this book we often use the vector notation $\mathbf{x} = (x, y, z)$ in three dimensions (and $\mathbf{x} = (x, y)$ in the plane). By a *domain* or *region* we mean an open set $D$ (a set without its boundary). An important example of a domain is the ball $\{|\mathbf{x} - \mathbf{a}| < R\}$ of center $\mathbf{a}$ and radius $R$, where $|\cdot|$ denotes the euclidean distance.

More precisely, a *boundary point* of any set $D$ (in three-dimensional space) is a point $\mathbf{x}$ for which every ball with center at $\mathbf{x}$ intersects both $D$ and the complement of $D$. In this book we denote the set of all boundary points of $D$ by bdy $D$. A set is *open* if it contains none of its boundary points. A set is *closed* if it contains all of its boundary points. The *closure* $\overline{D}$ is the union of the domain and its boundary: $\overline{D} = D \cup$ bdy $D$.

We will not repeat here the definitions of continuity and (partial) differentiability of functions, which are similar to those for one variable.

**First Vanishing Theorem.**    Let $f(\mathbf{x})$ be a continuous function in $\overline{D}$ where $D$ is a bounded domain. Assume that $f(\mathbf{x}) \geq 0$ in $\overline{D}$ and that $\iiint_D f(\mathbf{x}) \, d\mathbf{x} = 0$. Then $f(\mathbf{x})$ is identically zero. (The proof of this theorem is similar to the one-dimensional case and is left to the reader.)

**Second Vanishing Theorem.**    Let $f(\mathbf{x})$ be a continuous function in $D_0$ such that $\iiint_D f(\mathbf{x}) \, d\mathbf{x} = 0$ for all subdomains $D \subset D_0$. Then $f(\mathbf{x}) \equiv 0$ in $D_0$. (This theorem is analogous to the theorem in one dimension that a function is zero if its indefinite integral is.) Proof: Let $D$ be a ball and let its radius shrink to zero.

A function is said to be of *class $C^1$* in a domain $D$ if each of its partial derivatives of first order exists and is continuous in $D$. If $k$ is any positive integer, a function is said to be of *class $C^k$* if each of its partial derivatives of order $\leq k$ exists and is continuous.

The mixed derivatives are equal: If a function $f(x, y)$ is of class $C^2$, then $f_{xy} = f_{yx}$. The same is true for derivatives of any order. Although pathological examples can be exhibited for which the mixed derivatives are not equal, we never have to deal with them in this book.

The *chain rule* deals with functions of functions. For instance, consider the chain $(s, t) \mapsto (x, y) \mapsto u$. If $u$ is a function of $x$ and $y$ of class $C^1$, while $x$ and $y$ are differentiable functions of $s$ and $t$, then

$$\frac{\partial u}{\partial s} = \frac{\partial u}{\partial x} \frac{\partial x}{\partial s} + \frac{\partial u}{\partial y} \frac{\partial y}{\partial s}$$

and

$$\frac{\partial u}{\partial t} = \frac{\partial u}{\partial x}\frac{\partial x}{\partial t} + \frac{\partial u}{\partial y}\frac{\partial y}{\partial t}.$$

The *gradient* of a function (of three variables) is $\nabla f = (f_x, f_y, f_z)$. The *directional derivative* of $f(\mathbf{x})$ at a point $\mathbf{a}$ in the direction of the vector $\mathbf{v}$ is

$$\lim_{t \to 0} \frac{f(\mathbf{a} + t\mathbf{v}) - f(\mathbf{a})}{t} = \mathbf{v} \cdot \nabla f(\mathbf{a}).$$

It follows, for example, that the rate of change of a quantity $f(\mathbf{x})$ seen by a moving particle $\mathbf{x} = \mathbf{x}(t)$ is $(d/dt)(f(\mathbf{x})) = \nabla f \cdot d\mathbf{x}/dt$.

## VECTOR FIELDS

A vector field assigns a vector to each point. In two variables it is the same as two scalar functions of two variables, which can be written as $x = g(x', y')$ and $y = h(x', y')$. It can also be regarded as a "coordinate change" from primed to unprimed coordinates. The first-order partial derivatives form a matrix

$$\mathcal{J} = \begin{pmatrix} \dfrac{\partial x}{\partial x'} & \dfrac{\partial y}{\partial x'} \\ \dfrac{\partial x}{\partial y'} & \dfrac{\partial y}{\partial y'} \end{pmatrix},$$

called the *jacobian matrix*. The jacobian determinant (or just jacobian) is the determinant of this matrix, $J = \det \mathcal{J}$. In case the functions $g$ and $h$ are linear, $\mathcal{J}$ is just the matrix of the linear transformation (with respect to the coordinate systems) and $J$ is its determinant.

For example, the transformation from polar to cartesian coordinates $x = r \cos\theta$, $y = r \sin\theta$ has the jacobian matrix

$$\mathcal{J} = \begin{pmatrix} \dfrac{\partial x}{\partial r} & \dfrac{\partial y}{\partial r} \\ \dfrac{\partial x}{\partial \theta} & \dfrac{\partial y}{\partial \theta} \end{pmatrix} = \begin{pmatrix} \cos\theta & \sin\theta \\ -r \sin\theta & r \cos\theta \end{pmatrix}.$$

The jacobian determinant is $J = \cos\theta \cdot r\cos\theta + \sin\theta \cdot r\sin\theta = r$.

Any change of coordinates in a multiple integral involves the jacobian. If the transformation $x = g(x', y')$ and $y = h(x', y')$ carries the domain $D'$ onto the domain $D$ in a one-to-one manner and is of class $C^1$, and if $f(x, y)$ is a continuous function defined on $\overline{D}$, then

$$\iint_D f(x, y)\, dx\, dy = \iint_{D'} f(g(x', y'), h(x', y')) \cdot |J(x', y')|\, dx'\, dy'. \quad (2)$$

The size of the jacobian factor $|J|$ expresses the amount that areas are stretched or shrunk by the transformation. For example, a change to polar coordinates gives $|J(x', y')|\, dx'\, dy' = r\, dr\, d\theta$.

In three dimensions $\oint$ is a $3 \times 3$ matrix and the integrals in the formula (2) are triple integrals.

If **G** is a vector field in a simply connected domain in three dimensions, then $\nabla \times \mathbf{G} = 0$ if and only if $\mathbf{G} = \nabla f$ for some scalar function $f$. That is, **G** is curl-free if and only if it is a gradient. In two dimensions the same is true if we define curl $\mathbf{G} = \frac{\partial g_2}{\partial x} - \frac{\partial g_1}{\partial y}$, where $\mathbf{G} = [g_1, g_2]$.

## A.2 INFINITE SERIES OF FUNCTIONS

Given the infinite series $\sum_{n=1}^{\infty} a_n$, the *partial sums* are the sums of the first $N$ terms: $S_N = \sum_{n=1}^{N} a_n$. The infinite series *converges* if there is a finite number $S$ such that $\lim_{N \to \infty} S_N = S$. This means that for any number $\epsilon > 0$ (no matter how small) there exists an integer $\mathcal{N}$ such that $N \geq \mathcal{N}$ implies that $|S_N - S| < \epsilon$. $S$ is called the *sum* of the series.

If the series does not converge, we say it diverges. For instance, the series

$$\sum_{n=1}^{\infty} (-1)^{n+1} = 1 - 1 + 1 - 1 + \cdots$$

diverges because its partial sums are $1, 0, 1, 0, 1, 0, 1, \ldots$, which has no limit.

If the series $\sum_{n=1}^{\infty} a_n$ converges, then $\lim_{n \to \infty} a_n = 0$. Thus if the $n$th term does not converge or its limit is not zero, the series does not converge. This is the $n$th *term test* for divergence. The most famous subtle example is the series $\sum_{n=1}^{\infty} 1/n$, which diverges even though its $n$th term tends to zero.

If a series has only positive terms, $a_n \geq 0$, then its partial sums $S_N$ increase. In that case either the series converges or the partial sums diverge to $+\infty$.

To every series $\sum_{n=1}^{\infty} a_n$ we associate the positive series $\sum_{n=1}^{\infty} |a_n|$. If $\sum_{n=1}^{\infty} |a_n|$ converges, so does $\sum_{n=1}^{\infty} a_n$. If $\sum_{n=1}^{\infty} |a_n|$ converges, we say that $\sum_{n=1}^{\infty} a_n$ is *absolutely convergent*. If $\sum_{n=1}^{\infty} a_n$ converges but $\sum_{n=1}^{\infty} |a_n|$ diverges, we say that $\sum_{n=1}^{\infty} a_n$ is *conditionally convergent*. This book is full of conditionally convergent series.

**Comparison Test.** If $|a_n| \leq b_n$ for all $n$, and if $\sum_{n=1}^{\infty} b_n$ converges, then $\sum_{n=1}^{\infty} a_n$ converges absolutely. The contrapositive necessarily follows: If $\sum_{n=1}^{\infty} |a_n|$ diverges, so does $\sum_{n=1}^{\infty} b_n$. The limit comparison test states that if $a_n \geq 0, b_n \geq 0$, if $\lim_{n \to \infty} a_n/b_n = L$ where $0 \leq L < \infty$, and if $\sum_{n=1}^{\infty} b_n$ converges, then so does $\sum_{n=1}^{\infty} a_n$.

### SERIES OF FUNCTIONS

Now let's consider $\sum_{n=1}^{\infty} f_n(x)$, where $f_n(x)$ could be any functions. A simple example is the series

$$\sum_{n=0}^{\infty} (-1)^n x^{2n} = 1 - x^2 + x^4 - x^6 + \cdots,$$

which we recognize as a geometric series with the ratio $-x^2$. It converges absolutely to the sum $(1 + x^2)^{-1}$ for $|x| < 1$, but diverges elsewhere. The best known series are the power series $\sum_{n=1}^{\infty} a_n x^n$. They have special convergence properties, such as convergence in a symmetric interval around 0. However, most of the series in this book are not power series.

We say that the infinite series $\sum_{n=1}^{\infty} f_n(x)$ *converges* to $f(x)$ *pointwise* in an interval $(a, b)$ if it converges to $f(x)$ (as a series of numbers) *for each* $a < x < b$. That is, for each $a < x < b$ we have

$$\left| f(x) - \sum_{n=1}^{N} f_n(x) \right| \to 0 \qquad \text{as } N \to \infty. \tag{1}$$

That is, for each $x$ and for each $\epsilon$, there exists an integer $\mathcal{N}$ such that

$$N \geq \mathcal{N} \quad \text{implies} \quad \left| f(x) - \sum_{n=1}^{N} f_n(x) \right| < \epsilon. \tag{2}$$

It is always expected that $\mathcal{N}$ depends on $\epsilon$ (the smaller the $\epsilon$, the larger the $\mathcal{N}$), but $\mathcal{N}$ can also depend on the point $x$. For this reason it is difficult to make general deductions about the sum of a pointwise convergent series, and a stronger notion of convergence is introduced.

We say that the series *converges uniformly* to $f(x)$ in $[a, b]$ if

$$\max_{a \leq x \leq b} \left| f(x) - \sum_{n=1}^{N} f_n(x) \right| \to 0 \qquad \text{as } N \to \infty. \tag{3}$$

(Note that the endpoints are included.) That is, you take the biggest difference over *all* the $x$'s and *then* take the limit. It is the same as requiring that (2) is valid where $\mathcal{N}$ *does not depend on* $x$ (but still depends on $\epsilon$, of course).

**Comparison Test.** If $|f_n(x)| \leq c_n$ for all $n$ and for all $a \leq x \leq b$, where the $c_n$ are constants, and if $\sum_{n=1}^{\infty} c_n$ converges, then $\sum_{n=1}^{\infty} f_n(x)$ converges *uniformly* in the interval $[a, b]$, as well as absolutely.

**Convergence Theorem.** If $\sum_{n=1}^{\infty} f_n(x) = f(x)$ *uniformly* in $[a, b]$ and if all the functions $f_n(x)$ are continuous in $[a, b]$, then the sum $f(x)$ is also continuous in $[a, b]$ and

$$\sum_{n=1}^{\infty} \int_a^b f_n(x) \, dx = \int_a^b f(x) \, dx. \tag{4}$$

The last statement is called *term by term integration*.

Term by term differentiation is considerably more delicate.

**Convergence of Derivatives.** If all the functions $f_n(x)$ are differentiable in $[a, b]$ and if the series $\sum_{n=1}^{\infty} f_n(c)$ converges for some $c$, and if the series of derivatives $\sum_{n=1}^{\infty} f_n'(x)$ converges *uniformly* in $[a, b]$, then $\sum_{n=1}^{\infty} f_n(x)$

converges uniformly to a function $f(x)$ and

$$\sum_{n=1}^{\infty} f_n'(x) = f'(x). \tag{5}$$

A third kind of convergence, not mentioned here, plays a central role in Chapter 5.

## A.3   DIFFERENTIATION AND INTEGRATION

### DERIVATIVES OF INTEGRALS

We frequently deal with integrals of the general form

$$I(t) = \int_{a(t)}^{b(t)} f(x, t)\, dx. \tag{1}$$

How are they differentiated?

**Theorem 1.**   Suppose that $a$ and $b$ are constants. If both $f(x, t)$ and $\partial f/\partial t$ are continuous in the rectangle $[a, b] \times [c, d]$, then

$$\left(\frac{d}{dt}\right) \int_a^b f(x, t)\, dx = \int_a^b \frac{\partial f}{\partial t}(x, t)\, dx \qquad \text{for } t \in [c, d]. \tag{2}$$

The idea of this theorem is simply that

$$\frac{1}{\Delta t}\left[\int_a^b f(x, t + \Delta t)\, dx - \int_a^b f(x, t)\, dx\right] = \int_a^b \frac{f(x, t + \Delta t) - f(x, t)}{\Delta t}\, dx,$$

from which one passes to the limit as $\Delta t \to 0$.

For the case of integrals on the whole line, we have the following theorem.

**Theorem 2.**   Let $f(x, t)$ and $\partial f/\partial t$ be continuous functions in $(-\infty, \infty) \times (c, d)$. Assume that the integrals $\int_{-\infty}^{\infty} |f(x, t)|\, dx$ and $\int_{-\infty}^{\infty} |\partial f/\partial t|\, dx$ converge uniformly (as improper integrals) for $t \in (c, d)$. Then

$$\frac{d}{dt} \int_{-\infty}^{\infty} f(x, t)\, dx = \int_{-\infty}^{\infty} \frac{\partial f}{\partial t}(x, t)\, dx \qquad \text{for } t \in (c, d).$$

For the case of variable limits of integration, we have the following corollary.

**Theorem 3.**   If $I(t)$ is defined by (1), where $f(x, t)$ and $\partial f/\partial t$ are continuous on the rectangle $[A, B] \times [c, d]$, where $[A, B]$ contains the union of all the intervals $[a(t), b(t)]$, and if $a(t)$ and $b(t)$ are differentiable functions on $[c, d]$, then

$$\frac{dI}{dt} = \int_{a(t)}^{b(t)} \frac{\partial f}{\partial t}(x, t)\, dx + f(b(t), t)\, b'(t) - f(a(t), t)\, a'(t). \qquad (3)$$

**Proof.** Let $g(t, a, b) = \int_a^b f(x, t)\, dx$, where $t$, $a$, and $b$ are regarded as three independent variables. Then $I(t) = g(t, a(t), b(t))$. By the chain rule, $I'(t) = g_t + g_a a'(t) + g_b b'(t)$. The first term $g_t = \partial g / \partial t$ (with constant $a$ and $b$) is given by Theorem 1. The other partial derivatives are $g_b = f(b, t)$ and $g_a = -f(a, t)$. The identity (3) follows. [There is also an analog of this theorem for $a(t) = -\infty$ or for $b(t) = +\infty$.] □

For functions of three (or two) variables we have similar theorems. The analog of Theorem 1 is the following.

**Theorem 4.** Let $D$ be a fixed three-dimensional bounded domain and let $f(\mathbf{x}, t)$ and its partial derivative $(\partial f / \partial t)(\mathbf{x}, t)$ be continuous functions for $\mathbf{x} \in \overline{D}$ and $t \in [c, d]$. Then

$$\left(\frac{d}{dt}\right) \iiint_D f(\mathbf{x}, t)\, d\mathbf{x} = \iiint_D \frac{\partial f}{\partial t}(\mathbf{x}, t)\, d\mathbf{x} \qquad \text{for } t \in [c, d]. \qquad (4)$$

We omit the multidimensional analog of Theorem 3, which is not required in this book.

## CURVES AND SURFACES

A curve may be intuitively visualized as the path continuously traced out by a particle. More precisely, a *curve C* in space is defined as a continuous function from an interval $[a, b]$ into space. Thus it is given by a triple of continuous functions

$$x = f(t), \qquad y = g(t), \qquad z = h(t) \qquad \text{for } a \le t \le b. \qquad (5)$$

The triple of functions is called a parameterization of the curve. It is really the *image* of the functions in $\mathbf{x}$ space, which corresponds to our intuitive notion of a curve. For instance, the two parametrizations

$$x = \cos\theta, \qquad y = \sin\theta \qquad \text{for } 0 \le \theta \le \pi$$

and

$$x = -t, \qquad y = \sqrt{1 - t^2} \qquad \text{for } -1 \le t \le 1$$

represent the same semicircle in the plane (traced counterclockwise). Every $C^1$ curve can be given the natural arc length parameter defined by $ds/dt = |d\mathbf{x}/dt|$.

If the three coordinate functions are of class $C^1$, we say that the curve $C$ is of *class $C^1$*. The curve $C$ is *piecewise $C^1$* if there is a partition

$a = t_0 < t_1 < \cdots < t_n = b$ such that the coordinate functions are $C^1$ on each closed subinterval $[t_{j-1}, t_j]$; at each endpoint $t_j$ the functions and their first derivatives are computed as one-sided limits.

A *surface* $S$ is defined as a continuous function from a closed set $\overline{D}$ in a plane into three-dimensional space. Thus it too is given by a triple of functions

$$x = f(s, t), \qquad y = g(s, t), \qquad z = h(s, t) \qquad \text{for } (s, t) \in \overline{D}. \qquad (6)$$

This triple of functions is a *parametrization* of $S$, the variables $s$ and $t$ are the *parameters*, and $\overline{D}$ is the *parameter domain*. Of course, it is the *image* of the functions in three-dimensional space which corresponds to our intuitive notion of a surface. For example, the unit sphere $\{|\mathbf{x}| = 1\}$ is parameterized by spherical coordinates:

$$x = \sin\theta \cos\phi, \quad y = \sin\theta \sin\phi, \quad z = \cos\theta \quad \text{for } 0 \leq \phi \leq 2\pi, \ 0 \leq \theta \leq \pi.$$

The parameters are $\phi$ and $\theta$; they run over the rectangle $[0, 2\pi] \times [0, \pi]$, which is the parameter domain $\overline{D}$.

A surface may also be given as the *graph* of a function $z = h(x, y)$ where the point $(x, y)$ runs over a plane domain $\overline{D}$. Thus it is given parametrically by the equations $x = s, y = t, z = h(s, t)$ for $(s, t) \in \overline{D}$.

A surface is of *class* $C^1$ if all three coordinate functions are of class $C^1$. The union of a finite number of overlapping surface is also considered to be a surface. A surface is *piecewise* $C^1$ if it is the union of a finite number of $C^1$ surfaces that meet along their boundary curves. For instance, the surface of a cube is piecewise $C^1$, as it consists of six smooth faces. Alternatively, a surface can be described implicitly, as stated in the following theorem, which can be proved using the implicit function theorem [F1].

**Theorem 5.** Let $F(x, y, z)$ be a $C^1$ function of three variables (defined in some spatial domain) such that its gradient does not vanish ($\nabla F \neq 0$). Then the level set $\{F(x, y, z) = 0\}$ is a $C^1$ surface. That is, local $C^1$ parameterizations like (6) can be found for it.

## INTEGRALS OF DERIVATIVES

In one variable we have the fundamental theorem of calculus. In two variables we have the following theorem.

**Green's Theorem 6.** Let $D$ be a bounded plane domain with a piecewise $C^1$ boundary curve $C = \text{bdy } D$. Consider $C$ to be parametrized so that it is traversed once with $D$ on the left. Let $p(x, y)$ and $q(x, y)$ be any $C^1$ functions defined on $\overline{D} = D \cup C$. Then

$$\iint_D (q_x - p_y) \, dx \, dy = \int_C p \, dx + q \, dy. \qquad (7)$$

The line integral on the right side may also be written as $\int_C (p, q) \cdot \mathbf{t}\, ds$, where $\mathbf{t}$ is the unit tangent vector field along the curve and $ds$ is the element of arc length. If $(p, q)$ is the velocity field of a fluid flow, the line integral is the circulation of the flow. If, for instance, $D$ is an annulus, then bdy $D$ is a pair of circles traversed in opposite directions.

A completely equivalent formulation of Green's theorem is obtained by substituting $p = -g$ and $q = +f$. If $\mathbf{f} = (f, g)$ is any $C^1$ vector field in $\overline{D}$, then $\iint_D (f_x + g_y)\, dx\, dy = \int_C (-g\, dx + f\, dy)$. If $\mathbf{n}$ is the unit outward-pointing normal vector on $C$, then $\mathbf{n} = (+dy/ds, -dx/ds)$. Hence Green's theorem takes the form

$$\iint_D \nabla \cdot \mathbf{f}\, dx\, dy = \int_C \mathbf{f} \cdot \mathbf{n}\, ds, \tag{8}$$

where $\nabla \cdot \mathbf{f} = f_x + g_y$ denotes the divergence of $\mathbf{f}$.

In three dimensions we have the divergence theorem, also known as Gauss's theorem, which is the natural generalization of (8).

**Divergence Theorem 7.**  Let $D$ be a bounded spatial domain with a piecewise $C^1$ boundary surface $S$. Let $\mathbf{n}$ be the unit outward normal vector on $S$. Let $\mathbf{f}(\mathbf{x})$ be any $C^1$ vector field on $\overline{D} = D \cup S$. Then

$$\iiint_D \nabla \cdot \mathbf{f}\, d\mathbf{x} = \iint_S \mathbf{f} \cdot \mathbf{n}\, dS, \tag{9}$$

where $\nabla \cdot \mathbf{f}$ is the three-dimensional divergence of $\mathbf{f}$ and $dS$ is the element of surface area on $S$.

For example, let $\mathbf{f} = (x, y, z) = x\mathbf{i} + y\mathbf{j} + z\mathbf{k}$ and let $D$ be the ball with center at the origin and radius $a$. Then $\nabla \cdot \mathbf{f} = \partial x/\partial x + \partial y/\partial y + \partial z/\partial z = 3$, so that the left side of (9) equals

$$\iiint_D 3\, dx\, dy\, dz = 3\left(\frac{4}{3}\pi a^3\right) = 4\pi a^3.$$

Furthermore, $\mathbf{n} = \mathbf{x}/a$ on the spherical surface, so that $\mathbf{f} \cdot \mathbf{n} = \mathbf{x} \cdot \mathbf{x}/a = a^2/a = a$ and the right side of (9) is $\iint_S a\, dS = a(4\pi a^2) = 4\pi a^3$.

As a reference for this material, see [MT] or [EP], for instance.

## A.4  DIFFERENTIAL EQUATIONS

**Existence Theorem for ODEs.**  Consider the system of ODEs

$$\frac{d\mathbf{u}}{dt} = \mathbf{f}(t, \mathbf{u}) \tag{1}$$

together with the initial condition $\mathbf{u}(a) = \mathbf{b}$. Let $\mathbf{f}(t, \mathbf{u})$ be a vector function of a scalar variable $t$ and a vector variable $\mathbf{u}$ of class $C^1$ in a neighborhood of the point $(a, \mathbf{b})$. Here $\mathbf{u} = (u_1, u_2, \ldots, u_N)$ and $\mathbf{f} = (f_1, f_2, \ldots, f_N)$. Then there exists a unique solution $\mathbf{u}(t)$ of class $C^1$ for $t$ in some open interval containing $a$. That is, there is a unique curve in the $(t, \mathbf{u})$ space that passes through the point $(a, \mathbf{b})$ and satisfies (1).

## REGULAR SINGULAR POINTS

A linear second-order ODE has the form

$$a(t) u'' + b(t) u' + c(t) u = 0. \tag{2}$$

Its solutions form a two-dimensional vector space. A point where one of the coefficients is infinite or where $a(t) = 0$ is called a *singular point*. Suppose that the point in question is $t = 0$. The origin is called a *regular singular point* if the quotient $b(t)/a(t)$ behaves no worse than $t^{-1}$ and the quotient $c(t)/a(t)$ behaves no worse than $t^{-2}$ near $t = 0$. Examples include the Euler equation

$$u'' + \frac{\beta}{t} u' + \frac{\gamma}{t^2} u = 0 \tag{3}$$

and the Bessel and Legendre equations discussed in Chapter 10.

The main theorem about regular singular points states that near a regular singular point the solutions of (2) behave roughly like the solutions of the Euler equation. To be precise, assume that the limits

$$\beta = \lim_{t \to 0} t \frac{b(t)}{a(t)} \qquad \text{and} \qquad \gamma = \lim_{t \to 0} t^2 \frac{c(t)}{a(t)} \tag{4}$$

exist. Assume furthermore that $tb(t)/a(t)$ and $t^2 c(t)/a(t)$ are analytic near $t = 0$; that is, they have convergent power series expansions in powers of $t$. (For instance, they could be arbitrary polynomials.)

Let $r$ and $s$ denote the two roots of the quadratic equation

$$x(x - 1) + \beta x + \gamma = 0, \tag{5}$$

known as the *indicial equation*. Then the solutions of the simple Euler equation (3) are $C t^r + D t^s$ for arbitrary $C$ and $D$, except in the case $r = s$. The exponents $r$ and $s$ are called the *indices*.

## Theorem.

(a)  If $r$ and $s$ do not differ by an integer, then all the solutions of (2) have the form

$$C t^r \sum_{n=0}^{\infty} p_n t^n + D t^s \sum_{n=0}^{\infty} q_n t^n \tag{6}$$

with arbitrary $C$ and $D$. (This includes the case that $r$ and $s$ are complex.)

(b)  If $s - r$ is a nonnegative integer, then the solutions have the form

$$Ct^r \sum_{n=0}^{\infty} p_n t^n + (Cm \log t + D) t^s \sum_{n=0}^{\infty} q_n t^n \tag{7}$$

for some constant $m$ and with arbitrary constants $C$ and $D$. In case $r = s$, we have $m = 1$. All these series converge in at least a neighborhood of $t = 0$.

The same theorem is true in a neighborhood of any other point $t_0$ if the powers of $t$ are consistently replaced by powers of $(t - t_0)$ in (4), (6), and (7). For examples, see Sections 10.5 and 10.6 of this book, or see [BD].

## A.5  THE GAMMA FUNCTION

The logarithm can be defined by an integral. So can the gamma function, although in a different way. It is

$$\Gamma(x) = \int_0^{\infty} s^{x-1} e^{-s}\, ds \qquad \text{for } 0 < x < \infty.$$

The integral converges because the exponential makes it small as $s \to \infty$. As $s \to 0$, the factor $s^{x-1}$ is integrable because $x > 0$.

The gamma function is a generalization of the factorial. In fact, it has the following properties.

$$\Gamma(x + 1) = x \Gamma(x). \tag{1}$$

$$\Gamma(n + 1) = n! \qquad \text{if } n \text{ is a positive integer.} \tag{2}$$

$$\Gamma(\tfrac{1}{2}) = \sqrt{\pi}. \tag{3}$$

To prove (1), we integrate by parts:

$$\Gamma(x + 1) = \int_0^{\infty} s^x e^{-s}\, ds = -s^x e^{-s}\Big|_0^{\infty} + \int_0^{\infty} x s^{x-1} e^{-s}\, ds = 0 + x\Gamma(x).$$

Now $\Gamma(1) = \int_0^{\infty} e^{-s}\, ds = 1$, so that $\Gamma(2) = \Gamma(1) = 1$, and so on, and (2) follows by induction. Third, substituting $s = r^2$, we have

$$\Gamma(x) = 2 \int_0^{\infty} r^{2x-1} e^{-r^2}\, dr.$$

When $x = \tfrac{1}{2}$, $\Gamma(\tfrac{1}{2}) = 2 \int_0^{\infty} e^{-r^2}\, dr = \sqrt{\pi}$ from Exercise 2.4.6.

The gamma function on the "half-integers" is given by the formulas

$$\Gamma(\tfrac{1}{2} + n) = (2n - 1)(2n - 3) \cdots (5)(3)(1) \cdot 2^{-n} \sqrt{\pi} = \frac{(2n)! \sqrt{\pi}}{n! 2^{2n}} \tag{4}$$

where $n$ is a nonnegative integer. They are proved by induction using (1) and (3):

$$\Gamma\left(n + \tfrac{1}{2}\right) = (2n - 1)2^{-1}\Gamma(n - \tfrac{1}{2}) = (2n - 1)(2n - 3)2^{-2}\Gamma(n - \tfrac{3}{2}) = \cdots$$

$$= (2n - 1)(2n - 3)\cdots(5)(3)(1) \cdot 2^{-n}\Gamma\left(\tfrac{1}{2}\right)$$

$$= \frac{(2n)!}{(2n)(2n - 2)\cdots(4)(2)}\frac{\sqrt{\pi}}{2^n} = \frac{(2n)!\sqrt{\pi}}{n!2^{2n}}.$$

Another useful identity is

$$\Gamma(x) = \frac{2^{x-1}}{\sqrt{\pi}}\Gamma\left(\frac{x}{2}\right)\Gamma\left(\frac{x + 1}{2}\right), \tag{5}$$

which can be used to give another derivation of (4).

The formula for the *surface area of the sphere* $\{x_1^2 + x_2^2 + \cdots + x_n^2 = R^2\}$ of radius $R$ in $n$ dimensions is

$$A_n = \frac{2\pi^{n/2}R^{n-1}}{\Gamma(n/2)}. \tag{6}$$

For instance, $A_2 = 2\pi R$, $A_3 = 4\pi R^2$, $A_4 = 2\pi^2 R^3$, and $A_5 = (8/3)\pi^2 R^4$.

The identity (1), $\Gamma(x) = \Gamma(x + 1)/x$, can obviously be used to extend the definition of the gamma function to negative numbers, except for the negative integers. Indeed, we first use (1) to define $\Gamma(x)$ for $-1 < x < 0$, then for $-2 < x < -1$, and so on. It follows from Theorem A.3.2 that $\Gamma(x)$ is a differentiable function for all $x$ except the negative integers.

# REFERENCES

[AS] M. ABLOWITZ and H. SEGUR, *Solitons and the Inverse Scattering Transform*, SIAM, Philadelphia, 1981.

[Ak] N. I. AKHIEZER, *The Calculus of Variations*, Harwood Academic Publishers, 1988.

[AJS] W. AMREIN, J. JAUCH, and K. SINHA, *Scattering Theory in Quantum Mechanics*, Benjamin-Cummings, Menlo Park, CA, 1977.

[Bl] D. BLEEKER, *Gauge Theory and Variational Principles*, Addison-Wesley, Reading, MA, 1981.

[Bo] F. BOWMAN, *Introduction to Bessel Functions*, Dover Publications, New York, 1958.

[BD] W. BOYCE and R. DiPRIMA, *Elementary Differential Equations and Boundary Value Problems*, Wiley, New York, 8th ed., 2004.

[BF] R. BURDEN and J. FAIRES, *Numerical Analysis*, 3rd ed., PWS, Boston, 8th ed., 2004.

[CB] J. BROWN and R. CHURCHILL, *Fourier Series and Boundary Value Problems*, McGraw-Hill, New York, 7th ed., 2006.

[CL] E. CODDINGTON and N. LEVINSON, *Theory of Ordinary Differential Equations*, McGraw-Hill, New York, 1955.

[CF] R. COURANT and K.O. FRIEDRICHS, *Supersonic Flow and Shock Waves*, Springer, New York, 1999.

[CH] R. COURANT and D. HILBERT, *Methods of Mathematical Physics*, 2 vols., Wiley-Interscience, New York, 1991.

[CS] A. CONSTANTIN and W. A. STRAUSS, Exact steady periodic water waves with vorticity, *Comm. Pure Appl. Math.* 58 (2004), pp. 481–527.

[Dd] R. DODD, J. EILBECK, J. GIBBON, and H. MORRIS, *Solitons and Nonlinear Wave Equations*, Academic Press, New York, 1984.

[DM] H. DYM and H. McKEAN, *Fourier Series and Integrals*, Academic Press, New York, 1972.

[Ed] E. R. EDMONDS, *Angular Momentum in Quantum Mechanics*, Princeton University Press, Princeton, NJ, 2nd ed., 1960.

[EP] C. EDWARDS and D. PENNEY, *Calculus and Analytic Geometry*, Prentice Hall, Englewood Cliffs, NJ, 6th ed., 2006.

[Ev] L. C. EVANS, *Partial Differential Equations*, Amer. Math. Soc., Providence, 1998.

[Fd] H. FLANDERS, *Differential Forms with Applications to the Physical Sciences*, Dover, New York, 1989.

[Fi] N. FILONOV, On an inequality between Dirichlet and Neumann eigenvalues for the Laplace operator, *St. Petersburg Math. J.* 16 (2005), pp. 413–416.

[Fl] W. FLEMING, *Functions of Several Variables*, Springer-Verlag, New York, 1977.

[Fo] G. FOLLAND, *Introduction to Partial Differential Equations*, Princeton University Press, Princeton, NJ, 2nd ed., 1996.

[Fy] R. P. FEYNMAN, *Lectures on Physics*, Addison-Wesley, Reading, MA, 1965.

[Ga] P. R. GARABEDIAN, *Partial Differential Equations*, Wiley, New York, 1964.

[HJ] R. W. HALL and K. JOSIĆ, The mathematics of musical instruments, *Amer. Math. Monthly*, 2001.

[Hu] T. J. R. HUGHES, *The Finite Element Method*, Dover Publications, 2000.

[IJ] G. IOOSS and D. JOSEPH, *Elementary Stability and Bifurcation Theory*, Springer-Verlag, New York, 1981.

[Ja] R. D. JACKSON, *Classical Electrodynamics*, Wiley, New York, 3rd ed., 1998.

[Jn] D. S. JONES, *Methods in Electromagnetic Wave Propagation*, Oxford University Press, Oxford, UK, 2nd ed., 1994.

[Jo] F. JOHN, *Partial Differential Equations*, Springer-Verlag, New York, 4th ed., 1995.

[Js] R. S. JOHNSON, *A Modern Introduction to the Mathematical Theory of Water Waves*, Cambridge U. Press, Cambridge, UK, 1997.

[Ka] M. KAC, Can one hear the shape of a drum?, *Amer. Math. Monthly*, vol. 73, 1966, pp. 1–23.

[Ki] H. KIELHÖFER, *Bifurcation Theory*, Springer, New York, 2004.

[Kr] I. KREYSIG, *Advanced Engineering Mathematics*, Wiley, New York, 7th ed., 1992.

[LL] E. H. LIEB and M. LOSS, *Analysis*, Amer. Math. Soc., Providence, 1997.

[MOS] W. MAGNUS, F. OBERHETTINGER, and R. SONI, *Formulas and Theorems for the Special Functions of Mathematical Physics*, Springer-Verlag, New York, 1966.

[MT] J. MARSDEN and A. TROMBA, *Vector Calculus*, W. H. Freeman, San Francisco, 5th ed., 2003.

[Me] R. E. MEYER, *Introduction to Mathematical Fluid Dynamics*, Dover Publications, 1982.

[MF] P. M. MORSE and H. FESHBACH, *Methods of Theoretical Physics*, 2 vols., McGraw-Hill, New York, 1953.

[MI] P. M. MORSE and K. U. INGARD, *Theoretical Acoustics*, McGraw-Hill, New York, 1968.

[Ne] A. C. NEWELL, *Solitons in Mathematics and Physics*, SIAM, Philadelphia, 1981.

[PW] M. PROTTER and H. WEINBERGER, *Maximum Principles in Partial Differential Equations*, Prentice Hall, Englewood Cliffs, NJ, 1967.

[RS] M. REED and B. SIMON, *Methods of Modern Mathematical Physics*, 4 vols., Academic Press, New York, 1972–1979.

[RM] R. D. RICHTMEYER and K. W. MORTON, *Difference Methods for Initial-Value Problems*, Wiley-Interscience, New York, 2nd ed., 1967.

[Sa] G. SANSONE, *Orthogonal Functions*, R. E. Krieger, Melbourne, FL, 1977.

[Se] R. SEELEY, *Introduction to Fourier Series and Integrals*, Benjamin-Cummings, Menlo Park, CA, 1967.

[Sw] G. SEWELL, *Numerical Solution of Ordinary and Partial Differential Equations*, Academic Press, New York, 1988.

[Sh] S. SHEN, Acoustics of ancient Chinese bells, *Scientific American*, April 1987, pp. 104–110.

[Sm] J. SMOLLER, *Shock Waves and Reaction-Diffusion Equations*, Springer-Verlag, New York, 1983.

[Sg1] G. STRANG, *Introduction to Applied Mathematics*, Wellesley-Cambridge Press, Cambridge, MA, 1986.

[Sg2] G. STRANG, *Linear Algebra and Its Applications*, Academic Press, New York, 3rd ed., 2003.

[St] H. L. STRAUSS, *Quantum Mechanics: An Introduction*, Prentice Hall, Englewood Cliffs, NJ, 1968.

[SV] W. A. STRAUSS and L. VAZQUEZ, Numerical solution of a nonlinear Klein–Gordon equation, *Journal of Computational Physics* 28, pp. 271–278, 1978.

[TS] A. TIKHONOV and A. SAMARSKII, *Equations of Mathematical Physics*, Dover Publications, 1990.

[TR] P. TONG and J. ROSSETTOS, *Finite Element Method*, MIT Press, Cambridge, MA, 1977.

[We] H. F. WEINBERGER, *A First Course in Partial Differential Equations*, Blaisdell, Waltham, MA, 1965.

[Wh] G. B. WHITHAM, *Linear and Nonlinear Waves*, Wiley-Interscience, New York, 1974.

[Zy] A. ZYGMUND, *Trigonometric Series*, Cambridge University Press, Cambridge, England, 3rd ed., 2003.

# ANSWERS AND HINTS TO SELECTED EXERCISES

## Chapter 1

### Section 1.1

2. (a) Linear; (b) nonlinear
3. (a) Order 2, linear inhomogeneous; (c) order 3, nonlinear
5. (a), (d), and (e) are vector spaces.
7. Linearly independent
9. Its dimension is 2, not 3.
10. The three functions $e^{-x}$, $e^{2x}$, $xe^{2x}$

### Section 1.2

1. $\sin(x - 3t/2)$
3. $u(x, y) = f(y - \arctan x)$ for any function $f$ of one variable.
5. $u = f(y/x)$
7. (a) $e^{x^2-y^2}$; (b) a sketch of the characteristics shows that the solution is determined only in the region $\{x^2 \le y^2\}$.
8. By either the coordinate method or the geometric method, $u = e^{-c(ax+by)/(a^2+b^2)} f(bx - ay)$, where $f$ is arbitrary.
13. $u = x + 2y + 5/(y - 2x) + \exp[(-2x^2 - 3xy + 2y^2)/5] f(y - 2x)$ where $f$ is arbitrary.

## Section 1.3

2.  $u_{tt} = g[(l - x)u_x]_x$ where $l$ is the length of the chain.
3.  $u_t = (\kappa/c\rho)u_{xx} - (\mu P/c\rho A)(u - T_0)$, where $P$ is the perimeter of the cross section, $A$ is its area, and $\mu$ is the conductance across the contact surface.
4.  $u_t = ku_{zz} + Vu_z$
5.  $u_t = ku_{xx} - Vu_x$
9.  $4\pi a^5$

## Section 1.4

1.  Try a simple polynomial in $x$ and $t$.
4.  The hottest point is $x = 5l/8$.
5.  $ku_z = -Vu$ on $z = a$
6.  (a) $u_1'' = 0$ in $0 \le x \le L_1$, $u_2'' = 0$ in $L_1 \le x \le L_1 + L_2$, together with four boundary and jump conditions. Hence $u_1(x) = ax + b$ and $u_2(x) = cx + d$. Solve for $a$ and $b$ using the four conditions. (b) $u_1 = 10x/7$ and $u_2 = 10(2x - 3)/7$

## Section 1.5

1.  Solve the ODE. The solution is unique only if $L$ is not an integer multiple of $\pi$.
2.  (a) Take the difference of two solutions and solve the ODE. Not unique. (b) Integrate the equation from 0 to $l$. The function $f(x)$ must have zero average.
5.  $u = f(e^{-x}y)$ where $f(0) = 1$

## Section 1.6

1.  (a) $\mathcal{D} = 3$, so it's hyperbolic. (b) parabolic
2.  Use the method of Example 2.
5.  $\alpha = 1, \beta = -4, \gamma = 1/\sqrt{3}$

# Chapter 2

## Section 2.1

1.  $e^x \cosh ct + (1/c) \sin x \sin ct$
3.  $(l/4 - a)(\sqrt{\rho/T})$
4.  As in the text, $u_t + cu_x = h(x + ct)$. Let $w = u - f(x - ct)$. Show that $w_t + cw_x = 0$ and then find the form of $w$.

6. Let $m(t) = \max_x u(x, t)$. Then $m(t) = t$ for $0 \leq t \leq a/c$, and $m(t) = a/c$ for $t \geq a/c$.

8. (b) $u(r, t) = (1/r)[f(r + ct) + g(r - ct)]$
   (c) $(1/2r)\{(r + ct)\phi(r + ct) + (r - ct)\phi(r - ct)\} + (1/2cr)\int_{r-ct}^{r+ct} s\psi(s)\,ds$

9. $\frac{4}{5}(e^{x+t/4} - e^{x-t}) + x^2 + \frac{1}{4}t^2$

11. $u = f(3x - t) + g(x - 3t) - \frac{1}{16}\sin(x + t)$

## Section 2.2

5. $dE/dt = -r\int \rho u_t^2\,dx \leq 0$

6. (a) $(1 - c^2\beta'^2)\alpha f'' + c^2\left(\alpha\beta'' + \dfrac{n-1}{r}\alpha\beta' + 2\alpha'\beta'\right)f' - c^2\left(\alpha'' + \dfrac{n-1}{r}\alpha'\right)f = 0$

## Section 2.3

1. $(0, 0)$ and $(1, T)$

4. (a) Use the strong maximum principle. (b) Apply the uniqueness theorem to $u(1 - x, t)$. (c) Use the equation preceding (4).

5. (a) At the point $(-1, 1)$

## Section 2.4

1. $2u(x, t) = \mathscr{E}\mathrm{rf}\{(x + l)/2\sqrt{kt}\} - \mathscr{E}\mathrm{rf}\{(x - l)/2\sqrt{kt}\}$

3. $e^{3x+9kt}$

6. $\sqrt{\pi}/2$

8. The maximum is $(4\pi kt)^{-1/2} e^{-\delta^2/4kt} = (1/\sqrt{\pi}\delta)se^{-s^2}$ where $s = \delta/\sqrt{4kt}$. This tends to zero as $s \to +\infty$.

9. $x^2 + 2kt$

12. (c) $Q(x, t) \sim \frac{1}{2} + \dfrac{x}{\sqrt{4\pi kt}} - \dfrac{1}{3\sqrt{\pi}}\left(\dfrac{x}{\sqrt{4kt}}\right)^3$

17. $v(x, t)$ satisfies the diffusion equation with initial condition $\phi(x)$.

18.
$$u(x, t) = \int_{-\infty}^{\infty} \exp\left[-\frac{(x - Vt - z)^2}{4kt}\right]\phi(z)\frac{dz}{\sqrt{4\pi kt}}$$

## Section 2.5

1. Take $\phi \equiv 0$ and $\psi \neq 0$.

## Chapter 3

### Section 3.1

2. $1 - \mathscr{E}rf[x/\sqrt{4kt}]$

3. $w(x, t) = (4\pi kt)^{-1/2} \int_0^\infty [e^{-(x-y)^2/4kt} + e^{-(x+y)^2/4kt}]\phi(y)\,dy$

### Section 3.2

2. At $t = a/c$, it is a truncated triangle. At $t = 3a/2c$, it has a plateau up to $x = 5a/2$ that drops down to zero at $x = 7a/2$. At $t = 3a/c$, it has two plateaus, on $[0, a]$ and on $[2a, 4a]$.

3. $f(x + ct)$ for $x > ct$; $f(x + ct) - f(ct - x)$ for $x < ct$

4. Same as the preceding answer but with a plus sign

5. The singularity is on the line $x = 2t$.

6. $u(x, t) = tV$ for $0 < ct < x$; and $u(x, t) = (at - x)V/(a - c)$ for $0 < x < ct$

9. (a) 4/27; (b) $-1/48$

### Section 3.3

1. $u(x, t) = \int_0^\infty [S(x - y, t) - S(x + y, t)]\,\phi(y)\,dy$
$$+ \int_0^t \int_0^\infty [S(x - y, t - s) - S(x + y, t - s)]\,f(y, s)\,dy\,ds$$

### Section 3.4

1. $\frac{1}{6}xt^3$

3. $(x + 1)t + \sin x \cos ct + (1/c^2)\cos x\,(1 - \cos ct)$

8. Compute $[\mathscr{S}(t)\psi]_{tt} = \frac{1}{2}c[\psi'(x + ct) - \psi'(x - ct)] = c^2[\mathscr{S}(t)\psi]_{xx}$.

9. Differentiating, $\partial u/\partial t = \int_0^t \mathscr{S}'(t - s)f(s)\,ds + \mathscr{S}(0)f(t)$. The last term vanishes. Differentiating again, we have

$$\frac{\partial^2 u}{\partial t^2} = \int_0^t \mathscr{S}''(t - s)f(s)\,ds + \mathscr{S}'(0)f(x, t)$$

$$= \int_0^t c^2 \frac{\partial^2}{\partial x^2}\mathscr{S}(t - s)f(s)\,ds + f(t)$$

$$= c^2 \frac{\partial^2}{\partial x^2} \int_0^t \mathscr{S}(t - s)f(s)\,ds + f(t) = c^2\frac{\partial^2 u}{\partial x^2} + f(t).$$

13. $u = x$ for $x \geq ct$; $u = x + (t - x/c)^2$ for $0 \leq x \leq ct$

14. $u \equiv 0$ for $x \geq ct$; $u = -c \int_0^{t - x/c} k(s)\,ds$ for $0 \leq x \leq ct$

# Chapter 4

## Section 4.1

2. $(4/\pi) \sum_{n \text{ odd}} (1/n) e^{-n^2\pi^2 kt/l^2} \sin(n\pi x/l)$

3. $\sum_{n=1}^{\infty} A_n \sin(n\pi x/l) e^{-in^2\pi^2 t/l^2}$

## Section 4.2

1. The eigenvalues are $\left(n + \frac{1}{2}\right)^2 \pi^2/l^2$ and the eigenfunctions are $\sin[(n + \frac{1}{2})\pi x/l]$ for $n = 0, 1, 2, \ldots$.

## Section 4.3

1. Draw the graph of $y = \tan \beta l$ versus $y = -\beta/a$. Do the cases $a > 0$ and $a < 0$ separately.

8. (b) Two eigenvalues if $la_0 + 1$ and $la_l + 1$ are negative and their product is greater than 1. One eigenvalue if $(la_0 + 1)(la_l + 1) < 1$. No eigenvalue if $la_0 + 1$ and $la_l + 1$ are positive and their product is greater than 1.

9. (b) $\tan \beta = \beta$; (d) no

12. (b) $[\sin \beta l][-\sin \beta l + \beta l] = (1 - \cos \beta l)^2$
    (e) Some of them are solutions of $\sin \gamma = 0$ and others of $\tan \gamma = \gamma$. The eigenfunctions are 1 and $x$ for $\lambda = 0$; $\cos(n\pi x/l)$ for $n = 2, 4, 6, \ldots$; and $l\sqrt{\lambda_n} \cos(\sqrt{\lambda_n}x) - 2 \sin(\sqrt{\lambda_n}x)$ for $n = 3, 5, 7, \ldots$
    (f) $u(x, t) = A + Bx +$ two series

13. $\lambda = \beta^2$ where $\tan \beta l = k/\beta c^2$ and $X(x) = \sin \beta x$

14. $\lambda_n = 1 + n^2\pi^2$, $u_n(x) = x^{-1}\sin(n\pi \log x)$ for $n = 1, 2, 3, \ldots$.

15. If $\lambda = \beta^2$, the equation for $\beta$ is

$$\frac{\rho_1}{\kappa_1} \cot \frac{\beta\rho_1 a}{\kappa_1} + \frac{\rho_2}{\kappa_2} \cot \frac{\beta\rho_2(l-a)}{\kappa_2} = 0.$$

16. $\lambda_n = (n\pi/l)^4$, $X_n(x) = \sin(n\pi x/l)$ for $n = 1, 2, 3, \ldots$

17. $\lambda = \beta^4$, where $\beta$ is a root of the equation $\cosh \beta l \cos \beta l = 1$. The corresponding eigenfunction is

$$X(x) = (\sinh \beta l - \sin \beta l)(\cosh \beta x - \cos \beta x) \\ - (\cosh \beta l - \cos \beta l)(\sinh \beta x - \sin \beta x).$$

18. (c) $\cosh \beta l \cos \beta l = -1$
    (d) $\beta_1 l = 1.88$, $\beta_2 l = 4.69$, $\beta_3 l = 7.85$, $\ldots$. The frequencies are $c\beta_n^2$.
    (e) For the bar $\beta_2^2/\beta_1^2 = 6.27$, while for the string the ratio of the first two frequencies is $\beta_2/\beta_1 = 2$. Thus, relative to the fundamental frequency, the first overtone of the bar is higher than the fifth overtone of the string.

## Chapter 5

### Section 5.1

1. $\pi\sqrt{2}/4$
4. $2/\pi + (4/\pi)\sum_{\text{even }n\geq 2}(1-n^2)^{-1}\cos nx$. Set $x=0$ and $x=\pi/2$. The sums are $\frac{1}{2}$ and $\frac{1}{2}-\pi/4$.
5. (a) $l^2/6 + 2(l^2/\pi^2)\sum_{n=1}^{\infty}[(-1)^n/n^2]\cos(n\pi x/l)$    (b) $\pi^2/12$
7. $-7\pi^4/720$
9. $\frac{1}{2}t + (\sin 2ct \cos 2x)/4c$
10. $E_n = [4l\rho V^2/(n\pi)^2]\sin^2(n\pi\delta/l) \sim 4\rho V^2 l^{-1}\delta^2$ for fixed even $n$ and small $\delta$.

### Section 5.2

1. (a) Odd, period $= 2\pi/a$; (b) neither even nor odd nor periodic
9. Zero
10. (a) If $\phi(0+) = 0$
11. The complex form is

$$\sum_{n=-\infty}^{\infty}(-1)^n\frac{l+in\pi}{l^2+n^2\pi^2}\sinh(l)\,e^{in\pi x/l}$$

14. $\frac{1}{2}l - (4l/\pi^2)\sum_{n\text{ odd}}(1/n^2)\cos(n\pi x/l)$

### Section 5.3

1. (a) All the multiples of $(1, 1, -2)$. (b) The coefficients are $\frac{4}{3}$, $\frac{5}{2}$, and $-\frac{11}{6}$
2. (b) $3x^2 - 1$
4. (a) $U - (4U/\pi)\sum_{n=0}^{\infty}[1/(2n+1)]e^{-(n+1/2)^2\pi^2kt/l^2}\sin[(n+\frac{1}{2})\pi x/l]$
   (c) $(4l^2/k\pi^2)|\log(\epsilon\pi/4|U|)|$
6. $2\pi ni$; yes
10. (b) $\cos x + \cos 2x$ and $\cos x - \cos 2x$
11. (a) The expression equals zero in these cases.

### Section 5.4

1. (a) Yes; (b) no; (c) yes
6. $\cos x$ on $(0, \pi)$, $-\cos x$ on $(-\pi, 0)$, zero at the points $x = -\pi, 0, \pi$
7. (c) Yes; (d) yes; (e) no
12. $\pi^2/6$
13. $\pi^4/90$
14. $\pi^6/945$

16. $a_0 = \pi$, $a_1 = -4/\pi$, $b_1 = a_2 = b_2 = 0$

19. (b) $\int_a^b X^2 \, dx = [(\partial X/\partial x)(\partial X/\partial \lambda) - X(\partial^2 X/\partial \lambda \, \partial x)]|_a^b$

   (c) Use $X(x, \lambda) = \sin(\sqrt{\lambda}\, x)$. Evaluate part (b) at $\lambda = (m\pi/l)^2$.


## Section 5.5

4. (d) $A = (2/l^2) \int_0^l (2l - 3x)\phi(x)\, dx$


## Section 5.6

1. (a) $1 + \sum_{n=0}^{\infty} A_n e^{-(n+1/2)^2 \pi^2 t} \cos[(n+\frac{1}{2})\pi x]$, $A_n = (-1)^{n+1} 4(n+\frac{1}{2})^{-3} \pi^{-3}$
   (b) 1

4. $u = c^{-2} kx(l - \frac{1}{2}x) + $ series

6. $\omega = N\pi c/l$, where $N$ is any positive integer, provided that $g(x)$ and $\sin N\pi x/l$ are not orthogonal.

7. No resonance for any $\omega$

9. $(k - h)x + h + \sum_{n=1}^{\infty} (2/n\pi)[(-1)^n k - h] \cos 3n\pi t \sin n\pi x$

10. $(l - x)^{-1} \sum_{n=1}^{\infty} a_n e^{-n^2 \pi^2 kt/l^2} \sin(n\pi x/l)$

13. (b) The difference $v = u - \mathcal{U}$ satisfies homogeneous BCs. Separate variables. Show that $v$ is the sum of a series each term of which decays exponentially as $t \to +\infty$.
   (c) $\mathcal{U}(x, t)$ blows up as $r \to 0$ and $\omega \to m\pi c/l$, where $m$ is any integer. If $r = 0$ and $\omega = m\pi c/l$, use the method of (11), (12) to see the resonance.


# Chapter 6

## Section 6.1

2. $[Ae^{kr} + Be^{-kr}]/r$

3. $AI_0(kr) + BK_0(kr)$, where $I_0$ and $K_0$ are associated Bessel functions

4. $u = B + (A - B)(1/a - 1/b)^{-1}(1/r - 1/b)$

6. $(1/4)(r^2 - a^2) - [(b^2 - a^2)/4][(\log r - \log a)/(\log b - \log a)]$

7. $(r^2 - a^2)/6 - ab(a + b)(1/a - 1/r)/6$

9. (c) $\gamma = 40$


## Section 6.2

1. $U = \frac{1}{2}x^2 - \frac{1}{2}y^2 - ax + by + c$ for any $c$

3. $[2\sinh(2\pi)]^{-1} \sinh 2x \cos 2y + x/2\pi$

## Section 6.3

1. (a) 4; (b) 1
2. $1 + (3r/a)\sin\theta$
3. $\frac{3}{4}(r/a)\sin\theta - \frac{1}{4}(r/a)^3\sin 3\theta$

## Section 6.4

1. $1 + (3a/r)\sin\theta$
5. (b) $u = \frac{1}{2}(1 - \log r/\log 2) + [(1/30)r^2 - (8/15)(1/r^2)]\cos 2\theta$
8. $aA\log(r/b) + B$
10. The first two terms are $(2r^2/\pi a)\sin 2\theta + (2r^6/9\pi a^5)\sin 6\theta$.

# Chapter 7

## Section 7.1

8. $(\nabla w_0, \nabla w_1) = +0.45$, $(\nabla w_1, \nabla w_1) = 1.50$, $c_1 = -0.30$
9. $c_1 = -0.248$, $c_2 = -0.008$
10. $c_1 = \frac{1}{2}$

## Section 7.4

1. 
$$G(x) = \begin{cases} x_0(l - x)/l & \text{for } 0 \le x_0 \le x \le l \\ x(l - x_0)/l & \text{for } 0 \le x \le x_0 \le l \end{cases}$$

6. (b) $\int_{-\infty}^{\infty} y[y^2 + (\xi - x)^2]^{-1} h(\xi)\, d\xi/\pi$
7. (a) $f(s) = A\tan^{-1}s + B$. (b) Use the chain rule. (d) $h(x) = \frac{1}{2}\pi A + B$ for $x > 0$, $-\frac{1}{2}\pi A + B$ for $x < 0$
8. (b) $\frac{1}{2} + \pi^{-1}\tan^{-1}((x-a)/y)$
   (c) $\frac{1}{2}(c_0 + c_n) + \pi^{-1}\sum_{j=1}^{n}(c_{j-1} - c_j)\theta_j$, where $\theta_j$ is the angle between the $y$ axis and the vector from $(a_j, 0)$ to $(x, y)$.
12. $-1/(4\pi|\mathbf{x} - \mathbf{x}_0|) + |\mathbf{x}^*|/(4\pi a|\mathbf{x} - \mathbf{x}_0^*|)$ for $|\mathbf{x}| > a$ and $|\mathbf{x}_0| > a$
13. A sum of four terms involving the distances of $\mathbf{x}$ to $\mathbf{x}_0$, $\mathbf{x}_0^*$, $\mathbf{x}_0^\#$ and $\mathbf{x}_0^{*\#}$, where * denotes reflection across the sphere and # denotes reflection across the plane $z = 0$
14. There are 16 terms, similar to those in Exercise 13. (Written in terms of the Green's function for the whole ball, there are eight terms.)
15. (a) Use the chain rule. (b) In polar coordinates the transformation takes the simple form $(r, \theta) \mapsto (r^2, 2\theta)$.
16. $\frac{1}{2}(A + B) + [(A - B)/\pi]\arctan[(x^2 - y^2)/2xy]$

17.  (b)

$$\int_0^\infty xg(\eta)\left[\frac{1}{(y-\eta)^2+x^2}-\frac{1}{(y+\eta)^2+x^2}\right]\frac{d\eta}{\pi}+\text{term with }h$$

22.  $C+\int_{-\infty}^\infty h(x-\xi)\log(y^2+\xi^2)\,d\xi$

24.  $u(x_0,y_0,z_0)=C+\iint h(x,y)\left[(x-x_0)^2+(y-y_0)^2+z_0^2\right]^{-1/2}dx\,dy/2\pi$

## Chapter 8

### Section 8.1

1.  (a) $O((\Delta x)^2)$; (b) $O(\Delta x)$
3.  $\left(\frac{2}{3}u_{j+1}-\frac{2}{3}u_{j-1}-\frac{1}{12}u_{j+2}+\frac{1}{12}u_{j-2}\right)/\Delta x$, using the Taylor expansion

### Section 8.2

1.  (a) $u_1^4=u_3^4=\frac{99}{64}, u_2^4=\frac{35}{16}$
2.  A wild result (unstable). Negative values begin at the third time step.
3.  (a) $\frac{29}{64}=0.453$; (b) 0.413
4.  $u_j^4=16, 14, 16, \frac{71}{4}, 14, \frac{29}{4}, 4, \frac{29}{4}$ (from $j=-1$ to $j=6$)
5.  $u(3,3)=10$
8.  (c) $u_j^1=0, \frac{9}{77}, \frac{3}{11}, -\frac{37}{77}, \frac{3}{11}, \frac{9}{77}, 0$ (from $j=0$ to $j=6$)
10. (a) Explicit; (b) $\xi=-p\pm\sqrt{p^2+1}$, where $p=2s(1-\cos(k\,\Delta x))$. One of these roots is always $<-1$.
11. (b) $\xi=1-2as[1-\cos(k\,\Delta x)]+b\,\Delta t$. So the stability condition is $s\le 1/2a$.
12. (a) 79.15; (b) 0.31
13. (a) $u_j^{n+1}=[2s/(1+2s)](u_{j+1}^n+u_{j-1}^n)+[(1-2s)/(1+2s)]u_j^{n-1}$
    (b) $(1+2s)\xi^2-4s\xi\cos(k\,\Delta x)+2s-1=0$. The roots $\xi$ may be real or complex but in either case $|\xi|\le 1$.
14. (b) $s_1+s_2\le\frac{1}{2}$

### Section 8.3

2.  (a) $u_j^2=0, 0, 0, 4, 2, -4, 2, 4, 0, 0, 0$
    $u_j^3=0, 0, 16, -16, -13, 38, -13, -16, 16, 0, 0$
    (b) $u_j^3=0, 0, 1, 2, 2, 2, 2, 2, 1, 0, 0$
    (c) Case (a) is unstable, while case (b) is stable.
3.  (b) $u=x^2+t^2+t$
5.  $u_j^3=u_j^9=2, 1, 0, 0, 0, 1, 2$. Reflection of the maximum value $(=2)$ occurs at both ends at the time steps $n=3, 9, 15, \ldots$, as it should.

7. The middle value at $t = 4$ is 35.

11. (b) $(u_j^{n+1} - u_j^n)/\Delta t + a(u_{j+1}^n - u_{j-1}^n)/2\,\Delta x = 0$

    (c) Calculate $\xi = 1 - 2ias\,\sin(k\Delta x)$, so that $|\xi| > 1$ for almost all $k$; therefore, always unstable.

## Section 8.4

2. At $n = 6$, the interior values are $\frac{63}{32}, \frac{111}{16}, \frac{15}{16}, \frac{63}{32}$.

3. At $n = 4$, one of the values is $447/64 = 6.98$.

4. 0, 48, 0, 0; 0, 16, 11, 24; 0, 5, 4, 0; 0, 0, 0, 0

5. From top to bottom the rows are 0, 0, 0, 0; 0, 4, 5, 0; 0, 11, 16, 0; 0, 24, 48, 0.

6. (d) The exact value is $-0.0737$

7. The values at the center are (a) $\frac{7}{16}$ and (b) $\frac{5}{16}$, while the exact value is $\frac{1}{4}$. The exact solution is $u(x, y) = xy$.

9. Using the starting values $u_{j,k}^{(0)} = 0$ at the nonghost points and appropriate values at the ghost points so as to satisfy the BCs, compute $u_{3,2}^{(1)} \sim -0.219$ and $u_{3,2}^{(2)} \sim -0.243$. The exact value at that point is $u(1, \frac{2}{3}) = -0.278$.

## Section 8.5

1. $\frac{1}{3}$

2. (a) $A = \frac{1}{2}|x_2 y_3 - x_3 y_2 + x_3 y_1 - x_1 y_3 + x_1 y_2 - x_2 y_1|$

   (b) $v(x, y) = |x_2 y_3 - x_3 y_2 + (y_2 - y_3)x - (x_2 - x_3)y|/2A$

# Chapter 9

## Section 9.1

1. Either $|\mathbf{k}| = 1$ or $u = a + b(\mathbf{k} \cdot \mathbf{x} - ct)$, where $a$ and $b$ are constants.

## Section 9.2

3. $ty$

6. (a) $(\pi R/r)(\rho^2 - (r - R)^2)$ if $|\rho - R| \le r \le \rho + R$, where $r = |\mathbf{x}|$.

   (b)

$$u(\mathbf{x}, t) = \begin{cases} At & \text{for } r \le \rho - ct \\ A\dfrac{\rho^2 - (r - ct)^2}{4cr} & \text{for } |\rho - ct| \le r \le \rho + ct \end{cases}$$

   and $u(\mathbf{x}, t) = 0$ elsewhere. Notice that this solution is continuous.

   (e) The limit is $[A/4c^2][\rho^2 - (\mathbf{x}_0 \cdot \mathbf{v})^2/c^2]$.

7.  (a) $u(\mathbf{x}, t) = A$ for $r < \rho - ct$, $u(\mathbf{x}, t) = A(r - ct)/2r$ for $|\rho - ct| < r < \rho + ct$, and $u(\mathbf{x}, t) = 0$ for $r > \rho + ct$.
    (c) The limit is $A\mathbf{x}_0 \cdot \mathbf{v}/2c^2$.

11.  $u = [f(r + ct) + g(r - ct)]/r$

12.
$$u = \begin{cases} -\dfrac{1}{4\pi r} g\left(t - \dfrac{r}{c}\right) & \text{for } t \geq \dfrac{r}{c} \\[2mm] 0 & \text{for } 0 \leq t \leq \dfrac{r}{c} \end{cases}$$

16.  (b) $u(\mathbf{0}, t) = At$ for $t \leq \rho/c$, and $u(\mathbf{0}, t) = A[t - (t^2 - \rho^2/c^2)^{\frac{1}{2}}]$ for $t \geq \rho/c$.

17.  $A\rho^2/2c^2$

### Section 9.3

7.  $u = \frac{1}{2}At^2$ for $r < \rho - ct$; $u \equiv 0$ for $r > \rho + ct$; $u = $ cubic expressions for $ct - \rho < r < ct + \rho$ and for $r < ct - \rho$.

9.  $u(r, t) = \int_0^t \int_{|r-ct+c\tau|}^{r+ct-c\tau} f(s, \tau) s \, ds \, d\tau / 2cr$

### Section 9.4

1.  $xy^2z + 2ktxz$

## Chapter 10

### Section 10.1

2.
$$u = \frac{64a^2b^2}{\pi^6} \sum_{m,n \text{ odd}} \frac{1}{m^3n^3} \sin\frac{m\pi x}{a} \sin\frac{n\pi y}{b} \cos\left[\pi ct\left(\frac{m^2}{a^2} + \frac{n^2}{b^2}\right)^{1/2}\right]$$

3.  $-\infty < \gamma \leq 3k\pi^2/a^2$

4.  (c) $\lambda_{mn} \sim n^2\pi^2 + (m + \frac{1}{2})^2\pi^2$ for large $m, n$

5.  (a) 2; (d) 4; (e) $\infty$

### Section 10.2

2.  $u = \sum_{n=1}^{\infty} A_n \cos(\beta_n ct/a) J_0(\beta_n r/a)$, where $J_0(\beta_n) = 0$, the $A_n$ are given by explicit integrals, and $\beta_n = a\sqrt{\lambda_{0n}}$.

4.  $u = Ae^{-i\omega t} J_0(\omega r/c)$, where $A$ is any constant.

5.  $u = B + 2B \sum_{n=1}^{\infty} [\beta_n J_1(\beta_n)]^{-1} e^{-\beta_n^2 kt/a} J_0(\beta_n r/a)$, where $J_0(\beta_n) = 0$.

## Section 10.3

4.  Separating $u = w(r, t)\cos\theta$, we have

$$w(r, t) = \sum_{n=1}^{\infty} A_n \cos\frac{\beta_n ct}{a} \frac{1}{\sqrt{r}} J_{3/2}\left(\frac{\beta_n r}{a}\right)$$

where the $\beta_n$ are the roots of $\beta J'_{3/2}(\beta) = \frac{1}{2}J_{3/2}(\beta)$ and

$$A_n = \int_0^a \frac{r^{5/2} J_{3/2}(\beta_n r/a)\, dr}{\frac{1}{2}a^2\left(1 - 2/\beta_n^2\right) J_{3/2}^2(\beta_n)}$$

5.  $u = B + 2(a/\pi)(C - B)\sum_{n=1}^{\infty}[(-1)^{n+1}/n]\, e^{-n^2\pi^2 kt/a^2}(1/r)\sin(n\pi r/a)$
6.  $t \simeq k^{-1}\log\left(\frac{16}{5}\right) \simeq 193.9$ sec
7.  (a) $C + (B/a)\left(3kt - \frac{3}{10}a^2 + \frac{1}{2}r^2\right)$
    (b) $-2Ba^2 \sum e^{-\gamma_n^2 kt/a^2}\sin(\gamma_n r/a)(r\gamma_n^3 \cos\gamma_n)^{-1}$ where $\tan\gamma_n = \gamma_n$
9.  $u(r, t) = \sum_{n=1}^{\infty} A_n e^{-n^2\pi^2 kt/a^2}(1/r)\sin(n\pi r/a)$, where
    $A_n = (2/a)\int_0^a r\phi(r)\sin(n\pi r/a)\, dr$.
10. $u = C + (a^3/2r^2)\cos\theta$
11. The solution $u(r, \theta, \phi)$ is independent of $\phi$ and odd in $\theta$. Separating variables, we find $r^\alpha P_l(\cos\theta)$, where $\alpha = l$ or $-l - 1$. We exclude the negative root. So the solution is $u(r, \theta, \phi) = \sum_{l\,\text{odd}} A_l r^l P_l(\cos\theta)$. The coefficients are found from the expansion of $f(a\cos\theta)$ in the odd Legendre polynomials.

## Section 10.5

2.  $J_{3/2}(z) = \sqrt{2/\pi z}\,(\sin z/z - \cos z)$
    $J_{-3/2}(z) = \sqrt{2/\pi z}\,(-\cos z/z - \sin z)$
7.  Show that $v$ satisfies Bessel's equation of order 1.
14. $u(r) = J_0(ikr)/J_0(ika)$
15. $u(r) = H_0^+(ikr)/H_0^+(ika)$
16. $u(r) = \sqrt{a/r}(J_{1/2}(ikr)/J_{1/2}(ika)) = a\sinh kr/(r\sinh ka)$
17. $u(r) = \sqrt{a/r}\,H_{1/2}^+(ikr)/H_{1/2}^+(ika) = ae^{-kr}/(re^{-ka})$
18. The eigenvalues are $\lambda = \beta^2$, where $\beta J'_n(a\beta) + hJ_n(a\beta) = 0$. The eigenfunctions are $J_n(\beta r)e^{\pm in\theta}$, where $\beta$ is a root of this equation.
19. $v(r, \theta) = \{J_n(\beta r)N_n(\beta a) - J_n(\beta a)N_n(\beta r)\}e^{\pm in\theta}$, where $\lambda = \beta^2$ and $\beta$ are the roots of $J_n(\beta b)N_n(\beta a) = J_n(\beta a)N_n(\beta b)$.

## Section 10.6

5.  $a_0 = \frac{1}{4}$, $\quad a_1 = \frac{1}{2}$, $\quad a_2 = \frac{5}{16}$, $\quad a_l = 0 \quad$ for $l$ odd $\geq 3$, and

$$a_{2n} = (-1)^{n+1} \frac{(4n+1)(2n-3)!}{4^n (n-2)!(n+1)!} \quad \text{for } n \geq 2.$$

6.  $u = \frac{1}{3} + (2r^2/3a^2)P_2(\cos\theta)$

7.

$$u = \frac{1}{2}(A+B) + \frac{1}{2}(A-B)\sum_{l \text{ odd}} \frac{(-1)^{(l-1)/2}(2l+1)(l-2)!!}{(l+1)!!}\left(\frac{r}{a}\right)^l P_l(\cos\theta)$$

where $m!! = m(m-2)(m-4)\cdots$

8.  Same as (10.3.22) except that $l$ is not necessarily an integer, and $P_l^m(s)$ is a nonpolynomial solution of Legendre's equation with $P_l^m(1)$ finite. The equation $P_l^m(\cos\alpha) = 0$ determines the sequence of $l$'s, and (10.3.18) determines $\lambda_l$.

## Section 10.7

3.  (b) $(2-r)e^{-r/2}$, $r\cos\theta\, e^{-r/2}$, $r\sin\theta\, e^{\pm i\phi}e^{-r/2}$

# Chapter 11

## Section 11.1

3.  (b) Because $w(1) = 0$, $\lim_{x\to 1} w(x)/(x-1)$ is finite; similarly as $x \to 0$. Use this to prove part (b).

## Section 11.2

1.  $A$ has the entries $\frac{1}{3}, \frac{1}{6}, \frac{1}{6}, \frac{2}{15}$. $B$ has the entries $\frac{1}{30}, \frac{1}{60}, \frac{1}{60}, \frac{1}{105}$. Hence $\lambda_1 \sim 10$ and $\lambda_2 \sim 42$.

2.  $\lambda_1 \sim 10$ and $\lambda_2 \sim 48$

3.  $\iint |\nabla w|^2\, dx\, dy = \pi^8/45$, $\iint w^2\, dx\, dy = \pi^{10}/900$, $Q = 20/\pi^2 = 2.03$, $\lambda_1 = 2$

4.  (a) $Q = 10$, $\lambda_1 = \pi^2 \sim 9.87$

6.  (a) $A$ has the entries $\frac{4}{3}, \frac{3}{2}, \frac{3}{2}, \frac{9}{5}$. $B$ has the entries $\frac{8}{15}, \frac{7}{12}, \frac{7}{12}, \frac{9}{14}$.
    (b) 2.47 and 23.6     (c) $\lambda_1 = \pi^2/4 = 2.47$ and $\lambda_2 = 9\pi^2/4 = 22.2$

7.  (a) The square of the first zero of the Bessel function $J_0$, which is 5.76

## Section 11.3

1.  Multiply the PDE by the eigenfunction and integrate.

## Section 11.4

1. (a) $(n\pi/\log b)^2$, $\sin(n\pi \log x/\log b)$
4. $u(x, y) = \sum_{n=1}^{\infty} A_n x^{-1} \sin n\pi x \cosh n\pi y$, where
   $A_n = 2 \operatorname{sech}(n\pi) \int_1^2 \sin n\pi x \, f(x)x \, dx$.
5. The eigenvalues are the roots of $J_0(\sqrt{\lambda_n}l) = 0$.

## Section 11.6

5. Using rectangles, $5\pi^2/16 < \lambda_1 < 9\pi^2/16$ and $\frac{\pi^2}{2} < \lambda_2 < \pi^2$.

# Chapter 12

## Section 12.2

2. $u(x, t) = V/2c$ for $|x - x_0| < ct$, and $u = 0$ elsewhere.
3. $u(x, t) = V/2c$ in a certain region with corners at $(x_0, 0)$ and $(0, x_0/c)$, and $u = 0$ elsewhere.
5. $u = f(t - r/c)/4\pi c^2 r$ in $r < ct$, and $u = 0$ in $r > ct$.
10. $H(t - r/c)(2\pi c)^{-1} \int_0^{t-r/c} [c^2(t - \tau)^2 - r^2]^{-1/2} f(\tau) \, d\tau$

## Section 12.4

1. $(4\pi kt)^{-1/2} \int_{-\infty}^{\infty} \phi(y) e^{-(\mu t + x - y)^2/4\kappa t} \, dy$
3. $e^{-mr}/4\pi r$
5. (c) $u = 2\pi z(r^2 + z^2)^{-3/2}$
6. $u(x, y) = \int_0^{\infty} \int_{-\infty}^{\infty} [\pi \sinh(k)]^{-1} f(\xi) \sinh(ky) \cos(kx - k\xi) \, d\xi \, dk$

## Section 12.5

3. $1 - \cos t$
5. $[\cos(c\pi t/l) - (l/c\pi) \sin(c\pi t/l)] \sin(\pi x/l)$
7. $1 + e^{-4\pi^2 kt/l} \cos(2\pi x/l)$

# Chapter 13

## Section 13.3

2. Write the solution as in (2) with $c_1 = c_2 = 1$. Then $u(x, t) = f(x - t) + G(x + t)$ for $x < 0$, and $u = H(x - t)$ for $0 < x < t$, and $u = 0$ for $x > t$. The jump conditions lead to the fourth-order ODE $G'''' + 2G''' + 2G'' + 2G' + G = -f''''(-t) - 2f''(-t) - f(-t)$, which can be solved.

7. $-e^{ikr_+}/4\pi r_+ + e^{ikr_-}/4\pi r_-$, where $r_\pm = [x^2 + y^2 + (z \mp a)^2]^{1/2}$. The reflected wave is the second term.

### Section 13.4

1. Let $\beta = \sqrt{\lambda + Q}$. Then $\beta$ satisfies $0 < \beta < \sqrt{Q}$ and $\tan \beta = -\beta + Q/\beta$. This equation can be graphed as in Section 4.3 to obtain a finite number of eigenvalues depending on the depth.

2. $\lambda = -Q^2/4$ is the only eigenvalue and $\psi(x) = e^{-Q|x|/2}$ is the eigenfunction.

### Section 13.5

9. $(2c)^{-1}e^{-t/2}J_0[(i/2c)\sqrt{c^2t^2 - x^2}]\,H(c^2t^2 - x^2)$
10. $J_0(2\sqrt{xy})$ for $(x, y) \in Q$

# Chapter 14

### Section 14.1

2. $u = [x/(t + 1)]^5$
3. $u = x/(1 + t)$
5. $[(4tx + 8t + 1)^{1/2} - 1]/2t$
7. $u = -t + \sqrt{t^2 + x^2}$ for $x > 0$, $u = -t - \sqrt{t^2 + x^2}$ for $x < 0$.
9. $\partial z/\partial t = -a(\phi(z))/[1 + ta'(\phi(z))\phi'(z)]$, $\partial z/\partial x = 1/[1 + ta'(\phi(z))\phi'(z)]$, $\partial u/\partial t = \phi'(z)\partial z/\partial t$, $\partial u/\partial x = \phi'(z)\partial z/\partial x$, etc.
10. $u = (1 - x)/(1 - t)$ in the triangle bounded by $t = 0, t = x$, and $x = 1$. Elsewhere, $u = 0$ or $1$. The shock begins at $t = 1$ and proceeds along the line $x - 1 = \frac{1}{2}(t - 1)$.
12. $(x + t + \frac{1}{2}t^2)/(1 + t)$

### Section 14.2

4. (a) $\pm \int_{f_2}^{f} [P(y)]^{-1/2}\,dy = x - x_2$
10. $u(x, t) = (3t)^{-1/3} \int_{-\infty}^{\infty} A[(x - y)/(3t)^{1/3}]\phi(y)\,dy$

### Section 14.3

2. A circular arc. Minimize $\int_0^1 \sqrt{1 + u'^2}\,dx + m\int_0^1 u\,dx$, where $m$ is a Lagrange multiplier.
4. $y = (x^3 + 11x)/12$
8. $(k/\pi)^{3/2}e^{-k|v|^2}$, where $k = 3/(4E)$. This is known as the Maxwellian.

# INDEX